Applied Electronic Design

D. Joseph Stadtmiller
Mohawk Valley Community College

Upper Saddle River, New Jersey
Columbus, Ohio

Library of Congress Cataloging in Publication Data

Stadtmiller, D. Joseph
Applied electronic design / D. Joseph Stadtmiller
p. cm.
Includes bibliographical references and index.
ISBN 0-13-094758-X
1. Electronic apparatus and applications—Design and construction. 2. Electronic circuit Design. I. Title.

TK7836 .S728 2003
621.381—dc21 2002029285

Editor in Chief: Stephen Helba
Acquisitions Editor: Dennis Williams
Editorial Assistant: Lara Dimmick
Production Editor: Steve Robb
Production Coordinator: Tim Flem, PublishWare
Copy Editor: Roberta Dempsey
Design Coordinator: Karrie Converse-Jones
Cover Designer: Jason Moore
Production Manager: Pat Tonneman
Marketing Manager: Ben Leonard

This book was set in Century Book by PublishWare. It was printed and bound by R. R. Donnelley & Sons Company. The cover was printed by Phoenix Color Corp.

Pearson Education Ltd.
Pearson Education Australia Pty. Limited
Pearson Education Singapore Pte. Ltd.
Pearson Education North Asia Ltd.
Pearson Education Canada, Ltd.
Pearson Educación de Mexico, S.A. de C.V.
Pearson Education—Japan
Pearson Education Malaysia Pte. Ltd.
Pearson Education, *Upper Saddle River, New Jersey*

10 9 8 7 6 5 4 3 2 1
ISBN 0-13-094758-X

To Francine, Jes, Joe, Jon, and Ginger

Preface

The purpose of this book is to better prepare students to enter the fast-paced world of electronics by applying the theoretical knowledge learned in their foundation courses on analog and digital electronics. It is a continuation of the premise that prompted my first book, *Electronics: Project Management and Design*, which discussed all the facets of a design project as completed in a business environment. *Applied Electronic Design* offers a different perspective as it discusses the design process and issues relating to a wide range of electronic applications. Following are the underlying strategies of the book:

- Expose students to the design perspective from a wide range of electronic applications
- Discuss the Six-Step design process, the development of design specifications, and the importance of concurrent engineering concepts
- Emphasize general design issues, such as manufacturability, quality, reliability, ease of use, and serviceability
- Show methods for dealing with ambient temperature variations, EMI radiation, and noise immunity

The book's main theme is the application of basic analog and digital electronics concepts to actual design problems. This experience will promote a deeper understanding of electronics, better problem-solving skills, and the development of design skills for electronics students.

I developed this text by combining my industrial and academic experiences. Also, comments and reviews from my first book, *Electronics: Project Management and Design*, indicated that there is a general need for capstone-oriented texts and more specifically that a book with a design focus is highly desirable. Reviewers of the early *Applied Electronic Design* manuscript helped to refine the approach used and the topical coverage of the final product.

The book is targeted as a textbook for applied electronics, design, electronic design, senior-project, and electronic project courses. It could also be used as a supplementary text for any electronics program. It is most applicable to the third and fourth year of four-year electronic programs but can also be used in the second year of many two-year electronics programs. It is designed to function completely on its own or side by side with *Electronics: Project Management and Design.*

The book can be segmented into two areas: Chapters 1 through 5 present the design process, and Chapters 6 through 12 cover seven different areas of electronic circuit applications. The coverage of the design process includes defining the design problem, overall design considerations, beginning the design, selecting components, and printed circuit board design. The last seven chapters comprise the following topics: power supplies, amplifiers, oscillators/clock circuits, control circuits, digital logic circuits, embedded systems, and telecommunications. Each of these chapters can be considered a mini-handbook on the topic covered.

Every attempt has been made to simplify the information presented in the book. The book is intended for use at many levels, but a basic knowledge of circuit theory and both analog and digital electronic components is assumed. Each subject area is discussed to a depth that promotes a broad understanding of the underlying design issue. Each chapter starts with an introduction that highlights the topics to be covered. Examples are provided wherever practical, and each chapter concludes with a summary and exercises.

Applied Electronic Design will surely help electronic students take the all-important first step into their professional careers by introducing design problems and applications that represent the major application areas in the electronics field. In many cases, it might also help students determine initial career directions.

Acknowledgments

I would like to thank all of my teachers, professors, and mentors who helped develop the foundation of my knowledge. Thanks also to those whose comments and suggestions regarding my first book helped develop the overall content of this book.

Thanks to Prentice Hall Sales Representative Donna Conroy for supplying me with much reference material, and Acquisitions Editor Dennis Williams for his ongoing direction and support. Steve Robb, the production editor; Tim Flem, the production coordinator; and Roberta Dempsey, the copy editor, were all key players in finalizing and producing this book. It was a pleasure to work with them, and I appreciate the results of their efforts on the final product.

Finally, I thank the following reviewers, whose comments contributed much to the content and character of the book: Don Abernathy, DeVry Institute of Technology–Dallas; Habib Rahman, Saint Louis University; Hesham E. Shaalan, Georgia Southern University; and Sidney Soclof, California State University, Los Angeles.

Contents

1 The Design Problem

Introduction

This book applies basic electronics knowledge to a variety of design problems that can be resolved with electronic circuits. The abilities to adapt, design, and create new processes and devices are the skills that promote human survival more than any others. Other hominid species that failed to develop these skills have long since perished while our ancestors persevered. Currently, as we experience the benefits and changes caused by the Internet and the continuing information revolution, we are again realizing the potential of an idea and its resulting design. Beginning as a communication scheme to connect research universities and the defense community, the computer and communications framework that resulted in the Internet began in the mid-1970s. While the computer revolution was causing rapid change and creating many new possibilities, the infrastructure for the Internet was being slowly developed. Roughly 15 years later the personal computer and the Internet would team up to change the way we work and live at a level that rivals the development of the wheel, the printing press, and the atomic bomb. Clearly, the computer, the Internet, and the resulting telecommunications revolution have the potential to change the world as much as any of these historic developments.

Our future as a people, and the future of our businesses and our world, will continue to depend on our ability to design new devices and create new technology. Our most severe technical challenges will involve energy efficiency, natural resource utilization, and pollution. Many of the solutions to these and other problems will be electrical and electronic products and systems, while many will include a combination of technologies. Communication; computing; industrial automation; energy development and efficiency; and medical, entertainment and defense applications of electronics technology will continue to demand creative solutions to design problems. This book will discuss a variety of common electronic design problems and how to approach their solutions as theoretical electronic and circuit concepts are applied in the real world of electronic design.

This chapter begins with a definition of concurrent engineering and discusses the design issues that most projects must address. Then a process called the Six Steps is reviewed as a method for solving design problems. The final topic is the development of design specifications. These represent the formal definition of the design problem for the development team. The specific topics covered in this chapter are as follows:

- Concurrent Engineering
- The Six Steps
- Developing Design Specifications

1–1 ▶ Concurrent Engineering

The design problem is unique because it involves solving a problem with the creation of some device or system. As we create the design, it is important to focus on all of its requirements. During the design process, it is natural to concentrate on a design's central functional requirements, while ignoring seemingly trivial issues such as manufacturability and customer ease of use. However, in the end a design's success will be determined by how well it addresses *all* of the initial requirements.

Concurrent engineering is a design philosophy that promotes the consideration of all the requirements of a design, beginning with the very first step of the design process through to its completion. The concurrent engineering requirements include topics such as reliability, quality, customer use, marketing and sales issues, and manufacturing and financial issues, in addition to the more obvious functional performance specifications. Concurrent engineering is a design process that improves on traditional methods. This process comprises the following strategies, which are summarized in Figure 1–1:

1. Initially develop complete specifications defining the project, and minimize changes to the specifications after the project is started.
2. Consider manufacturing, quality, customer use, field service, and disposal issues at the very beginning of the project.
3. Utilize a project team with representatives from all affected departments. Promote innovation in the team. Empower them and give them incentives to promote their success.
4. Develop detailed schedules with distinct project phases. Deliverables should be defined for the end of each phase, and these must be complete before moving onto the next phase.
5. Include a project verification stage in which the quality, performance, and reliability of the project is verified independently, and the project is utilized in the intended environment.

1. Develop complete specifications
2. Consider manufacturing, customer use, and quality up front
3. Utilize a multifunctional team
4. Develop detailed schedules
5. Include a project verification stage in the project schedule
6. Involve key suppliers
7. Envision the goal
8. Promote continuous improvement

▲ **FIGURE 1–1**
Concurrent engineering strategies

6. Involve key suppliers very early in the project.
7. Maintain a perspective of the overall project goal and all of the different design issues on a concurrent (simultaneous) basis.
8. Provide for continuous improvement by evaluating the performance of the project when complete and reflecting on what went well and what needs improvement.

When the principles and strategies of concurrent engineering are applied to the design process, good things happen. The most important result is the synergy gained from the team approach to design project management. Concurrent engineering is the application of good common sense to engineering design projects to rectify the situations that evolved as companies became large and departmentalized. The specific results of concurrent engineering are as follows:

1. The multifunctional team ensures that all design issues are addressed up front. Customer, manufacturing, quality, financial, and field service issues are considered equal to other design issues.
2. Design activities are scheduled in an orderly way, taking advantage of parallel paths wherever possible and resulting in minimal linear time for the project. The project team believes the schedule can be achieved.
3. When involved early in the process, key suppliers can provide new and different perspectives that can improve the project results.

1. Manufacturing, customer use, and quality are considered up front.
2. Project schedules are minimized.
3. Key suppliers provide new ideas.
4. Project delays are identified and addressed quickly.
5. Project delays from specification changes are minimized.
6. Project visibility is high.
7. Short ramp-up time to manufacture.
8. Quality is improved.
9. Creativity increases.
10. Team members feel ownership.

▲ **FIGURE 1–2**
Summary of the results of concurrent engineering

4. Project delays are identified and addressed as they are realized.

5. Delays that result from incomplete or changing specifications are eliminated.

6. The project visibility is very high throughout the company.

7. The ramp-up time for starting manufacturing is planned and minimized.

8. Quality is measured independently.

9. Creativity is maximized.

10. The team feels a joint ownership in the project and its results.

The results of concurrent engineering are summarized in Figure 1–2.

1–2 ▶ The Six Steps of Problem Solving

At this point in your academic career, you have probably solved many problems. Many of them have been homework problems, which you did quickly after skimming the problem write-up and then frantically paging through the book to find an appropriate formula. After plugging in the numbers and performing the calculations, you compared the answer with those in the back of the book. If the answer checked out, you moved on to the next problem. If not, you repeated the process with perhaps a different formula. Only after numerous failures to achieve the book's answer will the average student reread the problem to verify the facts given and confirm what is required in the form of an answer. Very often at this juncture comes the realization that the problem write-up is asking for something completely different from what the student thought initially. A simple change in direction now will often lead to the successful solution. And so it is, with the educational process and life, as the pace of our lives continues to go faster and faster.

The primary role of engineers, technologists, and technicians is to supply technical solutions to problems; design problems, operational problems, and failure problems. As you enter a career in the electronics industry, the pace will continue to be rapid. Today's industry requires superior solutions to a complex set of problems, many of them design problems. In this section we will examine a simple process for solving engineering design problems that works. The Six-Step process is a basic problem-solving process that has been around for a long time. The Six Steps that follow are simple, easy to use, and will solve many types of problems:

Step One: *Research* the problem by gathering information.

Step Two: Completely *define* the problem by studying it and listing as many facts as possible to fully define it.

Step Three: *Plan* the solution. Develop a plan, a list of steps to solve the problem.

Step Four: *Execute* the plan outlined in Step Three.

Step Five: *Verify* the results achieved in Step Four, making sure that they do solve the original problem.

Step Six: *Conclude.* Develop a set of conclusions, and note what is learned in the process.

Figure 1–3 shows a summary of the entire process. Let's look at some examples.

Example 1–1

This problem was posed to a product engineer whose job it is to support the phase-in process for new products as they are turned over to manufacturing. New product phase-in to manufacturing is the process of preparing the company to manufacture a product. The problem is to improve the manufacturability of a new capacitance fuel gauge system for small aircraft. This new product is symptomatic of one designed initially without using concurrent engineering concepts. The details relating to its manufacturability are being addressed as the product is being phased into manufacturing, instead of at the beginning of the project.

Step One: Research and gather information

Step Two: Define the problem

Step Three: Plan the solution

Step Four: Execute the plan

Step Five: Verify the solution

Step Six: Develop a conclusion

▶ **FIGURE 1–3**
The Six Steps of problem solving

The capacitance fuel gauge system uses capacitance sensors to determine the quantity of fuel remaining in a fuel tank. It includes a circuit board, a capacitance sensor, and interconnecting wires all mounted to a stainless steel plate that mounts the unit into the gas tank of small aircraft. The circuit board is totally potted (encapsulated, molded into an epoxy potting compound) to form the shape of small cube. Where the interconnecting wires exit this cube, there is a need to protect the wires from bending or, in engineering terms, provide strain relief for the wires. On the initial prototypes fabricated by engineering, a rubber boot, shaped like a small cylinder, has been glued onto the main body of the fuel gauge to perform the strain relief function. A better solution is needed, one that is more effective as a strain relief and more manufacturable. To solve the problem, the product engineer uses the Six Steps:

Step One—Research the problem: Gather information about the problem. The engineer reviews the assembly process and observes and experiences the installation of the fuel gauge system in an aircraft.

Step Two—Problem definition: To provide a functional and easily manufactured strain relief for the capacitance fuel gauge system. The engineer develops a list of facts about the problem.

1. The interconnecting wires needed mechanical strain relief.
2. The term "easily manufactured" meant a simple, fixturable process that didn't add significant assembly time to the product.
3. The current method was messy, time-consuming, and not reliable.

Step Three—Plan the solution: Develop a plan, a list of steps intended to solve the problem. The engineer has come up with the following steps for solving this problem:

1. Develop two alternative solutions that address all the issues brought up in the problem definition.
2. Prototype and test each of the solutions.

Step Four—Execute the plan: Simply perform the solution plan. The product engineer performs the solution plan.

Step Five—Verify the results: Check the results achieved, making sure that they solve the original problem. The product engineer reviews each solution with engineering and manufacturing and chooses the preferred solution.

After analyzing the problem, the engineer has devised a couple of solutions. One alternative involves changing the tooling (the mold for applying the potting compound) for the potted assembly to include the strain relief and the other utilizes an off-the-shelf strain relief. The tooling change is the most attractive solution because it requires no additional parts or assembly. The strain relief will become part of the main assembly and will be formed at the same time as part of the potting process. When both solutions were built up and tested, the tooling change was the selected alternative. The benefits of this solution were the following:

1. No additional parts were needed.

2. No additional labor time was needed.

3. The strain relief was very secure and functional.

Step Six—Develop a set of conclusions: Take note of what is learned in the process. What the product engineer learned from solving this problem was the advantage of designing all the requirements of a design into any tooled parts included in the design.

This next example discusses how a company might solve the problem of increasing sales and market share of one of its product lines by implementing the Six-Step process.

Example 1–2

A company in the music electronics business desires to develop and market a new innovative electronic guitar tuner. A project manager is selected and a multifunctional project team has been put together to study the project and, if feasible complete it. Here is how the project would be completed using the Six Steps:

Step One: The project team will gather all the information needed to consider this project. The exact information needed will include technical, market, and financial information required to determine if the project is feasible and can proceed to Step Two. Step One will be complete when there is a project proposal that includes all of the technical information gathered on guitar tuners, information on the market for these products, and financial data that estimates the cost and profitability of the project if it is completed.

Step Two: With management approval of the Step One proposal, the project team proceeds to Step Two, where the design specifications are completed. Design specifications define the problem for a product development project, and their development is discussed later in this chapter. The result of Step Two is a set of specifications for the guitar tuner that will allow the project team to determine a solution plan and to develop the product using concurrent engineering concepts.

Step Three: With the design specifications complete, the project team develops a solution plan that is better known as a project schedule for product development projects.

Step Four: Now the Project Team can begin implementing the project, which will begin with the preliminary design stage, in which initial design ideas will be generated, explored, and simulated. Next, components will be selected, procured, and breadboarded to test the design concepts. After testing and modifying the breadboarded circuits, the team will assemble and test prototype circuit boards and a prototype guitar tuner. Step Four includes the preliminary design, component selection, and breadboarding and prototype development, which all involve the implementation of the design.

Step Five: The project team will now verify the design to make sure that it meets the original specifications. Team members will perform product assurance tests on the guitar tuner as well as field tests and a financial analysis. If approved at the completion of Step Five, the guitar tuner will be ready to be released for sale to customers.

Step Six: This step will be completed some time after the project has been released, when the project's performance with regard to sales, profitability, quality, and customer use issues will be reviewed on a monthly basis. The results of the project will be summarized and reviewed to determine what improvements can be made on this product in the future, as well as on the way similar future projects will be managed.

In each of the examples shown, the Six-step process was applied to solve the problem. For solving a design problem, the Six Steps mandate that before beginning work on the design, information must be gathered, the design problem must be completely defined, and a plan or schedule must be created. After the design is complete, it must be verified and then modified for improvement while noting general conclusions. It is most important to understand the significance of Step Two, defining the problem. Failure to complete this vital step results in a higher number of poor design solutions than a mediocre performance on any other step. It is also important to emphasize the last of the Six Steps, developing a conclusion. All too often our desire to complete a design prevents us from taking the time to review and reflect on what we learned from the process. Developing a conclusion promotes the widely used concept of continuous improvement—the ongoing learning that should be part of our everyday life. The next section reviews how to define the design problem formally with design specifications.

1–3 ▶ Developing Design Specifications

What are specifications and why do we need them? Specifications are nothing more than a formal document listing the requirements of a project or product. They are a detailed list of the facts related to the problem definition for a design project. When you construct a house, you follow a set of specifications that define details that must be met upon the house's completion. The general contractor is responsible for satisfying these specifications, but all of the subcontractors (i.e., carpenters, electricians, and masons) must be aware of these requirements to ensure that all aspects of the specifications are fulfilled. Generally speaking, a specification is simply a non-trivial problem definition.

The other benefit of using a formal specification is that a format can be developed that will become a useful guide in completing specifications for similar future projects. This will help ensure that key aspects of a future project are reviewed before the design is started. When one small detail remains unidentified up from, it can result in the project being completely aborted. A better-developed, better-defined set of specifications is more likely to result in a final design that will function and meet the original need.

Before we discuss the development of specifications further, let's again look at what precedes and drives their development. First, there is the need for whatever the design project represents. This results from a desire to meet the strategic priorities set by the company. The strategic priorities drive the specific business reason for the project. During Step One, a significant amount of information is collected about a project and a proposal is put together. This information is developed to determine the viability of the project and provide for the development of the specifications. At this point the multifunctional team will get together and use the Step One proposal information to develop the specifications. This is a critical point for the application of concurrent engineering principles. The specifications must be developed jointly with all of the ultimate requirements of the project in mind.

Marketing representatives are the experts on how the project will be employed as well as the business aspects of its success. The marketing team members are usually not technical experts and are therefore not aware of the technical details, knowledge, and capabilities required to complete the project. Most people that have a house built for them are not experts on building houses, but they are most knowledgeable on what they want their house to be. As they sit down with their contractor to discuss their requirements, they often find that aspects of the home they desire will present difficulties to the contractor. These difficulties will likely increase the price of the house beyond what the owner is willing to pay. The specifications for the house are developed with this give-and-take attitude in mind.

A similar situation occurs in industry. For most new products a company's marketing department, while representing the customer, will define the general requirements for the new product. These requirements are usually called the "market specifications" and are included in the Step One proposal, along with the market/business aspects of the project. Many times the initial marketing specifications are impractical, "pie in the sky" requirements. It is important to discuss the initial market specifications and filter them down to something that is more "real world" before development begins. The marketing, engineering, finance, quality, and manufacturing departments will work together to help develop and review the design specifications. This will ensure that the product not only meets the customer's need, but that it can be developed and manufactured at a cost and quality level that will ensure its business success.

Let's review what might be included in a set of specifications for a new electronic product. In utilizing a general specification, development engineers will look for answers to the following types of questions:

What is the power source for the product being developed?

What is the range in power source voltage over which the product must function?

In what ambient temperature range will the product have to function?

How large can the product be physically?

What are the criteria for the appearance of the product?

What tolerance levels should be selected for the electronic components?

In addition to answering these types of questions, the specifications should also discuss how the product is to operate. In some cases it is possible to be specific up front in defining the product's operation. Yet in other more complicated situations, it may be impractical to state exactly how the product will function. In these cases, the statement of function is handled in a general way.

We can categorize specifications into the following Specification Format:

General Description

Performance

Power Input

Package

Environmental

Operation

Agency Approvals

Cost Specifications

Special Requirements

Following are detailed explanations of each area of the design specifications.

General Description: This features a general write-up for the project describing its purpose, the broad approach to development, and the environment of the end use and the end user.

Performance: This section of the specification deals with the quality or performance level of the project or product under development. To complete this section, it is necessary to identify the key outputs or end result of the project. Next, ideal conditions for parameters that will affect performance should be defined to provide a consistent basis for making the measurements. Then the acceptable range in variation of the outputs must be determined.

Example 1–3

Let's take the example of a digital thermometer. The primary output of the temperature indicator is the displayed temperature. The environmental parameters that affect the displayed temperature are the ambient temperature of the indicator and the quality of the power supplied to the indicator. Following are the ideal conditions for determining this accuracy:

1. The thermometer is connected to a specific temperature sensor that is exposed to a temperature within the operating indication range of the indicator.
2. The thermometer indicator should be maintained at an ambient temperature of 25°C.
3. The input power is within the specification range.

In this case we will say that the operating range of the indicator is 0° to 400°C. We'll define the acceptable accuracy of the thermometer for this application as ±0.5% of range. In degrees, then, the acceptable range of the thermometer display is ±2°C (0.005×400°C). To check this out, the sensor is placed in a temperature within the operating range of the indicator, say 200°C. The acceptable reading on the indicator would be between 198° and 202°C. This is a very simple case because there is only one end result or output from the device, the indicated temperature.

In other projects there may be many outputs with many different variables relating to each one. If we were developing a waveform generator to output a triangle wave, a sine wave and a square wave, we would list each of those outputs in our performance specifications. Each type of output would also have accuracy specifications relating to the frequency, amplitude and overall representation of the waveform. In other words, the specifications analytically define how perfect each waveform must be.

To develop the performance aspect of a project specification, follow these steps:

1. Identify measurement aspects of all of the key outputs of the project or product.

2. Identify the parameters that affect these outputs and their ideal conditions for taking the measurements.

3. Determine acceptable ranges for these parameters.

Power Input: This section specifies the power to be supplied to the device. In most cases this needs to be considered. In others, such as a software-only project, it may be omitted. Much of the time the power supplied is standard 115 V AC, 60-cycle power. Even in this case, it is important to note the range in amplitude and frequency that the device can expect to see and, most important, to verify that the device will operate over that range. Most devices powered from standard 115 V AC use a ±10% variation in voltage level and a range in frequency of 50 to 60 Hertz.

In some projects a device may be capable of accepting power from both AC and DC sources. Specifications should be listed for each case. The load current or power consumption of the device, both typical and maximum values, should be identified also. If the DC source is a battery, then battery size and expected life will be important issues.

To complete the power input section of the specifications:

1. Identify the type and range of all power inputs to the device or project, making sure to include all pertinent parameters, such as frequency and amplitude.
2. Indicate the power (volt-amps) or current requirements on the power input.

Package: This part of the specification relates to the mechanical aspects of the design. The purpose is to define the general package criteria of the project without completing the design. In some cases there already exists a specific package that the project will utilize; that should be stated here. Other times there may be no specific requirement for the package size, and that should be stated here also. Most often, however, specific criteria for the package design is necessary to successfully meet the design goals of the project. The end result of this section of the specifications will identify all the key criteria, completely defining the package design problem for the package designer. Here are the areas that should be addressed:

Mechanical Size Limits: The largest, smallest, or specific volume to which the package must conform.

Environmental Rating: The weatherproof rating for the package—how well it is sealed and its ability to withstand corrosive atmospheres

Shape: The product shape, if important, and any information that will allow the designer to determine the optimum shape

Material: Specific material requirements for the package or any guidelines that will help to determine the proper material for the package

Human Engineering Aspects: All aspects of human interaction that the project will employ must be considered in the design. Are there keys that will need depressed? Does the device mount on a wall? Ease of use in every aspect of the project should be considered, including unpacking, installing, using, servicing, and even recycling and disposing. As we enter the new millennium, we must increasingly appreciate, preserve, and reuse our natural resources.

Environmental: This area of the specifications defines key variables in the environment that are the result of the design process. These primarily include ambient temperature, humidity, vibration, shock, electro-magnetic interference (EMI) immunity, and generation.

Ambient Temperature: This is the range in temperature to which the product will be exposed under normal operating conditions. This is listed as the operating range. It is also necessary to point out the storage temperature range. This is the range of temperature to which the device would be exposed over a normal product life while not operating.

Humidity: Humidity can have a detrimental impact on a variety of products and as such is an important environmental factor. The range in humidity over which the product is expected to operate, expressed as a percent of relative humidity, should be noted here.

Vibration: The vibration levels to which the project will be exposed should be specified. This is usually done in terms of the amplitude of the force applied, the frequency of the vibration, and the direction that the force will be applied to the project.

Shock: The purpose of this specification is to define the one-time shock force that the device should be able to withstand. In most cases this will be the worst-case shock applied during shipment if the device were to be dropped from a certain height on its side or on its corners. Specifications for standard tests used by shipping container manufacturers are often included in this section. In some applications there may be other sources of significant shock forces that can be applied to the device. These levels should be identified to allow development of the appropriate shock specifications.

Electro-Magnetic Interference (EMI) Immunity: The design's immunity to electrical interference of many kinds is an important requirement in many applications. The difficulty here is the creation of a subjective statement that specifies the project's EMI immunity goals in a way that can be verified and tested. Designers are generally concerned with interference such as radiated electrical fields, induced magnetic fields, electrostatic discharges, and power line transients. You can try to develop a set of requirements for each one of these areas or use existing standards, such as MIL-STD-461 (limits) and 462 (test procedures) issued by the U.S. Department of Defense. This standard is very complete and stringent and covers both immunity as well as emission. It may be a simple matter to state in the design specifications "meet MIL-STD-461." However, the actual process of verifying and meeting these specifications is a difficult task. It is best to review the various noise standards and pick the areas most important for consideration in the subject design and include them in the specifications. At the same time you must be able to verify the performance by creating the environment included in the design specifications. This may mean purchasing of some sophisticated and expensive test equipment or using an outside testing laboratory.

Electro-Magnetic Interference (EMI) Emissions: As the operating frequencies and the volume of equipment in operation have rapidly increased, the radiation of electronic equipment has come under increased scrutiny. The best source for this requirement comes from the Federal Communications Commission (FCC), which has defined two levels of requirements and testing, Class A and Class B equipment. Military Standard, MIL-STD-461, also addresses this issue. An appropriate specification statement should be determined and included in the design specifications. To verify the performance, you will need to measure radiated EMI from the design accurately. This may require the purchase of equipment or the use of an outside testing laboratory.

Operation: All of the operational aspects of the device will be addressed in this section. The steps and requirements for operation should be listed, starting with applying power to the unit. All the variables that are provided for adjustment of operation should be shown and discussed. As mentioned earlier, in simpler projects it may be possible to completely define the way the device will operate in the specifications. In cases that are software-intensive with many key depressions and displays, the specific operation is something that may be developed as the project design unfolds. However, it is important to make sure that the hardware requirements will support all of the overall requirements for the project. When the operational requirements are implemented with software, a separate operational specification called "Software Specifications" is usually generated. This details all of the needed software operations and requirements.

Agency Approvals: This area deals with both "recognized testing agencies and recognized specifications" with which the design project is required to comply. These include agencies such as Underwriters Laboratories (UL) and the Canadian Standards Association (CSA). Approval by these testing agencies is increasingly important in markets for a variety of product classifications. The identification of any required agency approvals, along with the corresponding specification, are imperative before beginning the design process.

Cost Specifications: This is one of the most important sections and is often overlooked. This section will determine the potential for the financial success of the project. When considering all the technical details, it is easy to overlook that both the cost of development and the manufacturing cost are crucial to the project's success. As such, they must be identified as part of the specifications. Cost specifications include both project cost estimates and manufacturing cost goals.

Preliminary Project Cost Estimates: This is an estimate of the total cost to develop the project. This step should be saved for last so that a reasonable estimate can be developed from the specifications completed thus far. This estimate should include both direct dollar expenditures and the total man-hours needed to complete the project.

Manufacturing Cost Goal: The manufacturing cost goal includes all manufacturing costs for the product and is the maximum number for the design team. This number should be tied to the profitability of the product and its intended selling price as defined in the Step One project proposal. It should also be tied to the anticipated annual sales volume. It is a goal that, if met, will assure meeting the intended market price for the product. If the sales volume meets or exceeds the projected numbers, the product will be a huge success. The manufacturing cost goal should be itemized to include the following categories: total cost of purchased parts, the labor cost to manufacture, and manufacturing overhead. The projected volume at which these costs are to be achieved should be stated also.

Special Considerations: This is simply a miscellaneous category that is a good place to discuss any design criteria that doesn't fit in any of the other sections already covered.

Following is a sample specification for an electronic digital thermometer.

Digital Thermometer Specifications

General Description: To develop a digital thermometer for use with an RTD (Resistance Temperature Detector) sensor to measure and display the temperature at the sensor location. The input signal to the digital thermometer is the resistance variation of the RTD sensor. The digital thermometer will operate off of 115 V AC, 60 Hz power and will utilize a standard enclosure, available off the shelf. The thermometer will be used routinely indoors and wall mounted. The temperature will be displayed on three seven-segment red LEDs. The accuracy should be within 0.5°F over a maximum range of 0° to 200°F.

Performance Specifications:

Rated Conditions:

Ambient Temperature = 25°C

Power Voltage = 115 V AC

Indication Accuracy: ± 0.25% of the 200°F range

Enclosure Specifications:

Size: Maximum size of 6″ × 4″ × 2″

Shape: The enclosure will be a purchased component from standard enclosures available on the market. A simple rectangular volume is preferable.

Material: Plastic preferred. Metal acceptable.

Human Engineering Aspects: Following is a list of key requirements:

1. The device shall be easily installed and connected.
2. It should be fail-safe if connected incorrectly.

Environmental Specifications:

Ambient Temperature:

Operation: 32°F to 122°F

Storage: –30°F to 122°F

Humidity: 10% to 90% relative humidity, non-condensing

Vibration: The digital thermometer shall be operable in a vibration environment with a vibration frequency from 0.3 Hz to 100 Hz with amplitudes as high as 0.2 g.

Shock: The digital thermometer shall be capable of withstanding shock that will most likely occur during shipment and must therefore meet the requirements of appropriate ICC specifications.

EMI Immunity: The digital thermometer shall be capable of operation in an environment as follows:

Radiated Electrical Fields:

–1 V per meter from 150 kHz through 25 MHz

–10 V per meter from 25 MHz to 1 GHz

Induced Magnetic Field: 20 A at 60 Hz into the enclosure

Power Line: ± 500 V, 50 ns duration over 360°

EMI Emissions: This design will meet the FCC Class B specifications for emissions.

Power Input: 115 V AC, 50 Hz to 60 Hz, ±10% in amplitude. Current draw a maximum of 100 ma.

Operation: The digital thermometer has no real operational requirements other than the requirement for sensor break protection. The sensor break protection will flash the display when the sensor is out of range. The digital thermometer will simply display the temperature when sufficient power is applied. When the signal is out of range, the display will flash.

Agency Approval Requirements: UL 1092 and equivalent CSA specifications

Special Considerations: None

Digital Thermometer Cost Estimates:

Project cost: $10,000

Manufacturing Cost Goal:

Purchased Parts: $95

Labor Costs: $40

Manufacturing Overhead: $15

Annual Volume: 1000 units

After completing the specifications, it is important that all members of the design and overall project team meet to discuss them. Often the process of putting together the specifications generates questions that had not been anticipated by the project's originators. Any issues should be discussed, resolved, and implemented in the final specifications. The final specifications should be signed off, dated, and distributed to all parties.

During most projects, situations occur that promote some change to the specifications. This is natural simply because it is impossible to foresee everything. One of the principles of concurrent engineering is to minimize or eliminate specification changes during the project. Specification changes that occur after significant project activity has been completed are very detrimental to the timely completion of any design project. If specification changes must be made, they should be implemented formally in writing, noting the revision level and date. The design specification should be a formal document and controlled like any other engineering document, by maintaining revision levels and communicating changes to all affected departments. With the specifications complete and signed off, the design process can begin. The following points are important for emphasis as you proceed with any design project:

1. The Project Specifications are a complete definition of the design problem and as such they are the primary basis for measuring the successful completion of the project.
2. Specification changes should be avoided during the project. If changes must be made to the specifications, they must be discussed and approved by all affected parties and implemented formally in the specifications.

Summary

In this chapter we discussed how concurrent engineering principles promote the success of a design project. We also discussed the six-step process for solving design problems. Step Two of the six-step process involves the definition of the problem in the form of a design specification. The process for completing design specifications was discussed and a sample design specification was presented.

Design specifications define all of the technical and business requirements for the new design. They should include a general description of the project and define its performance, power, and package requirements. They will define the environmental conditions under which the project will operate as well as how it will operate. These environmental conditions include ambient temperature, humidity, vibration, shock, EMI noise immunity, and the amount of EMI-radiated noise allowed. The specifications will also include agency approval requirements, cost goals and any other special requirements.

Once design specifications are developed, it is important that they be recognized as a formal document that is agreed to and signed off by the project team. It is also important to minimize the changes made to the specifications during the project. If changes must be made, the specifications should be formally revised and reapproved by the project team.

When the project is complete, the design specification is the primary document used to measure the success of the design project. Of course the ultimate success of the project will be determined by how well the design functions at its intended task and its financial viability. That is why it is so important that the design specifications reflect the complete requirements of the design up front, for careful consideration by the design team.

Reference

Stadtmiller, D. J. 2001. *Electronics Project Management and Design.* Upper Saddle River, NJ: Prentice Hall.

Exercises

1–1 List all of the benefits that result from a complete definition of the problem as described in Step Two of the six-step process.

1–2 List all of the benefits that result from the conclusion stage of problem solving as described in Step Six of the six-step process.

1–3 You are given a full-wave bridge rectifier that has four connections, all of which are unmarked. Define the problem and develop a solution plan for determining the two AC and DC plus/minus connections.

1–4 You wish to determine the key parameters and connections for a transformer that includes six unmarked connections. Define the problem and develop a solution plan.

1–5 You are given an LED display that includes a decimal point. The device is completely unmarked. Define the problem and develop a solution plan to determine all of its ten connections. There are five dual-in-line connections on the top and bottom of the display. The display's individual LEDs are available wired in either common anode or common cathode configuration.

1–6 List all of the benefits that result from the development of complete design specifications.

1–7 List the general categories that should be included in the design specifications.

1–8 How are modifications to design specifications implemented and controlled?

1–9 Develop a list of the operational parameters that must be considered when designing a +5 V supply.

1–10 Consider the design of a sine wave generator. List the operational parameters that should be considered in this design project.

2 Design Considerations

Introduction

Any product or system resulting from the design process must ultimately function in the real-world environment of its intended application. It may be a pacemaker that provides the critical pulse of life on a continuous basis in whatever environment the patient might encounter, or it might be a ground fault circuit interrupter that meticulously monitors the current in both legs of an electrical circuit, breaking the circuit upon detection of a minute difference in these current levels. An example of a less critical application is a portable CD player worn by a jogger that must play music while bouncing vigorously, in high or low ambient temperature conditions. From the jogger's perspective, the CD player's unflinching operation under these conditions is no less important and is probably the primary reason for its purchase.

Not only must a design function correctly in the intended environment, but it must also be manufacturable in a practical manufacturing environment. Increasingly demanding customers or users require that products and systems be easy to set up and use. It is expected that the design will function reliably and exude quality over the span of its projected life. In the unlikely event that the product fails, it must be easy to service or replace or represent a low enough cost that it can be discarded in an environmentally conscientious way.

These are design issues that vary significantly from the type of product to its manufacturing environment and application. They are the focus topics of this chapter. While these issues were discussed briefly in Chapter 1, this chapter provides more insight into their consideration at the beginning of the design process. This chapter addresses the following topics:

- Ambient Temperature
- Electro-magnetic Compatibility

- Packaging and Materials
- Manufacturability
- Ease of Use
- Seviceability
- Quality and Reliability

2–1 Ambient Temperature

The effects of ambient temperature on a circuit can change the value of a signal, or it can cause a component to fail due to excessive ambient temperature levels. When the value of an analog signal changes, this usually affects the value of some parameter important to the accuracy and/or function of the system. Because of the large margin between 1s and 0s, digital systems are usually not that sensitive to ambient temperature-induced accuracy errors. However, the accuracy of digital systems operated on a marginal basis can be compromised by ambient temperature. Generally speaking, when considering circuit reliability, the higher the ambient temperature, the shorter the operating life of the circuit. The effect of ambient temperature on a component can be determined from its data sheet.

What is most important at this point is the identification of the circuit areas that will be most affected by ambient temperature. The effect of ambient temperature must be considered on each component used in these circuits. To determine the change in analog signal levels that result from ambient temperature changes, review the temperature drift values in the specifications for each component in the circuit. These are usually stated in terms of parts per million per degree centigrade (5 ppm/°C), percent change per degree, or simply volts per degree. The effect of ambient temperature on various passive components is discussed in Chapter 4 along with a description of how to use ppm specifications. The sum total of these effects can be determined analytically by applying the induced ambient temperature changes in component values to the equations for the circuit in question. Circuit simulators, discussed in Chapter 3, can also be used to simulate the ambient temperature effects on all of the circuit components and worst-case scenarios can be developed. To meet the requirements for the design, it may be necessary to select different types or higher-grade components that are less sensitive to ambient temperature. Other approaches are either to control the ambient temperature of the sensitive circuit or to measure ambient temperature and compensate the circuit accordingly. Figure 2–1 summarizes the methods for addressing ambient temperature variations.

To consider the effect of ambient temperature on the reliability of electronic circuits and components, first determine the maximum temperature ratings for all of the components. The most critical are those that will get the hottest, those that dissipate the most power. Voltage regulators, amplifiers, and power switching circuits usually require the most power. Component packages for these devices usually accommodate heat sinks. Heat sinks are metal components designed to mount on electronic component packages for the purpose of dissipating heat. Make sure that high-power components have the proper heat sink to disperse the power they are dissipating.

1. Determine circuit areas sensitive to ambient temperature variations that affect system accuracy.

2. Estimate and/or simulate the net results of ambient temperature variations on these circuits.

3. Resolve unacceptable performance by:

   A. Selecting components with lower temperature coefficients.

   B. Matching positive and negative temperature coefficient devices to minimize the effect of temperature variations.

   C. Monitoring and controlling the temperature of the sensitive circuit to a constant value.

   D. Measuring the temperature of the sensitive circuit and providing some means of temperature compensation to the circuit.

▲ **FIGURE 2–1**
Design for Ambient Temperature Variations

Next, estimate the rise in temperature expected inside the enclosure for the electronic device being designed. Mechanical engineers have more experience with this and can determine this analytically because of their background in thermodynamics. The temperature rise can be estimated experimentally by placing the amount of power (use an appropriate number of miniature lightbulbs) that the circuit will generate in a simulated enclosure fabricated from the design material intended. Measure the rise in ambient temperature that occurs when the circuit goes from a no-power state to full power. Once the rise in ambient temperature is determined, add that number to the ambient temperature range to which the device will be exposed (available from the project specifications). The resulting total is the range in the internal case temperature to which each component will be exposed. Depending on the reliability requirements for the design, each component's maximum operating temperature should be at least 10% less than the maximum temperature allowed. Components that exceed this value can possibly be replaced with those that have a higher maximum operating temperature. Other resolutions involve reducing the power consumption of the electronic device, revisiting the design of the enclosure (its size and material), or adding additional air flow or cooling to reduce the expected internal ambient temperature rise. Figure 2–2 summarizes the methods for addressing ambient temperature reliability.

1. Determine the maximum temperature rating for all components.
2. Determine the circuit components that will dissipate the most power.
3. Estimate the rise in temperature above ambient temperature for any enclosure that houses electronics.
4. Maintain component operating temperatures to at least 10% less than their maximum rating by:
   A. Utilizing proper heat sinks on all high power components.
   B. Selecting components with higher operating ambient temperature ratings.
   C. Providing for as much passive air flow as possible.
   D. Providing active air flow (blowers, fans) or cooling as required.

▲ **FIGURE 2–2**
Design for Ambient Temperature Reliability

2–2 ▶ Electro-magnetic Compatibility

Electro-magnetic compatibility (EMC) is the design of electronic devices that promotes compatibility with electro-magnetic waveforms. It includes both the ability of an electronic device to be immune from electro-magnetic interference (EMI immunity) and the minimization of any generated and radiated electro-magnetic waveforms (EMI emissions).

EMI Immunity

For EMI to be a problem, there must be a source for the noise, a circuit that is sensitive to it, and a means for coupling or connecting the noise to the sensitive circuit. The primary sources for EMI are the AC power grid, radio signals, microprocesser-based equipment, electronic switching circuits, inductive switchgear (relays and contactors), ignition systems, and arc welders. EMI can enter an electronic instrument through external wire connections (conductive EMI), or it can present itself as an electromagnetic signal (radiated EMI). EMI can also be generated within the electronic device that can affect its own operation.

The result of EMI noise presented to electronic circuits is realized either as an analog signal that fluctuates incorrectly or as a circuit that ceases to operate properly in some way. This effect can last for a moment or continuously until the circuit is reset or powered down. One example is a digital display that fluctuates momentarily due to a noise spike on the AC power line. Another is a microprocessor-based circuit that locks up (does not respond to key depressions) due to noise induced from a cell phone transmission.

Conductive EMI is electrical noise that is introduced to the electronic circuit by a circuit connection to an external device. Noise either already exists on the conductor, is picked up by passing through an electromagnetic field, or results from sharing the same power supply or ground with another conductor. The AC power grid contains many noise signals that result from its generation and the frequent switching of large inductive loads. Whenever current flows through a wire, a magnetic field is created around the conductor. When a current-carrying conductor is exposed to another electrical or magnetic field, current can be induced in the conductor from this field. When this current is unplanned and unwanted, it is called noise or EMI that is magnetically coupled onto the conductor. When two circuits share the same ground or power supply, the current flowing through one circuit affects the voltage supplied to the other. In this case it is important to realize that each conductor represents a resistance to the circuit (see Figure 2–3). If the resistance value is made small (by using a larger conductor), the amount of noise voltage induced from one circuit to the other is minimized. This method of inducing noise signals onto conductors can affect external conductors and is a major source for generating noise internal to an electronic device. Radiated EMI occurs when an electronic circuit is exposed to an electromagnetic field directly and some noise current is induced in the circuit.

The methods for promoting EMI immunity in electronic devices combine the following concepts:

1. Minimize the amount of conducted EMI that can enter the electronic circuit.
2. Minimize the amount of radiated EMI the electronic circuit can pick up.

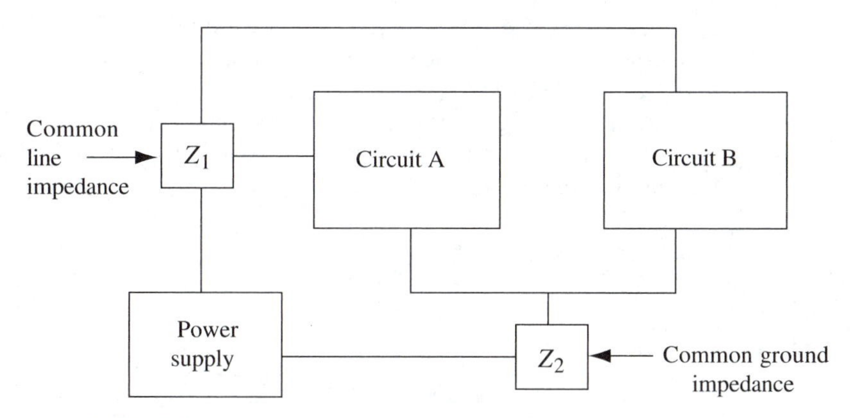

▲ **FIGURE 2–3**
Common ground and power supply

3. Minimize the amount of internal EMI created in the circuit.
4. Minimize the impact of any EMI noise that is presented to the circuit.

These are accomplished by implementing the following techniques:

1. *Elimination of the noise source.* If possible, the simplest and best approach is to eliminate the source of the noise completely. This can be done in a number of ways. If the source of the noise is an electromechanical contactor that is switching an inductive load, then the use of a transient suppressor, such as a metal oxide varistor (MOV) or an RC network placed across the contacts, can eliminate the source of the noise.
2. *Shielding.* The concept of shielding can be applied to conductors or complete electronic devices. It involves placing a conductive material around a conductor or device to pick up noise that would otherwise be induced onto it. The shield is grounded to drain off the induced noise signal with minimal impact to the conductor or device. Shielding works well in minimizing the affects of electrical fields over a wide range of frequencies. Shielding has a lesser effect on reducing noise generated by magnetic fields, especially at higher frequencies, because the current flowing in the shield can be induced on the shielded conductor unless the shield is made from a magnetic material. Shielded cable is available in a variety of configurations to apply the shielding solution to conductors. It is important to connect the shield to the zero signal reference for the signal being shielded. To shield circuits or devices, the entire device should be surrounded by the conductive shield and connected to the zero signal reference. Openings or gaps in the shield should be minimized.
3. *Grounding.* Grounding is the establishment of a zero signal point in the circuit. The basis for good grounding is to separate grounds for circuits where the possibility of noise voltages affecting either circuit exists. Ground loops should also be avoided. When a circuit is grounded in two places that are actually at different potentials, the result is a noise voltage that is the difference in the two ground potentials. This is known as a *ground loop.* Ground loops are also susceptible to noise from magnetic fields (see Figure 2–4).

   Good grounding practices involve keeping the grounds from different types of signals separate. The best example is a circuit that includes both analog and digital circuitry. In this case the analog circuit grounds and analog

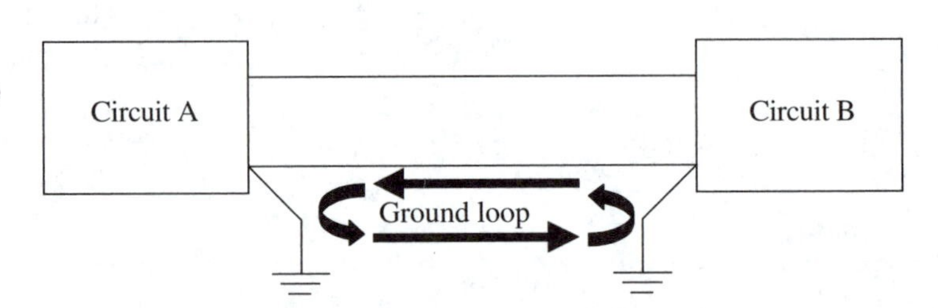

▶ FIGURE 2–4
Ground loop example

power supplies should be kept separate from digital circuit grounds and digital power supplies. In the case of long ground conductors, avoid the establishment of ground loops. For low-frequency signals, a single-point ground can be used, eliminating the ground loop that results from multiple ground connections.

4. *Impedance matching.* This approach is based upon matching the source impedance with the load impedance to create a balanced situation in a circuit. When this is done, induced noise voltages in the circuit become common mode noise as opposed to differential mode noise. The term *common mode noise* means that the noise is common to both the positive and negative sides of a signal. Differential mode noise is present on only one side of the signal. Common mode noise is easily removed from a signal by using a differential amplifier that amplifies only the differential signal and rejects the common mode signal.

5. *Noise filters.* To eliminate EMI from a circuit, noise filters can be used to attenuate noise from a signal. These are usually designed with capacitors but could be any combination of R, L, and C circuits. The values of the components are selected to attenuate the expected noise-voltage frequencies and pass the signal frequencies.

6. *Physical orientation.* When conductive or radiated noise sources exist, try to keep them as far away as possible from sensitive circuits. On circuit boards locate a switching power supply away from any analog signal amplifiers. In cables and on connectors, locate signals that can interact away from each other and place ground levels in between them. In the case of magnetic fields place circuit runs so that they are perpendicular with the magnetic field.

7. *Circuit isolation.* Transformers are used to isolate circuits for safety reasons, but there is also a benefit for eliminating noise voltages. When a ground loop exists between two circuits, a noise voltage is created between the two circuits if the grounds are at different potentials. Connecting an isolation transformer between the two circuits can break the ground loop (see Figure 2–5)

8. *Ferrite beads.* These are essentially a magnetic shield that serves to attenuate high frequencies without affecting DC or low frequencies. They are installed by passing a conductor through the ferrite beads. The result is a series RL circuit that serves as a high-frequency filter.

FIGURE 2–5
Isolation transformer example

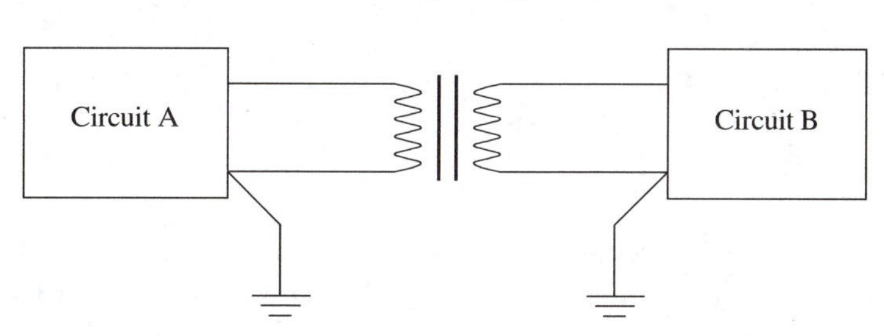

9. *Minimize the EMI effect.* There is always some level of EMI that will be presented to a sensitive circuit. In microprocessor circuits the result of EMI often causes the processor out of the normal program loop. It eventually locks up, unresponsive to any keyboard commands. To prevent this, a watchdog timer can be used to make this invisible to the user. A watchdog timer is a device that is reset at the beginning of every main program loop. The timer value is set at a time that is slightly greater than the maximum time to execute the main program loop. If the timer counts down before being reset at the beginning of the next program loop, then the processor must be out of the program loop. The watchdog timer invokes a reset and the processor will go to the beginning of the main loop. One other detail is that the RAM that contains parameters being used by the program must be preserved for this approach to work well. If the RAM is corrupted by EMI, then operation will not commence properly.

EMI Emissions

An electronic device can emit EMI as radiated waves or as interference that is conducted back into the power supply connections. In an effort to improve EMI immunity, many agencies have established standards to minimize noise emissions. By doing so they are taking the first step in promoting noise immunity—the elimination of noise at the source. The Federal Communications Commission (FCC), the Food and Drug Administration (FDA), and the European Committee (European Union's CE Standards) regulations all specify limits and test procedures to verify that all EMI emissions are held within their published regulations. These requirements are applied generally to any digital device that operates at frequencies of 10 kHz or higher. All of the steps listed previously to promote noise immunity can be utilized to minimize EMI emissions. There are additional measures that can be implemented when the printed circuit board is laid out. These are discussed in Chapter 5.

2–3 ▶ Packaging and Materials

There are a number of issues related to packaging and materials that are important for the electronic designer to consider. While mechanical designers will have primary responsibility for many of these design areas, these areas must be considered by the electronics design team as well. Following are the major packaging issues:

1. *Size and shape.* The overall size and shape of the package are indicative of the type of technology that should be utilized: SMT or through-hole technology.

2. *Material.* The enclosure material affects the EMI shielding, heat transfer capability, and agency approval requirements for the design. Even when plastic is the chosen enclosure material, EMI shielding can be accomplished

with the use of conductive paints applied to the inside surface of the enclosure. Metal enclosures will transfer heat to the outside much better than plastic, which is a consideration when estimating the expected heat rise in the enclosure. Finally, if agency approvals such as Underwriter's Laboratories (UL) or the Canadian Standards Association (CSA) are required, then any materials used should have the appropriate approval ratings. These ratings indicate the flammability of the material, and the safety standards of these approval agencies are a key concern.

3. *Enclosure seal.* The sealing capabilities of the enclosure also determine the amount of heat transfer that is possible from the enclosure's inside to its outside. There are various levels of standards for sealing and moisture-proof enclosures issued by the National Electrical Manufacturer's Association (NEMA). Examples are dustproof, waterproof under normal pressure, or waterproof under high pressure. These specifications might also dictate the need for what is called *conformal coating of the electronics* the entire immersion of an electronics assembly in a waterproof material. This is commonly done with varnishes and epoxies.

4. *Intrinsic safety.* In design applications in which there is extreme danger from any kind of electrical spark, there are requirements for what is called "intrinsically safe equipment." Chemical and petroleum processing plants are good examples of applications for these types of products. There are many levels of instrinsic safety specifications that limit any electronic switching below certain levels and contain any spark within the device.

2–4 ▶ Manufacturability

It is never too early to start thinking about a product's ease of manufacture, even though the lack of any physical form for the design limits visualizing many manufacturing design considerations. The circuit designer must become familiar with the needs and methods of manufacturing and experience the assembly process firsthand. It is also important to involve people in the manufacturing departments, such as manufacturing engineers, floor supervisors, and the assemblers, in the design process. It is the combination of all these sources of knowledge working with the designer that achieves the best results. As discussed in Chapter 1, teamwork and concurrent engineering principles promote the involvement of the manufacturing department early in the project and encourage a feeling of ownership by that department's members. They make the process work, and they will stand behind the product because they had a part in its development.

Manufacturing Process Definition

As the design develops, it is important to review and further define the manufacturing process. The manufacturing department leads this activity, but it needs much assistance from the design engineers. The definition of the manufacturing process involves the determination of the step-by-step assembly process. It also defines the assembly levels to be tested and the test procedures. As the various modules begin to take shape, make a list of the process steps that will be needed to assemble the product. Here are some key questions and issues that should be considered:

1. Can each module be completely tested on its own?
2. Can the final product be completely tested before being installed in its enclosure?
3. Only those manufacturing adjustments that are absolutely necessary should be included in the product.
4. Try to minimize and automate any testing and calibration that must be performed on the product.

There have been many changes in manufacturing ideologies over the last 20 years. The basis for this is a trend away from batch processing to a concept called *one-piece flow*. In batch processing, a large quantity of product is manufactured by assembling a quantity of each succeeding assembly level in different manufacturing areas, until the whole batch is completed. One-piece flow processing promotes the manufacture of a single-end product in a manufacturing cell until it is complete. This is also called *cellurized manufacturing*. Both processes are shown in Figure 2–6, which shows flowcharts for the manufacture of printed circuit boards (PCBs).

Take the assembly of a personal computer as an example. A company desires to manufacture 100 PCs in a week. Batch processing (see Figure 2–6) of these 100 units requires building 100 PC circuit boards on Monday. Testing and doing

▶ FIGURE 2–6
One-piece flow vs. batch processing

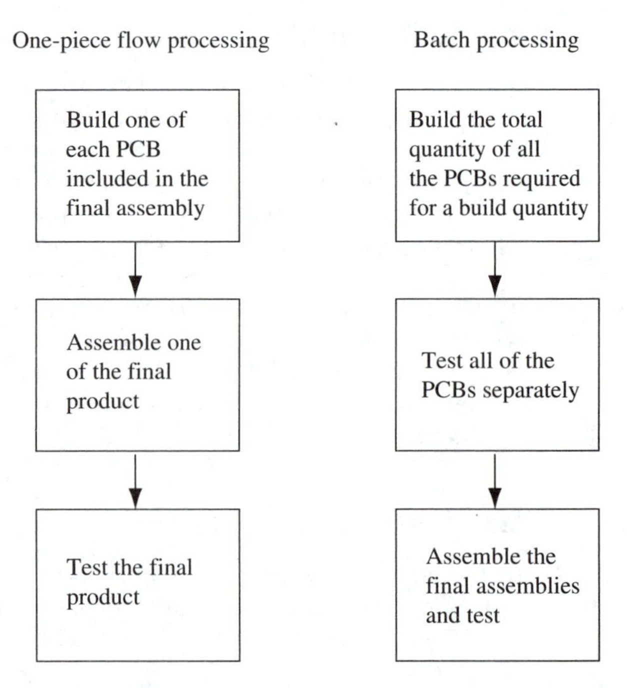

further assembly of those 100 circuit boards would occur on Tuesday. Final assembly of the 100 units would be performed on Wednesday. Then the final assemblies would be tested, calibrated, and burned-in on Thursday. The final assemblies would be packaged and ready for shipment on Friday.

A pure one-piece flow, cellurized manufacturing process (see Figure 2–6) dictates that one set of PC circuit boards be assembled, tested, and completed into a final assembly. Then another and another would be assembled until the 100 units are completed. All of the work would be completed in one manufacturing cell.

The benefits of one-piece flow processing are as follows:

1. Because each unit is completed shortly after all of the operations have been performed, process errors that occur are quickly identified and can be rectified before being repeated in other units.
2. Shipment lead times are reduced because units are available for shipment sooner than if they were batch processed.
3. One-piece flow represents a streamlined process in which the handling and movement of components and subassemblies around the plant are minimized.
4. All of the workers in the cell become cross-trained and are able to perform other cell operations. Because of this, they understand the complete assembly process and are more able to recognize problems in their infancy.

These are all significant benefits over the results experienced with batch processing. The problem with the one-piece flow concept arises with highly automated operations, where expensive equipment is required. Take the soldering operation for the PC example discussed previously. If batch processed, all of the boards would be assembled and then transferred to a wave-solder area for soldering. A pure one-piece flow process would require a wave-solder machine for each manufacturing cell. This is an expensive and space-consuming proposition that usually results in a compromise between batch processing and one-piece flow. The PC assembly process is just one example of the types of decisions that must be made. The manufacturing department or group and the design engineers must work together concurrently to determine the best methods for testing, calibration, and burn-in of all subassemblies and the final product.

Manufacturing Test and Calibration

When the manufacturing process is laid out, the stages in which assemblies will be tested are defined. These are all issues that will be developed later in the design process but should be in the background thought of the electronic designer as the initial design progresses. These stages will depend on the product and the type of process being used—batch processing or one-piece flow. Manufacturing test methods have also seen significant change in the last 20 years. These changes are similar to those discussed for manufacturing assembly.

To better understand this change in manufacturing test philosophy, let us explore the differences between batch processing and one-piece flow as they pertain to circuit board assembly and testing. The following represents a typical batch process circuit board assembly and test operation for a high-volume manufacturer.

Board Assembly and Test—Batch Processing

1. Incoming circuit board components are inspected before assembly.
2. Circuit boards are assembled in lots of 100 with auto-insertion machines. Some components are manually inserted before completing the assembly.
3. The assembled boards are sent to a wave-solder station, where they pass through the wave-solder process. The boards are cleaned and allowed to dry after soldering.
4. The quality department usually inspects each board or sample-inspects the boards for assembly errors and quality defects.
5. The inspected boards are sent to a test station, where an automated tester performs a series of preprogrammed tests on the particular circuit board.
6. If the board passes the tests, it proceeds to the next level of assembly. If not, a printout that includes information about the test failure is attached to the printed circuit board. The circuit board is sent to a test station, where a test technician will troubleshoot and repair the board. The repaired board is sent back to the automated test station for verification.
7. The final product is assembled using the circuit board under discussion in addition to other circuit boards that have experienced the same process. The final product is assembled and tested with a test fixture designed especially for the product. If the unit passes the final assembly tests, it is burned-in and readied for shipment. If not, the circuit boards that possess the malfunctions must be identified and corrected.

This process has been in use for many years. Following are the problems associated and experienced with this process:

1. Assembly or process errors are often repeated throughout an entire lot of boards before being detected during inspection or testing.
2. Because the process is segmented, there are time delays between the operations. There is generally poor communication between the people performing the different operations. Consequently, problem resolution and process improvement are not promoted.
3. Sometimes circuit boards pass their specific automated tests, but they will not function together with other circuit boards when assembled in the final product. Called *dynamic failures*, these result from the inability of the circuit board tests to verify the circuit board's complete operation.
4. In general, because so much time passes between the assembly of a circuit board and its assembly and test in the final unit, there is poor feedback to resolve quality issues as they develop.

In order to address and resolve these issues, cellurized manufacturing for circuit board assembly and test processes utilizing one-piece flow concepts were developed.

Board Assembly and Test—Manufacturing Cells

The following process describes cell manufacturing:

1. Incoming components are batch inspected, organized in assembly kits, and sent to the manufacturing cell.
2. The circuit boards are assembled and soldered within the cell. Small wave-solder machines have been developed for this purpose. If possible, all of the circuit boards included within a particular product are palletized (see Figure 2–7).
3. This means they are all attached to each other, assembled and tested as a unit, and then broken down into individual circuit boards for assembly. The individual boards are held together by a series of small laminate areas that are located between the individual boards. These areas are easily cut away or broken off to separate the boards.
4. Testing is completed with custom test fixtures within the cell. Only minimal testing is performed on the circuit boards until the final level of operation. This is easily accomplished using the palletized board approach. The testing philosophy applied here is to provide testing for a particular assembly only, when the cost of not testing the assembly is greater than the cost of testing it. In other words, an assembly should be tested only if the cost of finding a problem later is greater than the cost of testing the assembly.

The downside of cell manufacturing is realized when automated processes and equipment are applied. Because this type of equipment is expensive, it is hard to justify locating it in every cell, so two or more cells may share equipment by locating them in close proximity to it. The implementation of cell manufacturing has promoted the use of small, dedicated custom-designed test fixtures for testing assemblies and final products.

The test and calibration plan is developed as part of the overall manufacturing process and depends on the degree to which batch processing and cellular manufacturing techniques are applied. The test plan defines all of the assemblies that are tested along with the purpose for the tests.

Test Fixture Development

Once the test plan has been developed, test-fixture development can begin. Consider the test plan to be the specification for the design of the test fixture. When designing test fixtures, the design challenge is making quick and secure connections to points in the circuit where test measurements must be made. Measurement points that are available at the circuit board interconnections are easily accomplished by using a mating connector in the test fixture. For measurement points that are not accessible, a "bed of nails" approach must be used. A bed of nails is simply a test

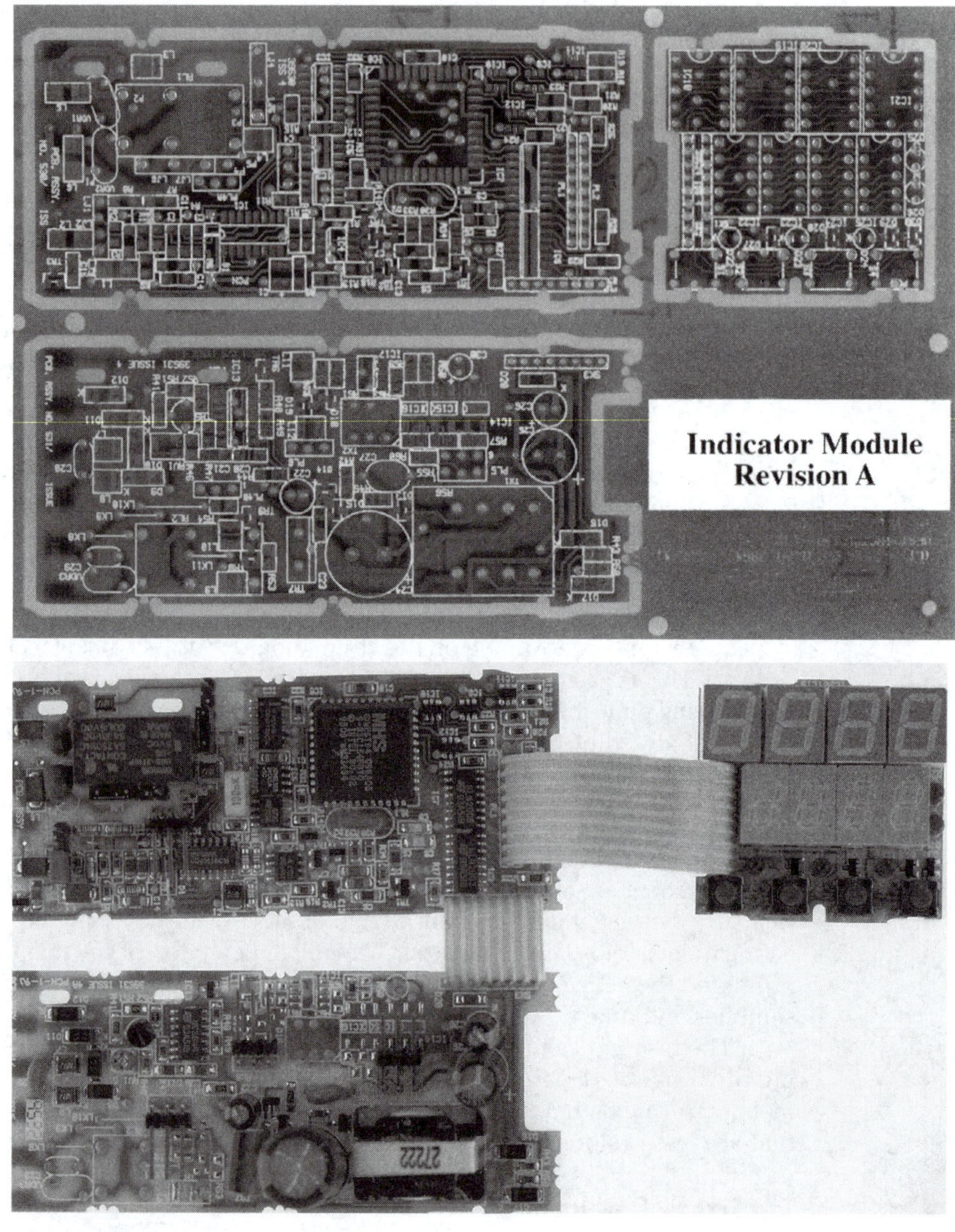

▲ **FIGURE 2–7**
Palletized circuit board

fixture that includes spring-loaded pogo pins (see Figure 2–8) that are located to make contact with specific pads on the bottom of the printed circuit board. There are two types of bed-of-nails fixtures. In one type the board is pressed down onto the bed of nails and mechanically held against the spring-loaded pogo pins. The other type is called a *vacuum fixture,* where air pressure pushes the pogo pins

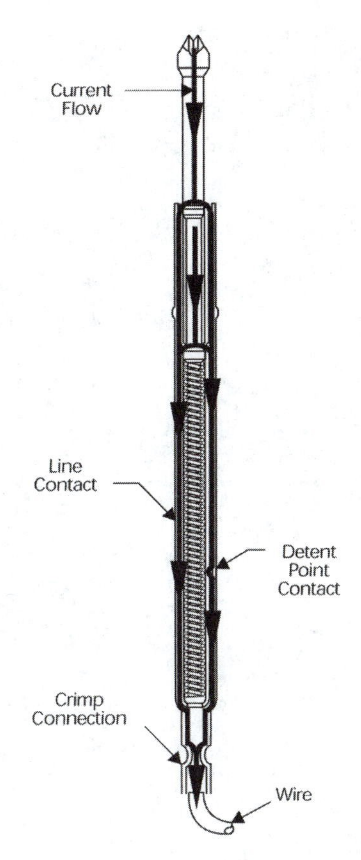

FIGURE 2–8
Spring-loaded test pin *(Courtesy of IDI Synergetix)*

against the board that is being tested. Companies that specialize in their design and construction usually develop vacuum fixtures. The design expertise and equipment to fabricate them are usually not available within most manufacturing companies.

Having determined a method for accessing test points, the fixture design can proceed. The test fixture is usually developed with an available programmable tester. Digital, analog, and combination circuit testers are available from a number of manufacturers (see Figure 2–9). Utilizing off-the-shelf automated test systems requires the development of high-level language programs that perform the tests. However, as mentioned previously, cellurized manufacturing has promoted the use of small custom-designed test fixtures for use in manufacturing cells. These custom fixtures can become significant design projects on their own and should be approached as a separate design subproject. Many of these are microprocessor-based designs, for which custom software is developed and circuit boards must be designed and laid out.

Test and Calibration Procedures

Along with any test or calibration fixture, a procedure must be developed to document the process to be followed while using the fixture. These procedures should be concise and accurate. Any changes made to these procedures should be formally

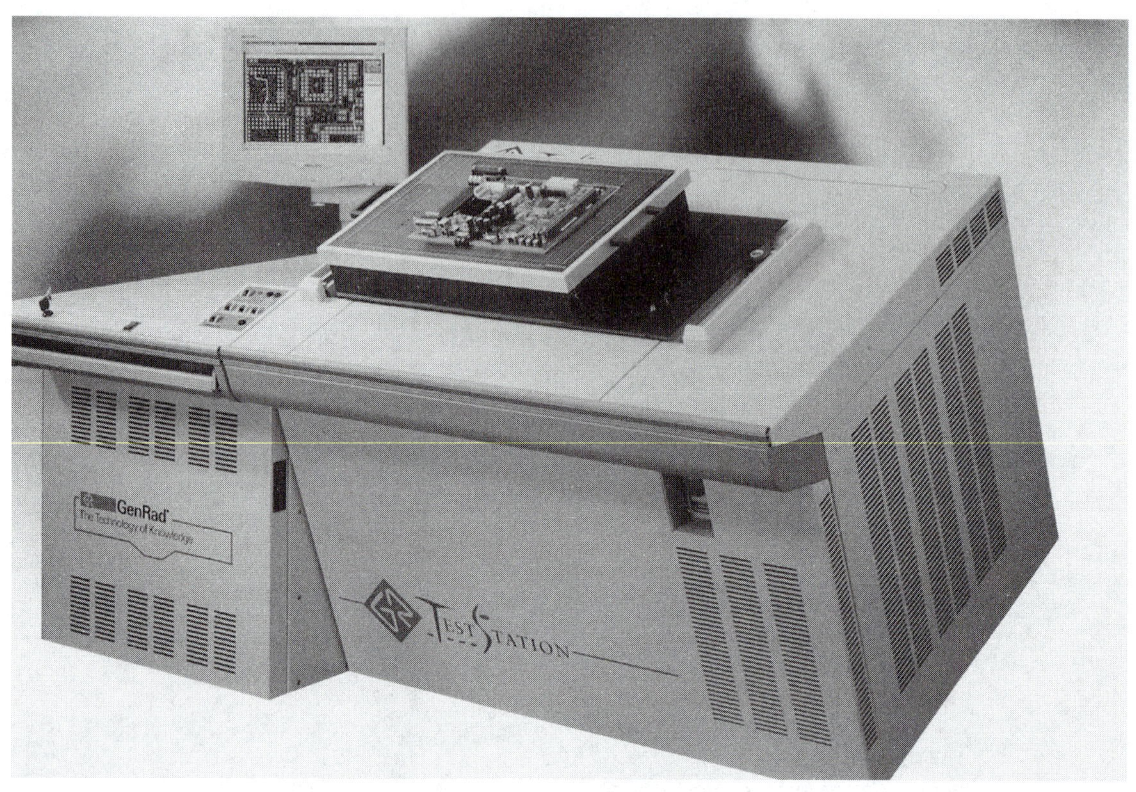

▲ **FIGURE 2–9**
Genrad tester *(Courtesy of Genrad, Inc.)*

controlled as any other engineering document. A test procedure is completed for each group of tests performed for an assembly and should reference the test fixture utilized. The test procedure should include the precise, step-by-step description of the tests to be performed in the sequence that they are to be completed. Completion of a test procedure often includes the compilation of test data and results for future reference. A blank sample of the test data sheet should be included in the test procedure. As with any manufacturing process, it is important to streamline all testing and calibration into a smooth process with minimal repetition of operations while maintaining accuracy and quality goals.

Burn-in

Another manufacturing process often used for electronic products is a process called *burn-in.* The purpose of the burn-in process is to minimize infantile failures. Infantile failures are premature failures that occur in weak components or connections. These failures happen shortly after a product is put in use. Many companies define infantile failures as those that occur between 1 and 30 days after a product is first used in an application. The goal of the burn-in process is to induce any weak components to fail before shipment to the customer, thereby causing a pending infantile failure to occur where it can be repaired without any negative impact on the customer.

There are many theories about burn-in and as many different practical approaches to its implementation. In order to induce pending infantile failures to occur, the product should be powered and operated in conditions that equal the customer's actual use. In order to accelerate the wear on weak components and accomplish burn-in quickly, the product is maintained at an elevated temperature. A more efficient burn-in is achieved if the temperature is cycled between room temperature, 25°C, and a temperature around 55°C. This causes the weak component to expand and contract as the temperature is cycled. If the lower temperature is set at 0°C instead of room temperature, the effectiveness of burn-in is further improved. The 0°C setting requires that the burn-in equipment have cooling capability. Finally, the on/off switching of power to the product is an additional stress that can often induce infantile failures.

Infantile failures are the result of weak or inferior components or connections. Many times components or connections are weakened by the manufacturing process. This is the case when a CMOS component is improperly handled and exposed to static electricity, thereby weakening it. Sometimes semiconductor components are weakened by exposure to severe temperature in a wave-solder process. Because components can be purchased pre-burned-in, many companies discontinue the burn-in process, neglecting the fact that the manufacturing process often weakens components.

Many companies believe in burn-in and perform it blindly, while others simply skip the process. The manufacturing department will always have a dim view of burn-in, which consumes much time and space and requires a significant capital investment. If a company has a solid manufacturing process that induces very few weak components, and it purchases top-quality components from quality suppliers and checks the process constantly, then burn-in may not be necessary. Otherwise, burn-in is the only final check to weed out weak components. A sure way to measure the need or effectiveness of burn-in is to monitor the number of infantile failures that occur in customer applications. The burn-in process should be considered at the very early stages of product design and reviewed throughout the project.

2–5 ▶ Ease of Use

"User friendly," a term coined in the 1980s, has been used—and overused—so much that people became tired of hearing it. The concept remains, though, and has increased in significance as a competitive advantage for all products. The primary reason for this is the increased use of microprocessors in products and the resulting increase in the number of features and options included within them. These variations in operation and features require the user to program or set up the product before use. The flashing clock readout on a VCR is a good example of the problem, as most people have a VCR, but few people can remember how to set its clock correctly after a power outage. Ease of use should be a goal for all products. Ease of use should include the sale, installation, set-up, programming, and use of the product. The best way to evaluate the ease of use of a product is to compare it to another competing design. Take every opportunity to compare the ease of use of products and make it a practice to evaluate every product's ease of use. The following areas are a measure of the ease of use of a product:

1. Requirement for an operator's manual. Can the majority of the most common operations be completed without the use of an operator's manual?
2. Quantity and clarity of the required steps. How many steps are required to complete a particular operation, and how clearly are these steps presented?
3. Amount of information to be remembered. How many items in the product's operation must be committed to memory in order to operate it?

To summarize, a product that is easy to use is one that can be operated almost intuitively without an operator's manual. If possible, the product should prompt you through the process. The steps required to set up and use the product should be minimal and clearly presented. Finally, the amount of information about the product required to perform common operations should be minimal and always noted on the product somehow.

2–6 ▶ Serviceability

A product's service requirements should be stated in its specifications. This is an area marketing people tend to gloss over so the service engineering group and the design engineers need to add focus to this topic. Serviceability can range from complete field repairability (repairable to the component level) to no serviceability at all (a throwaway unit). Very seldom are today's products repairable to the component level. Most assemblies are so small and packed together so intricately that disassembly usually damages the product. If the circuit board uses surface mount technology (SMT, explained in Chapter 3) components, it is usually impractical to replace those components. A typical intermediate position is to provide field reparability down to the board level, where circuit boards are changed out and the failed board is deemed not repairable and scrapped. Whichever the case, it is important to consider the serviceability of the product more seriously at this time. If the unit must be repairable to some degree, it must use packaging hardware that will allow disassembly and reassembly in the field. Many times the issues of ease of manufacturability will go against the ease of serviceability.

Example 2–1

A good example of the conflict between ease of manufacture and serviceability involved the method of mounting printed circuit boards to a plastic front panel on an industrial temperature controller. The initial design utilized threaded inserts pressed into plastic tabs that were part of a plastic front panel. The circuit boards were securely attached to the front bezel with screws. As a cost-saving measure, the manufacturing department desired to eliminate both the time-consuming insertions of the threaded inserts as well as the screwing operations. The design of the plastic front panel was changed to include plastic clips that would attach the circuit boards to the front panel. These design changes were integrated into the tooling for the plastic front panel and the change was implemented. The result of this cost-saving measure was a significant reliability and serviceability problem. The plastic clips did not hold the boards in place securely. This made the interconnections between the boards intermittent in high-vibration applications. The plastic clips also made it very difficult to remove the circuit boards once they were attached to the plastic front panel.

The result of this poorly implemented project was an increase in field failures. Also, customers and field service people were very unhappy with the new process for removing circuit boards from the front panel. At a great cost to the company, the design was changed back to the original design that utilized mounting screws.

2–7 ▶ Quality and Reliability

Quality as a measurement is a very broad term that is often made on a relative basis. It includes all aspects of the product, including those previously discussed: manufacturability, serviceability, and ease of use. Reliability on the other hand measures the performance and the time before failure.

Customer Quality Performance

There are a number of ways to measure the quality performance of a product from the customer's perspective. The first and most important is the customer's opinion about its quality. These measurements reflect the customer's expectations and therefore are subject to their point of view. How easy was the product to install? Was the operator's manual helpful and easy to understand? One customer might think that the operator's manual was comprehensive while another might think just the opposite, that the manual did not have enough information—or the right information. These measurements are best taken with a simple questionnaire that allows the customer to comment on the general appearance, function, and ease of use of the product. The performance of the product in these areas should be determined by taking an average of the respondent's replies, developing a rating for the product, and comparing the change in rating as attempts are made to improve weak areas.

The second method for measuring the quality of a product involves its failure and the reason for failure. From the customer's perspective, a failure is a failure. However, failures that occur sooner rather than later leave a less favorable impression. There are four time lengths, or levels of failure, that have important distinctions and are explained next. Figure 2–10 shows a summary of the four levels of failure.

Out-of-Box Failures

These are the most severe failures, because they cause the greatest amount of dissatisfaction for the customer. In these cases, as the name implies, the customer receives the product and unpacks it, and the product fails to perform right out of the box. Out-of-box failures can be induced by rough handling during shipment;

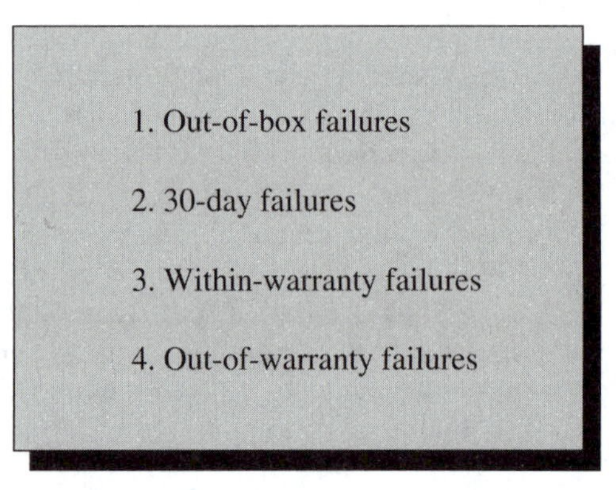

FIGURE 2–10
Summary of product failure types

otherwise they indicate a complete failure of the manufacturer's quality system. If the failure was induced by rough shipment, then a determination should be made as to the correct use of the packaging and any evidence of improper handling should be reviewed.

When the out-of-box failure is not due to shipment, then the exact nature of the failure must be determined. The most prevalent cause of these types of failures is incomplete testing of the entire assembly before shipment. Once identified, every effort must be made to make changes that prevent any reoccurrence of this particular problem. Another type of out-of-box failure may not involve a failure of the product at all. An example is when an operator's manual does not explain the operation of the product to the customer to the point where the customer believes it is not functional. In this case the product failure is caused by the operator manual.

Failures Within 30 Days of Receipt

These failures are called *infantile* or *premature failures*. The product is received by the customer, set up, and verified for use. After a period of less than 30 days, the product fails. This is a very special case, because the product did function at the customer location and now it has failed. This type of failure is usually caused by a component failure or a weak electrical connection. These failures can occur after routine use or when excessive strain is placed on the product. It is therefore important to determine the failed mechanism, be it a component or connection. The application of the product should be reviewed with the customer for any specifications that are being exceeded. If the failure occurs under normal use of the product, it is a true infantile failure. The only method available to minimize infantile failures is with the burn-in process discussed previously. Infantile failures are almost as aggravating to the customer as out-of-box failures.

Failures Within the Warranty Period

These are similar in nature to infantile failures except that the product operates for a longer period of time (greater than 30 days) but less than the specified warranty period. Of course, some percentage of the product will always fail before the warranty has expired. The actual percentage level of product failures under warranty significantly affects the product's success. These failures are usually due to a weak component, connection, or a design problem. Data indicating the root cause of the failure should be maintained and monitored.

Failures Outside the Warranty Period

These failures are the least bothersome for both the customer and the supplier. From the customer's perspective, that depends on the amount of time after warranty expiration that the failure occurs. A failure that occurs one week after a one-year warranty period is viewed much differently than a failure occurring one year after a one-year warranty expiration. Although these failures are not a financial burden to the manufacturer, they do represent a problem to the customer. Therefore, the reasons for these failures should be determined. Data regarding them should be kept and reviewed regularly while efforts are made to reduce the overall mean time between failure (MTBF, see next section) for the product. These four categories of failures represent four different aspects of the reliability of the product and the quality process that has been put in place to ensure it.

1. Out-of-box failures are a measure of how well a supplier checks out, packages, and ships the product to the customer.
2. 30-day failures are a measure of infantile failures, which can be affected by burn-in.
3. Warranty failures indicate the level of failures within the warranty period excluding 30-day failures. These are usually the result of inferior or misapplied components or other design problems.
4. Non-warranty failures are used in conjunction with all other failures to determine the overall MTBF for the product. These are the result of inferior or misapplied components, other design problems, or the limited life of the product.

Product reliability can be enhanced by a combination of design improvements, test fixture and procedure changes, product packaging, information enhancements, and modified burn-in.

Reliability Projections

When a design is completed, there are always questions about its reliability. The reliability of a design is determined on average by how long it functions properly in the intended application without failure. In industry, the reliability of a design is measured by a term called the *mean time between failure* (MTBF). Statistics theory defines the arithmetic mean as an average. The MTBF is the average time between failures of a product design. There are two ways to project the reliability of a design: accelerated life testing and statistical mathematical reliability projections.

Accelerated Life Testing

Accelerated life tests expose a design to conditions in which failure-prone areas are stressed to accelerate their wear. Take the example of a control relay that is normally cycled on and off five times a day. Accelerated life testing would cycle the relay on and off five times an hour. The net effect of 1 day of testing would simulate 24 days in the intended application. To properly set up an accelerated life test, the design should be reviewed to determine all of the expected failure areas. The test developed should include some component that addresses each of the expected failure-prone areas. Continued exposure to high ambient temperatures is the only way to accelerate wear on many internal electronic components.

Statistical Reliability Projections

The statistical projection of reliability of an electronic design is accomplished by the use of reliability data that is available for many components. This reliability data has been developed with military specifications developed for the purpose of projecting reliability. As discussed previously, military equipment is subject to the most rigorous performance, environmental, and reliability specifications. The data supplied in military specifications lists a mean time between failure for components that is determined by the power, voltage, current, and ambient temperature at which the component is operated. The MTBF for each component utilized in a design is totaled. The grand total is divided by the number of components to determine projected MTBF for the complete design. Software is now available that will take all of the components included within a design and project reliability while considering the operational data supplied. Manual methods for completing these calculations can be cumbersome for a complicated design.

Designers become aware of these are issues as they gain experience in a particular type of design application. Any application can have its own unique set of problems and priorities. What is most important is that circuit designers and software developers become intimately aware of a design's manufacturing and application environment. This can be accomplished by working on the manufacturing floor, going on trips to experience product applications, keeping abreast of trade periodicals and attendance at trade shows. Every attempt should be made to foster these types of activities. Finally, communication amongst engineering, manufacturing, service and quality is critical to the design's ultimate success.

Summary

In this chapter we discussed how to consider and address the most important general design issues that are confronted in most design projects. The issues that most affect the operation of the design are ambient temperature and EMI immunity. The affects of ambient temperature can be minimized by the selection of better grades of components, by temperature compensation, or by temperature control. EMI immunity is generally improved by minimizing the radiated, conducted, and internally generated EMI induced in the product.

The design issues that most affect the end user of the design and/or its manufacture and support are package technology (through-hole or SMT), manufacturing process selection, ease of use, serviceability, quality, and reliability. These issues can gain the most leverage from the concurrent engineering approach described in Chapter 1. In this chapter we discussed the criteria for selecting either through-hole or SMT circuit board processes. We defined the two main variations of manufacturing processes, batch processing and one-piece flow, and how they affect the assembly and testing of a product, as well as their strengths and weaknesses. Methods of automated testing electronic designs were reviewed as methods for making test point connections with either connectors or a bed-of-nails. The manufacturing process called *burn-in* was presented as a method of reducing infantile failures in a product design. Measurements and methods for improving the ease of use and serviceability of a design were reviewed.

Finally, the quality and reliability requirements for a design were discussed. Quality was defined from the customer's perspective in very basic terms while reliability was explored a little more deeply. This chapter presented four basic levels of reliability measures as it defined out-of-box, 30-day, warranty, and non-warranty failures. The significance of each category was discussed as well as methods for improvement. The reliability of design is generally measured in mean time between failures (MTBF). The MTBF can be projected with accelerated life testing or by using statistical methods.

Reference

Stadmiller, D. J. 2001. *Electronics Project Management and Design.* Upper Saddle River, NJ: Prentice Hall.

Exercises

2–1 Name the general types of problems that are experienced by electronic systems when they are exposed to high ambient temperatures.

2–2 What remedies are possible when an electronic component in a system is exposed to ambient temperatures that exceed the maximum temperature ratings of the device.?

2–3 List the four major packaging and material issues discussed in this chapter.

2–4 List the general concepts for maximizing EMI noise immunity.

2–5 In your own words, define what is meant by the term *ground loop.*

2–6 Discuss the main difference between conducted and radiated EMI.

2–7 Why are the grounds from analog and digital circuits kept separate as much as possible?

2–8 What are the two basic types of manufacturing processes from which to select when determining the manufacturing process for a new product?

2–9 Define what is meant by the term batch processing and discuss its advantages and disadvantages.

2–10 Define what is meant by the term one-piece processing and discuss its advantages and disadvantages.

2–11 What is meant by the term *palletized board assembly?* What are the advantages of using the palletized board concept?

2–12 Define what is meant by an *infantile failure.*

2–13 What is the process called *burn-in* and what is its purpose?

2–14 What parameter is used to describe the reliability of a design?

2–15 What are the two primary ways of projecting the reliability of a design?

2–16 List all of the ways discussed in this chapter that the customer quality level of a product can be measured.

2–17 Explain the difference between 30-day failures and out-of-box failures. Which type is more critical and why?

2–18 How can infantile failure levels be reduced?

2–19 How can out-of-box failures be reduced?

3 The Preliminary Design

Introduction

As in life, the first steps made in the design process are often the most difficult ones. Most design problems are complex and actually represent many different design problems all wrapped up into one. The tendency of many first-time designers is to try to design a complete solution all at once. It is extremely difficult for the mind to solve many complex problems simultaneously, so it is necessary to break down the problem into its basic functional components or modules. It is also important to plan the solution (Step Three of the Six Steps) in a logical manner.

As students, we become used to solving problems by analyzing well-defined circuits. Facing a new design problem with a clean sheet of paper is a revealing moment for the beginning design engineer. This chapter will discuss beginning the design process and developing a preliminary design by addressing the following topics:

- Divide and Conquer
- Preliminary Design Issues
- Enhancing Creativity
- The Initial Design
- Circuit Simulation Software
- Breadboarding

3–1 Divide and Conquer

The best way to start the process is to break down the design problem into smaller blocks or modules, these being the next largest functional modules contained within the overall design. Look at this like a gift-wrapped present that when unwrapped reveals another fully wrapped gift, and then another, and so on. A complex project sometimes has many sub-levels.

Example 3–1

Consider the problem of designing an audio tape deck to play back and record standard audiocassette tapes. In this example the top-level design problem is broken down into subprojects. In order to define the functional requirements inside the tape deck, think about the functional requirements from the operator's point of view. The top-level operational requirements of a standard tape deck are as follows:

1. Power on/off
2. Insert/eject cassette tapes
3. Play tapes
4. Fast forward
5. Rewind
6. Pause
7. Record
8. Stop
9. Adjust record levels
10. View record levels

The operational requirements listed can be categorized into the following common functional circuits in a block diagram:

1. *Tape Deck Control.* This module will control turning the tape deck on and off, all tape deck movement, insertion and ejection of tapes, and the selection of Play/Record.
2. *Audio Signal Playback.* This module includes the audio playback head and amplifiers to the output.
3. *Audio Signal Record.* The audio record head, the signal from the record inputs, the adjustment of the input signals and the display of the record levels reside in this module.
4. *Power Supply.* All of the modules listed previously will need a variety of DC power, which should be centrally developed and supplied by this module to the various circuits.
5. *The Tape Transport.* The cassette tape and the complete mechanism.

The functional requirements for the audio tape deck design have been broken down into five modules that can be approached as separate designs. See Figure 3–1 for the actual block diagram.

After subdividing the project, Step One and Step Two should be re-applied to each module. The level to which this occurs depends on the complexity of the

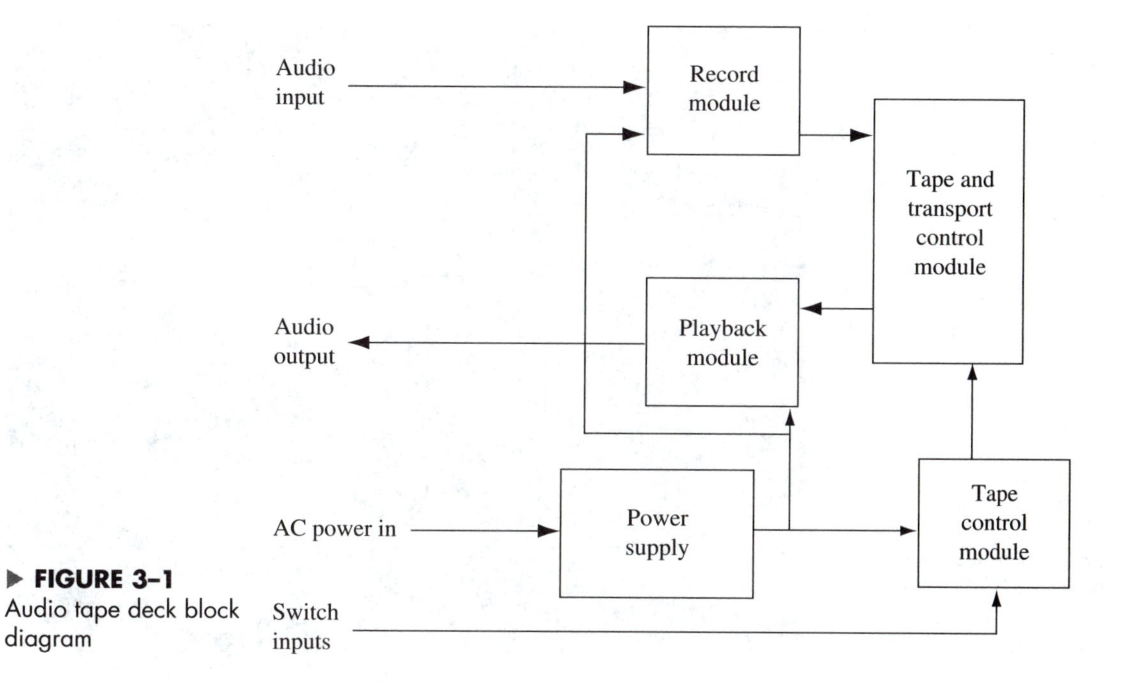

▶ **FIGURE 3–1**
Audio tape deck block diagram

module. In some projects the modules are highly complicated circuit boards, and it is appropriate to develop a complete specification for each. For other projects the requirements can simply be listed as they are designed.

3–2 ▶ Preliminary Design Issues

After the design problem has been broken down into modules, there is more research to be done before actually designing the circuits. This is necessary because even though data was collected and the design problem was defined for the overall design problem, it has now been broken down into smaller modules. More details about the design of these smaller modules must be determined. In a sense we are starting the Six-Step process all over again as we attempt to solve the smaller design problem of the submodules. The research process described in Step One of the Six Steps should be repeated for the submodule designs. If the module is complex, then Step Two should be completed for the module by developing design specifications.

Technology Selection

Another important choice to be made is the package technology that will be used for the project's electronic components. This is important for many reasons that will become obvious as we discuss them. The primary decision to be made is whether to use through-hole or surface-mount electronic package technology. Through-hole technology (see Figure 3–2) is the package technology that requires

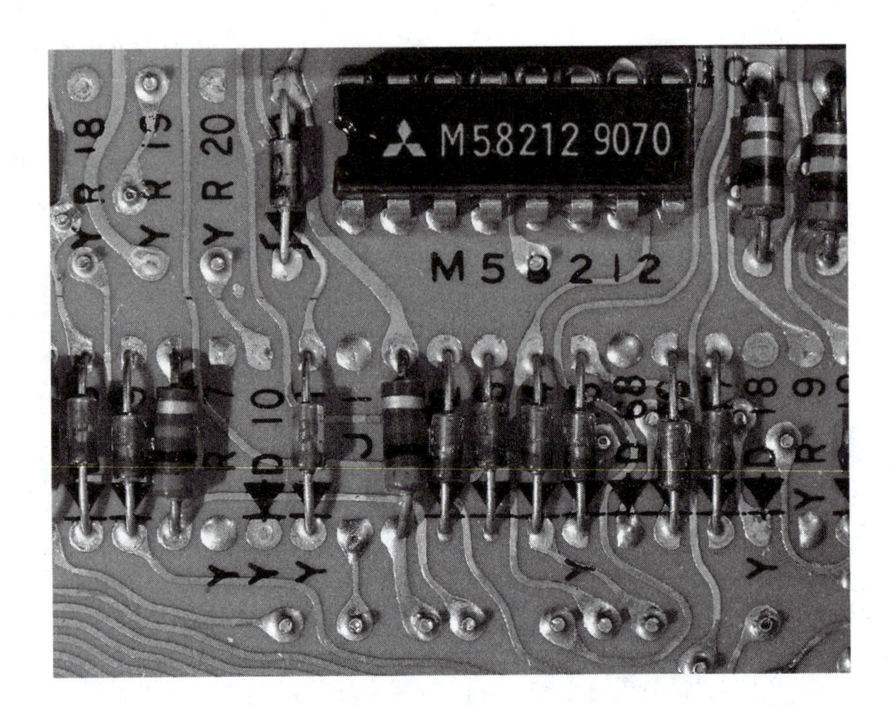

▶ FIGURE 3–2
Through-hole technology *(Alfred Pasieka/Science Photo Library/Photo Researchers, Inc.)*

a hole in the middle of a circuit pad located on the PCB for mounting the component. With surface mount technology (see Figure 3–3), the component solders directly to a pad that is on the surface of the board. Because a hole is not required in the PCB for SMT, the lead spacing can be made much smaller. This allows the reduction of the component sizes as well. The decision between through-hole and surface mount technology is an easy one because of their key differences and their impact. To make this decision, first pose the following questions:

1. Do the product specifications call for very low costs at very high-volume levels?
2. Does the product require an extremely small physical size?
3. Does the company that will manufacture the circuit boards have SMT manufacturing equipment and capabilities?

If the answer to any one of these questions is yes, then SMT should be strongly considered. Otherwise, through-hole technology should be utilized. To get a better understanding of this question, let's review the advantages and disadvantages of SMT:

Advantages:

Very small package sizes are possible.

Low assembly costs are possible in high volume through automated assembly.

▲ FIGURE 3–3
Surface-mount technology

Disadvantages:

SMT boards are difficult to breadboard and prototype.

Components have longer lead times and larger minimum buy quantities.

A large investment in equipment and know-how is required.

These boards are not easily repaired.

The criteria for this decision have been presented. Unless one of the unique advantages of SMT is required (small size or low costs @ high volume) or the company already possesses SMT capabilities, it will not be cost effective to make the investment required for SMT. This situation will change when SMT components cost significantly less than through-hole components, and/or SMT equipment prices continue to decline.

Many times SMT is utilized and all the components needed for the product are not available in SMT packages. This causes a difficult problem because both technologies must then be utilized, which detracts from one of the key SMT advantages: automated assembly. In these cases the SMT components are installed first with an automated pick and place machine. Then through-hole components are inserted manually as a secondary operation. SMT packages are unavailable when the power level is too high to be dissipated in an SMT package or when the volume level is too low for an SMT package to be developed. When using both through-hole and SMT package types, it is advisable to keep the application of the "mixed" technologies as localized as possible. If possible, try to keep all the SMT components on one circuit board. When both technologies are used on the same board, place all through-hole components on one side and SMT parts on the other side.

Manufacturing Cost Budget

At this point it is a good idea to consider the manufacturing cost of the design. A manufacturing cost budget should be developed for each of the functional modules. This is done by dividing up the total manufacturing cost goal for the product among each module, packaging/enclosure costs and total assembly costs. As the design develops, this cost budget should be used as a tool to measure the performance of the design in meeting the cost goals. The total cost of each module will be determined by adding the total cost of the parts and components within that functional module.

3–3 ▶ Enhancing Creativity

Most people are not exposed to a study of creativity in either high school or college. Nevertheless, it is a critical aspect of many careers, especially engineering. Creative thinking is a most useful process, yet it is largely undeveloped in the academic and industrial world. Creative thinking is enhanced when both information gathering and problem definition occur before starting the creative process.

Being creative involves the use of the subconscious, that part of our minds that is a mystery to us because we can't directly access it. Think of the mind's conscious and subconscious parts as being like a computer operating in a multitasking environment, where the background operating system controlling all of the basic functions compares to our subconscious mind and the specific applications programs represent the conscious mind. The applications programs don't appear to directly access the operating system software, but nevertheless, the operating system is there, operating and controlling all critical operations. The operating system in this case actually does a lot of work for the applications programs that are running and is an integral part of their results.

Compare your subconscious mind to the operating system software in your computer. It is a powerful and necessary part of your mind and it can do much good work for you. It differs from software in that it has tremendous creative power. Some of its best work is in the area of producing ideas, novel off-the-wall ideas that can solve problems in unique ways. However, the subconscious mind has some of the same limiting factors as a computer operating system. It has many applications programs running at the same time (i.e., making sure you are breathing and speaking and that your heart is beating) and it has a hard time being creative during busy periods or after a busy day. The best way to put your creative subconscious mind to work is to feed it with all the information and a problem definition and then assign it the task of coming up with some ideas. Relax and let your subconscious mind perform its magic.

Most of us have been troubled by a problem, then "slept on it," only to wake up with a solution right on the tip of our tongue. Many experts believe that the subconscious mind is our most valuable creative tool. When planning daily activities, schedule the review of particularly difficult problems toward the end of the day. Tell your subconscious mind—almost in jest—to figure out a solution. Then relax, sleep on it, and collect the ideas in the morning. The next time you are faced with a significant problem, try giving your subconscious mind a special assignment while you go take a nap. It has been said that the best work flows forth effortlessly. This is true, especially if the preliminary groundwork is done to set the stage.

After your subconscious mind has had time to review the problem, let your mind ramble as you write down the fresh ideas that come to your mind. It is important not to be too critical at this point; turn your discriminating, editing mind off. Now spread out the different ideas and let your mind focus on one idea and see what else your mind comes up with that is an offshoot of the original idea. Let your mind ramble again like this for each of the differing root ideas.

The end result should be a list of different and unique ideas. Many of them will be somewhat ridiculous, but others will surprise you. At this point you should evaluate the list of ideas with only a semi-critical outlook and attempt to combine the positive aspects of one with another. If things don't fit together, try turning one idea inside out, upside-down, or invert it. Play with the ideas until they work for the situation or you are convinced that they won't work. It helps to involve more than one person. Actually, the more people involved, the merrier, and the better chance for a novel idea.

After evaluation, if you don't have an acceptable idea, take note of what you have learned in the process and go back through the cycle again. Figure 3–4 shows a list of the steps to enhance creativity.

Example 3–2

This is an example of brainstorming often used at seminars on improving creativity. With about 20 people in the room, the group is asked to estimate how many different facts they anticipate being able to list about an average pencil. The group discusses this quickly and estimates that they will be able to list between 30 and 40 unique facts about the pencil. The session starts at one end of the room as each person in order is asked to state a unique fact about the pencil. Each fact is written and numbered on a marker board. The brainstorming proceeds quickly for about two complete cycles through the group. Then things slow a bit but still progress. At the end, the group may complete almost six cycles.

1. Gather information.
2. Define the problem.
3. Relax, let your subconscious mind work on the problem.
4. Collect the ideas by listing all of them.
5. Categorize the initial ideas.
6. Try combining or modifying the initial list of ideas to create more ideas.
7. Edit the list of ideas by excluding the ones that obviously do not apply to the problem.

FIGURE 3–4
Steps to enhance creative problem solving

One group developed a list of 114 facts about an average yellow pencil before being unable to continue. The process revealed the power of brainstorming, as one unique idea uncovered a number of other facts about the pencil. Each unique fact fed the process a little longer until, finally, the group could not continue. This exemplifies the creative power that is possible by combining the random thought processes and ideas of a number of people.

3–4 The Initial Design

We are now ready to proceed with the electronic design, which should proceed in a logical sequence with the input section first. The power supply designs should be completed last, if possible. Many times the designs of the different functional modules will proceed in parallel with different designers assigned to each module. The input and output of each of the functional blocks must be known in order to begin the design. You can usually begin sketching circuits on scrap paper. As the design becomes somewhat firm, you can make a neater sketch on paper with a light grid background.

Example 3–3

Continuing Example 3–1 let's complete the preliminary design of the tape deck control module of the audio tape deck. As shown in Figure 3–5, the inputs to the tape deck control module are the stop, play, record, fast forward, and reverse but-

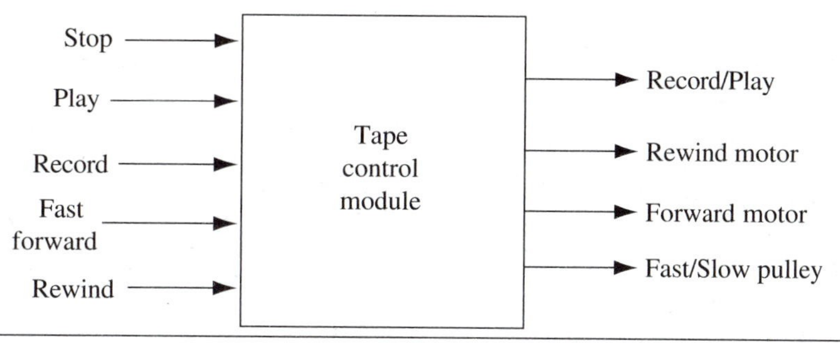

Stop	Play	Record	Fast Forward	Rewind	Record/Play	Rewind	Forward	Fast/Slow
1	x	x	x	x	0	0	0	0
0	0	0	0	1	0	1	0	1
0	0	0	1	0	0	0	1	1
0	1	0	0	0	0	0	1	0
0	1	1	0	0	1	0	1	0

▲ **FIGURE 3–5**
Tape control module block diagram and truth table

tons. The necessary outputs to the tape transport are digital signals that represent the following actions: record/play (1 = Record, 0 = Play), rewind motor, forward motor, and fast/slow speed pulley (1 = Fast, 0 = Slow).

This is a typical logic design problem in which the input and output relationship needed can be shown on a truth table. The truth table for the input and output relationship is also shown in Figure 3–5, which shows that there are four specific input codes that will result in a specific output code. For all other input codes the output code will be all zeros. Using logic circuit design methods, the circuit shown in Figure 3–6 was developed.

3–5 ▶ Circuit Simulation Software

Circuit simulation software has become increasingly effective and easy to use over the last 15 years. Circuit simulators allow the circuit designer to draw a circuit schematic and simulate circuit operation on a personal computer. The simulation allows the examination of any voltage or current in the circuit as a function of time. Circuit simulators provide a very quick and easy way to evaluate a particular design concept without access to expensive test equipment or having to procure parts and assemble a breadboard.

The first widely accepted circuit simulator was SPICE2, which was developed at the University of California–Berkeley in the mid-1970s after having modified its original SPICE program. SPICE stands for *Simulation Program with*

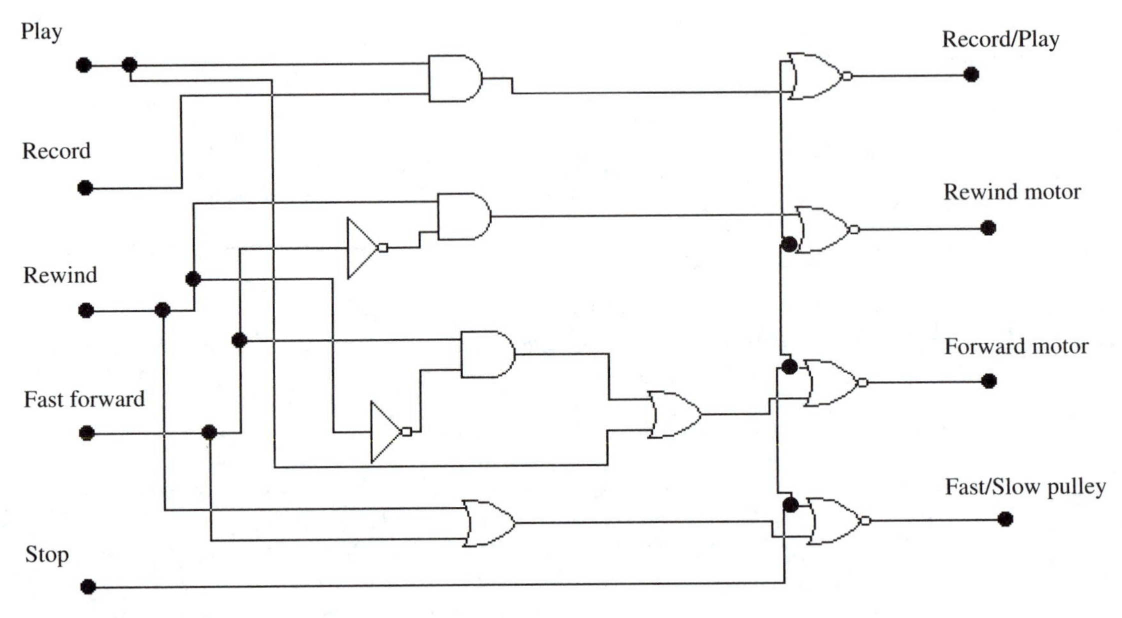

▲ **FIGURE 3–6**
Circuit solution

Integrated Circuit Emphasis. SPICE, SPICE2, and the most current Berkeley version, SPECE3F5 (commonly called BSPICE) are the recognized standard for analog circuit simulation. They were developed with public funds, so the software is in the public domain and available to U.S. citizens. XSPICE, a custom version of SPICE made for the U.S. Air Force, includes special modeling subsystems. PSPICE is a commercial version of SPICE developed by the MicroSim Corporation to operate on personal computers. PSPICE was followed by Electronics Workbench and many other commercially available software circuit simulators that can be run on personal computers. The key competitive differences between these simulators are as follows:

1. The number of device models available—in other words, how many of the available electronic devices have circuit simulation models
2. The allowable circuit complexity. This is usually realized as a limit to the number of components, connections, or circuit nodes.
3. The types of analysis available
4. Functional complexity and ease of use
5. Ability to output schematic to circuit board layout software

The importance of these differences depends on your perspective. Electronics students may be interested in just the basic analysis function, but in industry it is important to have a widely functional software circuit simulator that interfaces directly with circuit board layout programs.

Circuit simulators function by having a computer model for all electrical and electronic components. The circuit designer keys in the schematic diagram for the circuit to be analyzed using the available component models and specifying any parameter values. If a model does not exist for a particular device, one will have to be generated. All currently available circuit simulators allow the schematic to be drawn on the screen, whereas on older SPICE simulators the schematic was defined by specifying the devices to be connected between various circuit nodes. The schematic must be complete in every respect, including all power supplies and ground connections. A wide range of DC and AC power supplies are available as well as input signal sources. The circuit on the screen is almost the equivalent of a software breadboard. The designer then selects the voltages and currents to be analyzed. Some simulators simply plot a graph of voltage/current over time while others allow the connection of a software-driven DVM or oscilloscope to a particular point in the circuit to display the waveform as it would be seen on a real oscilloscope. The circuit simulation described thus far can be classified as basic DC and or AC circuit analysis, nothing more or less than could be accomplished on a circuit breadboard. However, it can be completed more quickly and without any physical components or test equipment. The following discussion covers the many types of analysis provided by state-of-the-art circuit simulators, which provide functions and features that are often impossible to perform on a breadboard.

Transient Analysis

Transient analysis is the determination of instantaneous changes that occur at a circuit node after power up or some other starting point. Transient analysis provides the circuit simulator with a significant functional advantage over breadboards. The analytical treatment of transient analysis often involves rigorous mathematical operations and the definition of boundary conditions. On a laboratory breadboard transient conditions can be difficult to simulate and usually requires a digital scope, storage scope, or some other data-recording device. Circuit simulators accomplish this task with relative ease and control, allowing the precise variation in initial conditions and analysis start and stop times. The transient analysis function provides accurate plots of voltage and or current over the specified time period.

Fourier Analysis

Fourier series analysis is another key circuit simulation tool. Fourier theory says that any non-sinusoidal periodic function can be described by a DC component with some number of sine and cosine functions. With this type of analysis, it is possible to determine the sine and cosine components that make up a complex waveform that exists at any particular circuit node. This information provides the circuit designer with the harmonic frequencies present in a signal as well as their relative amplitude. This can help to filter out unwanted signals by determination of their frequency and their possible source. In many circuit simulators the total harmonic distortion (THD) can also be calculated with the Fourier analysis function.

Noise Analysis

Circuit simulators can also simulate the various types of noise generated by components: thermal noise, shot noise, and flicker noise. Thermal noise is caused by the temperature and its induced effect on the interaction of electrons and ions in a conductor. Shot noise results from the discrete nature of electrons (there can be one electron or two electrons flowing in a circuit, but not 1.5 electrons) flowing in a semiconductor and is the most significant cause of transistor noise. Flicker noise is present in BJTs and FETs at low frequencies. When noise analysis is utilized on circuit simulators, the total noise present at a particular node resulting from these three types is calculated and recorded.

Distortion Analysis

Distortion results when an electronic device such as an amplifier fails to duplicate an input waveform correctly. The causes of distortion are nonlinear gain of a circuit or relative phase variations. Distortion caused by nonlinear gain is called *harmonic distortion,* while phase-induced distortion is known as *inter-modulation distortion.* Both types of distortion can be determined for a circuit and plotted as a function of frequency for a particular circuit node.

DC Sweep Analysis

As discussed in Chapter 1, in the development of specifications, variations in the DC power supply value are always an important consideration affecting circuit accuracy. Many circuit simulators provide a DC sweep analysis function that provides analysis of selected voltages or currents, as one or two DC supply values are varied. When selecting this type of analysis, the designer specifies the particular DC supplies to be varied and the circuit node being analyzed, as well as the beginning and ending supply values and the increment steps. The results will indicate the effect of the DC supply variations on the voltage and current of the particular node.

Sensitivity Analysis

Sensitivity analysis serves to determine the component variations that will most affect the accurate function of a circuit. DC sensitivity analysis includes the variation of all component values, one at a time, in order to determine which component has the greatest impact on a critical circuit voltage value. AC sensitivity analysis, on the other hand, varies the value of just one component, providing the critical variations of the component along with its impact on the circuit.

Parameter Sweep Analysis

The sensitivity analysis just described is usable for determining which component affects the circuit accuracy the most. Parameter sweep analysis provides for the variation of any component parameter value over a range and in increments specified by the user. Semiconductor components will have a number of parameter values that can be varied as compared to passive components, which will have few.

Temperature Sweep Analysis

Chapter 2 discussed the importance of considering ambient temperature effects on circuit performance. Temperature sweep analysis provides a helpful tool in the determination of ambient temperature sensitivity at very early stages in the design process. During this process, circuit operation of selected nodes is recorded for different ambient temperatures. The parameter value for all components that vary with temperature are changed accordingly and the impact on the circuit function is plotted.

Transfer Function Analysis

A transfer function mathematically describes the operation performed on an input signal by a functional circuit block to its output. Circuit simulators can analyze and determine the transfer function for a particular circuit. The designer specifies the inputs and outputs of the circuit to the transfer function analysis feature and then analyzes and determines the transfer function and input and output impedance of the circuit.

Worst-case Analysis

Worst-case analysis is an extremely useful design tool. Often during the design process, it is desirable to know the maximum and or minimum voltage for a particular circuit node. Worst-case analysis can provide this by making sensitivity analysis runs for each component and then plotting the maximum and minimum values found over the course of the sensitivity runs. This is critical information when determining accuracy specifications and selecting component tolerances.

Monte Carlo Analysis

Monte Carlo analysis involves the statistical probability of the variation in the parameter value of a circuit component. In other words, it uses the probability distribution function of a parameter value change to determine its value and the ultimate effect of the circuit node. Each parameter value is selected at random over the range specified using the selected probability distribution function (usually Guassian).

Software Circuit Simulators

Current software circuit simulators offer powerful design simulation tools, and there are many to choose from. There are a number of circuit simulators currently available that offer a wide range in features. There are three distinct levels of performance and application that are apparent: academic, medium-performance, and professional.

Academic Circuit Simulators

These circuit simulators usually are low priced (around $200) and are marketed primarily to high schools and colleges that offer technical and electronic programs. They represent lower-level analysis functions but are a very beneficial educational tool, primarily due to the ease with which circuits can be built and analyzed. While they may interface with other circuit board layout software, they have no or minimal circuit board layout capabilities themselves.

Medium-performance Circuit Simulators

These usually offer medium- to high-level circuit simulation and analysis and often include mixed signal simulation (analog and digital circuits combined) and programmable logic design. The suppliers usually have functionally limited student versions available for under $200, but the fully functional versions cost between $800 and $1000. There may be additional circuit board layout software that can interface directly with these circuit board simulators offered by the same supplier (at an additional price of between $1000 to $2000), or other circuit board layout software can be used directly. The circuit board layout software for the medium-level circuit simulators usually does not match the performance of the professional circuit board layout packages.

Professional Performance Circuit Simulators

These circuit simulators are usually the most costly, available for around $2000 to $4000. They provide optimum circuit simulation, including mixed signal and programmable logic design, combined with professional circuit board layout capabilities. Circuit board layout features include multilayer circuit boards, surface mount components, powerful auto-routing, and 3D renderings of the final board.

As with all software, circuit simulators are rapidly changing as new features are offered and new suppliers enter the market. There is a significant learning curve required to take advantage of the key features on new products and software versions. This is similar to what occurred with mechanical CAD software and printed circuit board layout software. The market is still developing, but what is desired most is a circuit simulator that is accepted and used widely in industry, that functions directly with professionally accepted circuit board layout software, and that is available as a limited-function student version at a reasonable cost.

3–6 ▶ Breadboarding

Circuit simulators provide powerful functions and are extremely beneficial, but they do fall short in some areas. For example, very few simulators take into account the actual maximum ratings for components. So a circuit might function flawlessly in the simulation but could fail when powered up because it exceeds some component rating. Consequently, it is important at some point to build up a real physical circuit before proceeding too far with the overall design project.

Breadboarding is the process of constructing an experimental schematic circuit from the actual components selected for the design. The purpose of breadboarding is to test the preliminary design of a circuit or module. The breadboard is usually temporary in nature and is more easily modified than a printed circuit board. There are many different methods for completing a breadboard. The method selected depends on the goals of the prototype phase and the complexity and circuit technology of the circuit to be breadboarded. The breadboarding technique discussions that follow do not include all breadboarding methods but do include the methods most viable today.

Solderless Breadboard

The solderless breadboard is one that most electronics students have had experience with in the laboratory. This breadboard consists of groups of circuit connection holes called *points* that are all located on a 0.1″ grid. These circuit points are arranged in groups that are connected together underneath the plastic surface of the breadboard assembly. Horizontal rows of five points are connected together and are arranged into a vertical column. A center channel that is provided for mounting a standard dual in-line package integrated circuit separates two such columns. The integrated circuit straddles the center channel, making connections to connecting points on each side of the channel. With the integrated circuit mounted in place, the laboratory breadboard provides four connection points for each pin on the integrated circuit. Other components, such as resistors and capacitors, can also be mounted across the center channel. There are also separate vertical columns where the points are connected vertically instead of horizontally. These are commonly used for the purpose of bussing power supply voltages and ground to various circuit points. There are a variety of solderless breadboards available, so to be sure about which points are connected to which, use your multimeter to verify the connection scheme of the breadboard you are using.

To construct a circuit with the laboratory breadboard, components are inserted into the connection points so as to implement all the connections shown on the schematic. This is accomplished by connecting two components to the same connection group (i.e., five points that are tied together) if they are electrically connected. The remaining circuit points are connected using 22- to 30-gauge wire that is stripped back about 3/16″ to 1/4″. Precut and pre-stripped wire is available in various lengths for use in these breadboards. The solderless breadboard is limited in the current that can flow in the connections and the wire typically used to make them. The inherent capacitance of the connections and potential crosstalk between conductors also limits the frequency of operation of the solderless breadboard system.

Universal PCB Breadboard

Universal PCB breadboards (see Figure 3–7) are printed circuit boards that have holes all located on a 0.1″ grid system. There are many varieties of these universal boards. In general, some of the holes in the grid system have copper pads to allow for soldering, and some of these pads are connected together. The variations come from the location of the holes that have copper pads and from the combinations of copper pad connections. Some versions are more suited to large numbers of discrete components, and others are geared toward integrated circuit applications. PCB-edge connections are provided on some models. Whichever variety is used, there are areas where the configuration of the universal circuit board requires modification with a hand drill to cut away circuit points that are connected.

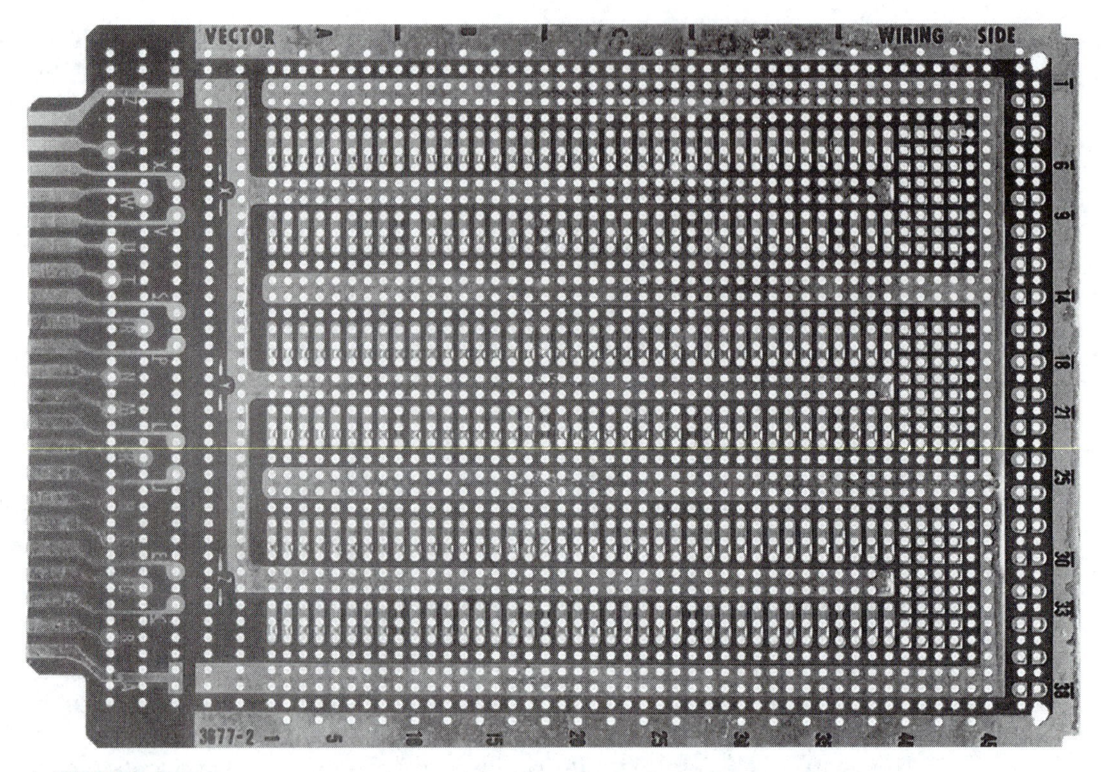

▲ **FIGURE 3–7**
Universal breadboard

Prototyping a circuit with the universal circuit board method is very tedious. After selecting the proper universal circuit board variety, the components are soldered in place. Next, the circuit connections shown in the schematic must be implemented either by soldering wires that connect all the circuit points or soldering together circuit pads with a continuous conductor to make a circuit board run. Finally, some areas of the copper connected pads may need to be disconnected. This can be accomplished by cutting away the connecting copper runs with a razor-blade knife or a hand drill fitted with a slicing tool. If carefully planned out and neatly implemented, the result can be very close to the eventual printed circuit board in function, physical size, and layout. This method can be combined with wire-wrap methods and SMT technology (to be discussed next). Figure 3–8 shows an example of a breadboard circuit that was developed with a universal breadboard scheme. Both the top and bottom views are shown. Notice how the circuit can be made to closely resemble the eventual printed circuit board.

Surface-mount Technology

Surface-mount technology provides unique challenges in breadboarding circuits. Again, there are many choices to be made. The most obvious is to breadboard the circuit with the through-hole equivalents rather than the actual SMT components.

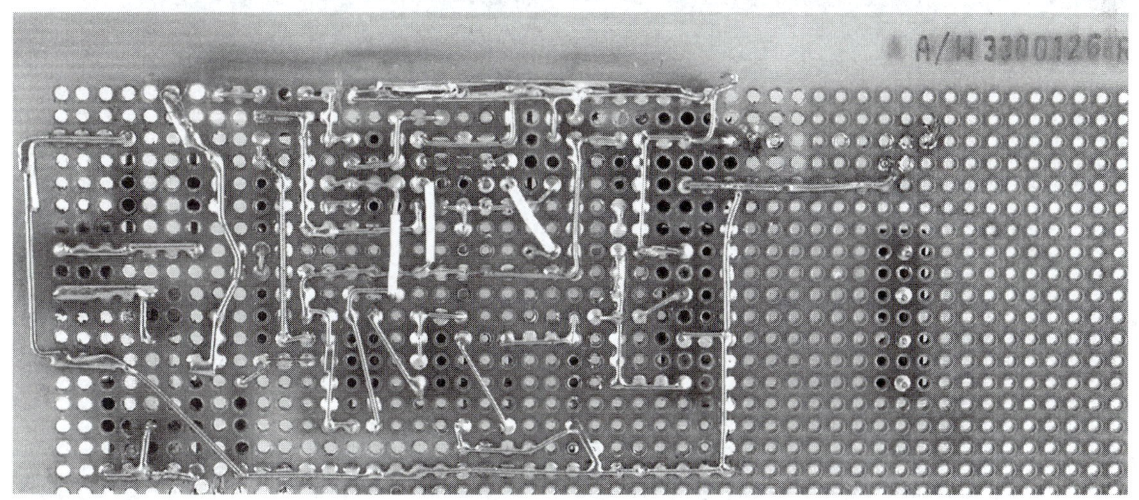

▲ **FIGURE 3–8**
Prototype example using Universal breadboard

If possible, it is best to avoid using of SMT components at the breadboard stage. Most SMT components are available in a through-hole package. If not, the SMT component can either be mounted on an SMT socket or on an SMT "carrier" board. Figure 3–9 shows an example of an SMT socket on a prototype board with extra area allotted for other through-hole circuitry. Figure 3–10 shows an example of an SMT carrier board that can be soldered into a universal breadboard prototype. SMT carrier boards allow the mounting of the SMT components where sockets may not be available or practical. The SMT carrier board can then be mounted to a universal printed circuit board and wired in the circuit with solder connections that go between the two boards. Soldering the SMT component to the carrier board

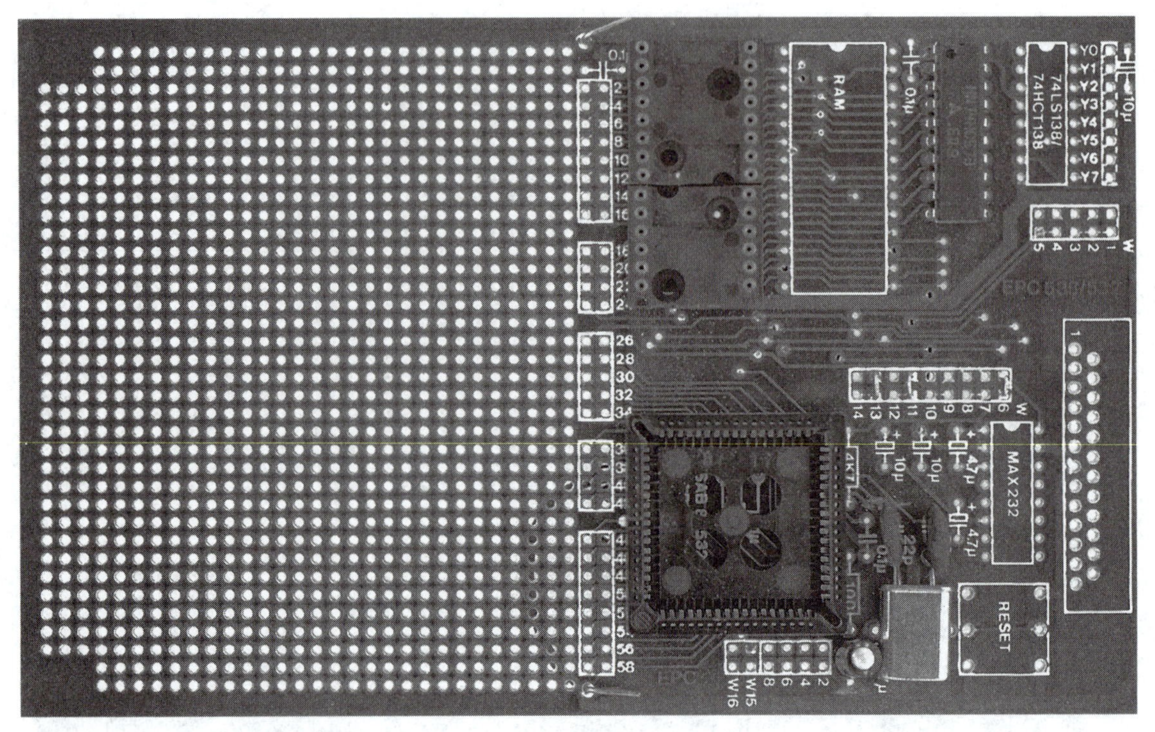

▲ FIGURE 3–9
SMT socket board

may take some advanced soldering skills, as the space between connections is small. Use an extra-fine tip for these applications and apply a thin coating of solder to the SMT pads.

It is best to build the breadboard with standard through-hole resistors and capacitors instead of SMT chip resistors and capacitors. The SMT chip components are extremely small, making them hard to handle and solder.

Wire-wrapping

Wire-wrapping is a solderless technique in which circuit connections are made by small wire connections that are tightly wrapped around pins called *wire-wrap pins*. Wire-wrap wire is generally 28- or 30-gauge wire. Wire-wrap pins, which feature right angle corners, are incorporated into integrated circuit sockets or part carriers or used as individual wire-wrap posts. Figure 3–11 shows an example of a wire-wrapped circuit. The wrapping is accomplished with a device called a wire-wrap gun or wire-wrap tool. To develop a breadboard utilizing the wire-wrap process, wire-wrap sockets are utilized for all integrated circuits, and part carriers are used for nonintegrated circuit-type devices such as resistors, capacitors, diodes, and transistors. The wire-wrap sockets and part carriers are lightly glued to a perforated phenolic board. Circuit connections are made point to point, as follows:

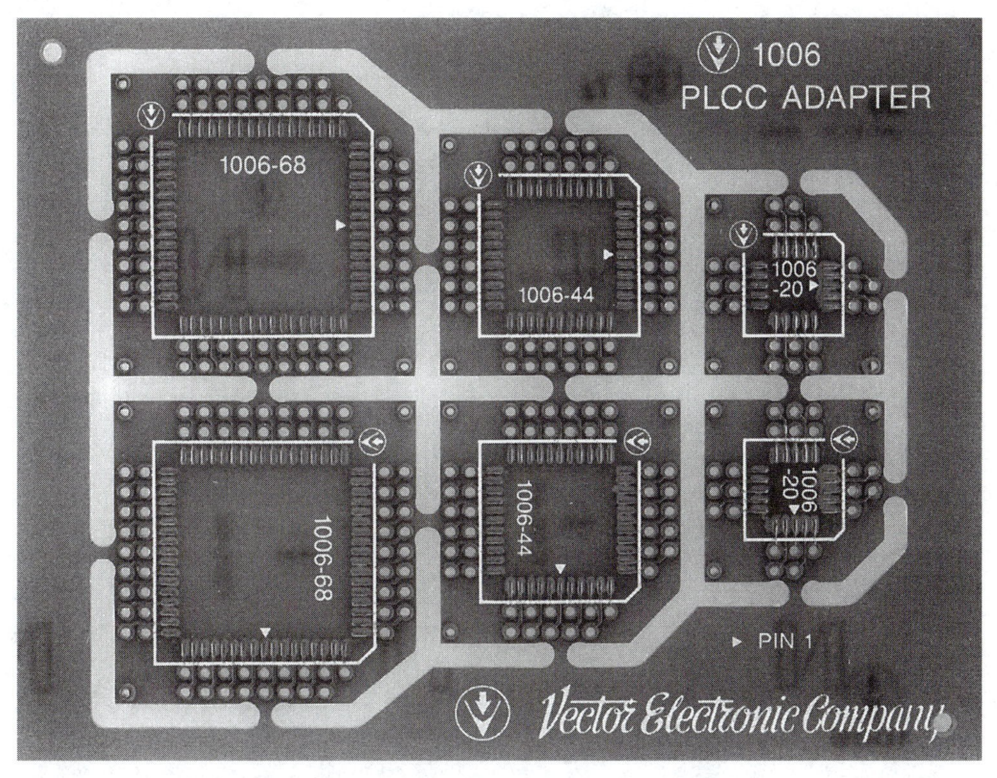

▲ **FIGURE 3–10**
SMT carrier board

1. Strip back the wire-wrap wire about 3/8″.
2. Insert the bare-wire end into the outer hole on the wire-wrap gun and bend the wire at a right angle with the axis of the wire-wrap gun.
3. Insert the center hole of the gun over the wire-wrap post to be connected, making sure that the end of the gun is flush with the back of the phenolic board.
4. Squeeze the lever on the wire-wrap gun and the bare wire will be wrapped clockwise around the wire-wrap pin.

Proceed to the other end of the connection and repeat the process. A reliable connection is actually made by the wire and the square edge of the wire-wrap pins as the wire is wrapped around the pins.

The Nonbreadboard

There are many times when the breadboarding process is actually implemented with a printed circuit board. This occurs most often with microprocessor-based digital boards, where the time and complexity of breadboarding is very high. In

▲ **FIGURE 3–11**
Wire-wrap application

this case the engineer ends up with only one wire-wrapped board, which is of questionable value after it is complete and made operational. Usually each software engineer will need a circuit board for software development, so numerous breadboards are often required in the early project stages. This situation promotes going directly to a printed circuit board. In this case the circuit board schematic is checked out and sent directly to the circuit board layout person, usually an electrical designer from the drafting department. (Circuit board layout will be discussed in detail in Chapter 5.) The circuit board is laid out and prototype quantities are ordered with a priority lead-time. When the circuit boards have been received, they are carefully built up and tested one section at a time. When functional problems are encountered, the boards are analyzed and corrected with a combination of "cuts" and "jumpers," component value changes, and additional components. In this case the first generation of the printed circuit board becomes the breadboard. The breadboard stage is not necessarily skipped, but it is replaced with the first version of the printed circuit board. It is desirable to make the prototype board a printed circuit board when one or more of the following is true:

1. The circuit is a complicated circuit board with embedded software.
2. The design is close to being a standard design, like well-known bus-oriented structures for microprocessors. In other words, the schematic has a good chance of being functional.
3. Testing of the prototype is not possible until some basic software is developed, and that is not scheduled to be complete for a couple of weeks.
4. More than one prototype board is required as soon as possible.
5. The prototype board must be reliable.
6. Engineering and technician time is at a premium and drafting layout time is more available.

The high degree of error checking that results from schematic capture software and circuit board layout programs has promoted the use of the nonbreadboard technique. When using complementary schematic capture and board layout packages, the most significant benefit is that the artwork will equal the schematic exactly. If the schematic is correct, then the artwork will be also. With this kind of accuracy, it is a fact that the actual printed circuit board has a higher chance of being equal to the schematic than any breadboard. This does not mean that the prototype board will function as required, which is why breadboards are still favored in many cases. The breadboard is more readily changed or just scrapped when major design problems develop.

Breadboarding Methods

Whichever method of breadboarding is selected, a schematic circuit is put into some physical form. To accomplish this accurately, an organized and methodical process must be implemented. The following procedure produces very good results:

1. Organize all the breadboarding materials and tools on a bench-top area to be used for the duration of the process.
2. It is imperative to start with a neat and orderly schematic. The schematic should be complete and include wire connections for all components that are breadboarded. All of the components on the breadboard should have a component number (e.g., R_1, C_1, and the like). The schematic should be laid out in an easy-to-understand format with inputs on the left-hand side and outputs on the right-hand side.
3. Make two copies of the schematic. One copy will be the breadboard master copy and should be initialed, dated, and filed with all breadboard documentation. As the breadboarding proceeds, use a highlighter marker to highlight each connection as it is made. When this process is used, the most common error, missing connections, is eliminated.

4. Start out wiring power to all points on the board where it is required. Make sure that the power and ground runs are made from heavier wire. Then follow the schematic, left to right, wiring the circuit in logical groups. For example, wire up all the power supply components and check out the power supply. Then wire all data connections from one component to the next component. This tends to minimize errors, because an error, such as wiring to an incorrect pin, will be noticed when an attempt is made to make the correct connection to that pin.

5. Make the connections as short and as neat as possible. Keep in mind to provide easy access for later inspection and the connection of test leads.

6. Many times there are integrated circuits where multiple components are located on one chip. Be sure to note which component on the integrated circuit is used for which schematic function by noting the pins and using a component subdesignation on the breadboard master schematic. Take the example of a Dual 4 Input NAND gate that is given a component designation of U_1. There are two 4 Input NAND gates on U_1, one will be designated $U_{1\text{–}A}$ and the other $U_{1\text{–}B}$. Also be sure to properly terminate all unused integrated circuit components as required in their specifications. Unused TTL gates should have their inputs tied high (+5 V), for example.

7. Make a sketch of the assembly showing the relative location of the components with their designation.

By performing this procedure, you may be surprised to find that the circuit will work the first time—that is, if you remember to turn on the power. Of course, the schematic design must be a "working" design to begin with.

Summary

In this chapter we discussed the steps for completing the preliminary design. The steps are listed here as a summary of the preliminary design process:

1. Divide the design problem into smaller modules.

   Complete a block diagram.

2. Define the design problem relating to these modules.

   Develop submodule specifications.

3. Do research on areas relating to each design problem.

   Gather more information.

4. Develop a cost budget.

   Develop cost goals for each submodule.

5. Apply creative thinking!

   Use your subconscious mind, and let your ideas flow.

6. Complete the preliminary design.
   Complete a preliminary design schematic.
7. Perform software simulation.
   Simulate the preliminary design.
8. Complete breadboard and evaluate.
   Breadboard and test the circuit.

When these steps are completed, the preliminary design is complete. A schematic diagram exists that defines the general components selected for the design and it shows their interconnection. The design has been simulated and basic breadboard testing has been completed. The next chapter discusses the selection of the actual components to be used in completing the formal design, the prototype printed circuit board.

Reference

Stadmiller, D. J. 2001. *Electronics Project Management and Design.* Upper Saddle River, NJ: Prentice Hall.

Exercises

3–1 What is the key advantage of utilizing surface mount technology?

3–2 True or false: Surface mount technology is always cheaper to manufacture than through-hole technology. What are the issues that determine which is cheaper?

3–3 What is the key disadvantage of using surface mount technology?

3–4 What is a manufacturing cost budget and how is it developed?

3–5 List the two steps to be completed before trying to utilize creative thinking to solve a problem.

3–6 List the competitive differences between the various circuit simulation software packages.

3–7 List the shortcomings of circuit software simulators in general when compared to the breadboard alternative.

3–8 Explain the concept of transient analysis as described in the section on circuit simulation software.

3–9 Explain the concept of DC sweep analysis as described in the discussion of circuit simulation software.

3–10 Does temperature sweep analysis performed by circuit simulation software do the same general function as ambient temperature testing in an environmental test chamber?

3–11 Explain the process and the result of sensitivity analysis performed on a circuit node by a circuit simulator.

3–12 Explain the difference between parameter sweep analysis and sensitivity analysis as performed by circuit simulators.

3–13 List the four different methods discussed in this chapter for completing breadboards. List one primary advantage and disadvantage of each.

3–14 Explain what is meant by the term *non-breadboard.*

4 Component Selection

Introduction

In Chapter 3 we discussed the preliminary design process. At this point the design is on paper and the circuits have been computer simulated, breadboarded, and tested. The design includes a variety of active and passive components. Before we start to lay out the printed circuit, we must select and procure the final components for use in developing the circuit board layout and constructing a prototype. Whether you are completing a two- or four-year electronics program, you have seen many resistors, capacitors, and inductors used in circuits. You should possess a solid understanding of their function in a circuit. The discussion that follows summarizes the important aspects of selecting passive components and discusses the many different types available. More importantly, the advantages and disadvantages of each type are reviewed. A method for selecting the right type of component for a specific application is presented. The selection of active components is also discussed in a general way. (Appendix A includes reference information on both passive and active components.) The specific topics of this chapter are as follows:

- Resistors
- Variable Resistors
- Capacitors
- Inductors
- Transformers
- Switches and Relays
- Connectors
- Selecting Active Components

4–1 ▶ Resistors

Once we have determined the nominal design value for a resistor in a circuit schematic drawing, we must convert that information into an actual physical resistor that we can connect in the circuit. To select a resistor, the value, the type (material and construction), tolerance, power rating, and temperature coefficient must be defined. The available types of fixed resistors are described in the following list:

1. *Molded carbon (carbon composition) resistors.* These are a common older type of resistor formed by molding carbon, insulating filler, and a resin binder. They are low cost and feature 5% and 10% value tolerances, a wide variety of power ratings, and a negative temperature coefficient of anywhere from –200 ppm to –500 ppm/°C. Example 4–1 describes a simple way to handle coefficients stated in parts per million (ppm).
2. *Carbon film resistors.* These have replaced molded carbon resistors as the most widely used fixed resistor. They are constructed by depositing a resistive carbon film material onto a ceramic rod. They offer smaller size and more stability than the molded carbon resistor and are still low cost. Carbon film resistors come in 1%, 5%, and 10% tolerances with negative temperature coefficients of –200 ppm to –500 ppm/°C and a variety of power ratings.
3. *Metal film resistors.* These resistors utilize the same construction method described for the carbon film resistors but metal film (nickel chromium) is deposited on the ceramic rod instead of the carbon film. Metal film resistors come in 1%, 5%, and 10% tolerances and a wide variety of power ratings and feature a positive temperature coefficient of less than +100 ppm/°C.
4. *Metal oxide resistors.* Tin oxide is deposited on a glass rod to form these resistors. The stability is very high. Metal oxide resistors come in 1% and 2% tolerances and power ratings of 1/5 W and 1/2 W and feature a positive temperature coefficient of less than +60 ppm/°C.
5. *Wire-wound resistors.* These are formed with resistive wire wrapped around an insulating rod. Wire-wound resistors typically have both higher power ratings, and very low resistance values are available. A wide range of values and power ratings are available at medium to high cost. Their temperature coefficient is positive and less than 100 ppm/°C.

Example 4–1

The temperature coefficient of many components is given in a format called parts per million (ppm). This ppm format requires some thought when first trying to apply it. A simple way to use any coefficient given in ppm form follows:

1. Divide the nominal value of whatever the variable in question is by the number 1,000,000. This will be the number of millions that the nominal value represents.

2. Multiply the number of millions calculated in step 1 of this example by the ppm value given.

A practical example would be to determine the resistance variation of a 10,000 Ω carbon composition resistor with a temperature coefficient of –200 ppm/°C from 25°C to 55°C. Follow the steps listed:

1. Divide

   10,000 Ω/1,000,000 = 0.01 million Ω

2. Multiply

   0.01 million Ω × 200 ppm/°C = 2 Ω/°C

3. Multiply

   2 Ω °C × 30°C (increase in temperature from 25°C to 55°C)
   = 60 Ω change

Since the temperature coefficient is negative, the 10,000-Ω resistance decreases by 60 Ω to a value of 9,940 Ω.

Resistor Selection

The criteria for selecting the actual resistor for use in a circuit depends upon the nominal resistance value, the level of stability, the tolerance required, the acceptable temperature variation, the power level, and cost. The following process is suggested to select all fixed resistors:

1. Determine which types of resistors have the value and power rating required.
2. Determine the minimum temperature coefficient that is acceptable for the application.
3. Chose the lowest-cost and most available resistor type that will meet the requirements of the circuit.

Figure 4–1 is a resistor comparison chart.

Thick Film Networks

These networks are fabricated with the same technology as metal film and carbon film resistors but are incorporated in a package that allows for multiple resistors of the same value. The packages include dual-in-line and single-in-line as well as SMT packages. For example, you can purchase eight 1000-Ω resistors in one dual-in-line package, and the resistors can all be connected on one side to the same pin or kept completely separate.

Resistor Type	Range of Values	Tolerance	Range of Temperature Coefficient	Power Rating	Cost Factor x = .02	Primary Advantage
Carbon composition	All standard 5% values	+/- 5% to +/- 10%	–200 to –500 ppm	1/8, 1/4, 1/2, 1, 2 W	1x	Cost and size
Carbon film	All standard 5% values	+/- 5% to +/- 10%	–200 to –500 ppm	1/8, 1/4, 1/2 W	1x	Cost and size
Metal film	All standard 1% values	+/- 1%	+100 ppm	1/4 W	2x	Tolerance and temperature coefficient
Precision metal oxide	All standard 1% values	+/- 1% to +/- 2%	+60 ppm	1/4, 1/2 W	4x	Temperature coefficient
Power metal oxide	All standard 5% values	+/-5%	+100 ppm	1/2, 3 W	15x	High power and temperature coefficient
Precision wire-wound	All standard 1% values	+/- 1%	+100 ppm	1, 2, 3 W	100x	High power and tolerance
Power wire-wound	All standard 5% values	+/- 5% to +/- 10%	+200 ppm	1 W on up	75x	High power

▲ **FIGURE 4–1**
Resistor comparison chart

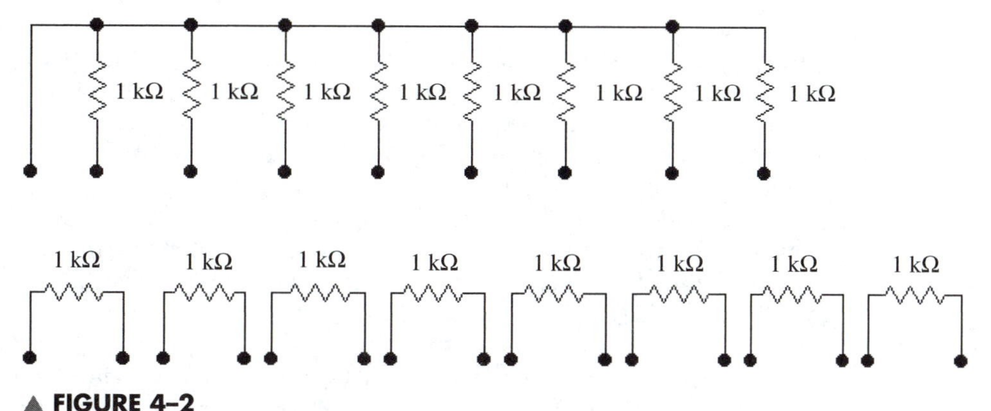

▲ **FIGURE 4–2**
Dual-in-line network configuration

Figure 4–2 shows a schematic diagram of these two configurations. These are handy in situations in which eight of the same value resistors are needed, such as for pull-up resistors on an 8-bit data bus.

Power Resistors

You can classify power resistors as anything over 1 W. These are typically wire-wound or metal film resistors that have the same characteristics as other wire-wound or metal film resistors, except that the packages are physically larger and designed to dissipate heat. Figure 4–3 shows examples of different power resistor packages.

4–2 ▶ Variable Resistors

Variable resistors are also called *potentiometers* and are simply resistors that can be adjusted to many values. These can be full-sized potentiometers or smaller trimming potentiometers called *trimpots*. They come in a variety of packages that can

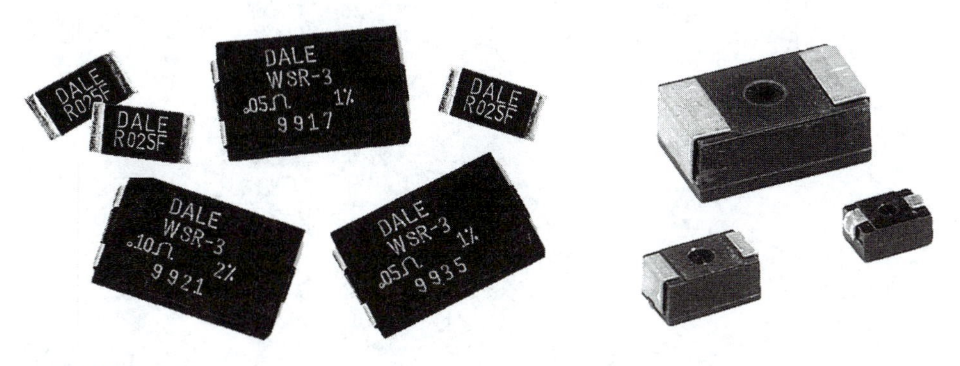

▲ **FIGURE 4–3**
Power resistor packages *(Courtesy of Vishay Intertechnology, Inc.)*

be circuit-board or panel mounted. Variable resistors are available that are adjustable with circular or linear motion. The circular adjustment types can be single or multi-turn. The multi-turn devices provide a more precise angular adjustment of the resistance value. Inside the potentiometer is either a wire-wound or a continuous film type of resistive element. The wire-wound type is similar to wire-wound fixed resistors and is fabricated by winding fine resistance wire around an insulating bobbin. Each end of the bobbin is connected to a terminal. The resistance between the two end terminals is the maximum resistance of the variable resistor. A third terminal, called the *wiper arm*, is moved along the surface of the resistive windings to achieve the variable resistance value. Wire-wound potentiometers perform much like their fixed resistor counterparts. They are stable and have a good temperature coefficient but are plagued by resolution issues inherent in their construction. As the wiper arm moves across the resistance wire wrapped around the insulating bobbin, the resistance measured will increase in steps as the wiper goes from contact with one "turn" to the next "turn." The number of turns (the number of times the wire is wrapped around the bobbin) will determine the resolution (the smallest resistor value change) of the variable resistor.

Continuous-style potentiometers are fabricated with a thick film composition consisting of metal film on a ceramic substrate. These variable resistors use materials such as cermet, carbon composition, carbon film, and metal film for the thick film compositions. Because they are constructed of a continuous length of conductive material, these potentiometers greatly improve on the resolution problems experienced with wire-wound variable resistors. However, they are limited to smaller power ratings.

Example 4–2

In the op amp circuit in Figure 4–4, determine the proper connection of the potentiometer so that the gain of the circuit increases with a clockwise adjustment of the variable resistor. The gain of the circuit increases as the value of the potentiometer increases. It is important to understand the correct wiring of potentiometers so that they will function properly, which means providing the proper

▶ FIGURE 4–4
Op amp adjustment circuit

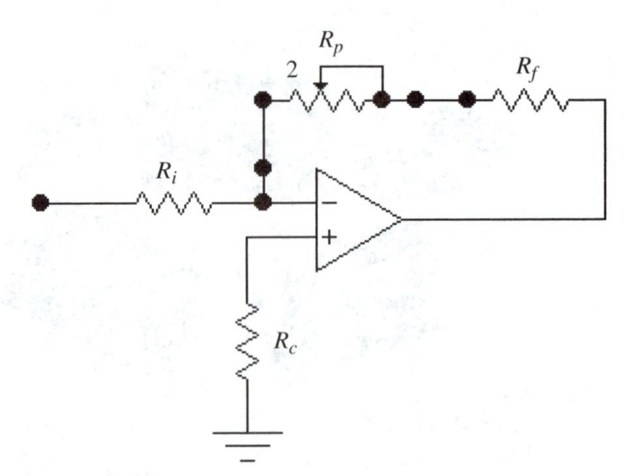

adjustment for the desired direction of adjustment. Potentiometers are three-terminal devices that have their total resistance between end terminals 1 and 3 and a variable resistance between terminal 2 (the wiper) and either of the other terminals. The direction of rotation (or the linear direction for linear adjustment types) that increases the resistance between terminals 2 and 1, and 2 and 3, is the critical point. Notice that the circuit shown has the second terminal connected to one of the end terminals. This is commonly done when the desired effect is to function simply as a variable resistor. In this case there are two common connections to the potentiometer. When a voltage divider function is required, all three connections to the potentiometer are separate.

To solve this problem, we must determine the end terminal (pin 1 or pin 3 of the potentiometer) to which the wiper arm (pin 2 of the potentiometer) should be connected. The solution will come from the answer to one question: When the potentiometer is turned clockwise, between which set of terminals does the resistance increase? This question can be answered by taking resistance measurements on a sample potentiometer or by remembering that for all variable resistors, the resistance between wiper terminal 2 and terminal 1 increases as the potentiometer is turned clockwise. To resolve our example problem, let's list what is known about the problem:

1. We desire the gain to increase when the potentiometer is adjusted clockwise.
2. The gain increases when the potentiometer resistance increases.

To resolve the problem, determine the two terminals where the resistance increases when the potentiometer is adjusted clockwise. The answer is between terminals 1 and 2. To which end terminal should wiper terminal 2 be connected? The answer is terminal 3. Connecting terminals 2 and 3 together shorts terminal 2 to the terminal 3 end. Since the resistance increases between terminal 2 and terminal 1, when the potentiometer is turned clockwise, the desired performance will result.

4–3 ▶ Capacitors

Capacitors are probably the second most utilized passive electronic component, and there are many variations from which to choose. When you are selecting capacitors, your goal will be to find a capacitor with the following:

1. The proper capacitance value
2. An acceptable tolerance rating
3. The proper working voltage
4. A temperature coefficient that is acceptable for the application
5. In some cases insulation resistance, quality factor, and dielectric absorption will also be important
6. Smallest size and cost available

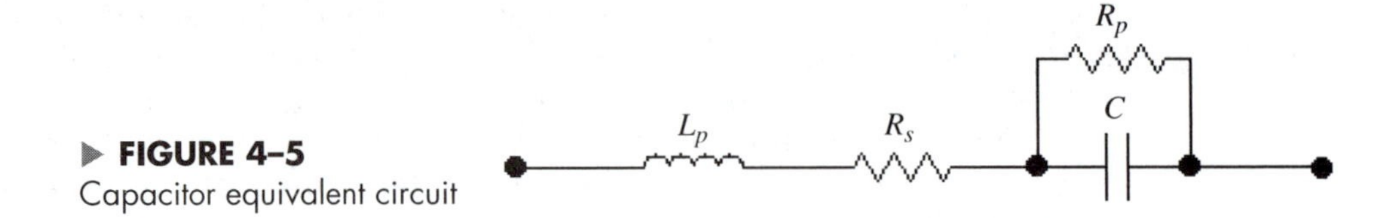

▶ **FIGURE 4–5**
Capacitor equivalent circuit

The physical model for a capacitor is shown in Figure 4–5. Inductance (L_P) and series resistance (R_S) are shown in series with a parallel capacitance (C) and resistance (R_P). The series resistance is a result of the resistance of the leads, plates, and any contact points. The resistance, in parallel with the capacitance, represents the leakage resistance that occurs through the insulation material around the plates. The capacitance shown is the true capacitance of the capacitor.

Insulation Resistance

The insulation resistance limits the capacitor's ability to completely block DC current as, theoretically, it should. That is why it is also called *DC leakage current.* The insulation resistance represents the ability of a capacitor to hold a charge for a period of time. It is an important parameter for capacitors used in integrator, sample hold, and peak detector circuits, which must hold a charge for a long period of time.

Equivalent Series Resistance, Dissipation Factor, and Quality Factor

The equivalent circut shown in Figure 4–5 can be converted to an equivalent series resistance (ESR) and a capacitance (C) as shown in Figure 4–6. The ESR is determined by calculating the equivalent series impedance for the parallel RC network at a given frequency, and adding the resistance component to the series resistance R_S, shown previously. L_P is negligible at low to medium frequencies. The dissipation factor (DF) is a measure of ESR/X_C. As such it is a measure of the AC loss of the capacitor. It is a unitless number that is expressed as a percent. The lower the DF number, the less loss is dissipated by the capacitor at a certain frequency. The quality factor or Q is simply 1/DF. It is the inverse of the dissipation factor. The DF and Q factors are important when precision operation at a certain frequency is required and when sampling a signal. Resonant frequency, precision filters, and sample and hold circuits are a few examples of circuits that require high Q and low DF.

Dielectric Absorption (DA)

Dielectric absorption is the phenomenon that allows a capacitor, which is quickly discharged and open circuited, to recover some of the charge that was discharged. In this case the charge is actually absorbed by the dielectric. In applications such

▶ **FIGURE 4–6**
ESR circuit

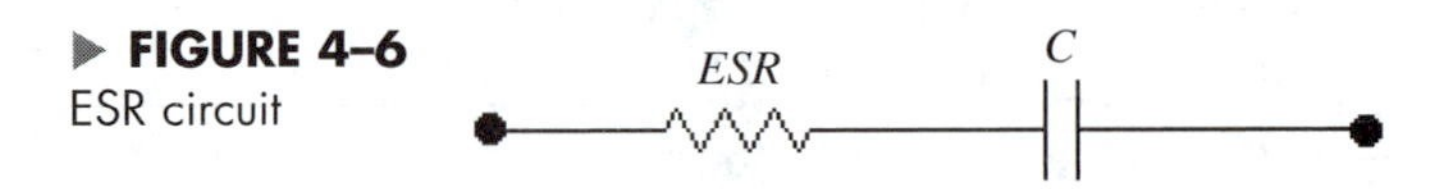

as sample and hold circuits, this recovered charge will add to the signal the next time it is sampled, causing an error. Dielectric absorption (DA) is an important consideration in sampling, timing, and high-speed switching circuits. The characteristic is expressed as a ratio and given as a percent. The higher the percent of DA rating for a capacitor, the greater the amount of DA effect.

Capacitor Types

To meet the circuit's requirements for capacitors, they are selected from many types of construction using different dielectric materials. The key is to find the desired performance characteristics in the smallest package at the least cost.

Ceramic Capacitors

These are widely used because they have a wide range of available values, and they are relatively small and inexpensive. The high dielectric constants available with the ceramic materials used in these capacitors result in large capacitance values for their size. The most common types of construction available for ceramic capacitors are the disc, multilayer, and chip variations. Figure 4–7 shows a typical disc type ceramic capacitor. The performance characteristics have the following ranges:

Capacitance range: 1 pF to 1μF

Working voltage: 25 V to 30 kV

Tolerance: ±5% to as high as +50% and –20%

Temperature coefficient: ±15% over temperature range

Relative size: Small

Relative cost: Inexpensive

Typical use: As medium- to high-frequency bypass, coupling, and filter capacitors

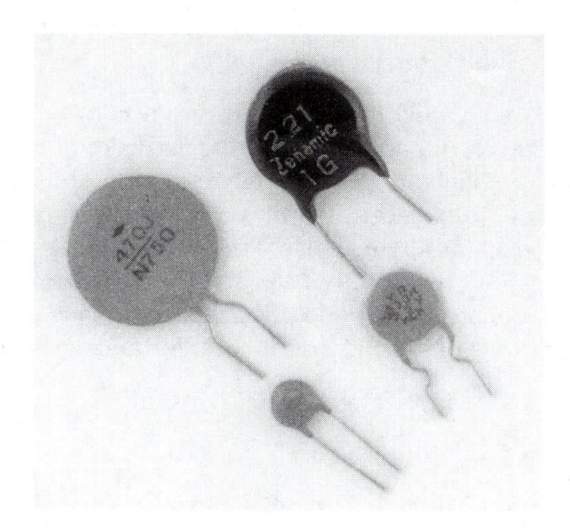

▶ FIGURE 4–7
Ceramic capacitors

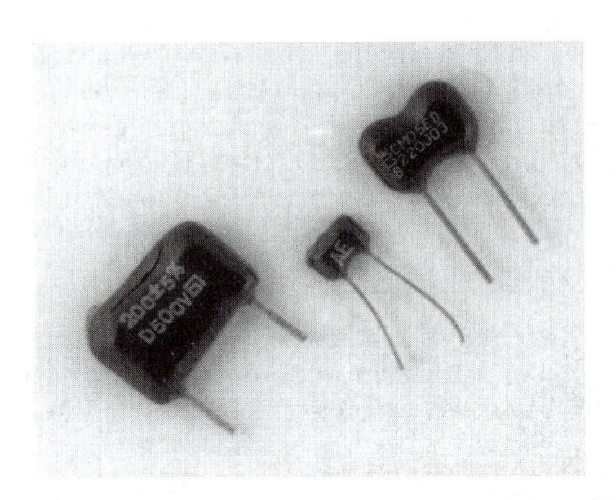

▶ **FIGURE 4–8**
Mica capacitors

Mica Capacitors

Mica capacitors (see Figure 4–8) offer superior performance when compared to the best quality ceramic capacitors when used at frequencies above 200 MHz. They are available in small capacitance values with tight tolerances, high working voltages, and stable temperature characteristics. Their performance characteristics have the following ranges:

Capacitance range: 2.2 pF to 0.01 µF

Working voltage: 50 V to 5 kV

Tolerance: ±0.5% to as high as ±20%

Temperature coefficient: +200 ppm/°C

Dissipation factor: 0.02% to 0.1%

Relative size: Medium to large

Relative cost: Medium

Plastic Film Capacitors

These are a category of many different plastic film dielectric materials that have similar performance characteristics. Plastic film capacitors (see Figure 4–9) are characterized by high working voltages, more stable temperature characteristics, low dielectric absorption, and a higher Q factor. These premiums result in their larger size and higher cost. The most common types of dielectric used to make plastic film capacitors are polyester, polypropylene, polystyrene, polyethylene, polycarbonate, teflon, and mylar. The following performance characteristics are the broad range of specifications for all film capacitors:

Capacitance range: 20 pF to 500 µF

Working voltage: 30 V to 10 kV

Tolerance: ±1% to as high as ±20%

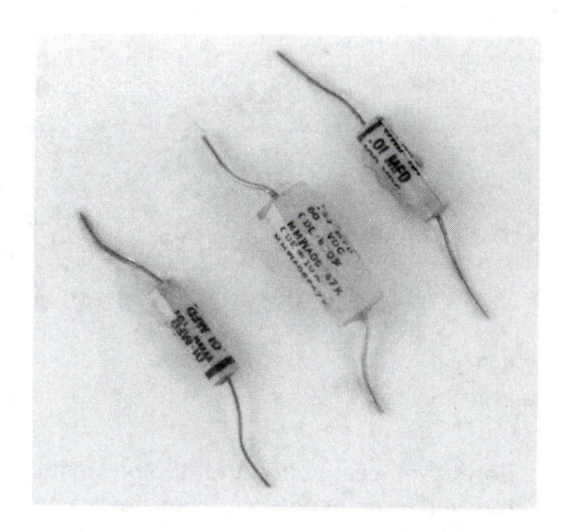

▶ **FIGURE 4–9**
Film capacitors

Temperature coefficient: ±2.5% to ±10% over temperature range

Relative size: Medium to large

Relative cost: Medium to high

Typical use: In low- to medium-frequency ranges where better than average capacitance tolerances, stability, temperature coefficients, and Q factors are required

Next, the individual characteristics that are unique to the different types of dielectric materials used in plastic film capacitors are discussed.

Polyester: These are considered as medium performance plastic film capacitors. They are designed for mounting on printed circuit boards and are available in values from 0.001 μF to 2.2 μF. Polyester capacitors feature low inherent inductance, tolerances of ±5%, ±10%, and ±20%, and a temperature coefficient ±10% change over the rated temperature range. Their size tends to be larger than other plastic film capacitors and they are available at medium cost. They are typically used as coupling capacitors.

Polypropylene: These capacitors (see Figure 4–10) feature very low dielectric absorption (0.001% to 0.02%) in a wide range of values (10 pF to 0.1 μF) and they are inexpensive. Their disadvantages are large case size, an inability to withstand high temperatures, and high inductance. For lower temperature applications, these would be a good choice for sample and hold circuits.

Polystyrene: Polystyrene capacitors offer the highest performance of the plastic film capacitors. They are available in ranges from 10 pF to 0.1μF at tolerances from ±1% to ±10%. Their stability is excellent with a temperature coefficient of around ±1%. The dissipation factor is low, which means the Q factor is high. They also feature a very low dielectric absorption factor.

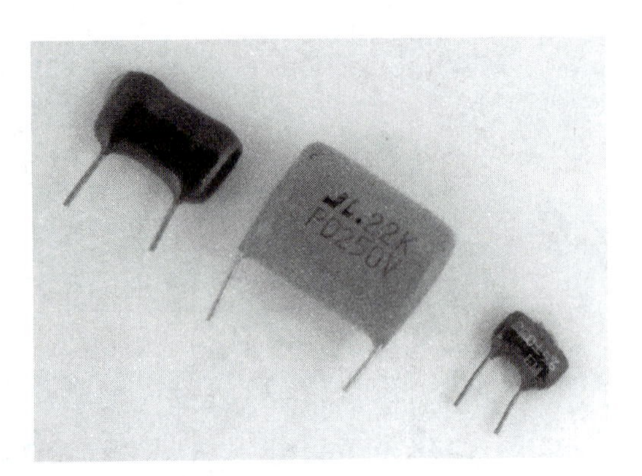

▶ **FIGURE 4–10**
Polypropylene capacitors

Polystyrene capacitors tend to be smaller than other plastic film capacitors and more expensive. They are typically used in filter networks, tuned circuits, and other precision charging circuits in the low- to medium-frequency range.

Polyethylene: These are specialized capacitors designed to suppress transients and noise from the input power connections to many industrial and commercial products. They are available in values from 0.001 μF to 1 μF and temperature coefficients of ±10%. The primary applications are as filter capacitors placed across the primary AC voltage lines.

Polycarbonate: These offer performance almost as good as polystyrene capacitors at a smaller size. If a high-quality capacitor is required in the smallest size, a polycarbonate capacitor may be a good choice. They are available in ranges from 10 pF to 0.1 μF and temperature variations on the order of –2.5%, +1% over the useable temperature range. Their cost is in the medium to high range. Applications include filter networks, tuned circuits, and other precision charging circuits in the low- to medium-frequency range, where small size is an overriding requirement.

Mylar: Mylar capacitors are the general purpose capacitors of the plastic film types. They are available in ranges from 0.001 μF to 0.22 μF and working voltages up to 100 V DC. Their tolerances and temperature coefficients are on the order of ±10%. They are available in small to medium sizes and low to medium cost.

Electrolytic Capacitors

Electrolytic capacitors are designed to achieve high capacitance values in as small a size as possible. The resulting design tradeoffs give electrolytics a lower range of working voltage, and they are polarized. Polarization means that one plate is made positive and the other negative. This polarization is due to the fact that the plates are made of different materials, unlike standard capacitor types. Reversing the polarity of the voltage applied to electrolytic capacitors will destroy them if the voltage is large enough. There are two dielectrics used to make electrolytic capacitors: aluminum and tantalum. Aluminum electrolytics are more common, physically larger, and have a wider range of values. The general range of capacitance values for electrolytic capacitors values goes from 0.1 µF to 5700 µF. Working voltages go from 10 V DC to 500 V DC, depending on the value and type. Tolerances are typically –10% to +50% and temperature coefficients are not usually listed. Their common use is as power supply rectifier filters, so the actual value of capacitance seldom requires high accuracy.

Aluminum Electrolytics: These feature a cylindrical package with either axial or radial leads (see Figure 4–11). The radial lead is usually preferred for circuit board mounting. There are variations in marking the polarized leads. Some manufacturers will mark the plus lead accordingly. Many others mark the negative lead with a large band with a negative sign embedded in it. There are various grades of aluminum electrolytics available that primarily determine the expected operational life. Their selection will involve using the lowest working voltage and smallest package size combined with the desired package style, grade, and capacitance value.

Tantalum Electrolytics: These electrolytics are typically smaller and restricted to a narrower range of capacitance and working voltage values. Capacitance values range from 0.1 µF to 100 µF and voltage ratings from 10 V DC to 35 V DC. There are two different packages: cylindrical with axial leads and teardrop shape, as shown in Figure 4–12. The teardrop shape is typically used in circuit board applications.

Capacitor Selection

With all the different types of capacitors to choose from and all the parameters to consider, capacitor selection may seem overwhelming. Yet, with a little experience, the more typical selections become routine. Most of the time the value and function required will result in the use of ceramic disc or aluminum electrolytic

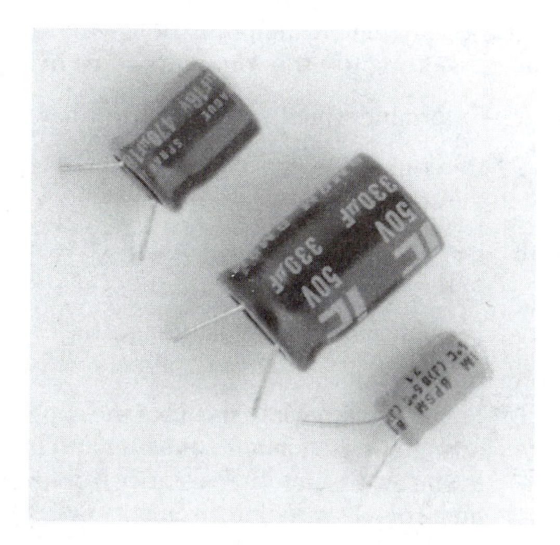

▶ FIGURE 4–11
Electrolytic capacitors

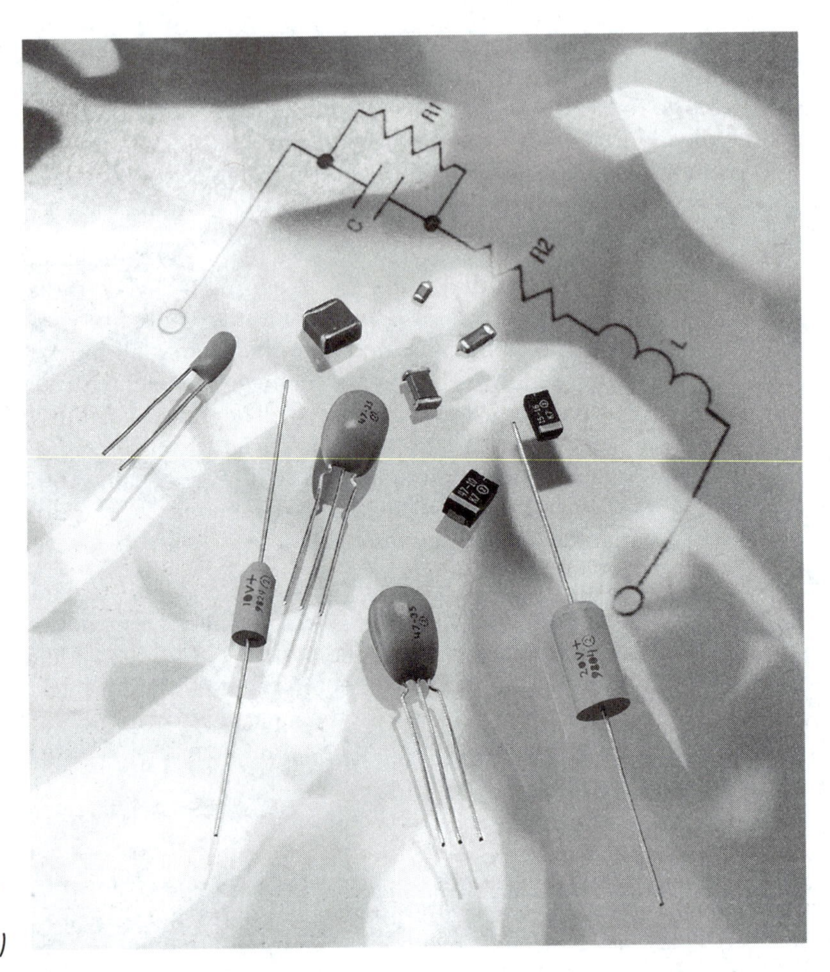

▶ FIGURE 4–12
Tantalum capacitors
(Courtesy of Vishay Intertechnology, Inc.)

capacitors. The nontrivial applications will require the most effort. To help in selecting capacitors, use the following steps in conjunction with Figure 4–13:

1. Determine which capacitor types have the value required.
2. Determine the capacitor types that have a working voltage sufficient for the application.
3. Determine those capacitor types that will function over the frequency range of the application.
4. From the application, determine the capacitor types that have an acceptable capacitance tolerance combined with the temperature coefficient.
5. Review any special aspects of the application, such as operation or function at specific frequencies (high Q and low DF), data conversion at low frequencies (insulation resistance), or accurate sampling, timing circuits, and high-speed switching circuits (low DA). Review the insulation resistance,

dissipation and quality factor, and dielectric absorption as required by the application and select the capacitor type that will meet these needs.

6. Review the sizes available of all the remaining capacitor types that meet all the criteria discussed so far.
7. Determine the cost and availability of all the capacitors reviewed in step 6 above. Make the decision based upon the combination of size, cost, and the other key factors defined in steps 1 through 5.

Back-to-back Electrolytics

When large capacitance values and small size are required for voltages that vary both positive and negative, the polarized plates of electrolytic capacitors are a significant limitation. To get around this, it is possible to place two polarized capacitors in series, back-to-back, with the negative plates connected together. Doing this achieves the effect of one nonpolarized capacitor equal to the series equivalent of the two polarized capacitors. Remember that two capacitors in series act like two resistors in parallel when determining the equivalent capacitance. There are some limitations of this practice that relate to tantalum capacitors. These limitations regard the equivalent capacitance value of the back-to-back capacitors and the fact that as the signal level across the capacitor increases, the equivalent capacitance will vary from the value calculated. Also, many manufacturers recommend against using "wet anode" tantalum capacitors in the back-to-back configuration.

Feed-through Capacitors

These are special capacitors for suppressing unwanted signal noise in the form of radio frequency interference (RFI). The purpose of feed-through capacitors is to shunt unwanted noise to ground at the entry or exit of a grounded metal enclosure or cavity. One plate of the feed-through capacitor makes contact with the feed-through conductor and the other capacitor plate is connected to both sides of the enclosure or cavity wall. These are intended for high-frequency applications and the range of capacitance values usually used is between 0.01 µF and 2 µF.

Example 4–3

On many circuit schematics, you will notice the placement of two bypass filter capacitors in parallel with values such as 100 µF and 0.1 µF. The 100 µF capacitor is usually an aluminum electrolytic type while the 0.1 µF value is usually a ceramic disc capacitor. The purpose of both of these capacitors is essentially the same: to bypass any frequency higher than DC, removing them from the power supply output. Why then are both capacitors required? Why can't the 100 µF capacitor filter perform this function alone? In analyzing this situation we will use the equation $X_C = 1/(2\pi fC)$ for capacitive reactance. The larger the value of the capacitance, C, the smaller X_C is at a given frequency, f, which is what is desired: a low impedance value that will bypass the frequency component to ground. The 100 µF capacitor

Capacitor Dielectric	Range of Values	Tolerance	Temperature Coefficient	Range of Working Voltage
Ceramic	1 pF to 1 µF	+/-5% to +/- 20%	+/- 15%	25 to 30 kV
Mica	2.2 pF to 0.01 µF	+/-0.5% to +/- 20%	200 ppm/°C	50 to 50 kV
Plastic film types	20 pF to 2.2 µF	+/- 1% to +/- 20%	+/- 2.5% to +/- 10%	30 to 10 kV
Polyester	0.001 µF to 2.2 µF	+/- 5% to +/- 20%	+/- 10%	30 to 10 kV
Polycarbonate	0.001 µF to 2.2 µF	+/- 1% to +/- 10%	+/- 10%	30 to 10 kV
Polystyrene	10 pF to 0.1 µF	+/- 1% to +/- 10%	+/- 10%	30 to 10 kV
Polypropylene	10 pF to 0.1 µF	+/- 1% to +/- 10%	+/- 10%	30 to 10 kV
Polyethylene	0.001 µF to 1 µF	+/- 1% to +/- 10%	+/- 10%	30 to 10 kV
Mylar	0.001 µF to 2.2 µF	+/- 1% to +/- 10%	+/- 10%	30 to 10 kV
Aluminum electrolytic	0.1 µf to 10,000 µF	–10% to +50%	NA	10 to 500 V DC
Tantalum electrolytic	0.1 µF to 100 µF	–10% to +50%	NA	10 to 35 V DC

▲ **FIGURE 4–13**
Capacitor selection chart

presents a much lower (100 times lower) capacitive reactance at any frequency when compared to the 0.1 µF capacitor. This reinforces the question: Why is the 0.1 µF capacitor necessary? The answer comes from the equivalent model of the capacitor discussed earlier in this section. The equivalent series resistance of the aluminum electrolytic capacitor and the ceramic capacitor also change with frequency. This limits their effectiveness as capacitors as frequency varies. The aluminum electrolytic is not effective above the range of 10,000 Hz. The ceramic capacitor is effective over a frequency range of roughly 1000 to 1 MHz. So neither

Frequency Range	Dielectric Absorption	Quality Factor	Size	Cost	Primary Advantage
High frequency	High	Low	Small	Low	Size and cost
High frequency	High	High	Medium	Low	Tolerance and temperature coefficient
Low to medium	Low	High	Medium to large	Medium to high	High quality factor and low dielectric absorption
Low to medium	Low	High	Medium to large	Medium to high	High quality factor, tolerance
Low to medium	Low	High	Medium to large	Medium to high	High quality factor, tolerance
Low to medium	0.001 to 0.02 %	High	Medium to large	Medium to high	Low dielectric absorption
Low to medium	0.001 to 0.02 %	High	Large	Medium to high	Low dielectric absorption
Low frequency	Low	High	Medium to large	Medium to high	High quality factor, tolerance
Low to medium	Low	High	Medium	Medium	General purpose plastic
Low frequency	High	NA	Small capacitance to size ratio	Low	Large capacitance for small size, polarized
Low frequency	High	NA	Small	Low	Large capacitance for small size, polarized

capacitor by itself can perform the complete needs of the circuit: bypassing all frequencies to ground. Both capacitors together function as a tag team, one taking over where the other one leaves off as the frequency increases.

Example 4–4

In this example we will select a capacitor for use in a sample-and-hold circuit (see Figure 4–14). The circuit requires a value of 0.01 μF and a 20 V rating. The temperature range for the circuit is 0°C to 85°C. A sample-and-hold circuit is

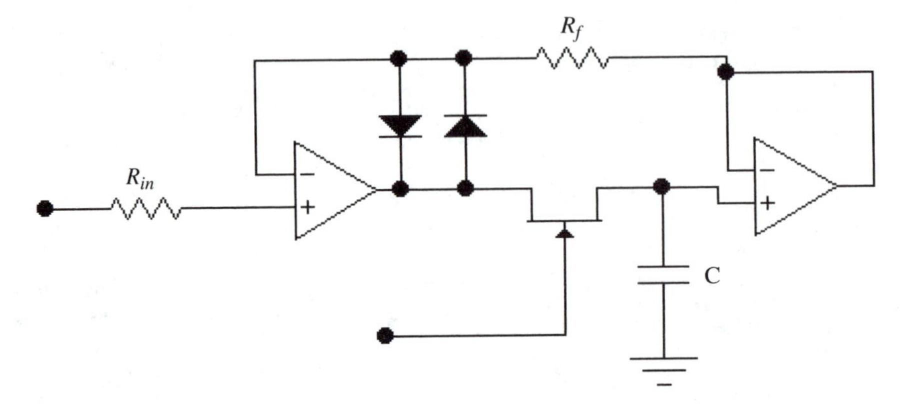

▲ **FIGURE 4–14**
Sample and hold circuit

often used with an A/D converter to sample the signal at one point in time and save it for the A/D converter to process. In sample-and-hold circuits a capacitor is connected to the input signal for a brief period of time so that the capacitor will charge up to the signal value. At the specified sample time, the input signal is disconnected and the A/D converts the sampled signal. In this case the dielectric absorption of the selected capacitor, as discussed earlier in this section, is critical. The correct capacitor will have a minimal DA factor. The capacitor types with the lowest DA factor are polystyrene, polypropylene, and teflon. Polystyrene will not be used because of its temperature limitations. Either polypropylene or teflon can be used for this application. Teflon has a little higher DA factor and will be higher in price. Polypropylene will have a lower DA factor and a larger size and will be cheaper.

Variable Capacitors

As with resistors, variable and trimming capacitors are available to provide for adjustable capacitance values. There are two basic types: air-variable and trimmer capacitors.

Air-variable capacitors are used to tune resonance circuits like those utilized in the front end of a radio receiver. Air-variable capacitors use an interleaved set of metal plates, with air as the dielectric. Capacitance variations of 1 pF to 200 pF are available. Trimmer capacitors are fabricated with mica, ceramic, and glass dielectrics and are used for fine-tuning a capacitance value. The key features of the variable capacitor types are:

Mica: Good stability and low temperature coefficient; good capacitance to size ratio and can handle moderate shock and vibration; low inductance and cost

Ceramic: High Q factor with predictable temperature coefficient; good capacitance to size ratio with low inductance; not usable in high shock and vibration environments; limited to 180° rotation

Glass: Possesses a high voltage capability and can be environmentally sealed, can handle many adjustments with smooth and nearly linear capacitance changes with rotation

4–4 ▶ Inductors

Inductors are the least-used passive electronic components because of their relative size and cost. At low to middle frequencies, the desired effect of the inductance can be achieved with a cheaper and smaller capacitive circuit. This is shown by the fact that we can construct both high- and low-pass filters from either resistor-capacitor circuits or resistor-inductive circuits. The reason that inductors are a more expensive approach for many circuit applications is that at low frequencies, relatively large inductor values are required. As frequency increases, inductors become a more viable approach in ripple reduction and other applications. In Chapter 6 we will see how inductors are combined with capacitors to form effective ripple reduction circuits in higher-frequency switching regulator circuits.

Inductors are used in filter and tuned circuits, as current limiters, and as ripple filters in power supplies. The three principal types of inductors have air, iron, or ferrite cores. The core is the material around which the conductor coil is wrapped. The conductive coil is covered with an insulating layer, usually a varnish, to prevent conduction between the adjacent coils. Magnetic core inductors offer a higher range of inductance values. There are also fixed and variable inductors. Figure 4–15 shows the schematic symbols that correspond to the core materials and fixed and variable symbols.

The equivalent circuit for an inductor is shown in Figure 4–16. R represents the coil resistance and C is the inherent capacitance of the winding. Inductors are classified by their Quality factor (Q) value that equals the ratio of the inductive reactance at a particular frequency to the winding resistance R. A high Q value means that the inductive reactance at that frequency is much larger than the coil resistance and therefore a small amount of power will be lost due to the coil resistance.

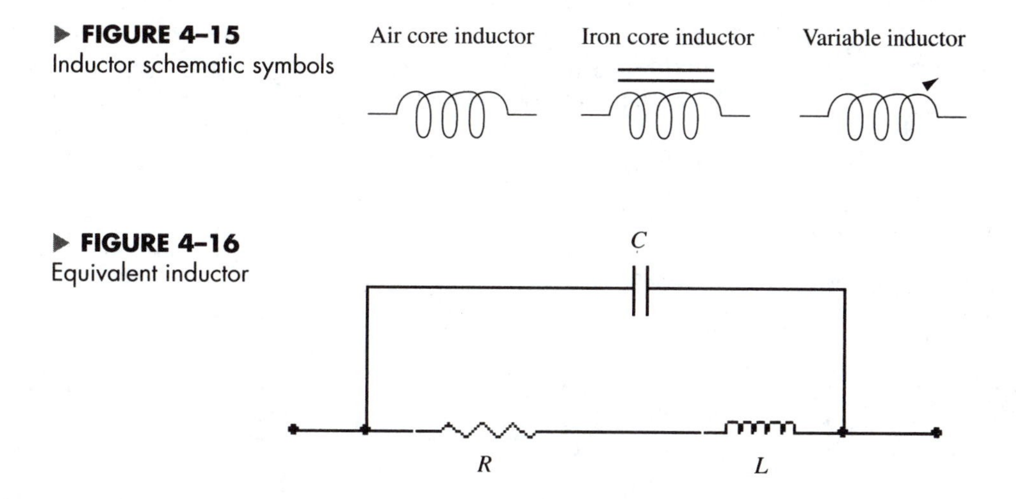

▶ **FIGURE 4–15**
Inductor schematic symbols

▶ **FIGURE 4–16**
Equivalent inductor

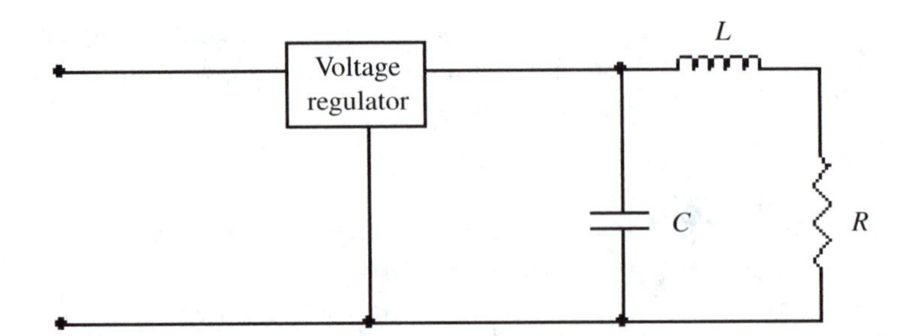

▶ **FIGURE 4–17**
Ripple reduction example

The more common applications of inductors are as follows:

Ripple Reduction: This application involves smoothing the ripple on the output of DC power supplies (see Figure 4–17). The critical parameters in this application are the minimum inductance value, DC current, working voltage, and maximum DC resistance.

Swinging Inductor: Swinging inductors (see Figure 4–18) are applied to the AC input of many power supplies to reduce the ripple input to the supply. The critical parameters in this application are both the minimum and maximum inductance values, DC current, working voltage, and maximum DC resistance.

Current Limiting: As current limiters, the value of inductance and the tolerance becomes important, along with the maximum AC current and maximum AC voltage drop for which the inductor is rated.

Tuned or Timing Circuits: These applications require tighter tolerances of inductance values and consideration of the inductor Q value, current, and DC resistance. Inductor Q value is ratio of the resistance of the inductor to the inductive reactance (R/X_L) at a specific frequency.

The range of inductor values available for the various types are as follows:

Air Core Inductors: Air core inductors are available from a few tenths of a microhenry to several hundred microhenries. These are usually used at high frequencies in the range of 100 kHz to 1 GHz. Very small air core inductors can be fabricated by a few loops of wire or loops on a printed circuit board.

▶ **FIGURE 4–18**
Swinging inductor example

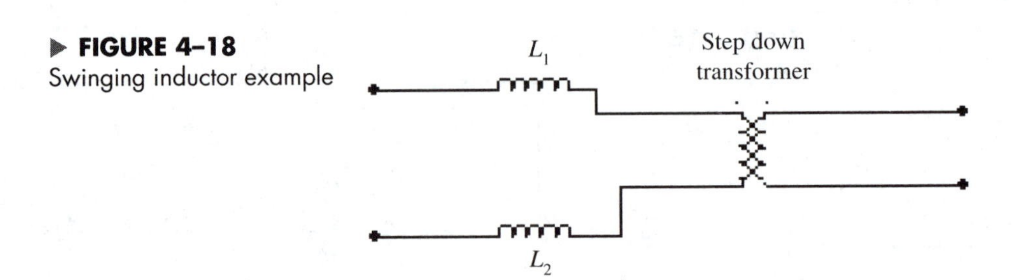

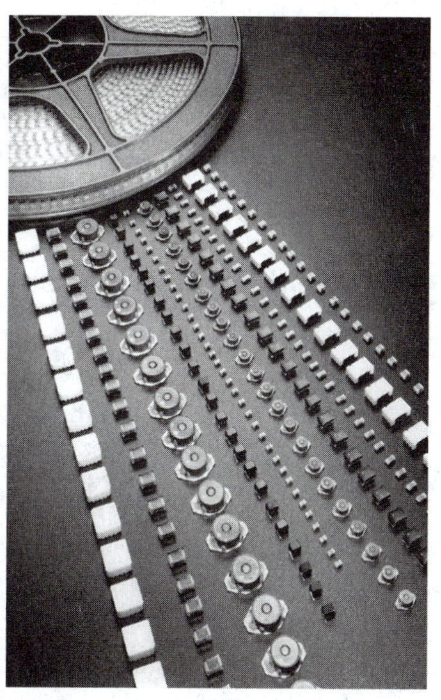

▲ **FIGURE 4–19**
Coilcraft inductor family *(Courtesy of Coilcraft, www.coilcraft.com)*

Iron Core Inductors: Iron core inductors are available from 0.1 to 120 mh with DC resistances ranging from a few to several hundred ohms. Maximum DC currents are on the order of 100 ma with Q values of about 50.

Variable Inductors: Adjustable magnetic core inductors are available that vary inductance by moving the magnetic core in or out of the coil with screw adjustments. Inductors come in a variety of sizes and shapes. Figure 4–19 shows some of the inductor packages available.

4–5 ▶ Transformers

A transformer consists of two separate coils that are wrapped around a closed magnetic circuit. They are used to step down, step up, isolate, and impedance match AC circuits. The ratio of the number of turns on the secondary to the primary turns is called the *turns ratio, n.* The voltage output from the secondary, V_{sec}, is equal to the primary voltage, V_{pri}, divided by the turns ratio. Conversely, the secondary current, I_{sec}, equals the turns ratio multiplied by the primary current, I_{pri}.

$$V_{sec} = V_{pri} \times n$$

$$I_{sec} = I_{pri} / n$$

There are a wide variety of transformer types based on primary and secondary voltages and currents. There are also many variations of primary and secondary windings. In either winding there may be a center tap that is a connection point in the middle of the winding. Center taps are used to create full-wave rectifiers with two diodes or combination plus-minus power supplies. In other cases a winding may actually be two equal windings that can be connected in parallel or series.

Example 4–5

In this example a power supply is designed for use as an industrial product that will operate off of 115 V AC or 230 V AC, depending on which is available to the user. To accomplish this, the product will have to be altered to change the turns ratio of the transformer used. This can be achieved by using a transformer that has two separate and equal windings on the primary side (see Figure 4–20). If 230 V AC operation is desired, then the two primary windings are wired in series. For 115 V AC operation, the windings are wired in parallel, which reduces the turns ratio and increases the current capability of the primary. The desired secondary voltage is 12 V. If the input voltage is 230 V AC, then the required turns ratio is 230/12 = 19.16. For 115 V AC, the turns ratio should be 115/12 = 9.58. Since a half a turn cannot be fabricated, a primary with two separate windings that are equal and 10 times the number of secondary turns will be the best solution. If the secondary has 10 turns, then each of the primary windings should have 100 turns. The turns ratio will vary from 20, when the two windings are wired in series, to 10 with the parallel connection.

The resulting option to the user will be a choice of how to wire the two primaries, in series or parallel. An even easier option is to include a switch in the power supply design that changes the series and parallel wiring of the primaries with the flip of that switch. The primary applications for transformers are discussed next, followed by the criteria for selecting transformers.

Step-down Transformers

This is the most common application of transformers and involves the step down of an AC power supply voltage to a smaller voltage AC voltage for use as a low voltage AC or DC power supply. In this case the turns ratio is greater than one, meaning that the primary has more turns than the secondary.

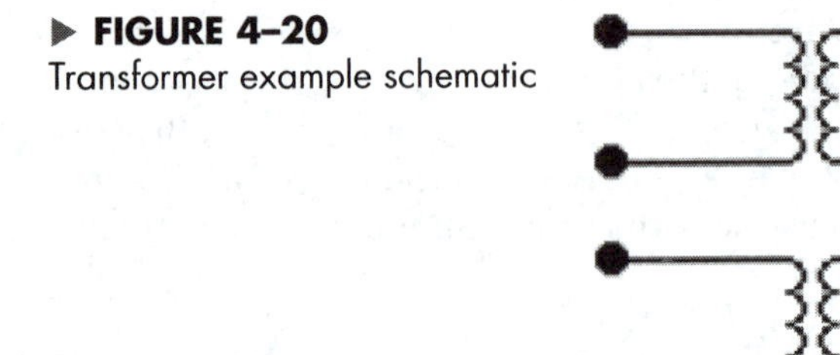

▶ FIGURE 4–20
Transformer example schematic

Step-up Transformers

These are utilized whenever it is necessary to increase the voltage level from the AC voltage level available. The best example of this is the development of 230 V AC when only 115 V AC is available. There are many other applications where the AC voltage must be increased slightly. Increasing 208 V AC to 230 V AC is another example. Step-up transformers have fewer turns in the primary than the secondary.

Isolation Transformers

These are used solely to isolate one AC voltage from another AC voltage. Transformers inherently accomplish this because there is no electrical connection between the primary and the secondary. The energy is transferred magnetically. The output voltage and current should be the same as the input voltage and current. The reason for their use is either to eliminate DC noise levels from a signal or to isolate the secondary circuit for safety reasons (see Figure 4–21). In selecting these transformers, the primary issue is finding a transformer that will handle the required power level.

Impedance Matching

Impedance matching involves the matching of a source impedance to the load impedance. This is desirable because maximum power is transferred only when the load impedance equals the source impedance. A transformer can achieve this because of the way the windings reflect the impedance value through the windings. The most common example is the typical 75 Ω source resistance of an antenna lead that is connected to the 300 Ω input resistance of a television receiver. An impedance matching transformer can be used to match these two resistances or, in other words, make the 300 Ω television load appear to the antenna as a 75 Ω load. The selection of impedance matching transformers is most dependent on the ideal turns ratio. The turns ratio can be determined by the formula where $R_{primary}$ equals the value of the source resistance and R_{load} is the actual load resistance. The proper power rating must also be determined by multiplying the maximum secondary output voltage times the output current. This will be the volt-amp requirement for the transformer.

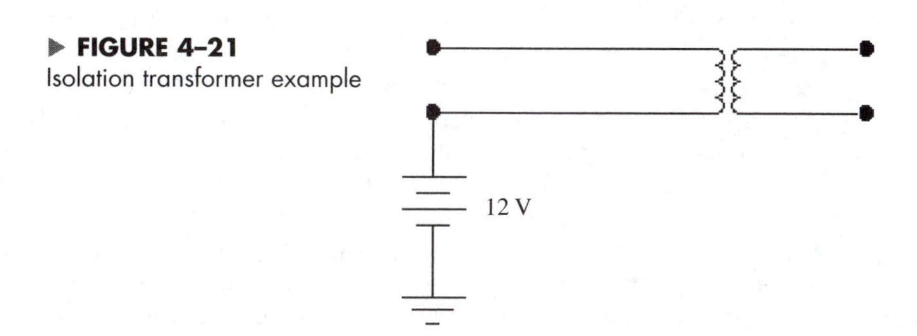

▶ FIGURE 4–21
Isolation transformer example

4–6 ▶ Switches and Relays

Switches are mechanical devices that switch electrical circuits. The two basic types of switches are maintained and momentary. Maintained switches maintain the switch position after they are engaged until they are mechanically repositioned. Momentary switches are spring loaded and revert back to their normal state after being released. Switch contacts are defined as common (C), normally open (NO), or normally closed (NC). The normal state of the contacts are the mechanical "off" state of the switch. Switches are classified by the terms *poles* and *throws*. Poles are merely the numbers of sets of contacts included in a switch. One set of contacts is needed for control of one circuit. If two or more circuits must be controlled, then additional sets of isolated contacts will be needed. The term *throws*, which is derived from the number of positions for old-fashioned "knife" switches, has been carried over to today's switches. A switch that makes or breaks only two points in a circuit is said to have a *single-throw contact.* A switch that makes or breaks one common contact to either a normally closed or a normally open contact is called a *double-throw contact.* To achieve more than two throws, some sort of rotary switch is required that will connect one common contact with potentially many other contacts. Switch selection is determined by mechanical size and style, contact arrangement, and the voltage and current rating of the contacts. Figure 4–22 shows a variety of switches.

Relays

Relays are electromechanical or solid-state devices that can best be defined as voltage-controlled switches. When a voltage is applied to the input, the output contacts change states. Electromechanical relays have been used in control circuits for many years. They continue to be utilized, in spite of the electronic alternative, because of their ease of use, versatility, contact ratings, and relatively low cost. They consist of a coil that, when energized, pulls in an armature, which changes the common contact—or switch position. These relays feature both AC and DC coils and contact arrangements up to four poles in single- or double-throw variations. The poles and throws specification for electromechanical relays is the same as described for switches. Devices called *contactors* are simply large electromechanical relays typically used in motor starter circuits. There are many different packages available for panel mounting, plug-in modules, or direct solder in circuit board variations. Selection of these relays involves the mechanical size and configuration, the coil voltage, the contacts arrangement (number of poles and number of throws), and the contact voltage and current ratings.

Solid-state relays are the electronic equivalent to electromechanical relays. They consist of an input circuit to which an input voltage is applied to change the state of the output contacts. The output contacts are an electronic switch that is either a transistor for switching DC circuits or a triac for AC applications. The input and output circuits are isolated by using opto-isolators. Opto-isolators use an LED and a photodiode to transmit a light signal that turns on the output when the input voltage is applied. Solid-state relays offer a significant improvement in switching life over electromechanical relays. Input voltages are limited to 5 V DC to 15 V

▲ **FIGURE 4–22**
Switch assortment *(Courtesy of C&K Switch Products)*

DC and only one contact arrangement is available: single pole/single throw (SPST). Care must also be taken to derate the current rating of solid-state relays when they are used at higher temperatures. To select solid-state relays, the input voltage, the output voltage being switched (AC or DC), and the voltage and current ratings of the contact are the key parameters.

Example 4–6

Determine the number of poles and throws for the switches and relays shown in Figure 4–23*.

4–7 ▶ Connectors

Connectors are an important aspect of any electronic design. Connectors involve any device that serves to make an electrical connection from one circuit element to another. There are generally three types of connector situations that occur in electronic design:

**Answers:* A. DPST B. SPDT C. SPST D. SPDT

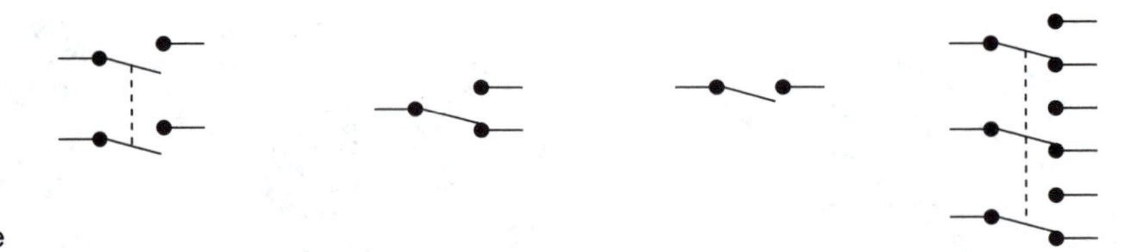

▶ **FIGURE 4–23**
Poles and throws example

Wire-to-wire

Wire-to-PCB (printed circuit board)

PCB-to-PCB

T key aspects that determine the applicability of a type of connector to a specific application are as follows:

1. Application type (wire-to-wire, wire-to-PCB, PCB-to-PCB; wire-to-wire could be flat cable, coaxial cable, or conventional wire)
2. Mounting (none, panel, printed circuit board)
3. Voltage across contacts
4. Current through contacts
5. Size
6. Number of contacts and spacing
7. Termination type (solder or crimp)
8. Environmental aspects (temperature, humidity, seal)
9. Contact resistance, which depends on the contact material; typical contact materials include beryllium copper, phosphor bronze, spring brass, and low-leaded brass. In higher-quality connectors, selective gold or silver plating is used on the specific contact areas. Low-cost connectors utilize an electro-tin plate.
10. Keyed or not keyed; keying prevents incorrect connection
11. Insertion force and pin alignment tolerance
12. Reliability (number of disconnects over life, design life)
13. Cost and ease of assembly

The most common types of connectors will be discussed next, along with their typical applications and advantages.

Printed Circuit Board Edge Connectors

Printed circuit board edge connectors are very cost effective because they form a one-piece connector that either connects a printed circuit board to some wires or to another printed circuit board. They rely on the printed circuit board to substi-

▶ **FIGURE 4–24**
Printed circuit board edge connector example *(Courtesy of Tyco Electronics)*

tute for the male portion of the connector. Nevertheless, when utilizing printed circuit board edge connectors, care must be taken to ensure that the design and quality of the printed circuit board will provide a secure connection. They are available in both crimp and solder terminations. Typical spacing of printed circuit board edge conductors are 0.156″ centers or 0.1″. Figure 4–24 shows examples of edge connectors.

Flat Cable Connectors

Flat cable connectors are designed to work with standard flat cable and form a termination that pierces and crimps the flat cable wire. This allows a very quick and low-cost connector assembly, which is its primary advantage. Flat cable connectors come in many configurations that include male pin and female receptacles as well as printed circuit board edge connectors. The flat cable connectors are connected to the flat cable by sandwiching the flat cable between the two sections of the connector and carefully pressing them together to pierce the insulation and form a crimp contact between the connector and the wire. The pressure must be applied in a consistent manner and is usually performed with a vice. Flat cable and the corresponding connectors come with many contact arrangements, which can be up to 64 contacts wide. Figure 4–25 shows examples of flat cable connectors.

D-Type Connectors

Connecting data input and output lines between two pieces of equipment is usually accomplished with some sort of D-type connectors. D-type connectors are terminated in a variety of solder lugs, including direct mounting to printed circuit boards. Standard D-type connectors include contact numbers of 9, 15, 25, and 37. Figure 4–26 shows examples of D-type connectors.

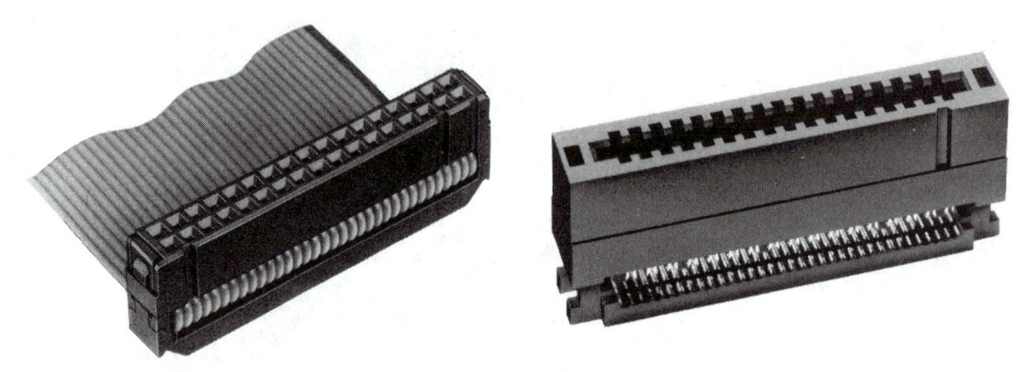

▲ **FIGURE 4–25**
Flat cable connectors *(Courtesy of Tyco Electronics)*

▲ **FIGURE 4–26**
D-type connectors *(Courtesy of Tyco Electronics)*

Coaxial Connectors

Coaxial connectors make connection on one axis and are usually restricted to connecting two wires: a signal and ground. These connectors (see Figure 4–27) are typically used for radio frequency (RF) and audio applications. The most popular RF coaxial connector is the standard BNC connector. However, there are many other popular varieties, such as SMA and SMB. Audio-type connectors are usually called phone plugs, phonojacks, and mini- and micro-plugs. The phone plugs are usually 1/4″ plugs and receptacles. Phono jacks are 3.18 mm in size while the miniplug is 3.58 mm and the microplug 2.46 mm.

Circular Connectors

These are usually multi-pin connectors of higher reliability. Many military-type connectors are of the circular type. (Military connectors are discussed in the next

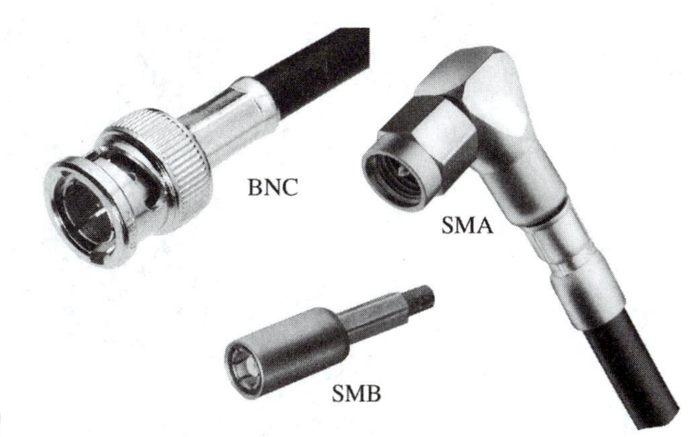

▶ FIGURE 4–27
RF Connectors *(Courtesy of Tyco Electronics)*

section.) A good example of a typical circular connector application is the keyboard connector on most personal computers. Figure 4–28 shows examples of circular connectors.

Military Connectors

Military connectors possess extremely high reliability and can withstand extreme environments such as moisture, ambient temperature, vibration, shock, and EMI. There are many different configurations and types and these are used whenever the requirements of the application justify their high cost.

Zero Insertion Force Connectors

Zero insertion force (ZIF) connectors are used whenever there are a high number of contacts—and the number of disconnects is also very high (see Figure 4–29). The primary purpose is to provide superior contact and preclude any damage to the connection point on either side caused by insertion. These connectors mate without any insertion force. The contact force is applied separately after insertion. The most typical application is EPROM programmers.

▶ FIGURE 4–28
Circular connectors *(Courtesy of Tyco Electronics)*

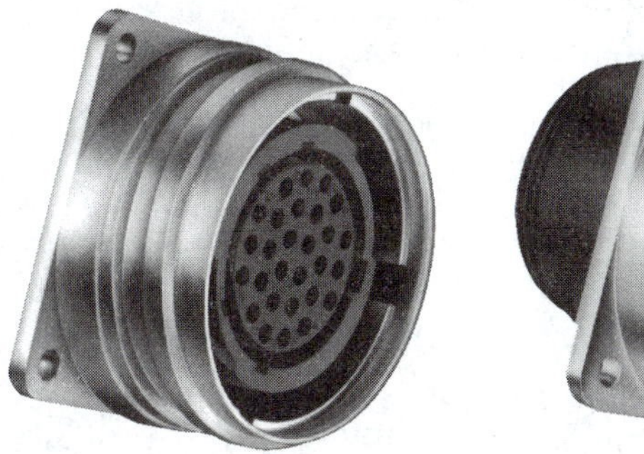

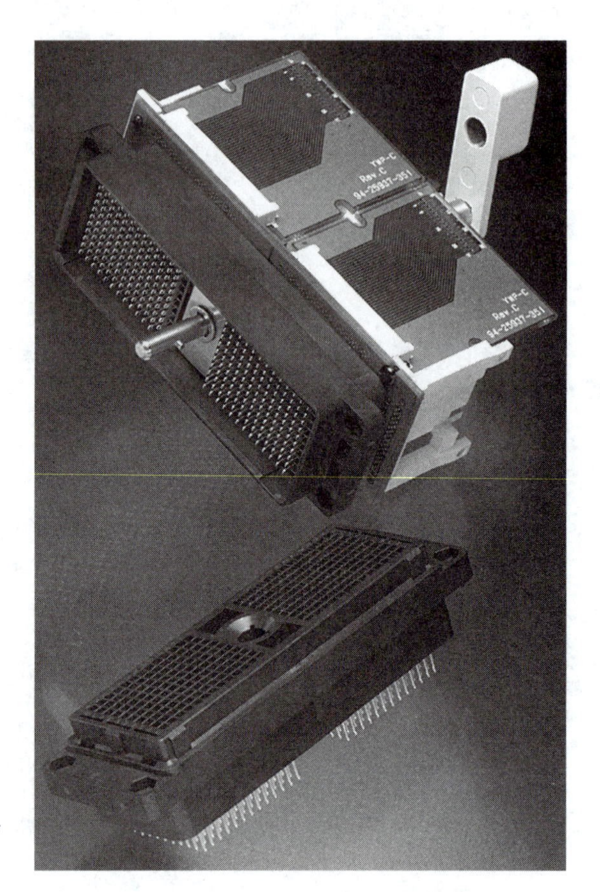

▶ **FIGURE 4–29**
ZIF connectors *(Courtesy of Tyco Electronics)*

4–8 ▶ Selecting Active Components

Active components are devices that are fabricated from semiconductor materials. The most commonly used active components are as follows:

Diodes: Rectifier diodes, signal diodes, shottky diodes, tunnel diodes, photodiodes, zener diodes, light emitting diodes (LEDs), varactor diodes

Transistors: Bipolar junction transistors (BJTs), field effect transistors (FETs), metal oxide semiconductor field effect transistors (MOSFETs)

Thyristors: Diacs, unijunction transistors, silicon-controlled rectifiers (SCRs), triacs

Analog Integrated Circuits: Voltage regulators, voltage references, comparators, op amps, DC to DC converters, A/D converters, D/A converters, multiplexers, filters, drivers, temperature sensors, special functions

Digital Logic Family Integrated Circuits: TTL logic, CMOS logic, ECL logic

MSI, VLSI Digital Integrated Circuits: Microprocessors, microcomputers, memories, programmable memories, programmable logic, drivers, interface circuits, encoders, decoders, multiplexers, and special functions

The selection and use of active electronic components is a broad area that is well covered in most electronic courses and texts. This section is a general review of active components, discussing their critical parameters and summarizing their selection.

The specifications for most active components start out with a section called "Absolute Maximum Ratings," where the maximum ratings for all parameters are listed. It is imperative to review every aspect of these ratings with the intended circuit application to ensure that these ratings are not exceeded. Exceeding these ratings usually means the component will malfunction and may sustain permanent damage. Depending on the application of the design, there should be at least a 10% safety factor between the maximum value of a parameter in a circuit and the absolute maximum value listed in the specifications. The key parameters of concern are usually:

Maximum power supply voltage

Maximum input voltage

Maximum differential input voltage

Maximum output current (source)

Maximum output current (sink)

Maximum operating temperature

Maximum voltage on any pin

Beyond the maximum parameters ratings, the determination of which active component fulfills the functional requirements of a circuit usually comes down to the functional capabilities of the circuit and the speed, tolerance, power consumption, temperature coefficient, reliability, complexity and ease of use of the component. These parameters are realized by a myriad of terms and pages of specifications that the data sheets for active components comprise. To complicate matters further, each semiconductor manufacturer may use slightly different terms for a particular parameter or signal line. This requires that the designer possess a solid understanding of the requirements for the component to be able to understand and sort out the various parameter symbols, names and their corresponding meaning and significance.

Data sheets usually include application hints. These should be reviewed completely as they are the key to the proper function of the active component in a circuit application. Application circuits are also included for many active components and these are often helpful in conveying what might not be so obvious from the rest of the specifications. Most active component data sheets require a lot of reading between the lines. Experience is the only way to develop this important skill. Fortunately, each semiconductor manufacturer has a staff of applications engineers that can be contacted for extra help and support.

Summary

In this chapter we have discussed the variety of components available and the process of their selection for use in an electronic circuit design. In each case the selection of a component involves its quality, size, cost, and labor requirements. This selection procedure is a critical part of the design process. It is important to remember the analogy of the "weak link in the chain" when considering circuit function and reliability. On the other hand, the financial objectives of the design cannot be ignored. This is why it is important to keep track of the cost goals of the project, continuously comparing the latest cost estimates, as they are determined.

The successful completion of any design involves the sound judgments made in the selection of components where the proper combination of function, quality, cost, and size are obtained. In the next chapter we will cover how to implement these components and the circuit schematic into a functioning printed circuit board assembly.

References

Harper, C. A., ed. 1977. *Handbook of Components for Electronics.* New York: McGraw-Hill.

Stadtmiller, D. J. 2001. *Electronics Project Management and Design.* Upper Saddle River, NJ: Prentice Hall.

Warring, R. H. 1983. *Electronic Components Handbook for Circuit Designers.* Blue Summit, PA: Tab Books.

Exercises

4–1 A 1% tolerance, 1 kΩ, metal film resistor is used in an application in which the ambient temperature will go from 0°C to 65°C. Assuming a temperature coefficient of 100 ppm/°C, what is the worst-case resistance range expected for this resistor value after combining the effects of the temperature coefficient and the resistor tolerance?

4–2 For each circuit shown in Figure 4–30, calculate the possible ranges of the voltage V_{OUT}. V_{IN} = 10 V.

4–3 In the circuits shown in Figure 4–31, calculate the smallest resolution values for the voltage V_{OUT}. V_{IN} = 10 V.

4–4 When selecting the actual resistors to be used in a circuit, what are the two key performance parameters that should be used to determine whether to select a carbon film or a metal film resistor? Use the resistor selection chart shown in Figure 4–1.

4–5 Which key design factors promote the use of a wire-wound resistor instead of a metal film resistor in a particular circuit?

4–6 If a design requirement can be accomplished by using higher tolerance resistor values or adjustable trimmer potentiometers, which is likely to be more cost effective?

4–7 Calculate the equivalent series resistance (ESR) for a 0.1 μF capacitor that has an insulation resistance (IR) of 100,000 Ω at a frequency of 60 Hz. Assume that series resistance and inductance equal zero.

4–8 Define in your own words the meaning of the term *dielectric absorption*. When is it an important consideration?

4–9 Define the terms *dissipation factor* and *quality factor*. How are they related?

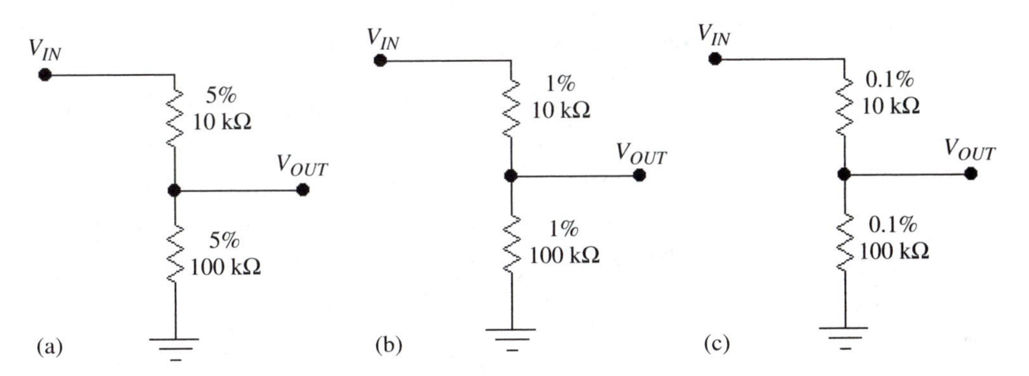

▲ **FIGURE 4–30**
1% and 5% voltage divider circuits (see Exercise 4–2)

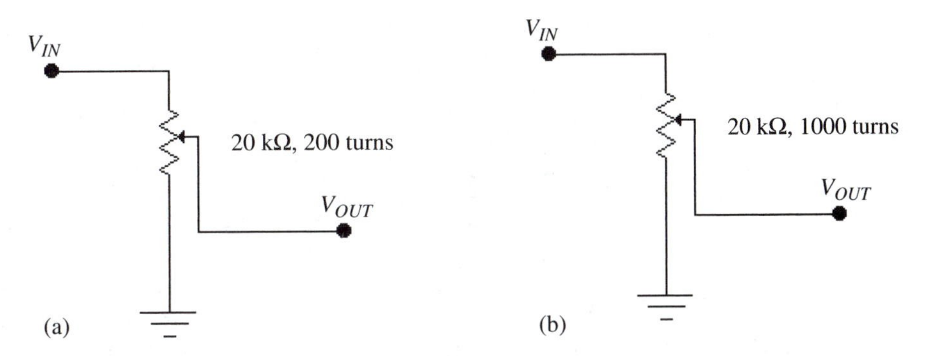

▲ **FIGURE 4–31**
Potentiometer voltage dividers with number of turns (see Exercise 4–3)

4–10 What type of capacitors would you select to perform power-supply decoupling of various integrated circuits on a printed circuit board?

4–11 A capacitor is to be used in a precision low-pass filter circuit for an audio signal. What are the key parameters of concern when selecting the capacitor to be used in this circuit? Based on using the best available capacitor to meet these requirements, which type of capacitor would you select?

4–12 A capacitor is to be used in a 60 Hz, sample-and-hold circuit that is connected up to an A/D converter. What is the most important parameter to be considered in the selection of the type of this capacitor?

4–13 A bipolar capacitor with a value of 10 μF is needed for a circuit. List the different types of capacitors and methods that you can use to achieve this overall capacitance value.

4–14 Show the schematic symbol for the following switches:

a. SPST maintained switch

b. SPST momentary switch

c. SPDT maintained switch

d. DPDT maintained switch

4–15 What are the three different types of applications for electrical connectors?

4–16 When replacing any type of electrical contact, which two electrical parameters must be known?

4–17 When selecting active components, list the maximum value parameters that should be considered.

5 Printed Circuit Board Design

Introduction

After the electronic design has been simulated, breadboarded and tested, the circuit is usually developed into a printed circuit assembly. The invention of the printed circuit board (PCB) was a significant factor in the development and growth of the electronics industry. Its development has been just as important as the transistor and the integrated circuit. Over the years the methods utilized to lay out and manufacture printed circuit boards have undergone significant change as large-scale integration and computer technology have been applied to the process. In order to develop a background for circuit board technology, the discussion begins with manual layout and taping methods and then proceeds to the current computer software process. The likelihood of laying out a PCB depends on the career path followed. Students seeking a technology degree are more likely to complete a board layout than students seeking an engineering degree. In any case there is a need to understand the process, as the PCB is a key component in any electronic design. The design engineer will review the PCB layout from an electrical perspective: grounding, component location, bypass capacitors, and the like.

The PCB is an important electronic component. Almost all electronic circuits are implemented with printed circuit boards. Because each circuit is different, the resulting PCB is a custom component designed specifically for a particular circuit. The primary function of the PCB is to make the electrical connections for the electronic circuit, but it also provides for mechanical mounting of the components and the board itself. The circuit conductors, while ideally viewed as 0 Ω conductors, in actuality possess impedance values that include resistance, inductance, and capacitance. In this chapter we will discuss the development and documentation of the printed circuit board as follows:

- Documentation accuracy
- Printed circuit board types

- General printed circuit board design considerations
- Printed circuit board development—manual
- Printed circuit board development—computer
- Printed circuit board documentation

5–1 Documentation Accuracy

At this point in the design project, there are two documents that define the circuit: the schematic diagram and the parts list. The accuracy of these documents is extremely important as we move into the next phase, because all of the drawings and documents developed later are based on these. Now is a good time to check and update the schematic diagram and the parts list.

It is good practice for whomever has design responsibility for an assembly, module, or system to maintain one set of all the drawings for the design, designated as "Design Master Drawings." The design master drawings—or simply "design masters"—are hard copies of the latest revision of all documents that define the design. They should be labeled in red as design masters with the date and initials of the responsible design engineer. The purpose of design masters is to accumulate all of the modifications to be made to the drawings in one well-assembled document as errors are found and problems resolved. Without drawings designated as design masters, you will soon find yourself with many copies, notes, and scraps of paper listing important changes that may or may not be passed on when final changes are made to the design drawings. Utilizing design master drawings will improve the accuracy of design documents and greatly improve the efficiency of the project engineer. At the beginning of the circuit board layout stage, the schematic and the parts list are checked and modified to correct any errors or implement changes made as part of the design simulation or breadboard testing.

5–2 Printed Circuit Board Types

Before discussing the layout methods, let us review the makeup of the printed circuit board and the different types currently available.

Circuit Board Laminates

Every printed circuit board starts out with what is called the laminate: the copper-clad material that is etched, plated, and drilled to complete the bare circuit board assembly. The laminate consists of a base material that has a copper foil applied to one or both sides. The typical base materials consist of paper, glass cloth, or glass mat combined with phenolic or epoxy resins. A base material and copper foil are combined to form a particular grade of laminate. Laminate grades have been established by NEMA (National Electrical Manufacturer's Association) and military specifications. NEMA type G-10 (MIL Spec type GE) is a very popular general-use grade of laminate made from glass cloth bonded with an epoxy resin. Other laminate grades include the NEMA prefix FR, which stands for "flame retardant." The FR grades are favored for use on PCBs used in products that must meet approval agency flame retardant requirements. NEMA type FR 4 is also very popular and is similar to the G-10 material with the addition of a flame-retardant epoxy. The mechanical strength, ease of machining, adhesion of the copper foil material, and ability to withstand changing environmental conditions determine the ultimate quality of a laminate material. Laminates are available with single-sided or dual-sided copper foils, and very thin laminates can be layered together to form multilayer circuit boards.

Printed Circuit Board Manufacturing Process

Traditionally, the printed circuit boards have been fabricated from what is called a subtraction process, the removal of copper from a laminate material. Processes that add copper runs have been developed and are becoming increasingly popular. The subtraction process is still predominate and is the process described here. Figure 5–1 shows a flowchart of the process. The basic subtraction process fabrication of a PCB involves the following:

1. The laminate is drilled and the holes are plated with a copper flash process.
2. The laminate material is thoroughly cleaned and dried.
3. The laminate is covered with a thin adhesive-backed sheet of material called a *photoresist* that is applied to the copper foil on the laminate. This photoresist material can be altered with the application of light to resist certain solvents called *developers*.
4. The photoresist material is exposed to fluorescent lights through the negative of the artwork to be etched. Where the negative allows the light to pass through, the photoresist is altered such that a developer solvent removes it from the laminate. Where the negative blocks the light, the photoresist material will not be sensitive to the developing solvent and remains on the laminate.
5. The laminate, with the developed photoresist material attached, is developed by placement in a developing solvent that strips away the sensitized photoresist material.
6. The result is the initial laminate with the remaining photoresist material covering the laminate where the copper runs are desired. The laminate is placed in a copper etching acid solution that strips away the exposed copper areas.

Single-sided Printed Circuit Boards

Single-sided printed circuit boards are the simplest variety, because the copper runs exist on only one side of the laminate. For through-hole technology components, the side of the board with the copper runs is called the *copper side*. The

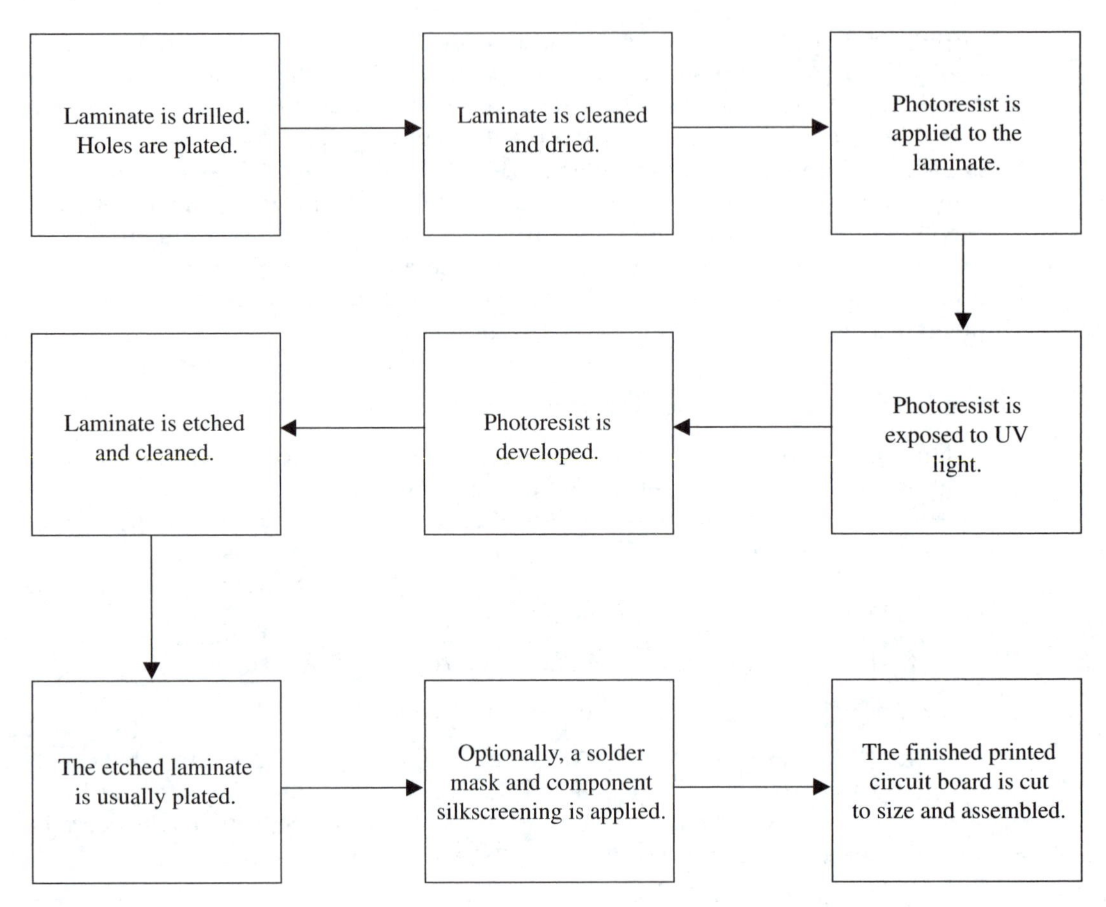

▲ **FIGURE 5–1**
Printed circuit board fabrication process flowchart

other side is designated the *component side.* For surface-mount technology circuit boards, the components and the copper runs are present on the same side. At the present time, single-sided boards are used only for very simple circuits not required to be a minimal size. Because all the connections must be made on a single side, the layout is more complicated, and more space is needed for the copper runs. Also, the holes that are drilled into the printed circuit for mounting through-hole components do not need to be plated through as on double-sided boards. This is because the connection of the component to the copper run can be assured by the solder connection. A single-sided printed circuit board will be cheaper but physically larger than the double-sided alternative.

Double-sided Printed Circuit Boards

The double-sided printed circuit board has copper runs on both sides of the board laminate. The circuit connections can be made much more easily, because there are two surfaces on which to make them. Double-sided boards also provide the ability to transfer a connecting run from one side of the circuit board to the other. This is accomplished through a hole in the printed circuit board called a via or feed-through. The via is a hole in the PCB whose only purpose is to transfer the connecting run from side to side. Consequently, components are not mounted in the via holes. The via hole must be copper plated to ensure connection between both sides of the board. Plating of the holes also improves the solder connection made when the components are installed into the board. Copper runs can be connected to either side of a component hole so circuit connections can be passed from one side of the board to the other through them as well. The double-sided PCB results in a smaller, denser circuit board.

Multilayer Circuit Boards

Multilayer circuit boards have additional thin laminates that provide circuit foils that can make additional circuit connections. Multilayer boards are utilized when complex circuit connections are required in a minimal space. Each layer is aligned with and sandwhiched between the outer layers of the circuit board. Plated-through via holes and component holes are used to transfer connections from one layer to another. If no connection is made at a particular layer, then the via or component hole is isolated from making a connection at that layer. The most typical application of multilayer boards today is the provision of two inner layers to make all the power supply connections. In this case one of the layers becomes power supply ground, and the other makes all the positive power supply voltage connections. This is an optimum situation for noise immunity, as the power supply ground layer is one large ground plane that serves as a ground shield for the entire board. Additionally, with the positive power supply voltage on one side and the ground layer on the other, the inner laminate material acts as a dielectric. With a dielectric between them, the ground plane and the positive supply circuit runs act like a distributed capacitor, providing very noise-free power to the entire circuit.

The four-layer variety of the multilayer printed circuit board is the most common circuit board style currently being utilized. This type of circuit board has two inner layers, one that provides power supply ground and another that supplies the nominal 5 V for most digital systems. The outer component and copper sides of the circuit board make all the other interconnections. These boards cost more than double-sided boards but provide exceptional noise immunity and better circuit densities. The increased utilization of plastic enclosures in the electronics industry has promoted the need for a shielding ground plane in place of the metal enclosures that had once accomplished this. The increased use of plastic electronic enclosures requires that the circuit boards themselves contain some shielding.

5–3 ▶ General Printed Circuit Board Design Considerations

In Sections 5–4 and 5–5 ahead, the specific method of circuit board layout for manual or computer methods is described. In this section the general considerations for printed circuit board design are discussed as they are applied to both methods.

Circuit Board Design Considerations

Following is a list of design considerations for circuit board design

1. *Connecting Runs.*
   a. In general, make all circuit runs (see Figure 5–2 for a summary) as short and as thick as reasonably possible while providing as much space between them as possible.
   b. Provide clearance on all sides of a printed circuit board. An area of about 3.8 mm to 10 mm wide (0.15″ to 0.40″) is recommended. Components and circuit runs should not be located in these areas. These clearance areas are needed to avoid interference with board handling fixtures, guidance rails, and alignment tools.
   c. All circuit pads should be larger than the connecting run to prevent the flow of solder away from the solder connection.
   d. Keep in mind the possibility of crosstalk between long adjacent runs. Insert a O-V potential run between them to minimize the potential for crosstalk.
   e. Try to keep signal and return runs together as they would be in a cable. The equal currents flowing in opposite directions will minimize the inductive effects.
2. *Circuit Board Perspective.* When developing circuit board artworks, negatives, and silk screens and while fabricating prototype boards, be aware of the proper perspective for the situation. (Is the view from the bottom or top side of the board?)
3. *Grounding and Shielding.* The information presented in Chapter 3 on grounding and shielding should be applied to the circuit board layout.

1. Make all circuit runs as short and as thick as reasonable.
2. Do not locate components within 0.15″ to 0.40″ from the edge of the circuit board.
3. Make circuit pads larger than the connecting runs.
4. Insert zero-potential runs between signal runs where crosstalk can occur.
5. Keep signal and return runs adjacent as they would be in a cable.

▲ **FIGURE 5–2**
Printed circuit board circuit runs—layout summary

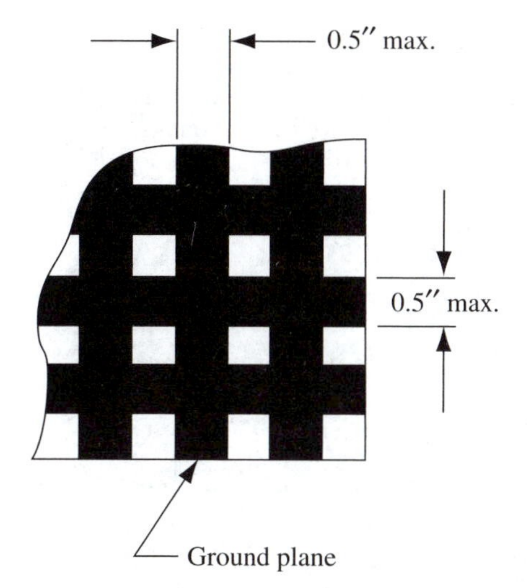

FIGURE 5–3
Crosshatch ground example

a. Ground plane: A ground shield should be utilized wherever possible. This means that the circuit board ground conductor should contain as much copper area as possible. The ideal situation is the one described earlier for multilayer boards, where one entire layer of the board is allocated to ground plane. In single- or double-sided circuit boards, this concept is implemented by making large portions of the circuit board available as a solid copper ground plane. Since most through-hole circuit boards are wave soldered, the ground plane is configured in a crosshatch scheme, as shown in Figure 5–3. This is done because large areas of solid copper absorb heat from the stream of wave solder, resulting in an uneven and lower solder temperature in that area of the board. This degrades the quality of the solder connection. The circuit board with a large exposed ground plane also has a tendency to warp because of the uneven absorption of heat caused by the large mass of exposed copper. The crosshatch ground plane resolves both of these issues.

b. Ground distribution: Ground should be separated and grouped into the following types: low-level analog signal grounds, low-level digital signal grounds, higher-level switched circuit grounds, and a chassis ground (enclosure or card-cage ground). These ground types should be connected at only one point. Within each type of ground, ground connections should be distributed to subgroupings of the appropriate circuitry in parallel as shown in Figure 5–4 and discussed in Example 5–1. This is done to preclude the effect of one long ground loop, where the current return flow from one area of a circuit can affect the operation of another circuit.

c. Guard rings: These are used with operational amplifier circuits to minimize leakage current that can occur at the input terminals to the op amp. The potential error, induced by this leakage current, increases dramatically when the signal source impedance is large. Guard rings are copper traces placed along each printed circuit board surface where the input terminals make contact. On a double-sided PCB with a through-hole

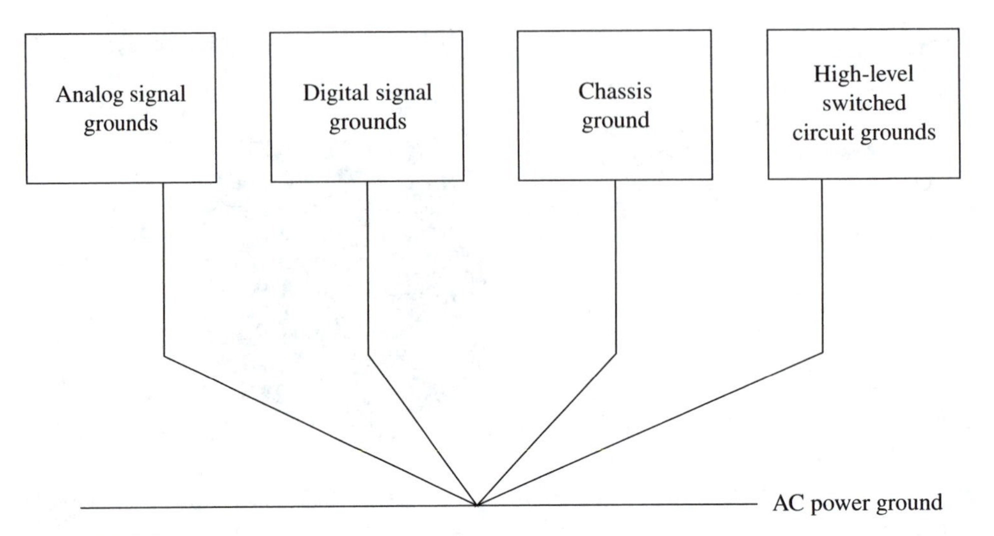

▲ **FIGURE 5–4**
Ground distribution system

package op amp, the guard ring should be placed on both sides of the board. The guard ring should circle around the sensitive op amp inputs and be connected to the same potential as the positive and negative inputs. See Figure 5–5 for examples of guard rings and their connections for various op amp circuits.

Example 5–1

A circuit schematic shows 15 digital integrated circuits to be laid out on a double-sided printed circuit board. The power supply ground connections are being planned for these integrated circuits. The problem is to determine a practical way

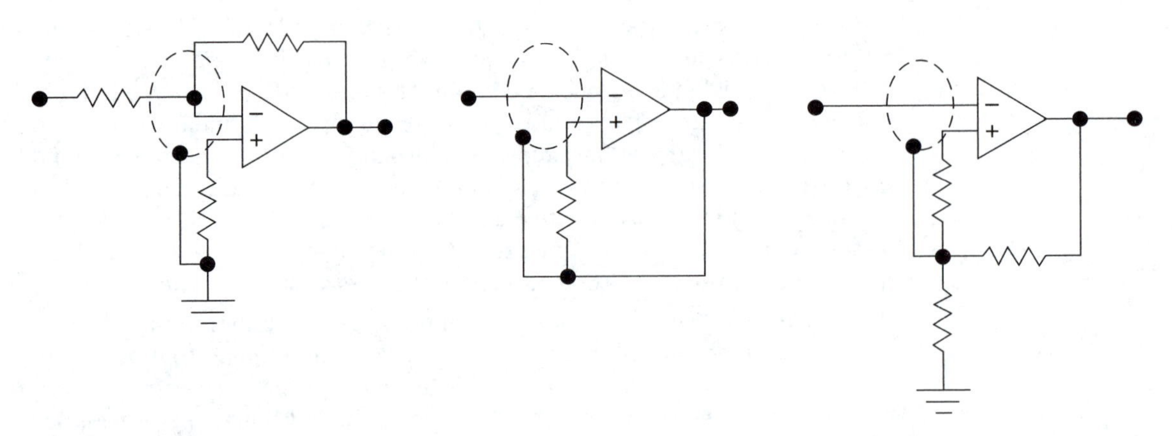

▲ **FIGURE 5–5**
Guard ring examples

of making these ground connections that will minimize the creation of a ground loop. According to the circuit board design considerations just discussed, ground connections of a similar type of circuitry should be grouped together and a separate ground connection should be made to subgroupings of that type of circuitry. In this example, all of the circuitry is low-level digital circuitry so these circuit grounds should all be kept together. One extreme approach is to have 15 individual parallel runs connecting to the common digital ground. This would require a large amount of circuit board area and make other connections very difficult. The other extreme is to provide one continuous ground connection to all the integrated circuits. A ground loop results when components, attached to the end of the loop, are at a higher ground potential than those at the beginning. Also, ground current flowing from the components at the end of the loop affects the ground level of those components at the beginning. The most practical solution is to break the 15 integrated circuits into three groups of 5 and provide parallel ground connections to each group of 5, as shown in Figure 5–6.

4. *Decoupling capacitors.* As power is distributed throughout a printed circuit board, the circuit runs exhibit some amount of inductive reactance. Inductive reactance will oppose a change in the current flow through the runs. At lower frequencies of operation, the inductive reactance is insignificant because the switched components have enough time to complete their switch transition. They can accommodate a delay in the availability of current caused by the inductive reactance. However, in high-

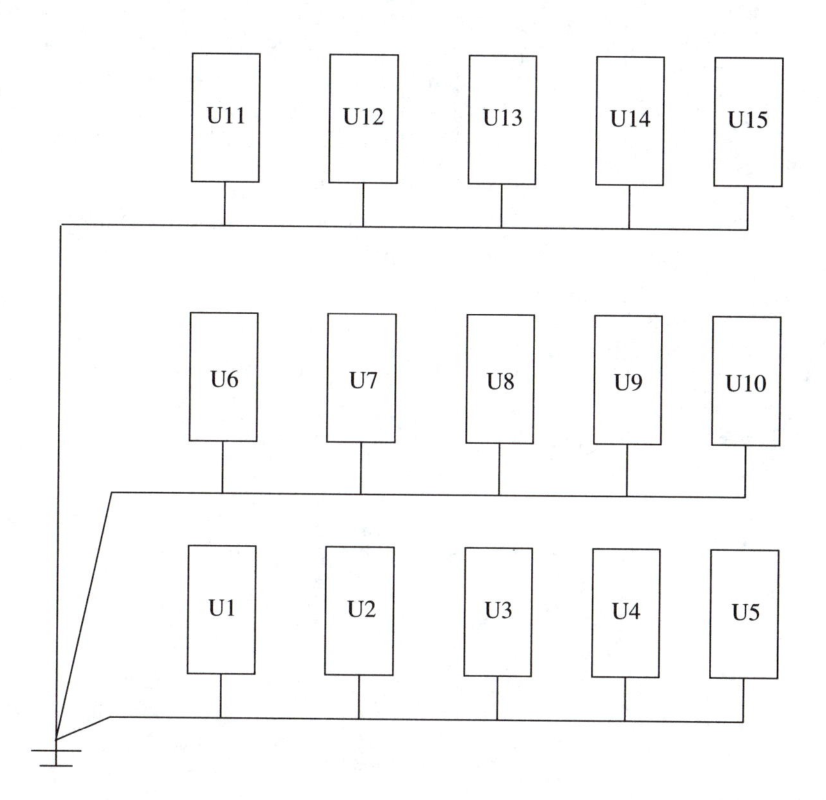

▶ FIGURE 5–6
Ground distribution example

frequency applications, such as typical digital circuits, the inductive reactance of the runs is more critical. It provides a delay of the additional current needed for the device to make the switch transition in time for the circuit to function properly. In these cases, a decoupling capacitor is used to counteract or decouple the power supply run from the effect of the inductive reactance. A 0.1 μF ceramic disc capacitor is typically used for this purpose. The 0.1 μF capacitor stores enough charge in reserve to supply the requirements of the switched component and enable it to switch in the required time. As a rule of thumb, one decoupling capacitor is used for every two integrated circuits. It is important to locate the capacitor as close to the component as possible with thick, short runs. Locating any decoupling capacitor on thin runs away from the component completely defeats its purpose (see Figure 5–7).

5. *Component placement and orientation guidelines.* The orientation and placement of components are important parts of any circuit board layout and varies depending on the type of package technology (through-hole technology or SMT) and the soldering process.

   a. Through-hole Guidelines: Through-hole components are usually either hand soldered or wave soldered. Wave soldering is the method most often used in a manufacturing environment, and it is the process in which component orientation can become important. Wave soldering is an automatic method of soldering in which liquid solder is continuously pumped through a spout to form a well-defined wave. The solder temperature can be tightly controlled over the surface of the wave. The circuit board to be soldered is passed over the wave and all the solder points on that side of the board are soldered. The advantages of wave soldering are as follows:

      1. Short solder times
      2. Reduced temperature distortion of the circuit board. This is because only a portion of the board is exposed to the wave at any time.

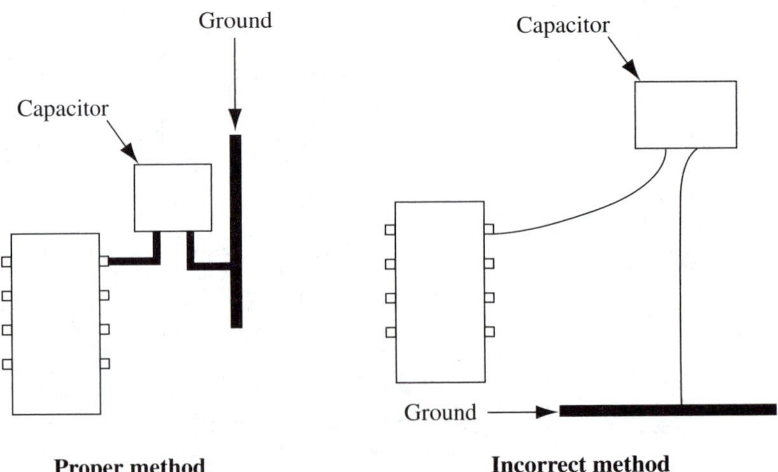

▶ FIGURE 5–7 Decoupling capacitor examples

3. There is a continuous flow of fresh solder returned to the wave. Any flux or other residue is filtered out of the process within the solder flow loop. If the circuit board is wave soldered, the orientation of the board as it flows through the wave should be determined. In determining which direction to pass the circuit board through the wave solder machine, the primary concern is the maximum-width circuit board that the wave-solder machine can process. If both dimensions of the circuit board are less than the maximum width for the solder machine, then the board can be passed through the wave in either orientation. If the machine can accommodate only one side of the board, then the longer dimension of the board must be parallel with the direction of flow, as shown in Figure 5–8. If neither dimension of the board fits through the wave, then either a larger wave-solder machine must be used, or the mechanical design must be redone to reduce the board size. This is not the type of information that one wants to learn about during the initial production run. When placing dual in-line packaging (DIP) through-hole packages, the orientation of the main body of the integrated circuits should be perpendicular to the intended flow of the wave solder. The connecting runs on the component side of the board should be in the direction of the wave flow. This is to preclude solder spilling over from one DIP connection to the next. The connecting runs on the component side of the board should be run perpendicular to the runs on the solder side of the board.

b. SMT Guidelines: SMT circuit boards require a lot more planning in their design. The land patterns or circuit pad sizes are critical for the completion of a reliable solder joint. Accordingly, the circuit pads used, either

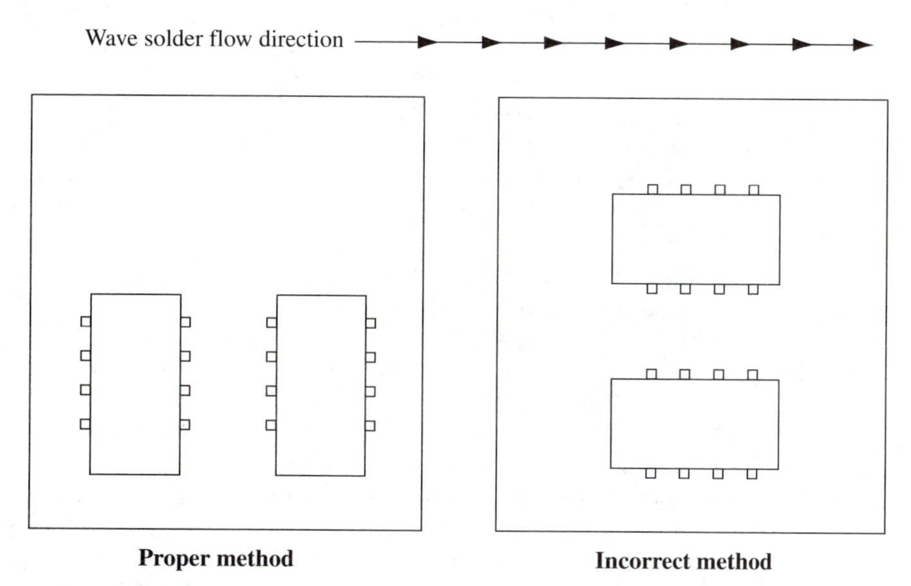

▲ **FIGURE 5–8**
Wave solder flow direction

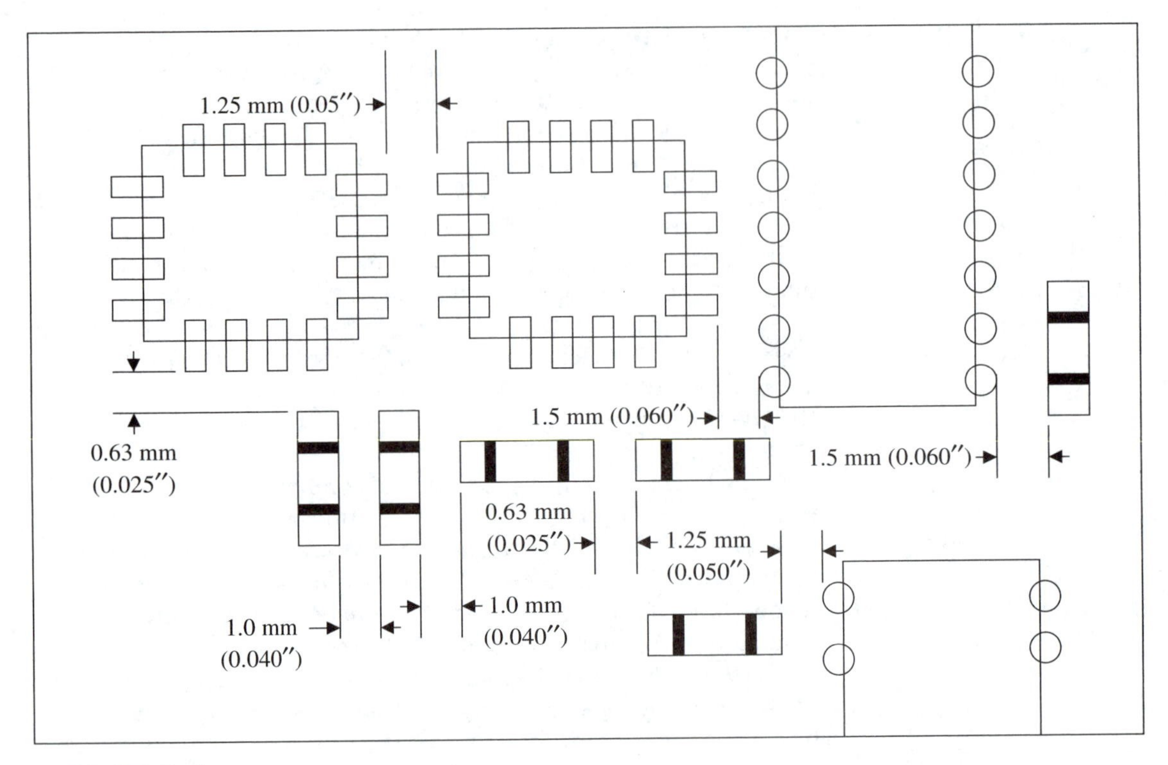

▲ **FIGURE 5–9**
SMT land pattern clearances

computer-generated or stick-on circuit "puppets," should be of the proper sizes as recommended by the Institute for Interconnecting and Packaging Electronic Circuits (IPC) in their standard IPC-SM-872. The clearances between the components, which provide for all manufacturing aspects of the circuit board, are shown in Figure 5–9.

SMT circuit boards can be soldered automatically in two ways: wave or paste-reflow soldering. Because SMT technology has a completely different set of processing issues, the component placement criteria are different. The square SMT packages present a special problem for wave soldering because leads are located on all sides of the package. It is impossible to place these components so that all pads are in the direction of the wave flow. For this reason a solder mask is recommended with most SMT circuits. A solder mask is a coat of epoxy resin covering the entire printed circuit board except where the pads will require soldering. Solder masks are used to eliminate solder bridging between adjacent conductors during wave soldering. Solder masks are currently used with most PCBs of either through-hole or SMT type, as circuit run densities have increased dramatically. Otherwise the placement of SMT components should be as follows:

1. All passive components should be mounted parallel to each other.

2. All integrated circuit packages should be mounted parallel to each other.
3. The longer axis of any integrated circuit package and that of passive components should be perpendicular to each other.
4. The longer axis of the passive components should be perpendicular to the direction of travel of the board through a wave-solder machine.

   If paste-reflow soldering is to be utilized, the placement of SMT components is not that critical.

6. *Tooling holes.* To provide for mechanical alignment on any parts placement or testing apparatus, a minimum of two (preferably three) unplated holes should be located in the corners of the circuit board. The actual hole diameters depend on the actual equipment being utilized, but they are generally between 2.5 mm and 3.8 mm (0.10″ to 0.15″). For SMT circuitry, optic targets are needed in addition to the tooling holes to orient and register the component pads to the center of the device. This is accomplished with fiducials, which are optical alignment targets that are silk-screened onto the board. Three fiducials should be placed on a known grid in the corners of the circuit board to form a three-point datum system. Figure 5–10 shows an example of the application of tooling holes and fiducials. This figure shows two different types and sizes of fiducials that can be used.

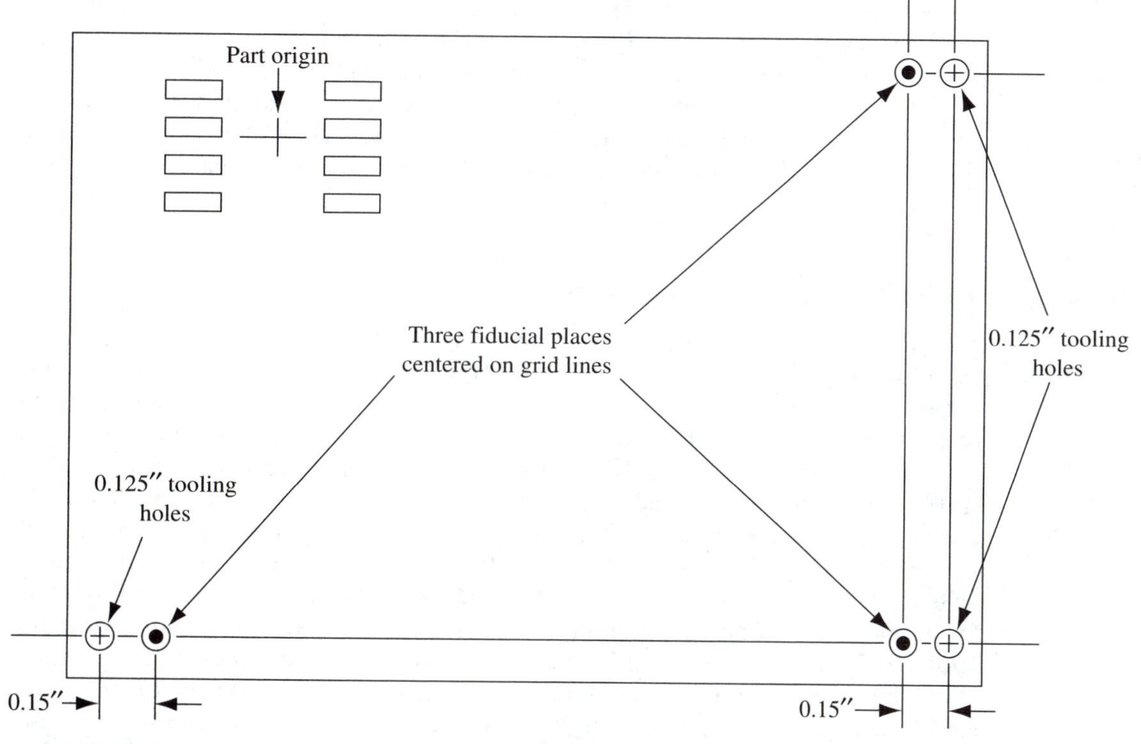

▲ **FIGURE 5–10**
Fiducial tooling hole example

7. *Consider circuit board testing requirements and the need for test points.* Review preliminary circuit board test plans to determine the circuit points that will require access during manufacturing and field-testing. Each point must be accessible for testing and provide a means for attaching meter and oscilloscope test probes. Test points are available as a standard component that can be soldered into the board. Test points are more critical on SMT boards because it is almost impossible to clip probes onto SMT components. With through-hole circuits, test leads can often be attached to component leads without the use of a purchased, assembled test point. If test points are not planned for, it is difficult for manufacturing and field service to test boards and will result in many artwork changes later in the project.

8. *Large or high-power components.* Large or high-power components also require specific attention to design details. It is important to ensure that large components are affixed to the circuit board with the appropriate mechanical strength to support their weight. Most large components designed for circuit board mounting have some means of mechanical mounting incorporated into their design other than the solder connections. Be sure to utilize the manufacturer's recommended circuit board mounting scheme. Do not rely on just the solder connections to mechanically hold a large component onto a printed circuit board.

   Components that utilize higher levels of power should utilize heat sinks when necessary. Be sure to determine this before laying out the circuit patterns by providing for the mechanical mounting of any heat sinks. When a component does not require a heat sink, but generates more than 1 W of power (i.e., a 2 W power resistor), be sure to consider its location relative to other temperature-sensitive components. Also, mount the device up off of the surface of the printed circuit board so that heat can radiate evenly in all directions.

5–4 ▶ Printed Circuit Board Layout—Manual

The manual circuit board layout method is one that has been used since the invention of the PCB. The manual layout method is seldom used in industry today, since the process has been, for the most part, replaced with computer software layout programs. However, there is a benefit to understanding this process before discussing the computer methods. Manual circuit board layout involves the use of layout templates to draw the components into position on a layout drawing that is usually a semitransparent medium of some kind. A mylar material is recommended, with one side having a matte finish for the layout drawing, because it erases cleanly and produces a strong, clear pencil image. The layout and the eventual artwork should be done over a precision grid background so that all the holes in the board can be located on the grid. A typical grid background is one with 1 mm spacing. The layout and the eventual taping are usually done at a scale of two times the actual size of the components, although standard 1X, 2X, and 4X templates and circuit puppets are available. A 4X scale, for example, would be used on small, dense circuits. Circuit puppets are adhesive-backed pads that conform to the var-

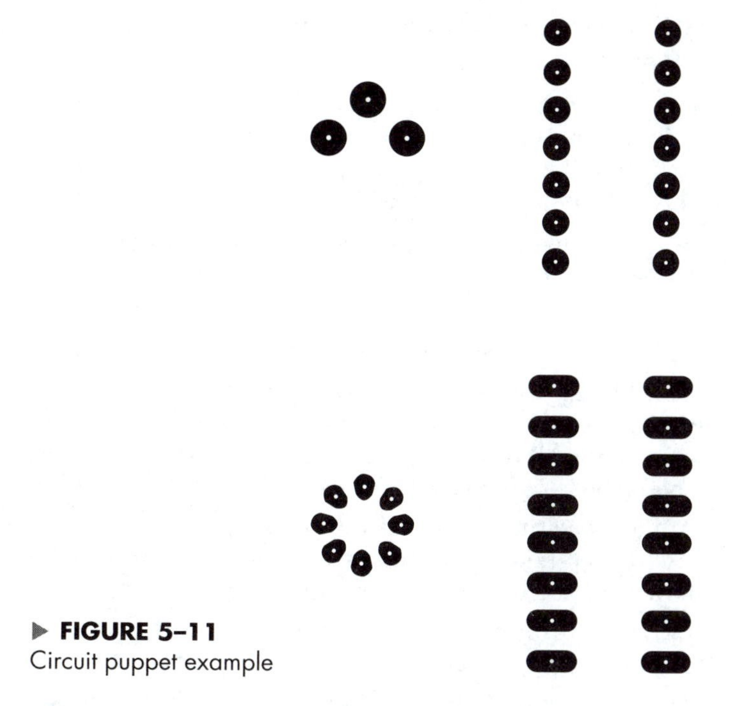

▶ **FIGURE 5–11**
Circuit puppet example

ious circuit components as shown in Figure 5–11. The layout is usually done as a positive. The dark areas define where the copper runs will be. Negative circuit puppets are available if one wishes to develop an artwork directly as a negative.

The Layout Drawing

Developing the layout drawing is the most difficult and critical part of the layout process. Its development will ultimately determine how well the circuit board functions and how easily it is manufactured, tested, and serviced. Completing the layout drawing involves selecting the location of all components on the board and defining the pattern of copper foil connections that will complete all the required connections. To complete the layout drawing, use the following process:

1. *Circuit board mechanical design.* Determine the desired type, size, and mounting configuration of the printed circuit board. This includes the length, width, thickness, and type (number of sides and layers). The choices for mounting usually include using mounting pads or standoffs, card cages, or other specialized hardware that affix to the board edge. These decisions must be made in conjunction with the overall mechanical design for the project as well as the selection of interconnection methods that are discussed below. Be sure the equipment used to process the board in manufacturing can accommodate the circuit board size selected.
2. *Board interconnection.* Determine the method and optimum location of all connections to the printed circuit board. The actual connecting method

must be selected and all relevant information about the connection must be determined.

3. *Component locations.* At this point the mechanical outline of the board and the location and space requirements for its mounting and interconnection have been defined. Next, select the ideal location of the components on the board while considering all of the following:
   a. Keep connections as short as possible. Keep components that have connections between them close together.
   b. Keep the functional blocks of the circuit together.
   c. Maintain an orderly flow of any signal from input to output.
   d. Make the thickness of the runs appropriate for the signal that they carry. Give special consideration to power supply and ground runs. These should always be as direct and as thick as possible. High current runs should be very thick.
   e. At the same time, consider the voltages present on adjacent runs and try to keep the runs as far apart as possible. Approval agencies often have spacing requirements for runs carrying voltages in excess of 30 V.
   f. Leave room for components that will dissipate a lot of power and consider the need for a heat sink.
   g. Attempt to keep component configurations as consistent as possible (i.e., have all integrated circuits going in the same direction, all polarized devices in the same orientation, resistors adjacent and in line, and so on).
   h. Provide access to testing for key circuit areas. Consider the eventual testing method that will be employed and the need for test points.
   i. Consider access to any adjustments or the need to remove any components from the board or the need to remove the board from the assembly.
   j. The board should present a professional and high-quality appearance.
4. *Layout copper runs.* The layout drawing should now include the desired location of all the components in addition to the complete mechanical profile of the board with mounting and interconnection hardware. The process of actually determining the connecting runs is the most difficult part of this process. The process usually takes a number of cycles, so it is recommended to place another layer of matte-finished mylar over the layout drawing that has been completed up to this point. This is done so as not to waste the efforts completed thus far, which occurs when the layout drawing must be redrawn after the first attempt to lay out the connections is abandoned for a better way. One way is to lay out the connections on the top layer of mylar. As the process evolves, simply replace the top layer and start over. The process is an iterative one that involves learning the best way to make the connections for a given circuit. The more experienced designer requires fewer iterations. At this stage, using a light table will make it easier to see through the different levels of mylar.

Draw in the connections with a simple line that will represent the actual circuit run. For double-sided boards, use a red pencil for one side of the board and a blue pencil for the other. For the most difficult connections, utilize a via hole to transfer a connecting run from one side or layer of the board to another. Via holes should not be overused. Each via requires that another hole be drilled in the board, adding a small cost to the board. Also, via hole connections are slightly less reliable than a solid copper run. It is best to make all power supply connections first and then proceed making the other connections.

5. *Layout design check.* After a number of attempts at completing the connecting copper runs, a successful layout drawing is complete. Before starting the taping process, it is important that whoever has design responsibility for the circuit board check all aspects of the layout design. This includes the mechanical size and shape, manufacturing and testing issues, and the location length and size of all circuit connections. This should be done before the taping process is started.

6. *The pad master.* After the layout design is checked, the layout is ready for taping. This involves the use of the circuit puppets, pads, and artwork tape to implement the pencil layout as a taped positive. The taping should be completed on what is called clear taping film. There are usually at least two layers (except for single-sided boards, where there is one) representing the layout so there must be a way to register or locate the sheets of taping film on top of each other. This is usually done with what is called a pin bar, where the pins in the pin bar line up with the taping film that is prepunched to the pin size and spacing on the pin bar. To align the different layers when not on the light table, crosshair puppets are added to each layer of the artwork to allow proper and accurate alignment.

   The taping is started for circuit boards with more than one side by generating what is called a pad master. The pad master is simply one layer of taping film that has pads marking the location of each hole that will be on the board. Generate the pad master by placing one layer of taping film over the layout and placing pads and puppets as appropriate over all of the via holes and mounting pads. Figure 5–12 shows an example of a pad master.

7. *Artwork taping.* The artwork is completed by placing another layer of taping film over the pad master and layout drawing. This layer of taping film represents the copper runs for one side of the circuit board. The copper runs are completed on the taping film by using special artwork tape available in many precision widths. Be sure that the tape overlaps the connecting areas; do not stretch the tape as it is applied. A layer of connecting runs is completed for each side of the board. A completed double-sided circuit board artwork will include a pad master and one layer of connecting runs each for the top and the bottom. A four-layer, multilayer board artwork consists of one pad master and four layers of connecting run taping films. Figure 5–13 shows one layer of circuitry for the same board pad master shown in Figure 5–12. Figure 5–14 shows Figure 5–13 aligned with Figure

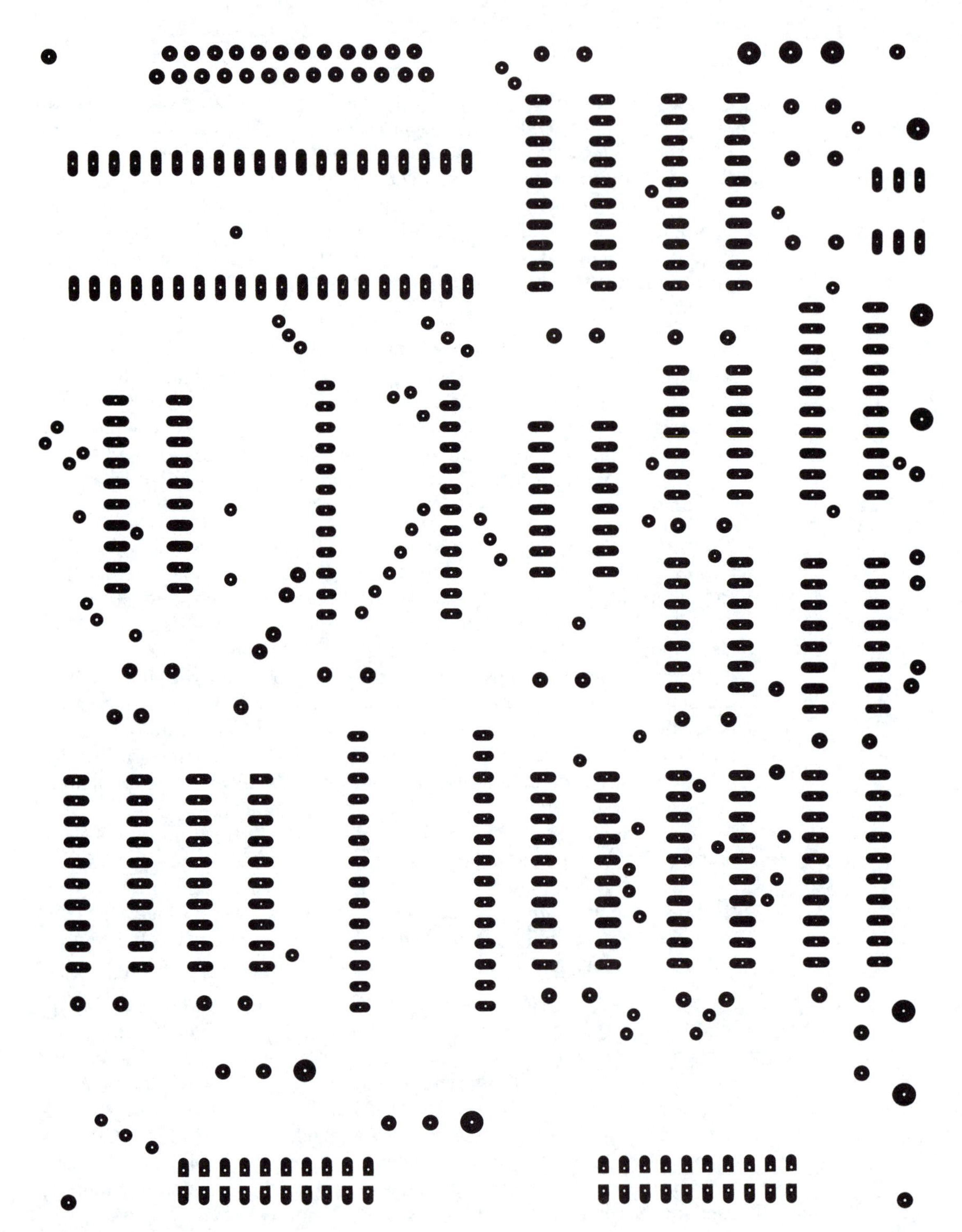

▲ **FIGURE 5–12**
Pad master example

▲ **FIGURE 5–13**
Single-layer taping example

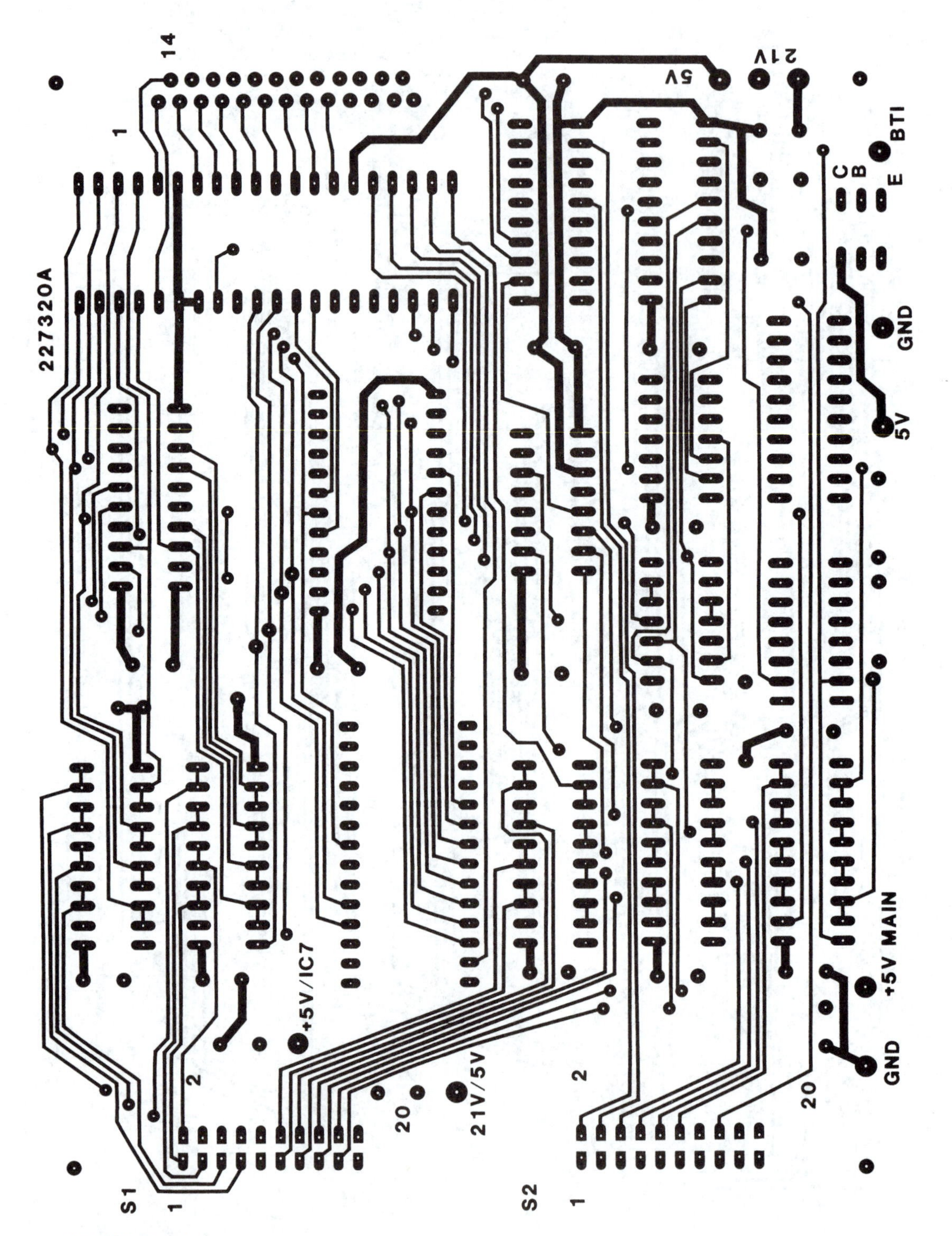

▲ **FIGURE 5–14**
One-side copper artwork (pad master and copper runs)

5–12. Figure 5–14 shows all copper areas on the top side of the circuit board. It represents the complete artwork for that side of the board.

8. *Artwork checking.* The artwork is now complete. If the board design is complex, there is a good chance that there are errors in the artwork. All it takes is one missing bit of tape or one forgotten hole. The detail included in a complex board is incredible. The possibility of an error increases proportionally with complexity. It is best to have two people check the artwork, one who reviews the schematic and another who checks the artwork. The board designer, for example, should review the schematic, while a competent person who is not the designer checks the artwork. This is because the board designer has spent many hours staring at this same artwork, and he or she will have difficulty getting the objective distance needed to see any errors in it. The person reviewing the schematic calls out each connection while the artwork checker verifies the connection. The schematic reviewer highlights each run as it is checked and continues until all of the runs have been verified.

9. *Artwork photography.* The completed and verified artwork is now ready to be photographed for reduction and developing a negative. The artwork is sent to a company that specializes in precision photography and reduction. The pad master is combined with the various layers, reduced, and photographed. This results in a 1:1 negative of each side or layer of the board. The negative and a document called a *fabrication drawing* or a "drill code" are sent to a circuit board manufacturer for fabrication. A drill code defines the mechanical aspects (i.e., size, laminate, plating, and hole sizes) of the circuit board.

5–5 ▶ Printed Circuit Board Layout—Computer

There are two computer methods that can generate circuit board artwork: custom software designed to lay out circuit boards or CAD software to draw the layout of the board. The latter method is identical to the manual process except that the computer is used as a drawing tool. Circuit board layout software packages are most often used today, so this discussion focuses on them. There are many software layout packages available and currently in use. The discussion that follows is general enough to describe the process but may not be entirely accurate when applied to a specific software package. As in the manual layout process, the schematic is the source of defining the circuit connections. The computer layout software requires what is called a schematic capture file to define the schematic. This is simply the schematic keyed into a schematic capture program, which is in a format compatible with the layout software package. All software layout programs have mating schematic capture software, and many accept other schematic capture files as well. When the schematic is created, it may be necessary to add or create a device library for components that may not be present in the standard device libraries in the software. The device library contains all of the pertinent information about the device, the type of component, the number of connections and the label for each connection, and the physical package definition. Make sure the device library is complete for all of the components on the schematic, because the package information is needed to begin the computer layout process.

The computer layout process begins with the same two steps as the manual layout process does: the layout of the mechanical outline and the interconnections. The result is an outline drawing that is the mechanical design of the board and its interconnections. Next, the components are positioned on the board layout by dragging them into position. The result is identical to the pad master drawing generated in the third step of the manual process. The circuit designer has to make a choice at this point whether or not to use the layout program's autorouter feature. An autorouter is a software routine that determines the path of the connections required by the schematic. It is analogous to the trial-by-error process described earlier in the manual layout method. The computer will attempt many circuit paths for making the connections and will choose the ones that its software intelligence determines are the best. The autorouter is a key part of a software layout package. The quality and quantity of the intelligence included in it determine its performance. The price of the software layout package is usually indicative of the amount of intelligence included and the quality of the autorouter, or, in other words, the higher the price the better the autorouter. The decision for using the autorouter is based largely on the experience one has with it. If you have no experience with a particular autorouter, then you can develop some through experimentation. Even the best autorouters are not perfect and cannot possibly possess all of the design criteria for a particular design nor the human insight to make design decisions. There are at least three areas where autorouters produce undesirable results:

1. Power supply and ground runs are not direct or large enough.
2. Connection runs are placed adjacent to areas where there is potential for picking up interference.
3. It utilizes too many via holes.

Many board designers have approached the autorouter decision by connecting up the power supply and ground connections manually. Then they engage the autorouter to make the rest of the connections and modify any undesirable results manually. This is probably the best way to use the autorouter function. There are occasions where an autorouter is unable to make all the connections, and these connections will have to be completed manually.

After completion of the computer layout process, whether it is accomplished manually or with an autorouter, the computer will save files that contain the artwork for each layer of the printed circuit board. The benefit of this process is that the artwork is guaranteed to be accurate and reflect the connections defined in the schematic capture file. If the schematic capture file is correct, then the artwork is as well. There is no need to check and verify that all the connections have been made as was necessary with the manual layout. However, it is still important for the design engineer to review the artwork to make sure that the design will meet the overall requirements for the circuit board. Another real benefit of software layout programs is that they generate the other drawings needed to fabricate and assemble the circuit board. Photography is unnecessary. The artwork and drill code files are simply sent to a circuit board fabricator on a floppy disc or any computer network—including the Internet. While software layout programs are very powerful and offer many benefits, they are somewhat difficult to learn and use. They are as sophisticated as most CAD drawing software programs. Every attempt is made to make the software as easy to use as possible, but there are simply too many complex features and functions that require knowledge and training to use them properly. These are the types of programs that require consistent use to develop expertise in them.

An ideal way to use the electronic design software tools available today—simulators and layout programs—is to have the design engineers perform schematic design on a simulator that is compatible with the layout software being used. When the breadboard and simulation is complete, the file containing the updated schematic is handed over to the board designer. Using the schematic in conjunction with the layout program, the board designer completes the layout and all relevant drawings. The result is a very fast and accurate circuit board development process.

5–6 ▶ Printed Circuit Board Documentation

With the artwork complete, it is time to put together the complete documentation package that will define both the unpopulated and the completely assembled printed circuit board. To define the unpopulated or bare printed circuit board, a document called a fabrication drawing or drill code must be generated.

The Fabrication Drawing

A fabrication drawing is required for each printed circuit board as it defines the board's mechanical requirements. The fabrication drawing, combined with all of the board artwork, completely defines the unpopulated circuit board. The fabrication drawing specifies the mechanical shape and dimensions of the PCB as well as the size and location of any holes. The fabrication drawing must also point out details such as the board laminate material, tolerances, plating, and other optional requirements, such as solder masks and silk-screening. Following is a detailed list of the issues that should be considered to note on the fabrication drawing:

1. The board laminate material
2. Requirements for a solder mask and silk screen
3. Reference to the artwork number and revision level
4. Reproduction of artwork tolerances on the circuit board
5. Plating specifications
6. All mechanical tolerances

Following is an example of the notes included on a typical fabrication drawing:

1. Material: FR4 glass epoxy, 1/16″ thick with 1 oz copper each side per Mil Spec #MIL-P-13949. All holes to be plated through with a minimum of 0.001″ thick copper. After plating the holes, the surface copper should have a total copper thickness of 2 oz.
2. Apply solder mask and silk screen per artwork drawings #12345678, Revision A.
3. For circuit artworks use drawing #12345678, Revision A.
4. Defects:
   a. Circuit run defects such as holes, nicks, and scratches shall not reduce the conductor width by more than ± 0.002″.
   b. Maximum allowable line reduction shall not exceed 0.005″.
5. Plating: The board should be plated with tin/lead 63/37 ± 5% plating to a total thickness of 0.0004″ to 0.0006″. All solder plated areas to be subject to hot oil solder reflow process.
6. Hole diameter tolerances: +0.005″, –0.002″.

Figure 5–15 is a fabrication drawing example.

Solder Mask

A solder mask is a coat of epoxy resin that covers the entire printed circuit board except where solder connections are to be made. The application of a solder mask is optional. Its purpose is to prevent solder from bridging over adjacent circuit runs and pads during automated wave-solder operations. Increasing circuit densities and the use of wave soldering has made the use of a solder mask very common. On through-hole-only technology boards, the artwork for the solder mask can be simply the pad master drawing discussed earlier. When surface-mount technology is used, the pads for all of the SMT components must be included on the solder mask artwork as well. In any case an artwork should be completed and labeled as “Solder Mask” for the subject fabricated printed circuit board. Most computer circuit board layout packages will generate a solder mask on request after the layout has been completed. The circuit board fabricator will apply the solder mask if called for in the fabrication drawing.

Silk-screening

Silk-screening is another optional process that involves the marking of the board with component and other reference numbers. Silk-screening is an aid to manufacturing and service personnel in the field and provides a professional appearance. If silk-screening is required, silk-screen artwork must be completed separately

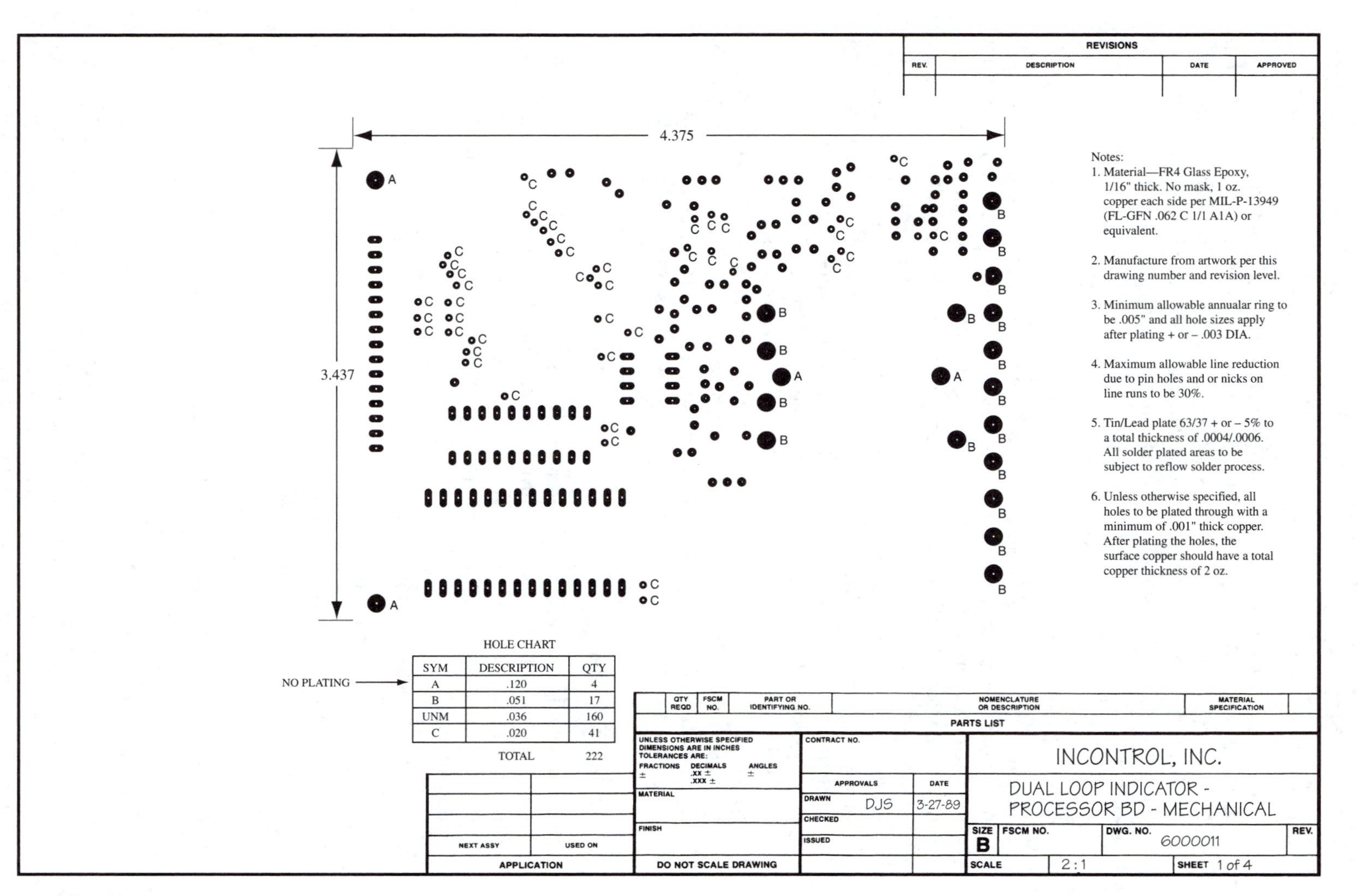

SYM	DESCRIPTION	QTY
A	.120	4
B	.051	17
UNM	.036	160
C	.020	41
	TOTAL	222

▲ **FIGURE 5–15**
Fabrication drawing example

that includes the outline and orientation of all component packages, their reference designations, and other pertinent information, such as test points and the like. Most computer layout programs will generate a silk-screen drawing automatically after the board layout is complete. The fabrication drawing must specify that the board is to be silk-screened and refer to the drawing number for the silk-screen artwork.

Solder Paste Screen

The solder paste screen is used as a screen for the application of solder paste to surface-mount technology circuit boards where either vapor phase or infrared reflow processes are utilized to make the solder connections. Therefore, a solder paste mask is necessary only for surface-mount circuit boards that will be manufactured with one of these reflow solder processes. The solder paste screen is similar to the solder mask, except that it includes only the surface-mount pads or connections on a given side of the circuit board. If through-holes exist on the circuit board, they are either vias or used for mounting through-hole components. In either case solder paste should not be applied to these holes. For that reason through-holes should not be included on the solder paste screen. Solder paste screens will be used by whomever assembles the printed circuit board. These artworks can be generated on request by most computer layout programs. The board fabricator will not have any need for the solder paste screen. The actual silk-screen itself will have to be ordered from a silk-screen fabricator per the solder paste screen artwork.

Assembly Drawing

The assembly drawing is a pictorial that will show the physical representation of the circuit board. The assembly drawing will show the location and reference designation of all components that are to be assembled to the board. It will include special notations about the assembly process and refer to the following other related documents:

1. Parts list and bill of material
2. Circuit board specifications
3. Circuit board artworks
4. Circuit board schematic
5. Circuit board assembly and test procedures

Figure 5–16 shows an assembly drawing example.

Parts List and Bill of Material

The circuit board parts list is essentially the same as any other parts list or bill of material. It should include the manufacturer and its part number, the company part number, reference designations, and the quantity used per assembly. The manufacturer's part number is the number assigned to the component by its supplier.

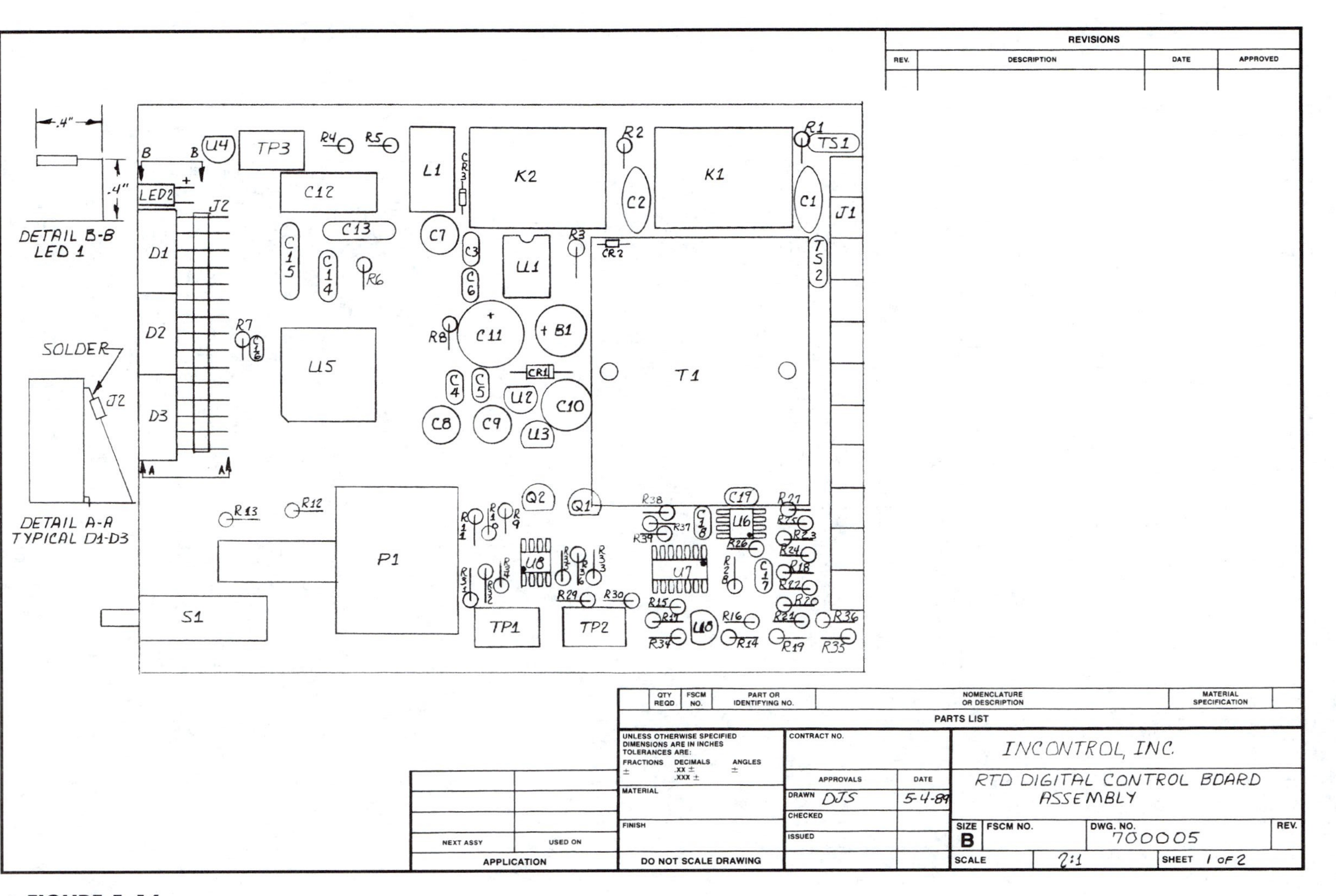

▲ **FIGURE 5–16**
Assembly drawing example

	Main Printed Circuit Board Assembly					
Item #	Description	Company Part Number	Manufacturer	Manufacturer Part Number	Component ID	Quantity
1	2.49 k ohm, 1% Metal Film, 1/4w Resistor	10000000	Res. Inc	2.49kX	R1,R13	2
2	110 ohm, 1% Metal Film, 1/4 w Resistor	10000001	Res. Inc	110X	R2	1
3	100 k ohm, 1% Metal Film, 1/4w Resistor	10000002	Res. Inc	100kX	R3--R6,R15	5
4	26.1 k ohm, 1% Metal Film, 1/4w Resistor	10000003	Res. Inc	26.1kX	R7	1
5	10.0 k ohm, 1% Metal Film, 1/4w Resistor	10000004	Res. Inc	10.0kX	R8,R9	2
6	430 k ohm, 1% Metal Film, 1/4w Resistor	10000005	Res. Inc	430kX	R11	1
7	5.11 k ohm, 1% Metal Film, 1/4w Resistor	10000006	Res. Inc	5.11kX	R10	1
8	1 M ohm, 1% Metal Film, 1/4w Resistor	10000007	Res. Inc	1MX	R12	1
9	470 k ohm, 1% Metal Film, 1/4w Resistor	10000008	Res. Inc	470kX	R14	1
10	1k ohm, 10 turn Trimpot	10000009	TP Inc.	101tp1	RZ	1
11	50 k ohm, 10 turn Trimpot	10000010	TP Inc.	502tp1	RG	1
12	10 k ohm,10 turn Trimpot	10000011	TP Inc.	102tp1	RR	1
13	.1 uF, 50 V, Ceramic Capacitor	10000012	Cap. Inc.	P1101	C1--C4	4
14	.01 uF, 50 V, Ceramic Capacitor	10000013	Cap. Inc.	P1102	C5	1
15	100 pF, 50 V, Ceramic Capacitor	10000014	Cap. Inc.	P1106	C6	1
16	.1 uF, 50 V, Mylar Capacitor	10000015	Cap. Inc.	M1101	C7	1
17	.22 uF, 50 V, Polypropylene Capacitor	10000016	Cap. Inc.	P3224	C8	1
18	.047 uF, 50 V, Polypropylene Capacitor	10000017	Cap. Inc.	P3473	C9	1
19	3300 uF, 16 V, Electrolytic Capacitor	10000018	Cap. Inc.	P5144	C10	1
20	330 uF, 16 V, Electrolytic Capacitor	10000019	Cap. Inc.	P5140	C11, C12	2
21	3 1/2 Digit LED A/D Converter	10000020	IC Inc.	TC7117CPL	U1	1
22	5 V .1 A Voltage Regulator	10000021	IC Inc.	LM320LZ-5.0	U2	1
23	5 V.1 A Negative Voltage Regulator	10000022	IC Inc.	LM340LAZ-5.0	U3	1
24	5 V 1 A Voltage Regulator	10000023	IC Inc.	LM340T-5.0	U4	1
25	Quad Operational Amplifier	10000024	IC Inc.	LM324P	U5	1
26	Voltage Reference- 2.5 Volts	10000025	IC Inc.	LM4040CZ-2.5	U6	1
27	115/20VAC,.3A Center Tapped Transformer	10000026	IC Inc.	MT2111	T1	1
28	Seven Segment .56" Red LED Display	10000027	IC Inc.	67-1463	D1--D4	4
29	Printed Circuit Board	10000028	PCB Inc.	NA	NA	1
30	Terminal Block -5 Position	10000029	Conn. Inc.	W5500	J1	1

▲ **FIGURE 5–17**
Parts list document example

The purchasing company assigns the company part number to the component. The component will be stocked and tracked by the company part number. Reference designations are numbers that tie a component from the schematic to the parts list and the assembly drawing. For example, a resistor is given a reference designation of R1 on a schematic. That same designation (R1) will be used to show the location of the resistor on the assembly drawing and to specify the resistor on the parts list. Figure 5–17 shows a parts list document example.

Summary

In Chapter 5 we have completed a general review of the printed circuit board: the materials used, types of construction, and their layout and fabrication. The layout of the PCB is as vital to its accurate and reliable operation as the quality of the electronic design. The circuit board layout is usually completed by people who specialize in this area. These specialists usually know the manufacturing process used by the company and are experts at using the chosen circuit board layout program. However, their knowledge and experitise in the theoretical function of a particular design is often limited. Therefore, it is important for all electronic designers to perform a complete review of the circuit board layout before the layout is etched in copper. It is most critical to review areas of the circuit where noise immunity and noise emissions are most significant: power supply, circuits, and signal grounding.

With the completion of this chapter, we conclude the general discussion of the electronic design process. For the remainder of the book, each chapter specializes on a particular area of electronic design and applies those concepts to specific design problems.

References

Coombs, C. F. 1988. *Printed Circuits Handbook*, 3rd ed. New York: McGraw-Hill.

Mardiguian, M. 1987. *Interference Control in Computers and Microprocessor-Based Equipment.* Gainesville, VA: Don White Consultants.

Ott, H. W. 2001. *Noise Reduction Techniques in Electronic Systems.* New York: Wiley.

Stadtmiller, D. J. 2001. *Electronics Project Management and Design.* Upper Saddle River, NJ: Prentice Hall.

Exercises

5–1 What is the purpose of using design master drawings?

5–2 What is the purpose of a via hole? What process is necessary to provide conductivity through a via hole from one side of a circuit board to the other?

5–3 What is a laminate, and what do the different grades of laminate indicate?

5–4 What is the difference between single-sided, double-sided, and multilayer circuit boards?

5–5 On which type of circuit board can components be placed on both sides of the circuit board?

5–6 Compare double-sided and multilayer printed circuit boards as far as cost, size, and noise immunity are concerned.

5–7 How are power supplies and ground usually distributed on multilayer circuit boards?

5–8 What is meant by the term *ground plane*, and why should it be designed in what is called a *crosshatched pattern?*

5–9 Why are the grounds from analog and digital circuits usually kept separate as much as possible?

5–10 What is a guard ring, and why is it used?

5–11 What is a decoupling capacitor, and why is it used?

5–12 When using wave soldering, why is it important to know the direction of the wave when determining component locations on a printed circuit board?

5–13 List and describe the documents and drawings required to completely document a printed circuit board.

5–14 Describe the purpose for a solder mask and what it consists of. How would you develop the artwork for a solder mask?

5–15 What is the purpose of silk-screening a printed circuit board? How would the artwork for one be developed?

5–16 List all of the drawings needed to specify a double-sided printed circuit board to a circuit board fabricator who supplies the bare printed circuit board. Assume that the board will have a solder mask and will be silk-screened.

5–17 Describe the purpose of a computer software board layout autorouter. What are the areas of performance that are a concern when using an autorouter?

5–18 Explain the difference between the two basic types of SMT soldering processes: wave soldering and reflow soldering.

5–19 What is the purpose of test points, and why are they even more critical on SMT circuit boards?

6 Power Supply Design

Introduction

The design of DC power supplies is a critical aspect of electronic circuit design. The study of power supply design is often viewed as tedious, and even trivial, when compared to elaborate large-scale integrated circuits and high-speed processors. As such, power supplies are often considered just a minor detail, to be worked out later on complex electronic design projects. But the simple fact is that almost every electronic circuit requires a power supply. An appropriate, efficient, and reliable power supply design will largely determine the level of successful operation and reliability of any electronic circuit. Even if power supply design is not in your future, you will be better able to work on a team with power supply designers when you have a basic understanding of their function and the competing design criteria.

Power supplies usually appeal to the analog-oriented individual, and their design is often delegated to someone who specializes in this area. As the performance of power supplies has increased over the years, their design has become more complex with the utilization of more switching circuits to decrease their size and cost. Many new integrated circuits have been developed that include elaborate functions that improve power supply reliability and efficiency and reduce their size. As we continue to deal with reduced energy supplies, the efficiency of power supplies becomes increasingly important. In this chapter we will discuss some of the many types of power supplies, their design concepts, and how to measure and compare their performance. The specific topics covered are as follows:

- Power supply specifications
- Linear DC power supplies
- DC-to-DC converters
- Switching power supplies
- Inverters

6–1 Power Supply Specifications

Before we discuss specific power supply circuits, we must first define the power supply design problem with a general set of requirements or specifications. Power supplies provide operating power for electrical and electronic devices by taking an input voltage and converting it to an appropriate output voltage. In order to properly design a power supply, we must define the input and its expected variations along with the output and the acceptable variations for it. The specifications that are developed next will identify all the key power supply design parameters. They are summarized in Figure 6–1.

Input: The input power to the power supply should be completely defined as AC or DC, and the expected frequency range, the nominal input voltage (RMS or peak voltages for AC), and the expected variation in the input voltage level should also be specified.

Output: The output required for the power supply is specified as AC or DC. If the output is an AC voltage, the required frequency range and the ac-

- Input voltage: DC or AC, frequency and range for AC, nominal input voltage and variation
- Output voltage: DC or AC, frequency and range for AC, nominal output voltage and allowable variation
- Dropout voltage: minimum input/output voltage differential
- Output current: continuous and maximum output current
- Ripple voltage: maximum % of allowable peak-to-peak ripple
- Efficiency: output power/input power
- Percent load regulation: % output change/% load change
- Percent line regulation: % output change/% line change
- Noise filter: attenuation shown in dB
- Generated noise: line and radiated, amplitude and frequency
- Failsafe features: current limiting, thermal overload
- Size and cost

▲ **FIGURE 6–1**
Key power supply specification parameters

ceptable variation must be noted. For both DC and AC supplies, the expected output voltage level must be identified along with its acceptable variation. Power supplies also have a maximum current rating that can be safely supplied on a continuous basis. For DC supplies the output voltage will always include some amount of variation around the nominal output voltage DC level called *ripple*. The amount of ripple voltage allowable must be defined in the specifications.

General Power Supply Requirements: These are design parameters that relate to the overall power supply design, such as input/output voltage, current, and ripple.

Dropout Voltage: The specs should include the minimum input to output voltage differential that will provide regulation of the output.

Efficiency: The efficiency of a power supply is measured by comparing the amount of output power supplied to the load to the total input power. The difference between these two numbers is the power consumed by the power supply. Efficiency is specified as a percent and is equal to the output power/input power.

Percent Load Regulation: Most power supplies provide regulated outputs, which means that they control the output to a specified value within the tolerance required. A typical 5-V DC power supply might output 5 V DC within a tolerance of ±1%. The acceptable range of output voltage for this supply is 4.95 to 5.05 V DC. When the supply is connected to a load, the output voltage will change slightly as the load is varied. The percent regulation is the percent of change in the regulated output voltage compared to the corresponding percent change in the load.

$$\text{Percent Load Regulation} = P_{OC}/P_{LC} \qquad (6\text{–}1)$$

where P_{OC} = the percent change in the regulated output voltage

P_{LC} = the percent change in the load.

Percent Line Regulation: This is defined as the output voltage change divided by the line voltage change expressed as a percentage.

Noise Filter: The input voltage presented to the power supply often includes noise signals that should be attenuated and filtered. The requirements for this are usually specified as dB levels of attenuation provided over a frequency range.

Generated Noise: Many switching types of power supplies generate noise as they regulate the output voltage. The magnitude of this noise voltage is of concern because of its effect on other electronic devices. It is regulated on certain product categories by the FCC and on all products imported into the European Union, which must meet CE electrical noise standards. These standards include the EMI fields that are radiated by the supply as well as the level of noise that is conducted back into the input power circuit. This input circuit is often the AC line voltage, which can be common to many other electronic devices and is why the noise voltage levels are of concern.

Failsafe Features: There are a variety of other special features that can be specified to make a power supply failsafe. Temperature measurement of the semiconductor regulating device and measurement of the load current or of changes in load current are the most common. In these cases the power supply circuit shuts down or limits the current when excessive load currents are drawn; this precludes damage to the power supply and load circuits.

Size and Cost: All power supply designs will have limits placed on their physical size and the cost of manufacture.

6–2 ▶ Linear DC Power Supplies

Linear power supplies are based on the concept of taking an input voltage and converting it to an appropriate level of DC voltage that exceeds the required output voltage. This voltage is then regulated down to the specified power supply output voltage. If the linear supply is a regulating supply, it monitors the output voltage and varies its resistance accordingly to provide the required output voltage.

There are two basic varieties of linear regulators, series and shunt regulators. Shunt regulators are represented by a series resistance with a regulating mechanism, usually a semiconductor, whose resistance is varied to draw more or less current in a circuit path that is in parallel with the load, thereby adjusting the output voltage to the required level. As the load resistance changes, the shunt regulator diverts more or less current through this parallel path (see Figure 6–2). The efficiency of the shunt type regulator is low when the load current is small (load resistance is high) because a large current is diverted through the regulator. This current isn't doing any work except regulating the output. When the load current approaches full load conditions for the power supply, the efficiency is much higher. A key advantage of the shunt regulator is the fact that when the load is a short circuit, the power supply will not be short-circuited.

Figure 6–3 shows a series regulator with the regulating mechanism in series with the load. The series regulator has the opposite characteristics as compared to the shunt regulator. The efficiency of the series regulator is low when the load

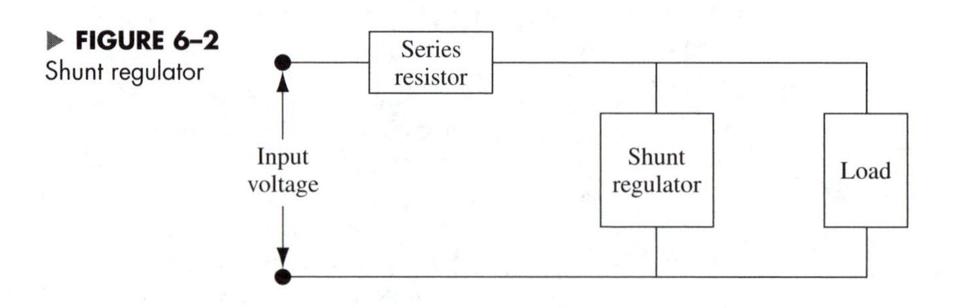

▶ **FIGURE 6–2**
Shunt regulator

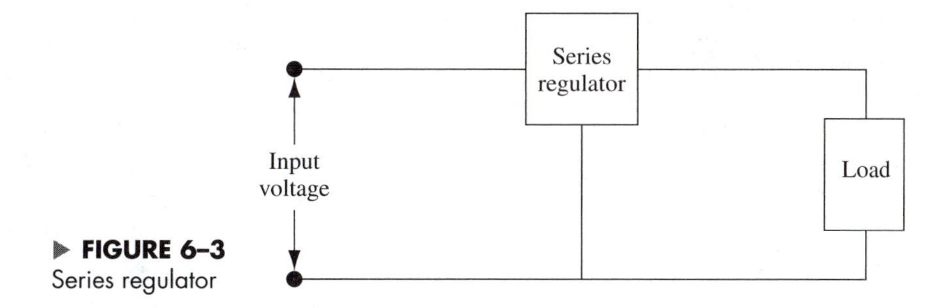

▶ **FIGURE 6–3**
Series regulator

current is at full load levels and high when the load current is low. When the load becomes shorted, excessive current is drawn from the supply. Series regulators are used much more often than shunt regulators. When a regulator is to operate very near full load condition most of the time and there is a possibility of the load becoming shorted, a shunt regulator becomes a viable option.

The block diagram for a general power supply is shown in Figure 6–4. A typical outlet voltage of 115 V AC at 60 Hz powers most power supplies. Accordingly most power supplies include a transformer to step down the AC voltage to a level closer to the ultimate output value. AC-powered linear regulators also require a rectifier circuit to convert the AC voltage to a pulsating DC voltage. DC-powered power supplies will not require the use of a transformer or a rectifier circuit. As shown in Figure 6–4, the primary components of a linear power supply are transformer, rectifier, filter, regulator, and load.

5-V DC Linear Series Regulated Supply

Let's begin by reviewing the detailed operation of a simple 5-V DC linear power supply that utilizes series regulation. See the schematic shown in Figure 6–5. Note that an input filter and output filter are not included in this circuit. The transformer steps the input AC voltage of 115 RMS V AC, 50 to 60 Hz down to a lower AC voltage at the secondary. A full-wave bridge rectifier converts the secondary voltage to pulsating DC with a peak value of 1.4 V less than the peak output of the secondary and a frequency that is twice the input voltage frequency. The rectifier filter capacitor smoothes out the pulsating DC. This produces a DC voltage with a small fluctuating component called the *ripple* riding on top of the DC voltage. The

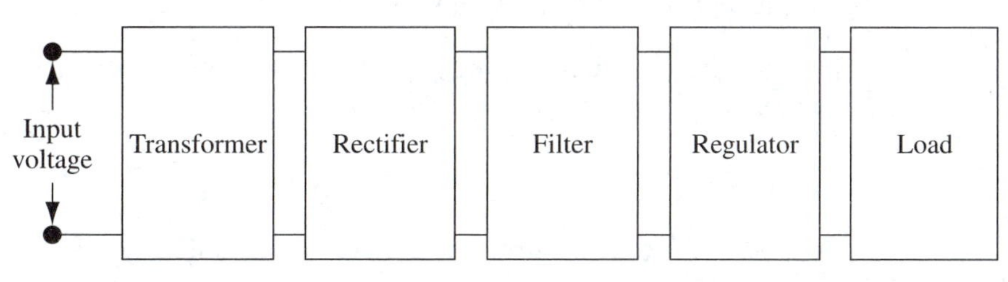

▲ **FIGURE 6–4**
Linear power supply block diagram

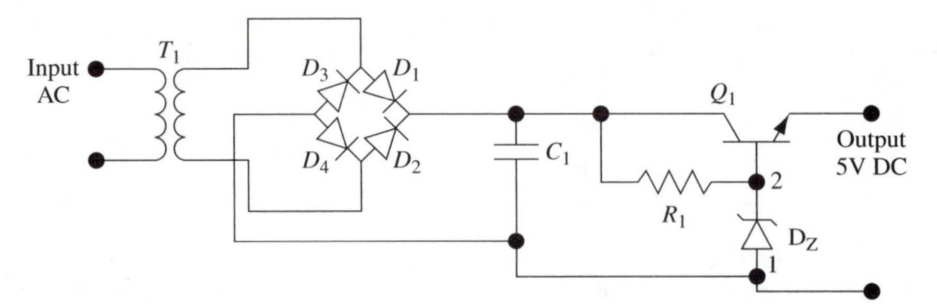

▲ **FIGURE 6–5**
5-V DC linear power supply with series regulation

rectified and filtered DC voltage is then regulated down to 5 V DC and connected to the load.

The regulator circuit is essentially an emitter-biased transistor circuit where the zener voltage maintains a constant base emitter voltage and the emitter voltage is controlled by the transistor to be the zener voltage plus the .7-V drop across the forward-biased PN junction that comprises the transistor base-emitter.

The basic design equations for this power supply are as follows:

$$V_{PSEC} = V_{PPRI} \times n \qquad (6\text{–}2)$$

where V_{PPRI} = the transformer peak primary AC voltage, V_{PSEC} = the transformer peak secondary voltage, and n = the transformer turns ratio

$$V_{RECT} = V_{PSEC} - 1.4\ \text{V} \qquad (6\text{–}3)$$

where V_{RECT} = the rectifier output voltage

Note: The 1.4 V is derived from the .7-V drop across each fix forward-biased silicon diode in the rectifier circuit.

$$V_{RIPPLE} \cong V_{RECT}/fC \qquad (6\text{–}4)$$

where f = the frequency of the pulsating voltage output from the rectifier and C = the value of the filter capacitor

Note: The frequency of the regulator output is equal to twice the frequency of the input frequency for all full wave rectifiers.

$$V_{RO} = V_Z + .7\ \text{V} \qquad (6\text{–}5)$$

where V_Z = the zener voltage of the zener diode and V_{RO} = the regulated output voltage to the load

Note: This equation is valid only if the zener is in its operating range.

I_O = the rated output current and is dependent on the maximum zener current for the zener to remain in regulation. The other circuit elements, primarily the transformer, diodes and transistor, must be selected to accommodate this current level on a continuous basis. Let's review a specific design example.

Example 6–1

For the power supply circuit shown in Figure 6–5, determine all component values and ratings for the supply to generate 5 V DC ±2% at 100 milliamps with less than 10% ripple voltage. The input voltage is 115 RMS V AC at 60 Hz and can vary in amplitude ±10%.

Solution

The solution of this design problem will require the determination of the following component parameters:

Transformer: primary voltage, secondary voltage, turns ratio and power rating

Rectifier Diodes: current rating, reverse breakdown voltage

C_1: capacitor type, value, voltage rating

Q_1: transistor type, voltage, current and beta

D_Z: zener voltage

R_1: resistor value, tolerance, type, and power rating

The peak input voltage V_{PPRI} is equal to 115/.707 which is 162.7 V. V_{PPRI} can vary ±10% or ±16.27 V, so the peak input voltage input to the primary can range from 146.4 V to 179 V. The 5-V DC output must be maintained within ±2% over this range of input voltage.

To maintain the nominal 5-V output, we will allot 3 additional volts for the series regulator to regulate down to the 5-V level. Therefore, the input voltage to the series regulator circuit, which is the rectifier output V_{RECT}, should be equal to a minimum of 8 V peak. Using the design equation $V_{RECT} = V_{PSEC} - 1.4$ V, V_{PSEC} can be found to equal 9.4 V.

Since $V_{PSEC} = V_{PPRI} \times n$, the turns ratio n is equal to V_{PSEC}/V_{PPRI} or 9.4/146.4 = .064. The primary must be able to handle a voltage as high as 200 V peak and the secondary could see a voltage as high as 20 V. The design equation that defines the zener voltage, $V_{RO} = V_Z + .7$ V is solved for $V_{RO} = 5$ V. The zener voltage $V_Z = 4.3$ V and a 1N749A is selected as the zener diode. The 1N749A has a zener voltage of 4.3 V with a zener test current of 20 mA; the knee current is 6 mA and the maximum current is 85 mA.

The ripple percentage is specified as a maximum of 10%. Ripple is stated as a percent of the nominal output voltage, or $V_{RIPPLE} = .5$ V. Solving, $V_{RIPPLE} \cong \text{LOAD}/fC$ with $I_{LOAD} = 100$ mA and f = 120 Hz. $C = 1666$ µF. The next highest standard capacitor value available is 1800 µF and the closest voltage rating that provides a reasonable margin of safety is 16 V.

The calculation of R_1 is accomplished by establishing a zener current that will be between the smallest (the zener knee current) and largest current (the maximum zener current) for zener regulation. There is a wide range of resistance values that will accomplish this as the nominal 9.4 volts supplied by the rectifier will provide the test current of 20 mA with a 9.4/20 mA = 470 Ω. The range of input voltage that the regulator circuit is expected to see is 8.46 to 10.34 V. With R_1 = 470 Ω the zener currents for this range in input voltage are 18 mA to 22 mA. Considering the possible current flow through R_1, it should be a .5-watt resistor.

There are many transistors that will function in this circuit. The key parameters are the maximum current and voltage ratings for the transistor. The regulator specifications call for a load current of 100 mA and the maximum voltage the transistor should see is 10.34 V. An NPN transistor such as a 2N3904 is selected with a maximum collector voltage of 40 V and a continuous collector current rating of 200 mA. The DC current gain for the 2N3904 ranges from 30 to 100.

Following is a summary of the selected components:

Transformer T_1: primary 115 RMS V AC, secondary voltage 6.6 RMS V AC, with a turns ratio of .064 and a power rating of 2 watts

Rectifier Diodes D_1–D_4, 1N4001 current rating = 1 amp, reverse breakdown voltage of 50 volts, or an equivalent bridge rectifier assembly

C_1: 1800 μF aluminum electrolytic, 16 volts

Q_1: 2N3904, $V_{CE} = 40\ V_{dc}$, I_C = 200 mA

D_Z: 1N749A, V_Z = 4.3 V

R_1: 470 Ω ±5%, carbon composition, .5 watts

Surge Resistor

When selecting the diodes to be used in the rectifier circuit, or a complete rectifier circuit, it is important to consider the surge current that will occur when the power supply is first turned on and the capacitors are all being charged. If the continuous current rating of the diodes is close to the maximum load current output from the power supply, a resistor called a *surge resistor* is often used to limit the flow of initial surge current (see Figure 6–6). The value of the surge resistor should be as small as possible, as it will lower the efficiency of the power supply and reduce the amount of voltage available to the regulator circuit.

Inductive Filters

From previous experience with inductors and capacitors, we realize that it is usually possible to achieve the same results of a capacitor filter with an inductor. This is true with the rectifier filter capacitor, whose job it is to reduce the ripple supplied to the regulator circuit. Theoretically, we could replace the filter capacitor completely with an inductor, as shown in Figure 6–7. However, the inductor required would be very large for a frequency as low as 120 Hz. It would be more practical to reduce the size of the capacitor by using both an inductor and capacitor, as shown in Figure 6–7b, but we will find that even this is not practical or necessary

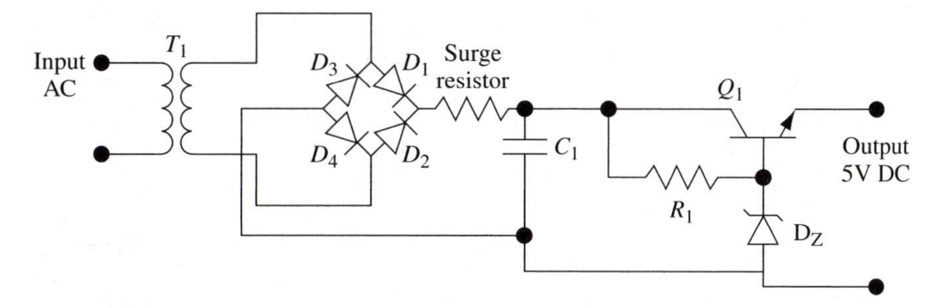

▲ **FIGURE 6–6**
Surge resistor example

with the integrated circuit regulators that we will discuss next. An inductor used as a filter is called a *choke*. The purpose of the inductor in the filter circuit is to have most of the AC voltage drop across the inductor at a particular frequency. Therefore, the relationship of X_L (inductive reactance) to X_C (capacitive reactance) at a given frequency will determine the AC voltage drop across the two circuit elements. From an AC perspective, the ripple output voltage of the filter is:

$$V_{RIPPLE} = X_C/(X_L \times V_{RECT}) \qquad (6\text{–}6)$$

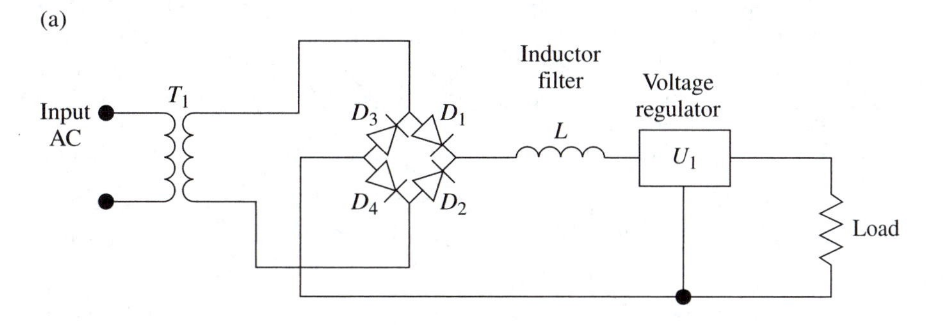

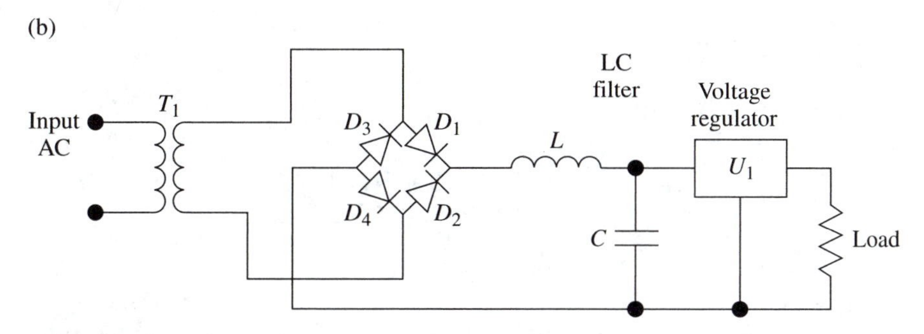

▲ **FIGURE 6–7**
Inductor rectifier filters

If we let $X_C = 1$ and $X_L = 20$ and $V_{RECT} = 8.4$ volts peak then the $V_{RIPPLE} = .42$ V, which is less than the specified .50 V. Calculating the resulting capacitor and inductor values that would achieve this at a frequency of 120 Hz (the frequency of the rectifier output voltage):

$$X_C = 1/(2\pi fC) = 1 \qquad C = 1/2\pi 120 = 1326\ \mu\text{F} \tag{6–7}$$

$$X_L = 2\pi fL = 20 \qquad L = 20/2\pi 120 = 26.5\ \text{mH} \tag{6–8}$$

The 26.5 mH inductor is still a very large inductance value, and the benefit of reducing the capacitor value such a small amount is not cost effective. We can conclude that using an inductive filter at a frequency of 120 Hz is not practical, especially with the ripple reduction capabilities of today's integrated circuit regulators. Later on when we discuss switching regulators, we will see that their higher frequency of operation provides for the effective use of inductor filters.

Integrated Circuit Regulators

The previous example was worthwhile because it was not only a review of a linear series regulator design, but it also provided example applications of selecting zener diodes and transistors. However, there are many integrated circuit linear regulators available that are popular and easier to use. The simplest varieties are called *three-terminal regulators*, and they perform functions identical to the regulator constructed from resistor R_1, zener diode D_Z, and transistor Q_1 in the previous example. In addition, they provide failsafe features such as current limiting and thermal shutdown. An LM309 linear regulator is a fixed output regulator that can replace the regulator circuit from the previous example directly. Following is a summary of the key specifications for the LM309:

Output Voltage: 5.05 V typical, minimum 4.8 to a maximum of 5.2 V

Output Current: up to 1 amp, depending on the package and heatsink provided. An LM309 in a TO39 package with no heatsink can provide 100 mA up to a temperature of about 50 °C.

Dropout Voltage: The input-to-output voltage difference, where the regulator ceases to regulate with further reduction of the input voltage

Ripple Rejection: 50 dB

Current Limiting: When the peak output current exceeds safe levels, the current is limited to these levels.

Temperature Shutdown: If the IC becomes overheated, the circuit simply shuts down its function as a regulator.

The drop voltage for most of these regulators is between 2 and 3 V. Later we will discuss special regulators that feature low dropout voltages.

Take note of the specifications when you consider the requirement for a heatsink. A heatsink is a thermally conductive device that attaches to a component for the purpose of absorbing heat and dissipating it away. The use of a heatsink is critical for the reliable operation of any component that conducts significant current levels. Heatsinks are often rated in °C/watt. This rating is indicative of the temperature that results when the heatsink is attached to a component that dissipates a certain wattage. A heatsink with a rating of 20 °C/watt will result in a 20 °C rise in temperature for each watt dissipated by the heatsink. The lower the °C/watt rating, the better the heatsink is at absorbing and distributing the heat.

Example 6–2

In this example we will modify the circuit of Example 6–1 to replace the regulator circuit with the LM309. As shown in Figure 6–8, the LM309 replaces the transistor, resistor and zener diode that Example 6–1's regulator circuit comprised. There are some considerations regarding filter capacitors.

Ripple Rejection

Let's examine the 50-dB ripple rejection stated in the specifications for the LM309. The ability of the LM309 to reject ripple is of course frequency-dependent. An examination of the data sheets for the LM309 indicates that the ripple rejection peaks at about 500 Hz (Figure 6–9). The data sheet also shows that the 50 dB specification is met by the device for frequencies ranging from 10 Hz to 100 kHz.

For this design the 50-dB specification means that the remaining ripple will be .003 (a 50-dB loss results in .3% voltage signal remaining) times the input ripple value. It appears like the large filter capacitor is no longer needed to reduce the ripple value. However, it is important to note that the input provided to the regulator must always be 2 to 3 volts higher than the regulated output voltage in order for the regulator to operate properly. So even though the regulator can reduce the ripple significantly on its own capacitor, C_1 is still needed to make sure that the voltage supplied to the regulator is sufficient for its reliable operation. In actual practice C_1 is specified at a value that will reduce the ripple to 10% before input to the regulator circuit. Capacitor C_1 combined with the ripple rejection capabilities of the integrated circuit regulator reduce ripple significantly and very practically.

Recalculating C_1 to provide 10% ripple from the nominal 8.4-V pulsating DC used previously sets the desired ripple at .84 volts. $V_{RIPPLE} \cong I_{LOAD}/fC$ with I_{LOAD} = 100 milliamps, f = 120 Hz, and C = 992 μF. The next highest standard capacitor value available is 1000 μF and the closest voltage rating that provides a reasonable margin

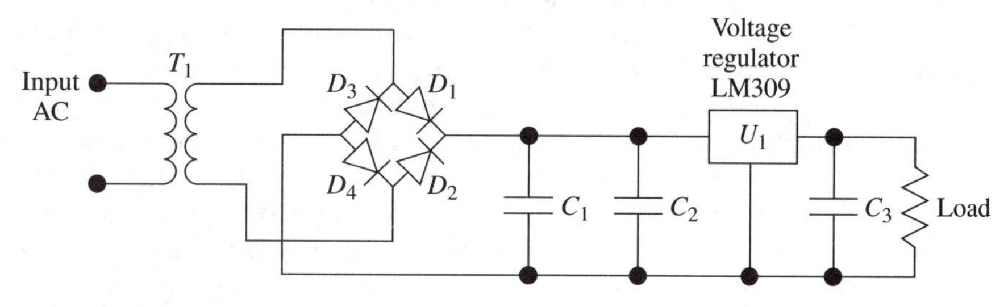

▲ **FIGURE 6–8**
LM309 regulator circuit

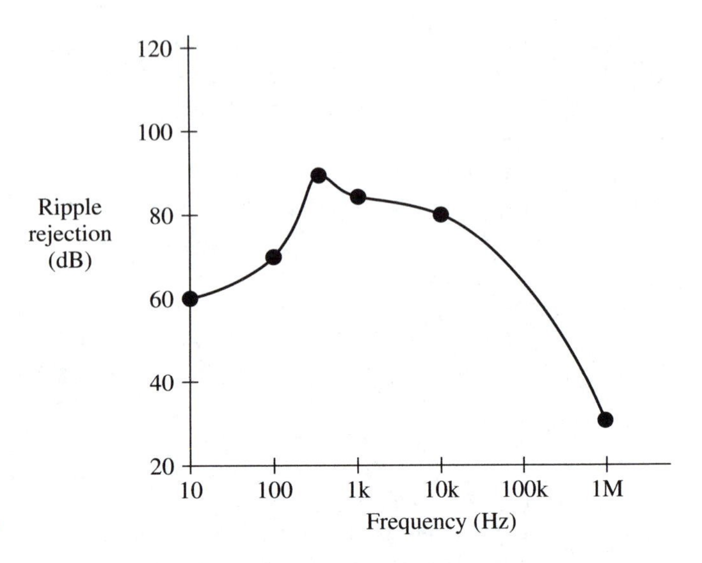

FIGURE 6–9
Ripple rejection bode plot

of safety is 16 V. Recalculating the output ripple value for the input ripple value of .84 volts, we have:

$$\text{dB} = 20\ \text{Log}_{10}\ (V_{OUT}/V_{IN}) \qquad (6\text{–}9)$$

$$-50\ \text{dB} = 20\ \text{Log}_{10}\ (V_{OUT}/.84\ \text{V})$$

$V_{OUT} = 2.65$ mV of output ripple voltage

C_1 can be reduced or increased in value, depending on the overall ripple requirements of the power supply. C_2, a 1-μF solid tantalum capacitor, is recommended when the regulator is located more than four inches away from the primary rectifier filter capacitor C_1. C_3, also a 1-μF solid tantalum capacitor, is not required to support the regulator's stability but is beneficial in improving the regulator's ability to respond to a transient load change.

Efficiency

The voltage drop across a regulator and its output current determine the power dissipated by a three-terminal regulator. The efficiency of these circuits will determined by how small this voltage drop can be maintained and still provide an input voltage high enough to provide for good regulation.

There are many other varieties of three-terminal fixed-output regulators that are similar to the LM309. They come in a variety of fixed voltage outputs, packages, and power ratings, but functionally they are the same. Current limiting and thermal shutdown features are usually included. In addition to positive fixed-voltage regulators, there are also negative fixed-output voltage regulators available to provide negative voltage supplies. The following list includes some popular three-terminal positive and negative voltage regulators with their basic specifications:

LM78LXX series: +5-V (LM78L05), +12-V (LM78L12), and +15-V (LM78L15) regulators at 100 mA

LM78MXX series: +5-V (LM78M05), +12-V (LM78M12), and +15-V (LM78M15) regulators at 500 mA

LM79LXX series: –5-V (LM79L05), –12-V (LM79L12), and –15-V (LM79L15) regulators at 500 mA

LM79MXX series: –5-V (LM79M05), –12-V (LM79M12), and –15-V (LM79M15) regulators at 500 mA

Adjustable Linear Regulator Circuits

Adjustable voltage regulators can be modified to regulate their output voltage over a range of output voltages. In this type of regulator circuit, the source of the reference voltage used to control the output is provided as an input to one of the regulator inputs. The terminal usually chosen for this purpose is the terminal usually connected to ground on fixed-voltage regulators. Integrated circuit adjustable voltage regulators are available in three-terminal packages. A popular series is the LM317 positive voltage and the LM337 negative voltage adjustable regulators. Their basic specifications are as follows:

LM317: output voltage range, +1.2 to +37 V, up to 1.5 amps output current

LM337: output voltage range, –1.2 to –37 V, up to 1.5 amps output current

A typical circuit application of the LM317 is shown in Figure 6–10. For this circuit:

$V_{OUT} = 1.25\ (1 + R_2/R_1)$

Combination ± Linear Power Supplies

For many analog circuits, ± power supplies are required to provide both positive and negative signals that have a common zero reference point called *analog*

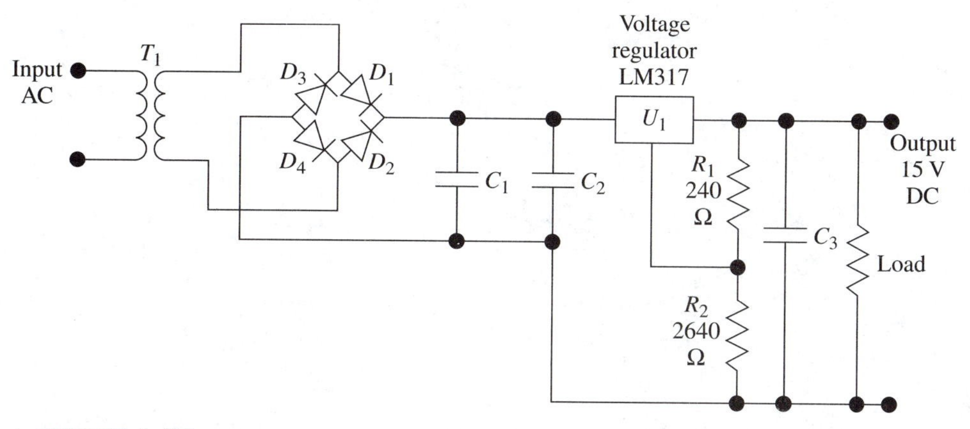

▲ FIGURE 6–10
LM317 adjustable regulator circuit

ground. A typical arrangement utilizes a center-tapped transformer and a full wave bridge rectifier connected, as shown in Figure 6–11. This circuit provides one-half of the secondary voltage minus the 1.4-V drop across the rectifier diodes to both the positive and negative rectifier voltage outputs.

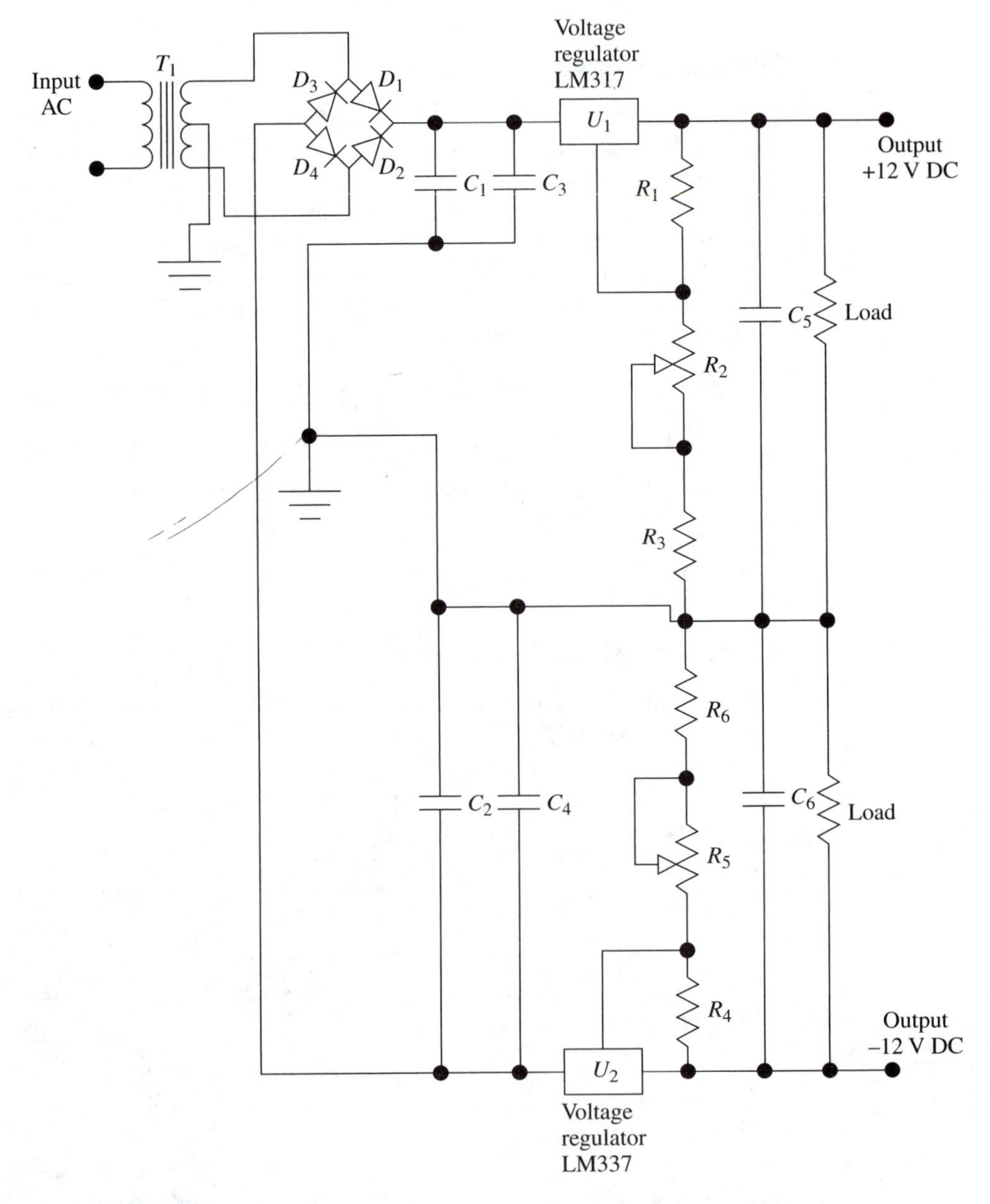

▲ **FIGURE 6–11**
±12 V DC adjustable regulated power supply

Example 6–3

Develop a dual power supply that will develop ±12 V that are adjustable over a range of 11.5 to 12.5 V with an output load current of 100 mA each. Ripple output should be less than .6 mV. The input power is 115 or 230 V AC ±10% at 50–60 Hz.

Solution

The LM317 and LM337 regulators are selected as the integrated circuit regulators for this power supply. The circuit schematic for this supply is shown in Figure 6–10. Since the regulated output is ±12 V the voltage supplied to the input must be ±15 V to allow adequate regulation. Therefore, the output of the secondary must be ±18 V when the input power is at its lowest possible value, 115 or 230 V AC minus 10%, which is 103.5 or 207 V AC. For the 115-V AC input situation, the turns ratio of the transformer must be ±18/103.5, which is 36/103.5 = .35. For 230 V AC the turns ratio must be ±18/207, which is 36/207 = .174. To accommodate either 115 V AC or 230 V AC, we can use two identical primaries with turns ratios of .35 relative to the secondary. When powering the supply with 115 V AC, the primaries will be wired in parallel, making the overall turns ratio .35. For 230 volts AC operation, the primaries can be wired in series to create a turns ratio of .174.

The rectifier filter capacitors should reduce the ripple at each half of the secondary by 10% or down to 1.8 volts. Using the formula $V_{RIPPLE} \cong I_{LOAD}/fC$:

$C = 100\text{ mA}/(120\text{ Hz} \times 1.8\text{ V}) = 463\ \mu\text{F}$
The next highest standard value capacitor available is 470 µF.

Typical ripple rejection by the LM317 and LM337 regulators is 65 dB; therefore, the ripple that is expected on the output voltage of this supply can be calculated as follows:

$$\text{dB} = 20\ \text{Log}_{10}\,(V_{OUT}/V_{IN}) \qquad (6\text{–}10)$$

$-65\text{ dB} = 20\ \text{Log}_{10}\,(V_{OUT}/1.8\text{ V})$
$V_{OUT} = .57\text{ mV}$

The data sheets for the LM317/337 recommend the use of an input filter capacitor of .1 µF (C_3 and C_4) and a solid tantalum output filter capacitor of 1 µF (C_5 and C_6). In order to calculate the output voltage adjustment resistors, the following equation is supplied by the data sheet for the LM317/337:

$V_{OUT} = 1.25\text{ V}\,(1 + R_2/R_1) + 50\ \mu\text{A}\ R_2$

The data sheet recommends that R_1 be 240 Ω. The range of adjustment for the output voltage was specified as 11.5 to 12.5 V. Therefore, when R_2 is near its minimum value, the output voltage should be 11.5 V, and it should reach 12.5 V when it is adjusted close to its maximum value. The output voltage equation is solved for both of these situations:

for $V_{OUT} = 11.5\text{ V} = 1.25\text{ V}\,(1 + \text{R}_2/240\ \Omega) + 50\ \mu\text{A}\ R_2$
$R_2 = 1949\ \Omega$

for $V_{OUT} = 12.5\text{ V} = 1.25\text{ V}\,(1 + R_2/240\ \Omega) + 50\ \mu\text{A}\ R_2$
$R_2 = 2140\ \Omega$

To meet this range of adjustment, R_2 must be about 1949 Ω at its minimum value and 2140 Ω. This can be accomplished with a fixed resistor in series with a trimmer potentiometer. A 1780 Ω 1% resistor (R_3 in Figure 6–11) is selected in series with a 500-Ω trimmer potentiometer (R_2 in Figure 6–11).This will provide more adjustment range than required, but the next smallest standard value potentiometer is 200 Ω, and that would not provide enough adjustment.

Following is a summary of the components selected for Example 6–3 as shown in Figure 6–10:

Transformer T1: primary 115/230 RMS V AC, secondary voltage 36 RMS V AC, dual primary with turns ratios of .35 and a power rating of 3.6 watts.

Rectifier Diodes D_1–D_4, 1N4001 current rating = 1 amp, reverse breakdown voltage of 50 V.

C_1 and C_2: 470 µF aluminum electrolytic, 50 V

U_1: LM317 Regulator

U_2: LM337 Regulator

R_1, R_4: 240 Ω, .5% metal film resistor

R_2, R_5: 500 Ω, 5% trimmer potentiometer

R_3, R_6: 1780 Ω, 1% metal film resistor

C_3 and C_4: .1 µF ceramic disc capacitor

C_5 and C_6: 10 µF solid tantalum capacitor

C_7 and C_8: 1 µF solid tantalum capacitor

Integrated Circuit Shunt Regulators

As mentioned previously, series regulators are used much more often than shunt regulators, which is obvious in any data book on power supply regulators. The key advantages of the shunt regulator are that its efficiency is highest at full load current and it also can handle a shorted load. These are the situations that promote the use of a shunt regulator:

1. The load is a consistent value that is very close to the full load of the regulator.
2. The regulator can be subject to a shorted load.

The LM431 is a three-terminal integrated circuit shunt regulator that is adjustable. If we connect this shunt regulator to the transformer and rectifier circuit of Example 6–2, we have the 5-V DC, 100 mA shunt regulator circuit shown in Figure 6–12.

Resistor R_S is calculated to develop a voltage drop of approximately 3 V to bring the voltage down to the 5-V level where the shunt regulator will maintain this value by adjusting the current shunted in parallel to the load.

R_S = 3 V/100 mA = 30 Ω (the closest standard value is 33 Ω)

$V_O = (1 + R_1/R_2)\ V_{REF}$

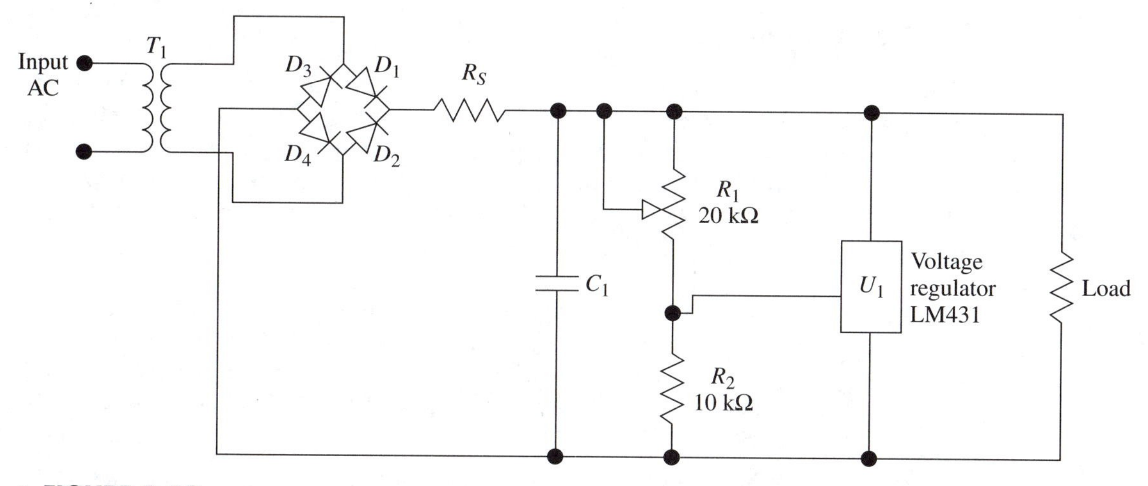

▲ **FIGURE 6–12**
5-V shunt regulator circuit

Letting R_1 = 10k Ω and $R_1 = R_2$ will create the desired operation. R_1 is adjusted to develop the 5-V output when connected to the load.

Low Dropout Voltage Regulators

These are a class of voltage regulators that feature a dropout voltage differential of less than 1 V. The actual range in variation of the dropout voltage differential for these regulators is from approximately .30 to .82 V. In general, these regulators promote better efficiency, because the power consumed by the regulator is a function of the output current times the voltage drop across the regulator. Also, when the input voltage supplied to the regulator comes from a battery, they allow operation of the circuit very close to the minimal battery voltage. Low dropout regulators also include the failsafe features discussed before, such as current limiting and thermal shutdown. Because these regulators are often used in battery-powered circuits, they also provide reverse battery protection and line transient protection. Following is a sampling of some popular low-dropout voltage regulators:

LM2940: fixed outputs of 5, 8, 12, or 15 V at 1 amp, .5-V dropout voltage

LM2926: fixed 5-V output at .5 amps, .35-V dropout voltage

LM2931C: adjustable output 3 to 29 V at 100 mA, .30-V dropout voltage

LM2951C: fixed outputs of 3.0, 3.3, 5 V or adjustable 1.24 to 29 V at 100 mA with .38-V dropout voltage

It is important to note that the previous discussion about integrated circuit regulators is general. Whenever using a specific integrated circuit in a design, consult the data sheet for that part and use all of the appropriate precautions and application information.

6–3 DC-to-DC Converters

DC-to-DC converters are a class of devices that convert an input DC voltage to another DC voltage. They are most often used to increase a DC voltage significantly or to generate a DC voltage of opposite polarity. Reducing DC voltage levels are typically accomplished with resistors, zener diodes, and/or voltage regulators in a manner similar to the methods discussed in Section 6–2. There are two basic types of DC-to-DC converters:

1. Switching-type (Step-up or Inverting) Regulators: those that utilize transistor switches to store energy in capacitors or inductors. The stored energy is used to step up the voltage or change its polarity. These regulators are actually a subset of the class of general switching regulators that are discussed in Section 6–4.
2. Push-pull/Flyback Regulators: These regulators convert a lower DC input voltage to AC and then use a step-up transformer to increase the AC voltage, which is converted back to a higher-level DC.

Many times when working on a circuit design, you'll find that there are requirements for special DC voltages to operate particular devices. Vacuum fluorescent displays are a good example. These bright, blue-green-colored displays utilize DC voltages on the order of 200 V for their power. The current requirements are very small. In order to supply this voltage, the circuit designer can use a separate transformer or secondary winding, to develop a larger AC voltage, and then rectify and regulate, as discussed in Section 6–2. An alternative is to use a DC-to-DC converter to step up an available, lower DC voltage to roughly 200 V DC.

Another common application of DC-to-DC converters is in predominately digital circuits that are powered with 5 V DC where a small section of analog circuitry is included that requires either a higher DC voltage and or a voltage with negative polarity. For example, let's say that a small op amp circuit is being added to a predominately digital circuit that has 5 V DC available and the op amp circuit requires = ±12 V DC for proper operation. DC-to-DC converters can be used to resolve both of these issues—generation of both +12 V DC and –12 V from the available 5 V DC. Of course, for this approach to be practical, the current requirements for the ±12 V DC supplies must be minimal.

Switching-type DC-to-DC Converters

The LM1577/2577 is a good example of an integrated circuit switching-type, step-up DC-to-DC converter, otherwise called a *"boost" voltage regulator.* The circuit shown in Figure 6–13 is an LM2577 IC used to convert a +5-V DC input voltage to +12 V DC at 800 mA. The operation of the LM2577 can be understood after a review of its block diagram, as shown in Figure 6–14.

The NPN transistor switch included in the LM2577 is switched at a frequency of 52 kHz. When the transistor is switched on, current flows from V_{IN} through the inductor L. The inductor opposes this attempt to change the current instantaneously by storing energy in its magnetic field, as the current increases over time at a rate of V_{IN}/L. See the simplified circuit in Figure 6–15. At the instant the transistor is switched off, the voltage across the load resistor will be determined by the current flowing through the inductor, multiplied by the quantity of the load resistance minus the voltage drop across the diode. The LM2577 controls this voltage by monitoring the output voltage across the load and adjusting the duty cycle (on time vs. off time) of the 52 kHz oscillator. If the voltage is lower than the desired output (+12 V DC, in this case), the duty cycle is increased. If the input voltage increases slightly, the duty cycle is decreased to bring it back to the desired value. The amplifier and comparator continually monitor the output voltage and make the appropriate changes to the duty cycle. The underlying concept for this circuit is that the output voltage is determined by the amount of energy stored in the inductor during the portion of the 52 kHz clock that the transistor is on. The output voltage is monitored, amplified by the error amplifier, and compared to the desired output. Then the duty cycle is adjusted accordingly.

Inverting, Switching-type DC-to-DC Converters

An inverting DC-to-DC converter is used to convert a positive DC voltage to a negative polarity. For situations where +5 V DC is already available and –5 V DC is also needed, an inverting, switching-type DC-to-DC converter can do the job if the current requirements are small enough. The LMC7660 is an example of this type of DC-to-DC converter. Inverting regulators of this type all use a similar technique to change the polarity of a DC voltage. This process uses two sets of switches that first charge up one capacitor with the input voltage. This charge is then passed on to a

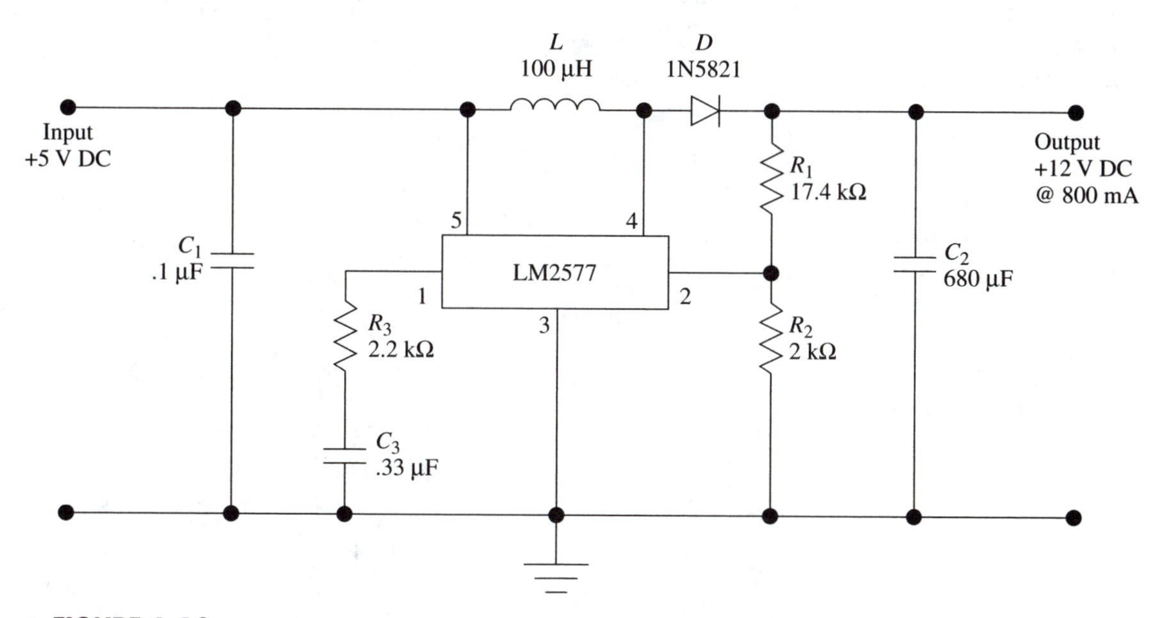

▲ **FIGURE 6–13**
+5-V to +12-V switching-type step-up DC-to-DC converter

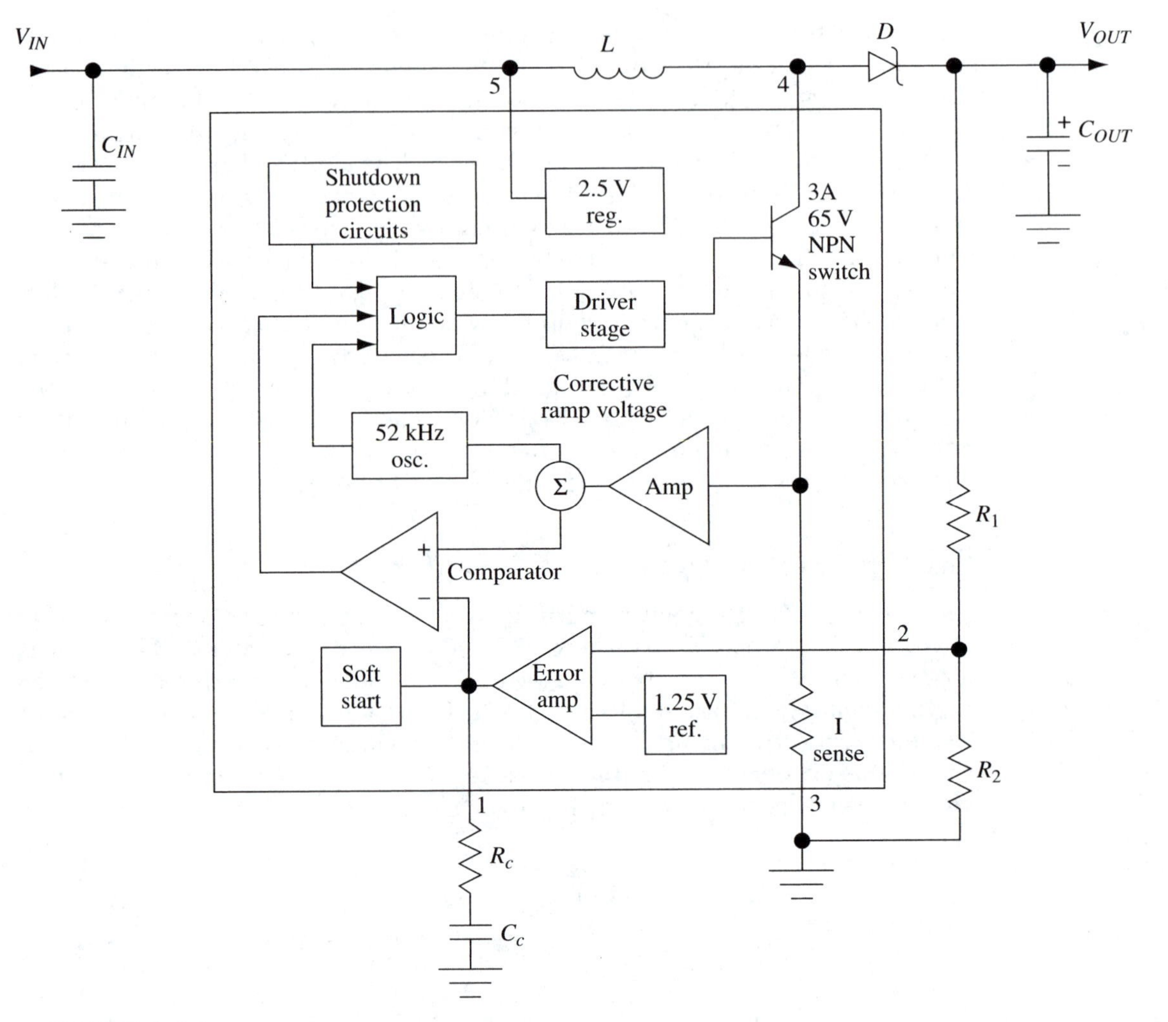

▲ **FIGURE 6–14**
LM2577 block diagram

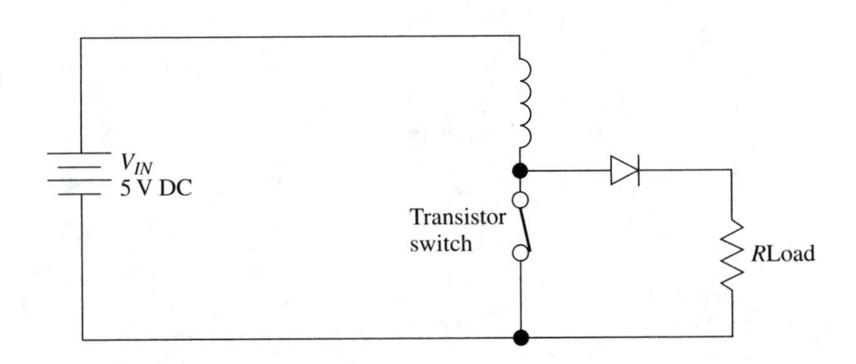

▶ **FIGURE 6–15**
Simplified LM2577 switching circuit

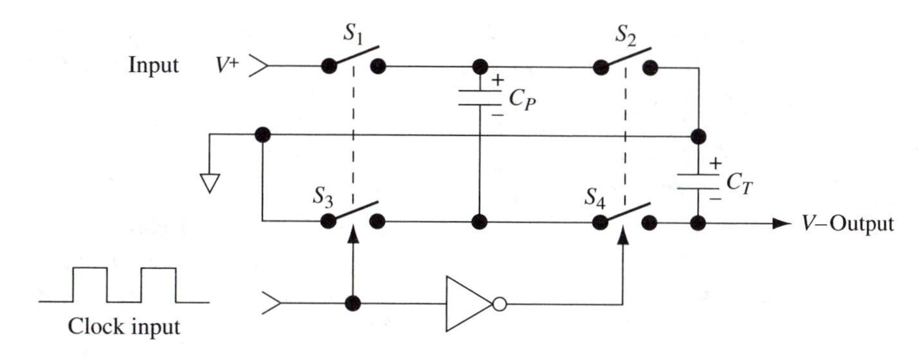

▲ **FIGURE 6–16**
Inverting, switching-type DC-to-DC converter functional diagram

second capacitor whose positive terminal is connected to ground, reversing the polarity of the output voltage (see Figure 6–16). The LM7660 has an internal oscillator that switches two sets of CMOS switches. The oscillator is connected directly to switch set *A*, then inverted and connected to switch set *B*. Therefore, when switch set *A* is closed, switch set *B* is open, and vice versa. The frequency of the oscillator is 10 kHz but can be reduced by placing a slow-down capacitor between pins 7 and 8. When a positive input DC voltage (in the range of 3 to 10 V for the LMC7660) is connected to *V*+ and switch set *A* is closed, the capacitor labeled *C*+ is connected to *V*+ and ground and therefore charges up to *V*+. Then switch set *A* is opened (which isolates capacitor *C*+ from *V*+ and ground) and switch set *B* is closed. This connects capacitor *C*+ in parallel with capacitor *C*–, whose positive terminal is connected to ground. *C*+ charges *C*–, initially only up to *V*+/2 but after a number of cycles the voltage across *C*– becomes equal to *V*+. However, the positive lead of *C*– is grounded, reversing the polarity of this voltage, the output of this inverting DC-to-DC converter.

Figure 6–17 shows an example application of the LMC7660 converting +5 V to –5 V. The maximum current this device can supply is 400 μA and power efficiency

▶ **FIGURE 6–17**
+5-V to –5-V DC-to-DC converter

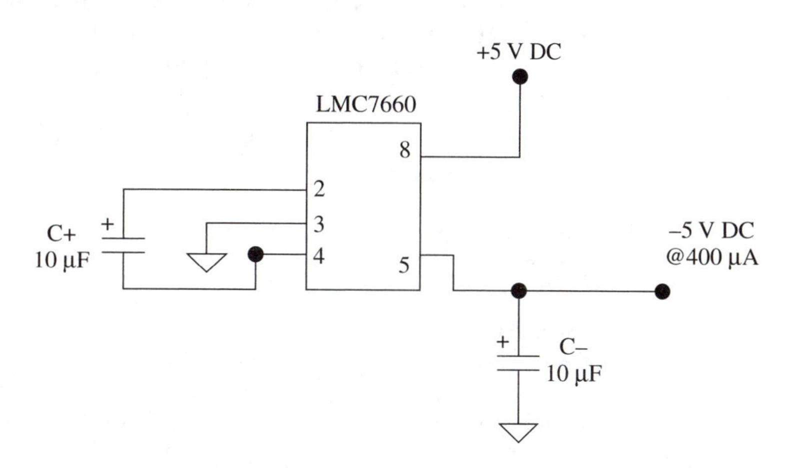

of this conversion is at least 90%. The efficiency can be improved by using a slow-down capacitor, which reduces the oscillator frequency and the quiescent operating current of the device.

Push-pull/Flyback DC to DC Converters

Push-pull/flyback DC-to-DC converters are a class of DC-to-DC converters that convert a DC voltage to AC (usually a square wave) and then use a transformer to increase, isolate, and otherwise modify the output DC voltage. They are usually employed to meet any of the following output requirements:

1. There is a large difference between the DC input voltage and the desired output voltage.
2. The current requirements of the converter output are significant.
3. There is a need for multiple DC output voltages.
4. The DC output voltages must be electrically isolated from the input voltage.

The switching-type DC-to-DC converters described previously do not provide isolation or the possibility of multiple output voltages. Also, because they store energy in the electric field of capacitors or the magnetic field of inductors, they cannot meet high voltage/current requirements without unacceptably high ripple levels. As the current and voltage requirements of DC-to-DC converters increase, so does the stored energy needed to support these increases. At some point the physical limitations of the largest practical inductor and capacitor are reached, and a different approach is needed.

There are two general types of DC-to-DC converters that are capable of higher voltage/current outputs: the push-pull converter and the flyback converter. Both types of DC-to-DC converters resolve the issues of higher voltage/current, isolation and multiple outputs by converting the DC voltage to an AC voltage. Then a step-up transformer is used to increase the voltage and/or change the polarity. This provides the opportunity for multiple outputs with the use of multiple secondary windings. The output voltage and current levels are limited only by the input and the transformer used. Isolation can be achieved by using a separate winding to feedback the output voltage that is measured and used to provide regulation.

Push-pull DC-to-DC Converters

The push-pull DC-to-DC converter consists of two transistors that are configured to operate much like a class B amplifier. One NPN-type transistor conducts current in one direction, while a PNP transistor conducts it in the other. A basic push-pull converter is shown in Figure 6–18. The input DC voltage causes each transistor to switch in succession, generating an alternating square wave input to the transformer primary that is transferred over to the secondary. The frequency of the square wave will depend on the inductive characteristics of the transformer. The turns ratio deter-

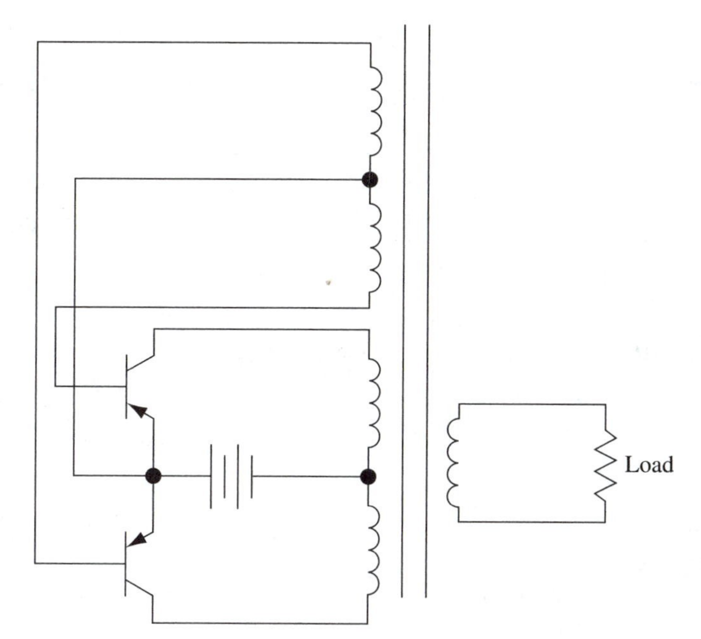

▶ **FIGURE 6–18**
Push-pull DC-to-DC converter

mines the secondary voltage which is in turn rectified, filtered, and regulated like the regulator circuits discussed in Section 6–2. Push-pull converters require matched transistors and the transformer winding must be wound to specifications that determine the proper inductance and output secondary voltage.

Flyback DC-to-DC Converters

Flyback converters eliminate the need for matched transistors and precision transformers by using just one transistor to develop a single polarity pulse waveform that is input to a transformer to be stepped up. The flyback converter is also called a *single-ended converter* because of the single polarity of the generated waveform. While the flyback converter simplifies the converter circuit design, the single polarity waveform fails to utilize the full power transfer capacity of the transistor and transformer. Nevertheless the flyback converter is practical in many situations. An example flyback regulator is shown in Figure 6–19 that utilizes the LM2587-12 integrated circuit flyback regulator. The circuit shown converts a voltage in the range of 4 to 6 V DC to ±12 V DC. A simplified functional diagram for the LM2587 is shown in Figure 6–20. The input DC voltage is connected to pin 5 of the LM2587 and regulated down to 2.9 V internally. The 2.9 V is used to power the internal circuitry, which includes an oscillator, amplifier and comparator. The circuits combine to switch the output transistor on and off appropriately to regulate the output voltage, which is fed back into the LM2587 on pin 2. The compensation connection adjusts the op amp's gain so that it is consistent over the operating frequency range of the regulator.

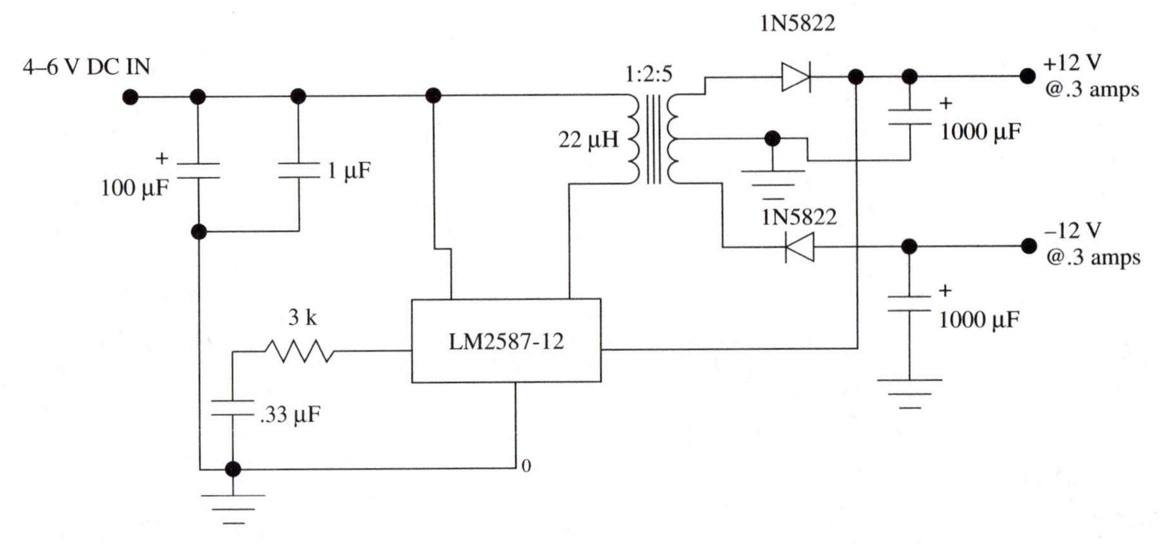

▲ **FIGURE 6–19**
LM2587 ±12-V flyback regulator circuit

Voltage References

A special type of DC-to-DC converter is used to generate precision DC voltages for use as a reference in an analog circuit. Reference voltages are used in analog circuits most often when signal conditioning is being performed. Signal conditioning is when we modify a signal's range, and/or level, to meet some other requirement. Let's say that we have an analog signal with a range of 0 to 2 V that must be converted to a range of 1 to 5 V, a commonly used standard in many instrumentation and control circuits. In this case we are changing the range and the level of the signal and would utilize a signal conditioner circuit to accomplish this. A precision DC voltage is needed to develop the 1-V offset of the 1- to 5-V output signal. A circuit called a *voltage reference* is used for this purpose and for many other applications.

Another common application of voltage references is to supply the reference voltage for analog-to-digital (A/D) converters or digital-to-analog (D/A) converters. In these applications, the voltage reference determines the range of analog values that correspond to the range of digital values. For example, if we supply an 8-bit A/D converter with a 2.5-V reference voltage, a 2.5-V analog input will result in a count of 255 decimal or 11111111 in binary. Conversely, when a digital input of 11111111 binary is input to a D/A converter with a voltage reference of 2.5 V, the analog signal output will be 2.5 V.

A voltage reference must supply a particular voltage value with low noise and a small temperature coefficient. It can be a circuit as simple as a zener diode or an integrated circuit that offers high precision and extremely low temperature coefficients. As was the case with voltage regulators, voltage references are available in a two-terminal shunt configuration or as series references with three or more terminals.

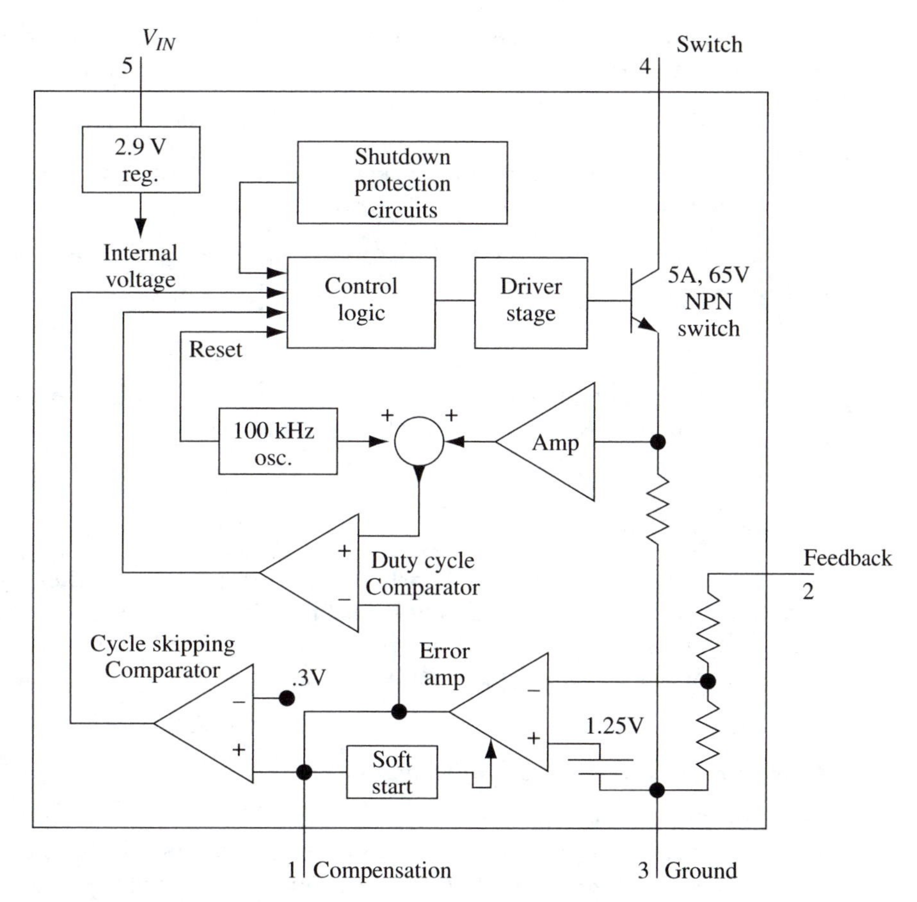

▲ FIGURE 6–20
LM2587 functional diagram

Before solid state voltage references were available, designers utilized various types of batteries that were developed for this purpose. The zener diode was the first semiconductor voltage reference; although it is not quite accurate or stable enough to be used for that purpose without additional circuitry. The voltage stability of the zener diode can be greatly improved by placing a rectifier diode in series with a zener diode. This combination, called a *reference diode,* can result in a temperature coefficient of less than 25 ppm/°C. Reference diodes are combined with op amps to provide voltage references that are highly accurate and offer low temperature coefficients.

Another device that operates in a manner similar to the zener diode is called the *bandgap reference.* Bandgap references are a direct result of integrated circuit technology and include a number of closely matched diodes that are fabricated on a silicon substrate. One of the diodes is connected in series with the combination of all of the other diodes in parallel. When identical currents drive the single and parallel diodes, the result is a stable voltage of 1.2 V across the circuit that exhibits an ideal temperature coefficient of zero.

When selecting a voltage reference the key design parameters, many of which are identical to those discussed previously for voltage regulators, are as follows:

1. Reference Voltage Value: the voltage reference required
2. Specified Reference Tolerance: the expected variation from the ideal voltage reference value
3. Temperature coefficient: the expected variation due to temperature changes
4. Maximum output current
5. Power consumption
6. Drop-out Voltage: the input/output voltage differential where the voltage reference ceases to provide the specified output reference value
7. Line regulation
8. Load regulation

Voltage reference circuits have been simplified greatly by the integrated circuit configurations currently available. But in any design it is important to pay attention to the details. Let's review a few of the varieties of the voltage references available and their typical applications.

A Two-terminal Bandgap Shunt Regulator

The ICL8069 is a very popular shunt type voltage reference that utilizes the bandgap technology discussed previously. It is a two-terminal device that is often shown schematically as a zener diode but offers much better voltage reference performance than a typical zener diode. The schematic shown in Figure 6–21 applies the ICL8069 as an adjustable voltage reference of 1.2 V or less. The output voltage of

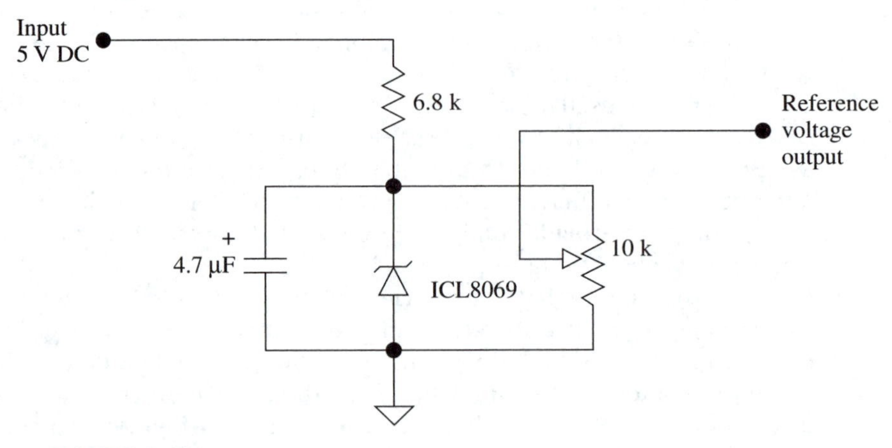

▲ **FIGURE 6–21**
ICL8069 adjustable voltage reference

the ICL8069 is 1.23 V, +20 mV or –30 mV. It is available with temperature coefficients of 10, 25, 50 and 100 ppm/°C. The maximum current that the ICL8069 can shunt is 10 mA in either the forward or reverse directions. This voltage reference is an inexpensive and accurate reference and is available with a low temperature coefficient (10 ppm/°C). However, it suffers the poor efficiency that is inherent with the shunt current reference approach.

A Three-terminal Series Voltage Reference

The MAX6120 is a good example of a three-terminal series type voltage reference that offers accuracy and temperature drift performance similar to the ICL8069 but offers much improved power efficiency. Like the ICL8069, the MAX6120 produces a 1.2-V reference voltage. It can accept input voltages that range from 2.4 to 11 V and offers a 1.2-V reference voltage ±12 mV with a temperature coefficient of 30 ppm/°C. Unlike the ICL8069 and other shunt regulators, the MAX6120 requires typically 50 µA to operate, independent of the input voltage. This offers maximum efficiency and promotes its use in battery-operated equipment and other power-sensitive applications. The typical application of the MAX6120 requires only one filter capacitor as is shown in Figure 6–22. Its maximum output power is 320 mW, which translates to a maximum output current of 320 mW/1.2 V or .266 mA.

Higher Voltage References

The MAX87X series is a group of voltage references that can supply references for 2.5 V (MAX873), 5.0 V (MAX875) or 10.0 V (MAX876). These references are based upon the three-terminal series references discussed previously, except that they have been expanded to include more elaborate functions and therefore require more pin connections. They utilize bandgap reference diodes as the primary reference generator. This voltage is amplified on chip to the desired higher reference value. The connections include the typical input voltage, ground, and output voltage connections that are combined with a temperature output signal, an output

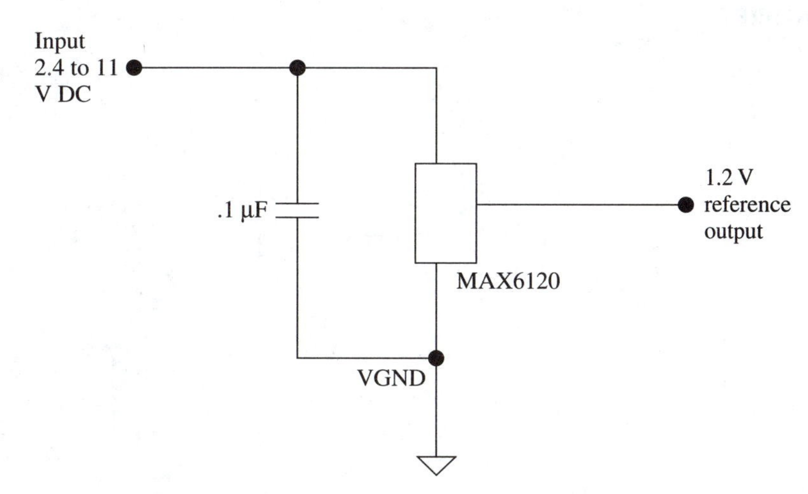

FIGURE 6–22 MAX6120 series voltage reference

adjust pin, and two test connections. The two test connections are for use only by the IC manufacturer and should be left unconnected. The temperature output signal labeled "TEMP" represents the temperature of the die and can be used to compensate the output voltage over temperature. The TEMP output changes at about +2 mV/°C and is about 608 mV at 25 °C. Since the MAX87X series has a negative temperature coefficient that is fairly linear from 25 to 60 °C, the TEMP output can be used with an op amp circuit to compensate the voltage reference for temperature variations. The output accuracy without trimming is roughly ±.5% and the output can be trimmed with a 100-kΩ potentiometer over 4% of the output voltage range. The temperature coefficient for the MAX87X series is only 7ppm/°C and the maximum power supply current is 280 μA. The maximum output current is 10 mA. An application circuit for the MAX87X family is shown in Figure 6–23.

Kelvin-sensed Outputs

At the high end of voltage reference performance is a class of devices called *Kelvin-sensed voltage references.* These devices use a commonly used technique called *Kelvin-sensing* that minimizes the effect of lead and other circuit resistance (connectors, etc.). Let's examine the three-terminal reference that supplies a reference voltage to the load shown in Figure 6–24. When we consider the lead resistance between the reference circuit location and the load, as symbolized by R_1 and R_2, we see that a voltage drop occurs across each resistor, creating a difference between the sensed and controlled output of the three terminal reference and the load.

A Kelvin-sensed output utilizes separate drive and sense lines to eliminate the error that results from the lead resistances shown as R_1 and R_2. In Figure 6–25 a Kelvin-sensed reference is connected to load with lead resistances R_1 and R_2 as in Figure 6–24. In this case, however, note the separate drive and sense lines available on the reference integrated circuit. What makes this circuit effective is the fact that the input resistance of the sense circuit is very high, so that the current that flows back to the reference through the sense lines is very low. Consequently, only a very

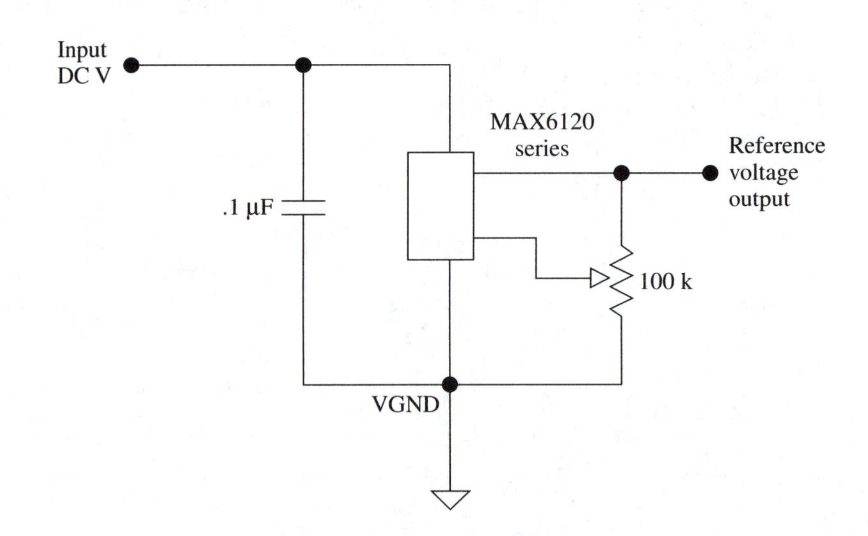

▶ FIGURE 6–23
MAX87X voltage reference

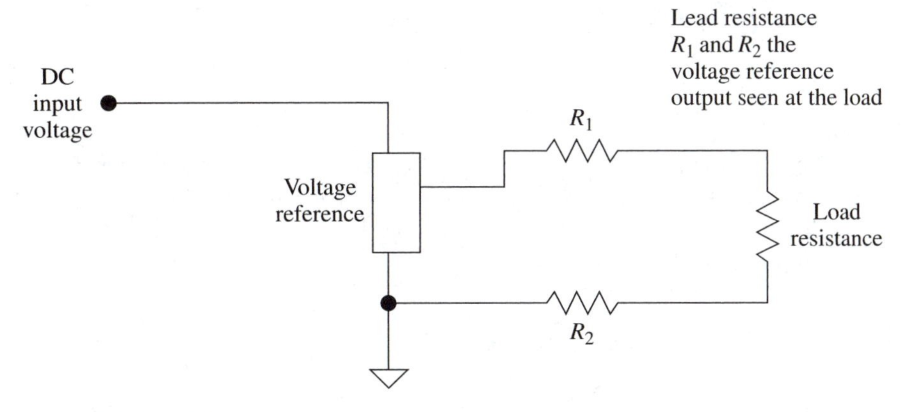

▲ **FIGURE 6–24**
Three-terminal regulator with lead resistance

small voltage is generated across the lead resistance of the sense lines (shown as R_3 and R_4). This means that the reference circuit is receiving a more accurate indication of the reference voltage present at the load and can thereby regulate this value accordingly.

The requirement for a voltage reference is usually the result of a critical circuit function on which the overall accuracy of the circuit is determined. Consequently, proper care must be taken in their selection and application. In the early days of electronics, circuit designers utilized bulky and expensive batteries as voltage reference that barely met their accuracy and temperature drift requirements. Today's technology provides accurate references with low temperature drift coefficients that include the following additional features:

1. Low quiescent current requirements
2. Low dropout voltages
3. Can drive capacitive loads

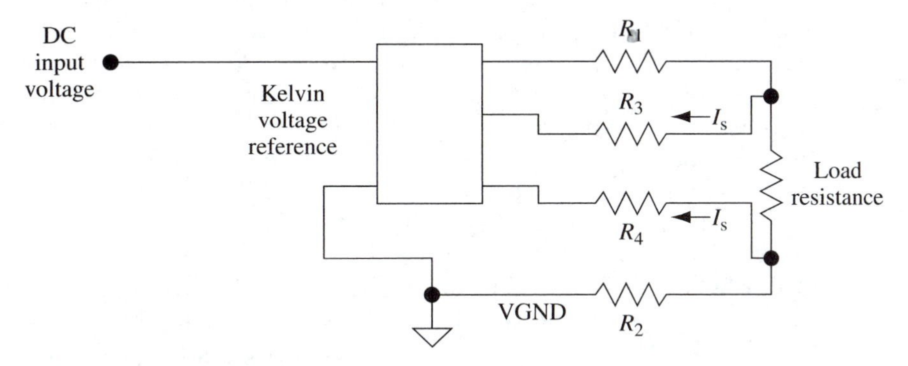

▲ **FIGURE 6–25**
Kelvin-sensing output voltage reference

4. Few external components required
5. Low cost
6. Small size

There are many varieties and choices to be considered in order to match the voltage reference to the cost and space requirements of the design. As with all other design decisions, it is best to make an error in the favor of higher quality and reliability, instead of the alternative.

6–4 ▶ Switching DC Power Supplies

While the linear DC power supplies discussed in Section 6–2 offer superior low-noise performance, they suffer from poor power efficiency. On the other hand, switching power supplies greatly improve power efficiency at the price of increased noise levels. Most digital circuits can accommodate higher noise levels and require higher efficiency as more and more circuitry is squeezed into smaller spaces. Consequently, switching power supplies are often used to generate +5 V DC for digital circuit applications and some less demanding analog circuits as well.

Recall that linear power supplies use the collector-to-emitter resistance of a transistor to attenuate an input voltage down to a regulated output voltage. The voltage drop across the transistor occurs continuously and represents a significant waste of power that results in the relative inefficiency of the linear power supply. Switching power supplies use a different approach. They step down the input voltage by switching the transistor on and off very quickly. The duty cycle of the switching is used to regulate the output voltage and the resulting waveform is smoothed out with inductive and capacitive filters. MOS transistors, which offer very low "on" drain-to-source voltages, are used in these types of supplies. The low "on" voltage across the transistor, combined with the fact that the transistor is switched on for only a portion of the switching cycle, result in significant improvements in power efficiency. Switching power supplies yield efficiencies in the area of 95%. When the switching frequency is made high enough, inductors can then be used for filtering in addition to capacitors, which helps to provide the necessary noise filtering.

Figure 6–26 shows the functional block diagram for a switching power supply. The switching transistor is switched on and off by the control/drive circuit in accordance with the measured value of the output voltage. The switching supply's output voltage is an input to the control/drive circuit, which compares this value with the desired output voltage. If the output voltage is greater than the desired voltage, the control drive circuit will take action to reduce the relative time that the switching transistor is on during the next cycle. For output voltages less than the desired voltage, the control/drive circuit will increase the on time of the switching transistor. The waveform that results at the input to the L-C filter is a pulsed waveform, whose maximum value equals the input voltage value while the minimum value is approximately zero. The L-C filter smoothes out this pulsed waveform and the output voltage is measured by the control/drive circuit, which controls the relative amount of on-time for the switching transistor. The catch diode shown in Figure 6–26 provides a return path for the current when the switching transistor

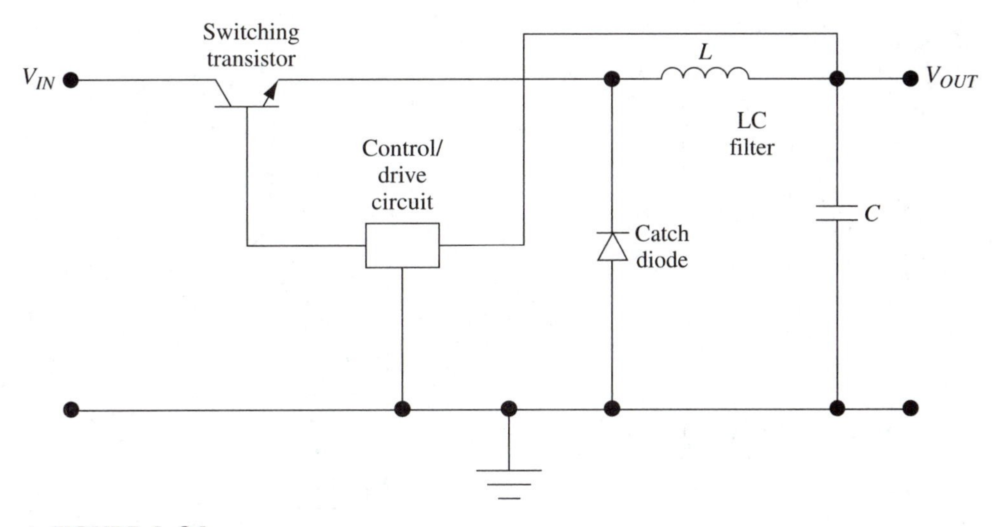

▲ **FIGURE 6–26**
Switching power supply block program

is off. There are two methods of controlling the switching transistor's relative on-time: using fixed on-time while the cycle frequency is varied or a constant frequency while the on-time is varied. The latter is the approach used most often and is commonly called pulse width modulation (PWM). PWM switching power supplies are used more often today than any other type of power supply.

Switching voltage regulators can be designed to operate in one of two modes: continuous or discontinuous. The current that flows through the inductor causes the operational differences between these modes. In the continuous mode, the current through the inductor never stops flowing during the operating cycle. When the current through the inductor drops to zero for a period of time during the normal operating cycle, the switching regulator exhibits discontinuous operation. Most often the continuous mode of operation is preferred because it offers better regulation and lower ripple. It does, however, require a larger inductor value than would otherwise be necessary for discontinuous operation. Many switching regulators can be operated in either the continuous or discontinuous mode. In this book we will only discuss examples of continuous-operation switching regulators.

The detailed design of a switching power supply can be easily accomplished with discrete components, combined with integrated circuit operational amplifiers for the control/drive circuit. However, there are a wide variety of specific switching regulator integrated circuits that have been developed for this purpose. Many of these ICs require only an input voltage, input filter capacitor and an output L-C filter to operate. Each family of switching regulator integrated circuits will usually offer a range of fixed output voltages and variable voltages, all available at a particular current level. There are also variations between these IC families such as efficiency, noise levels, and features such as thermal shutdown.

The LM2574 series is a good example of a family of switching regulators which offer fixed output voltages of 3.3 V, 5 V, 12 V, and 15 V as well as a variable output version. The maximum output current for the LM2574 is .5 amps. An input

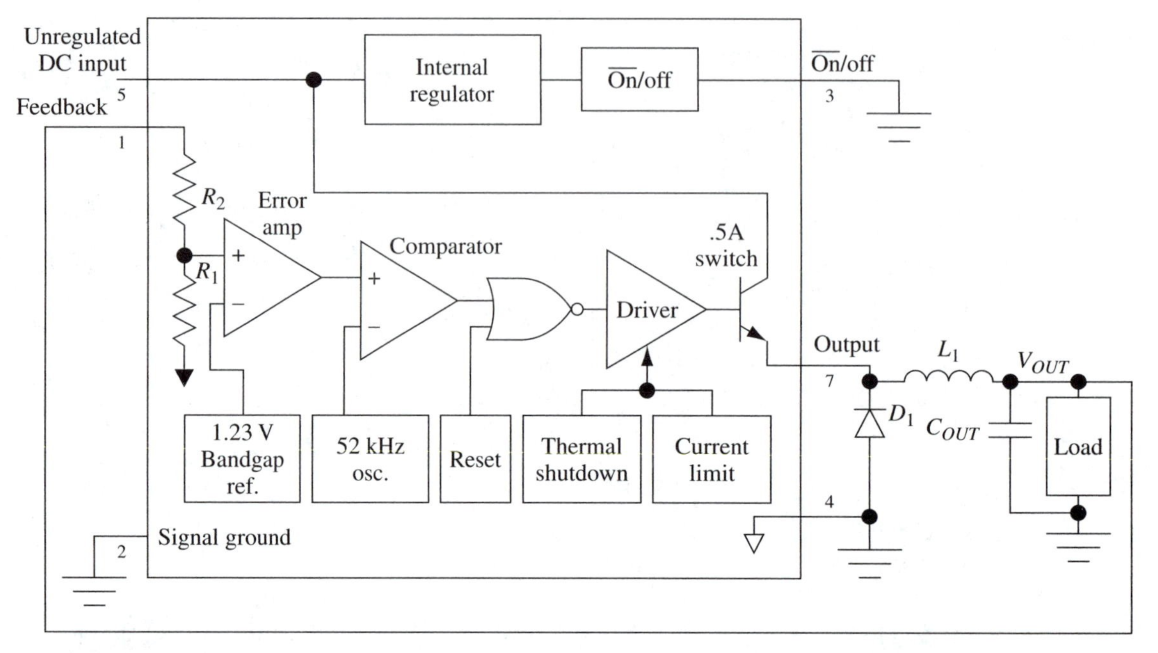

▲ **FIGURE 6–27**
LM2574 switching regulator block program

filter capacitor, catch diode, and output L-C filter are all that is needed to operate the LM 2574 family, switching regulator. The block diagram for the LM2574 is shown in Figure 6–27. This switching regulator operates at a fixed 52 kHz and uses PWM to control the output voltage.

Example 6–4

In this example we will design a 5-V switching power supply with .3 amp current output using the LM2574 family. The input voltage to the regulator is a DC voltage that can range from 8.5 to 10 V and the desired ripple value is 1% of the 5-V output or 50 mV.

Solution

The solution of this design problem requires the selection of the switching regulator IC, the input filter capacitor (C_{IN}), the catch diode (D) and the inductor/capacitor (L/C_{OUT}) that comprise the output filter circuit. See the circuit shown in Figure 6–28.

The fixed voltage output version the LM2574-5.0 is selected as the switching regulator IC. The input voltage range of 8.5 to 10 V is well within the 7- to 40-V input range requirements for the LM2574-5.0 switching regulator. The data sheets for these types of devices always include the design rules for their proper application. The steps outlined for the LM2574 are listed as follows:

1. Select the inductance value for the inductor using the chart shown in Figure 6–29. All of the values on the chart shown in Figure 6–29 assume continu-

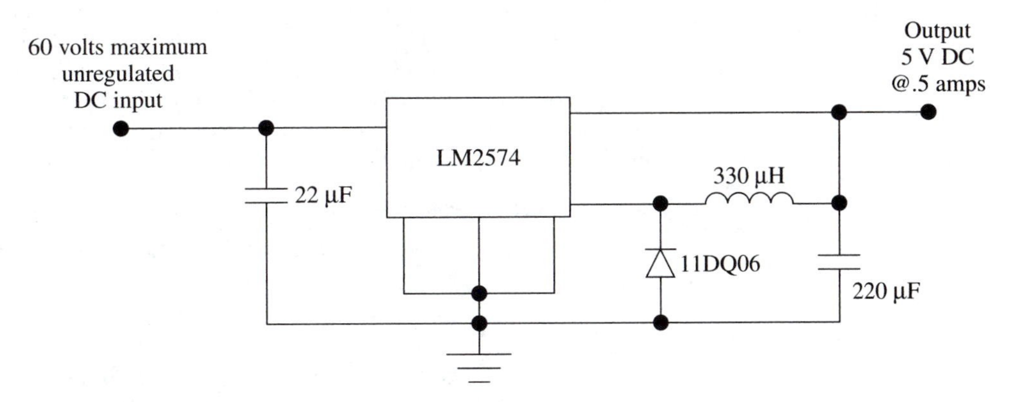

▲ FIGURE 6–28
LM2574-5.0 regulator circuit

ous operation of the switching regulator. The value of 220 µH is selected from the chart using the input voltage range of 8.5 to 10 V and the current requirement of .3 amps. The inductor selected shall be rated for operation at 52 kHz and should have a current rating of at least 1.5 times the load current (.3 amps), which is .45 amps.

2. The value of the output capacitor, C_{OUT} can be determined by the following formula:

   C_{OUT}: 13,300 ($V_{IN\text{-}MAX}/V_{OUT} \times L$) where L is given in µH and the resulting capacitor value results in µF.

   For this example $V_{IN\text{-}MAX}$ = 10 V, V_{OUT} = 5 V and L = 220 µH. C_{OUT} = 121 µF. The application hints for this IC indicate that capacitor values in the range of 100 µF to 330 µF will yield ripple values in the range of

▶ FIGURE 6–29
Inductor selection graph

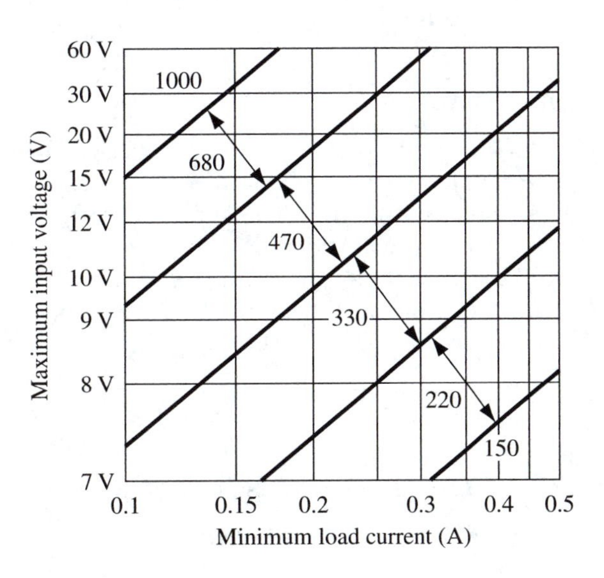

50 mV to 150 mV while larger values will reduce ripple to 20 mV to 50 mV. For this reason C_{OUT} is selected to be 470 µF. The voltage rating for C_{OUT} should be at least 1.5 times the output voltage or 7.5 V. An aluminum electrolytic capacitor with a value of 470 µF and rated at 26 V DC is selected for C_{OUT}.

3. The catch diode can be selected by using the following criteria:
   a. The current rating must be greater than 1.5 times the maximum load current.

      $1.5 \times .3$ amps = .45 amps

   b. The reverse voltage rating must be greater than 1.25 times the maximum input voltage.

      1.25×5 V = 6.25 V

   c. The diode should be a Schottky or fast-recovery type.

      A 1N5817 Shottky diode is selected with a reverse voltage of 20 V and a current rating of 1 amp.

4. To maintain stability, C_{IN} the input capacitor value must be at least 22 µF and have a voltage rating that will accommodate the maximum input voltage. A 22 µF electrolytic capacitor with a voltage rating of 16 V is selected for C_{IN}.

Switching regulators rapidly switch currents in many parts of the circuit that can cause problems when coupled with wiring inductance and ground loops. Consequently, these circuits are sensitive to the layout of the circuit. When breadboarding or laying out a printed circuit board artwork for a switching regulator, keep the leads to the input filter capacitor, catch diode, and output capacitor as short as possible. Use a single ground point or ground plane for all the ground connections for the circuit.

6–5 ▶ Inverters

An inverter is a device that converts DC power over to AC power, usually at a higher voltage level. In other words, an inverter performs the opposite function performed by a rectifier. This type of inverter is not to be confused with digital logic inverters. The most common inverter application meets the requirement to convert +12 V DC to 120 V sinusoidal AC, at 60 Hz. This is desirable for powering most domestic appliances from automotive or other low voltage DC power supplies. In order to accomplish this feat, the inverter circuit must perform the following tasks:

1. Conversion from DC to AC
2. Development of a fixed frequency
3. Simulation of a sinusoidal waveform
4. Step-up of the voltage

Conversion from DC to AC

The conversion from DC to AC is commonly performed by using the DC voltage to create a square-wave generator by driving two transistors in a typical push-pull arrangement. This was reviewed previously with the push-pull/flyback regulator in the DC-to-DC converter discussion in Section 6–3. See Figure 6–18. The circuit in Figure 6–18 requires some type of starting circuit to initiate oscillation. A self-starting inverter circuit is similar to the one shown in Figure 6–30.

When the DC voltage is connected, current flows through R to the base of both transistors Q_A and Q_B. Because there is always a slight difference between the characteristics of the two transistors, one will always turn on one before the other. If we assume that Q_A turns on first, then current flows through transistor Q_A to ground. While the current flow is changing, a voltage is induced in all secondary windings (N_C, N_{QA}, and N_{QB}) where the dotted end of each winding has a negative polarity. The positive voltage at N_{QA} keeps Q_A on while the negative N_{QB} voltage keeps Q_B off. The current level reaches its peak when the core saturates. At this point, because the current is not changing, the secondary voltages are no longer generated and transistor Q_A will shut off. During the previous time period when Q_A was on, the capacitor, C, maintains a negative voltage across N_{QB}, momentarily holding Q_B off. When Q_A switches off, the current flowing in N_A causes N_A to reverse polarity to maintain current flow. This turns Q_B on, which generates the other half of the AC cycle. The circuit exemplifies the

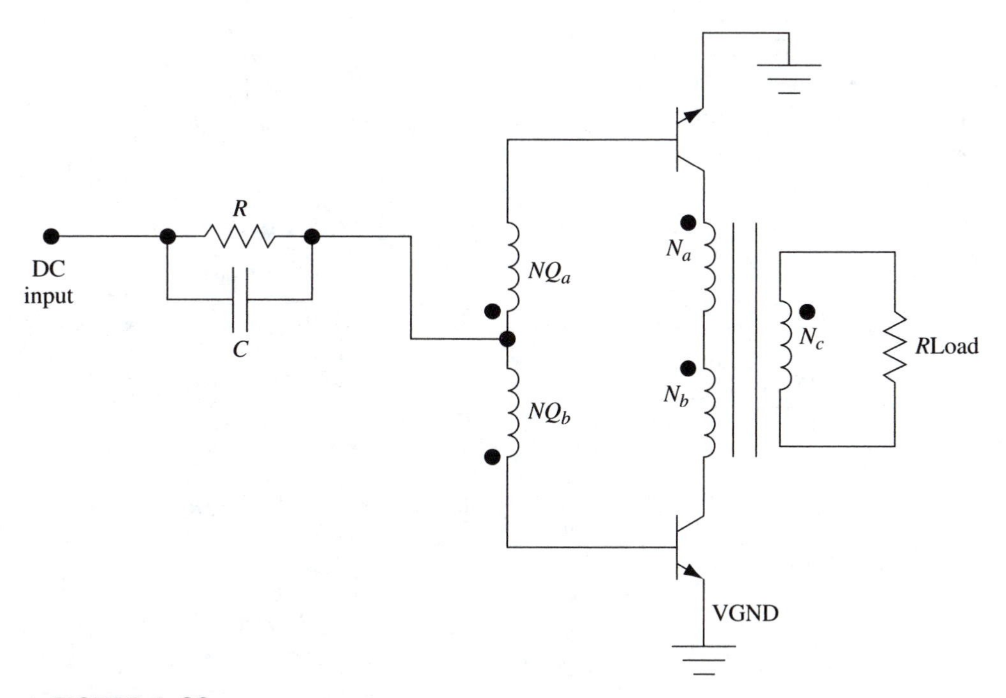

▲ **FIGURE 6–30**
Self-starting inverter circuit

concept of a self-starting push-pull inverter. The resistance, R, and the inductance of the transformer windings determines the inverter frequency.

The amplitude of the AC waveform seen at the load secondary N_C is determined by the turns ratio $N_C/(N_A$ or $N_B)$. The sinusoidal appearance of the waveform is dependent on the load impedance and can be improved by a circuit that matches the inverter output impedance to the load impedance.

There are many ICs available that combine a number of the functions required to build inverters. One of the simplest of these ICs is called simply *a dual output driver*, the CS3706. This IC possesses two output transistor drivers that can be configured to operate in the push-pull mode of operation. In other words, when one transistor is on, the other is off, and vice versa. The input to the IC is a TTL level digital signal that can be a square wave signal at the desired inverter frequency. Figure 6–31 shows the CS3706 connected as an inverter. The input to the circuit can be a square wave at the required inverter frequency. The amplitude of the AC voltage seen at the load secondary is determined by the turns ratio of the transformer. The sinusoidal quality of the AC waveform is determined by the switching frequency and its relation to the inductance of the transformer and the load impedance. There is a significant amount of detail regarding the application of inverters to provide precise sinusoidal waveforms to a wide range of load impedances.

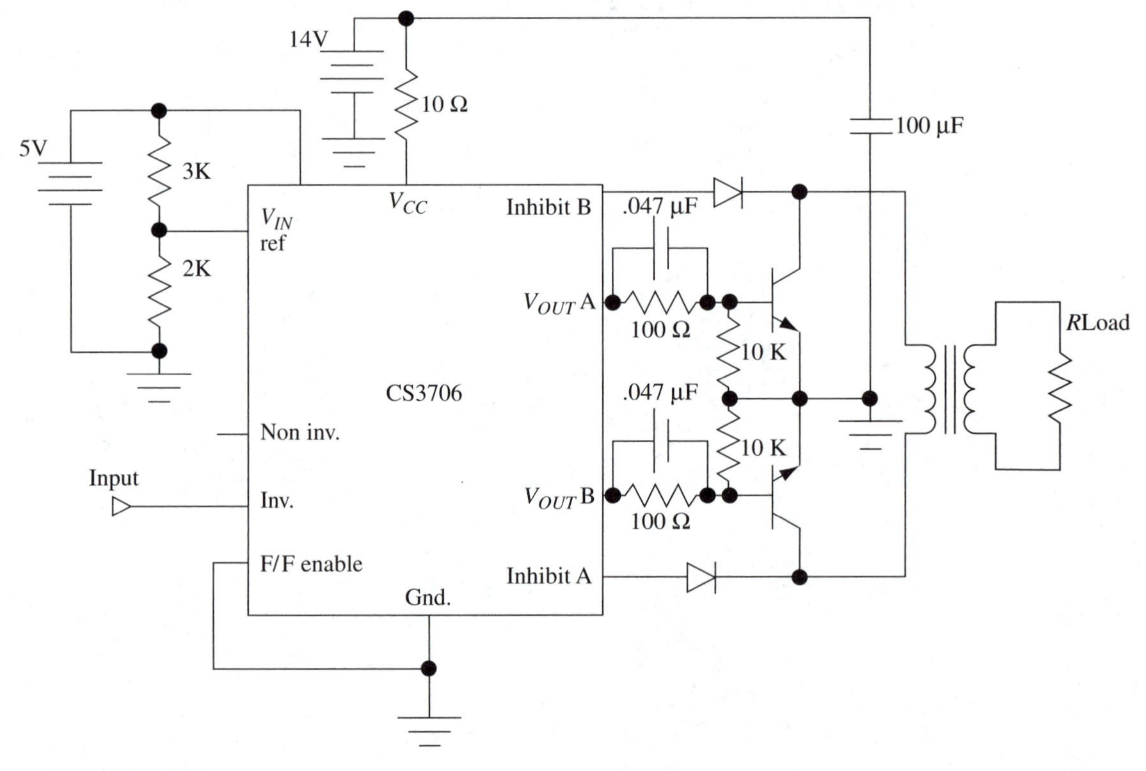

▲ **FIGURE 6–31**
CS3706 Inverter circuit

Summary

In this chapter we have reviewed a wide variety of power supply circuits. It should be apparent that power supply and converter circuits require strong circuit analysis capabilities. It should also be apparent that the development of many power supply ICs have simplified the task of the electronic designer significantly. This chapter is by no means a complete discussion of this topic but should be viewed as a starting point for the development of strong power supply and analysis capabilities.

Many of the circuits discussed have been the more simple circuits and applications to make the analysis readily understandable. However, the challenge will continue to be the improvement of power supply and converter circuits in the area of quality and efficiency. The quality of a power supply/converter circuit is determined by how accurately it develops the desired output, without noise, to a variety of load conditions. Its efficiency is determined by the amount of power it uses in the process. Rest assured that the products and applications of the future will require further improvements in both the quality and efficiency of power supply circuits. Based upon the history of electronic development, it appears that we will continue to find new ways of raising the bar of performance yet further.

References

Gottlieb, I. M. 1984. *Power Supplies, Switching Regulators, Inverters and Converters.* Blue Ridge Summit, PA: TAB Books.

Hnatek, E. R. 1981. *Design of Solid-state Power Supplies.* New York: Van Nostrand Rienhold.

Exercises

6–1 Explain the meaning of the term *dropout voltage.*

6–2 Explain the difference between the terms *percent line regulation* and *percent load regulation.*

6–3 A power supply outputs 5.2 V with a load current of 200 mA. When the load current changes to 220 mA, the output voltage falls to 5.15 V. Calculate the percentage load regulation.

6–4 A power supply with an input voltage of 20 V DC outputs 15 V DC. When the input voltage falls to 19.5 V, the output voltage dips to 14.9 V. Calculate the percent line regulation for this power supply.

6–5 A 620-Ω load is connected to a power supply that generates 6 V DC as an output. The power supply input is 115 V AC RMS, which draws 22 mA. Calculate the efficiency of this power supply.

6–6 Describe the function of a series regulator and an shunt regulator. What load conditions are the most efficient for each of these regulators?

6–7 Design a power supply using the circuit shown in Figure 6–5. Determine all of the component values and ratings for the supply to generate 5 V ±1% at 50 mA and with 10% ripple. The input voltage is 117 V RMS at 60 Hz and can vary in amplitude ±10% and 50–60 Hz.

6–8 Develop a ±12-V DC supply adjustable over the range of 11.25 to 12.75 V with an output load current of 75 mA. Ripple output should be less than .5 mV and the input power is 115 V AC, 50–60 Hz. Use the LM317 and LM337 regulators discussed in Example 6–3.

6–9 Why are inductors impractical to use as output ripple filters in power supplies that have an AC input frequency of around 60 Hz?

6–10 What are the primary functions of DC to DC converters? What is the primary difference between the switching and flyback type of DC-to-DC converters?

6–11 List the most important performance requirements for a voltage reference device.

6–12 Compare the performance of switching and linear supplies in all performance categories.

6–13 Compare the operation of DC to DC flyback converters to the operation of an inverter circuit.

6–14 What is the reason for using Kelvin sensing voltage references over a regular bandgap reference?

6–15 Develop an experimental procedure for determining the efficiency of any power supply. Detail and list each step.

7 Amplifier Design

Introduction

Even in today's digitally oriented world, there is still a need for amplifiers to increase the voltage and or current level of all types of signals. In spite of the fact that most signals to be stored or transmitted are now digitized, low-level signals must be amplified before being converted to digital in order to achieve reasonable resolution. Also, when digital signals are being converted to analog, their power levels must be increased further for many applications. Then there are the traditional applications of analog signals that are amplified without being digitized, transmitted or stored in the process.

Like every other aspect of electronic technology, amplifier design has changed significantly over the last 30 years. Sophisticated operational, power, and instrumentation amplifiers have been developed and implemented in integrated circuits. High-performance amplifiers are available in small packages at low prices. Like most things today, amplifier technology is available without having to create it. Consequently, fewer and fewer entry-level engineers understand the art of amplifier design and operation. This chapter will focus on developing an understanding of the functional requirements of amplifiers and the application of the current technology available. The particular topics to be covered are as follows:

- Amplifier performance
- DC amplifiers
- AC amplifiers
- Audio amplifiers
- Video amplifiers
- RF amplifiers

7–1 ▶ Amplifier Performance

The general definition of an amplifier is any device that increases the value of some input parameter to a higher level. In the electronic world, amplifiers are designed to increase the level of voltage and/or current of an input signal. An ideal voltage amplifier will have infinite input impedance and zero output impedance. It will accept an input signal and amplify it to the desired level uniformly over the entire range of amplitudes and frequencies possible for the input signal. Consequently, the ideal amplifier should completely reject any frequencies that are outside the possible frequency range of the input signal. An amplifier should have a consistent response time that provides for both an accurate reproduction of the input signal and a minimal delay in its reproduction.

The performance of a real amplifier, as compared to the ideal amplifier described previously, is indicated by the parameters shown in the following specifications, which are summarized in Figure 7–1:

Input Signal: Identifies the range of input signal amplitude that the amplifier can process

Input Impedance: The net input impedance seen by a source connected to the input

Output Signal: Indicates the range of output signal amplitude that the amplifier can supply

Output Impedance: The net impedance seen by a load connected to the output of the amplifier.

Gain: The range in gain that the amplifier is capable of, usually expressed in terms of dB

Bandwidth: The range in frequency that the amplifier can maintain a gain within 3 dB of a reference gain value

Response Time/Slew Rate: This parameter indicates how quickly the amplifier output can change. It is a measure of how well the amplifier can duplicate the input signal from the time perspective.

Distortion: The degree of unwanted, inaccurate signals present in the amplified signal. This is usually due to nonlinearities that exist in areas of amplifier operation. The most common measurement used for amplifiers is called *Total Harmonic Distortion (THD)*, or the distortion factor. This is the classification of the second and higher order harmonic distortion levels that are present in the output.

Noise Rejection: The ability of an amplifier to reject or attenuate input signals outside the range of the specified input signal.

Noise Level: The total noise level output from an amplifier that is either passed through or generated by the amplifier. This can be measured by applying a zero input signal to the amplifier input while replacing the input signal with its source impedance.

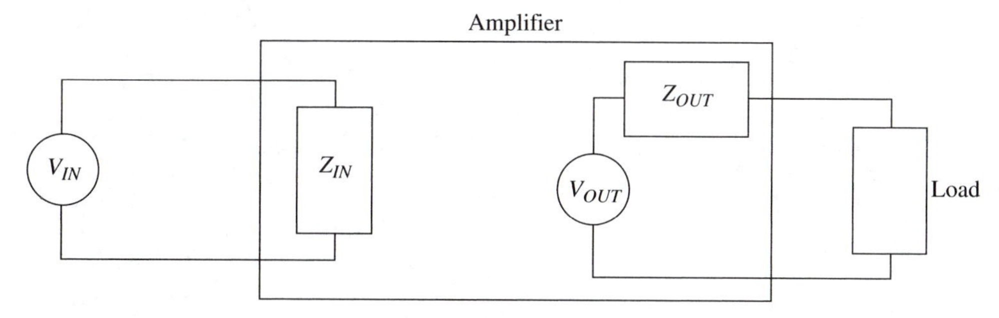

Amplifier specification parameters:

Input signal range	Response time/slew rate
Z_{IN}: Input impedance	Distortion
Output signal range	Noise rejection
Z_{OUT}: Output impedance	Noise level
Gain = V_{OUT} / V_{IN}	CMRR: Common Mode Rejection Ratio
BW: Bandwidth	Efficiency

▲ **FIGURE 7–1**
Amplifier specification parameters

Common Mode Rejection Ratio (CMRR): A parameter applicable to differential amplifiers that measures the degree to which signals common to both differential inputs are attenuated by the amplifier, usually indicated in dB

Efficiency: The output power, divided by the total input power; measures the overall power efficiency of an amplifier

Amplifiers are categorized in a variety of different ways: overall function, frequency range, signal level, and input configuration. Each of these categories are defined and discussed next, as well as a special category of amplifiers called *operational amplifiers.* All amplifiers utilize the concept of negative feedback to achieve gain and bandwidth performance needed to meet the intended application. Gain and bandwidth are opposing performance factors as higher gains result in lower bandwidths, and vice versa.

Amplifier Function

An amplifier can have one of the four following primary functions:

1. Voltage Amplification: a voltage controlled voltage source (VCVS)
2. Voltage to Current Converter or Transconductance Amplifier: a voltage controlled current source (VCIS)
3. Transresistance Amplifier: a current-controlled voltage source (ICVS)
4. Current Amplifier: a current-controlled current course (ICIS)

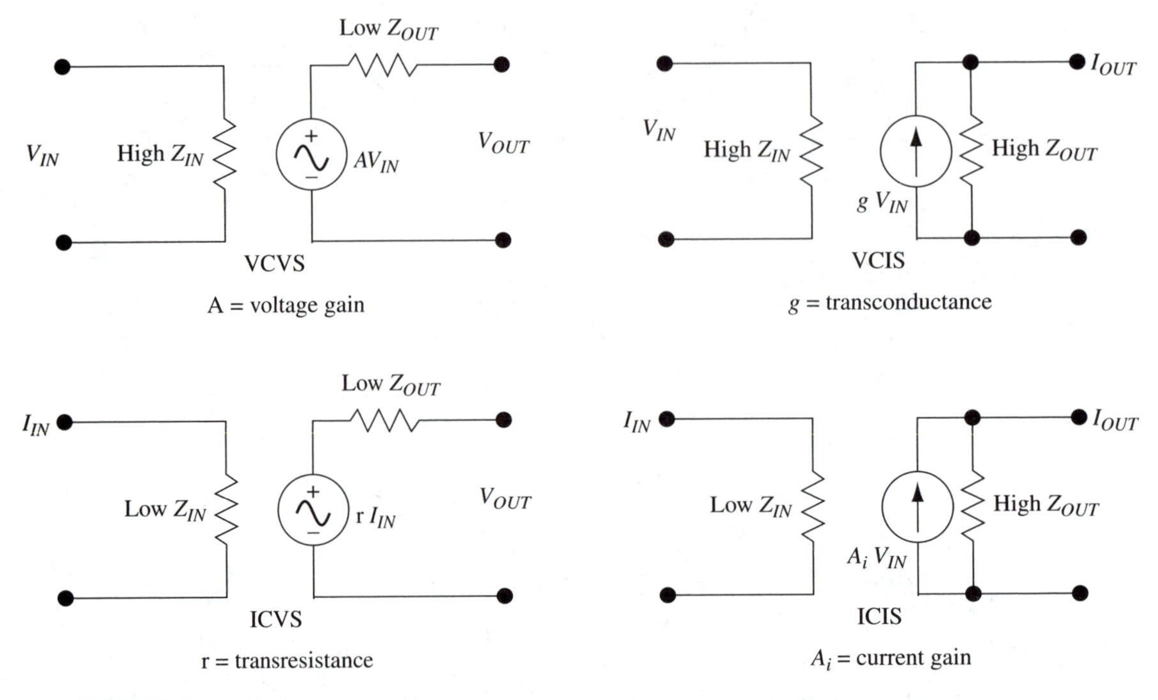

▲ **FIGURE 7–2**
Functional amplifier block diagrams

Figure 7–2 shows each of these four amplifier functions, along with the desired input and output impedances for each.

Amplifier Frequency Range

The major categories of frequency response are DC, telephony, audio, video, and RF amplifiers. DC amplifiers are used to amplify slowly changing signals from a variety of low-level signal transducers so that they can be indicated, recorded, and controlled. The most common applications are seen in industrial control environments where temperatures, pressures, and flow rates are continually monitored and controlled. Telephony applications include the transmission of voice signals only, which generally cover a frequency range of 100 Hz to 5 kHz. Audio amplifiers are used specifically to amplify signals that cover the human hearing range, roughly 20 Hz to 20 kHz, and are consequently subject to the high-fidelity requirements of our ears. Video amplifiers, also called *wide bandwidth amplifiers* because of their large bandwidth capabilities (20 Hz to 6 MHz), are used to transmit television and other video images. RF amplifiers that are used for most radio communications are designed to amplify signals with frequencies of 30 MHz to 4 GHz.

Amplifier Signal Level

The range of an amplifier's input signal level requires that a different set of design criteria be applied. A small signal amplifier, often called a *pre-amplifier*, must amplify a small signal while rejecting small signals outside the input frequency range while producing minimal distortion. Because the output voltage and currents are at a low level, the efficiency of the pre-amplifier is usually not a key concern. The output power of a pre-amplifier is generally less than 1 watt.

Power amps can have power outputs that range from 500 mW to hundreds of watts, and as a result they have design criteria that emphasizes their efficiency. Power amps are classified into categories that define their design. These categories are class A, B, AB, and C. You probably recall from introductory electronics courses that class A amplifiers establish a quiescent bias point in the middle of the transistor load line. This bias position precludes clipping of the input signal, but also promotes the inefficient operation of the amplifier. This is because when a zero signal input occurs with the class A amplifier, it still consumes power, yet no work is being performed. Class B amplifiers rectify this situation by locating the quiescent bias point at the transistor cutoff point. Operation at the cutoff point causes amplification of only half of the waveform, which is why class B amplifiers use two transistors. Class B amplifiers offer much greater efficiency when compared to class A amps because when the input signal is zero or small, the power consumed is small. Of course class B amplifiers suffer from something called *crossover distortion.* This phenomenon is caused because the cutoff point for two transistors is seldom exactly the same, so there is some distortion when the signal goes from positive to negative and vice versa.

Class AB amplifiers offer a compromise by moving the quiescent bias point slightly off the cutoff point, reducing efficiency slightly, but minimizing crossover distortion.

Class C amplifiers operate below cutoff and offer significant efficiency improvements, but they distort the signal greatly. Class C amplifiers are used to amplify pulsed waveforms where the frequency of the waveform and the presence of a pulse are the primary information carriers.

Amplifier Input Configuration

Amplifiers can be designed to accommodate single-ended or differential input configurations. Single-ended inputs are measured from a circuit common point, so the amplitude of the input signal is the input value referenced to this circuit common. On the other hand, differential input amplifiers have two input connections, neither of which is connected to a circuit common. The input signal processed by the differential amplifier is the difference between these two input points (see Figure 7–3). Differential amplifiers are useful when noise exists that is common to both signals. In this case the differential amplifier can be used to attenuate and practically reject what is called the *common mode noise signal.* The degree to which the amplifier can reject the common mode signal is called the *common mode rejection ratio (CMRR).*

(a)

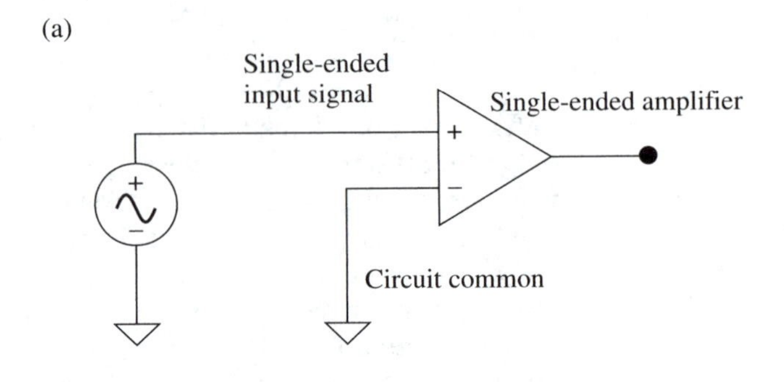

(b)

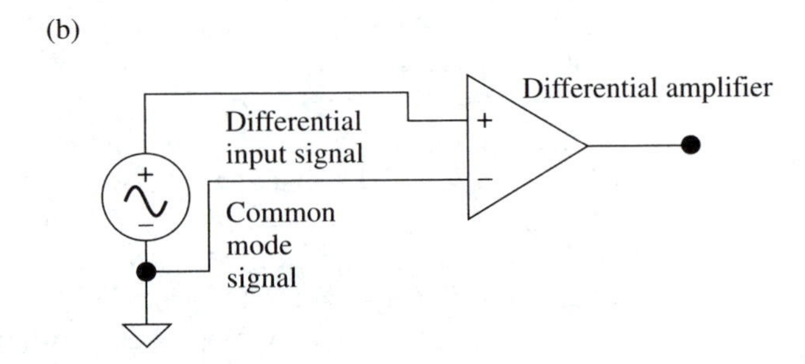

▶ **FIGURE 7–3**
Single-ended and differential input configurations

Operational Amplifiers

This is a special class of amplifiers that are well covered in most electronics curricula because they have become a primary building block in linear circuit applications. They were named *operational amplifiers* long ago when vacuum tube amplifiers, configured in this way, were used to perform mathematical operations in analog computers. Op amps feature independent positive and negative inputs and a very high open loop gain and provide the ability to develop almost any gain and function by connecting discrete components. Op amps can be configured as single-ended or differential and can perform any of the four amplifier functions (VCVS, VCIS, ICVS and ICIS). Op amps are often configured to function as DC amplifiers with closed loop negative feedback but are also commonly used as AC amplifiers. They can also be used open loop or with positive feedback to create a variety of oscillator and switching circuits.

7–2 ▶ DC Amplifiers

DC amplifiers are used to amplify static or slow-changing signals for the purpose of indication, storage, control, transmission, or to complete some mathematical operation. Integrated circuit op amps are used almost exclusively for these applications, because they offer high performance and are inexpensive and easy to use. This section will discuss the general application of op amps as DC amplifiers and will summarize the key parameters and concerns relative to this application.

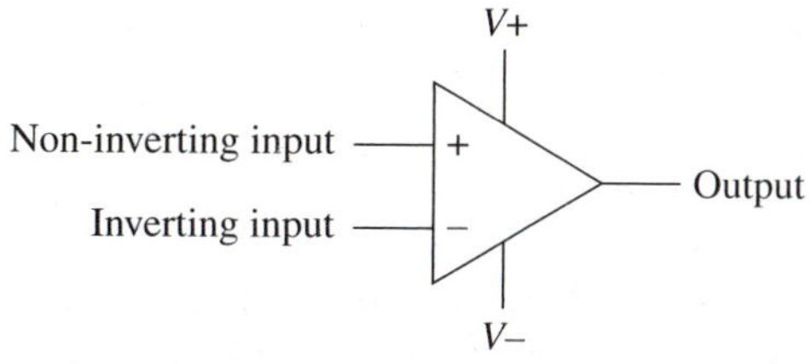

V+

Non-inverting input

Inverting input

Output

V–

FIGURE 7–4
Op amp schematic

The basic op amp, shown schematically in Figure 7–4, is a five-terminal device: the negative/inverting input, the positive/non-inverting input, an output and the power connections $V+$ and $V-$. The ideal op amp has an infinite gain, infinite input impedance, and zero output impedance and operates over an infinite bandwidth. Real op amps available as integrated circuits approach the ideal op amp definition in a practical sense. They offer a very large gain, high input impedance, low output impedance, and a wide operating bandwidth. The maximum output voltage possible for an op amp is within $V+$ and $V-$ that power it. Most op amps can only provide an output voltage over the range of $V+$ minus 2 V to $V-$plus 2 V. The maximum output current is specified on the data sheet for a particular op amp.

When used as an amplifier, the op amp is connected with negative feedback, realized by a resistor connected from the output terminal to the negative input. The connection of the input signal to either the inverting or non-inverting inputs determines whether the input signal will be inverted or not. Figure 7–5 shows a non-inverting op amp connected to $V+ = +12$ V and $V- = -12$ V. The circuit shown in Figure 7–5 is considered to be single-ended, because one of the input connections is shared with the common connection for the op amp circuit. The gain formula for the non-inverting, single ended amp circuit shown in Figure 7–5 is as follows:

$$V_O/V_I = \text{Gain} = 1 + R_F/R_I \tag{7–1}$$

$V_O/V_I = \text{Gain} = 1 + R_F/R_I$ $R_F = 100\ \text{k}\ \Omega$ and $R_I = 20\ \text{k}\ \Omega$ so the Gain = 5

If the input voltage ranges from 0 to 2 V, then the output voltage ranges from 0 to 10 V, which is just within the output range possible for the op amp powered by ±12 V.

FIGURE 7–5
Non-inverting single-ended amplifier

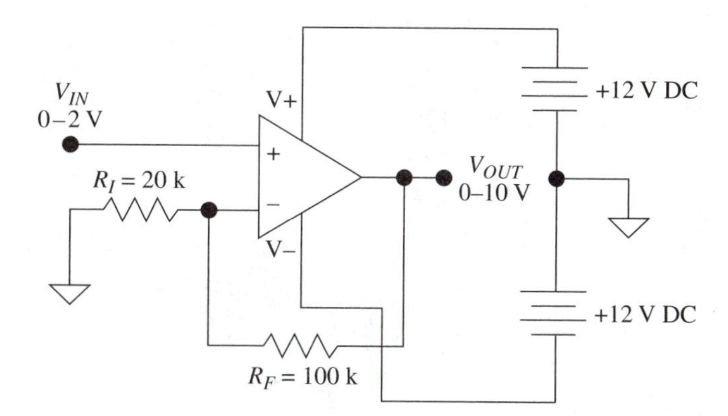

The input impedance seen by the input signal is given by the Formula 7–2:

$$Z_{IN\text{-}NI} = (1 + A_{OL}\, R_F/R_I)Z_{IN} \qquad (7\text{–}2)$$

A_{OL} is the op amp's open loop gain and Z_{IN} is the op amp's input impedance, both of which are available from the op amp data sheet. A review of this formula shows that the input impedance of the non-inverting amplifier is actually much greater than the input impedance of the op amp by itself.

The output impedance seen by the load connected to the non-inverting amplifier is given by the following formula:

$$Z_{OUT\text{-}NI} = Z_{OUT}/(1 + A_{OL}R_F/R_I) \qquad (7\text{–}3)$$

where A_{OL} is the op amp's open loop gain, Z_{OUT} is the op amp's output impedance, and both values are available from the op amp data sheet. This formula shows that the output impedance of the non-inverting amp is actually less than the output impedance of the op amp itself.

The operating bandwidth and the maximum current possible for the non-inverting amplifier are the values provided directly from the op amp data sheet.

A special case of the non-inverting amplifier commonly used is when $R_F = 0$ and $R_I = \infty$ as shown in Figure 7–6. This is called a *voltage follower* because the gain equals 1. It is often used because of the high impedance it presents to an input signal.

A single-ended inverting amplifier is shown in Figure 7–7. The primary functional differences between the inverting and the non-inverting amplifier are as follows:

1. The input impedance presented to the input signal of the inverting amp is usually much less than the input impedance of the op amp. The approximate formula for this is $Z_{IN\text{-}I} = R_I + R_F/A_{OL}$ and if $R_I > R_F/A_{OL}$ then $Z_{IN\text{-}I} = R_I$.
2. The output impedance of the inverting amp is approximately the data sheet value given for op amp's output impedance.
3. The Gain formula = $-R_F/R_I$, so the output is always the negative of the input.

▶ FIGURE 7–6
Voltage follower circuit

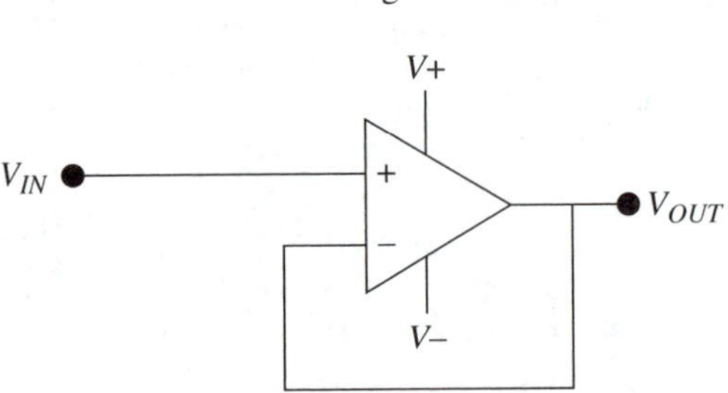

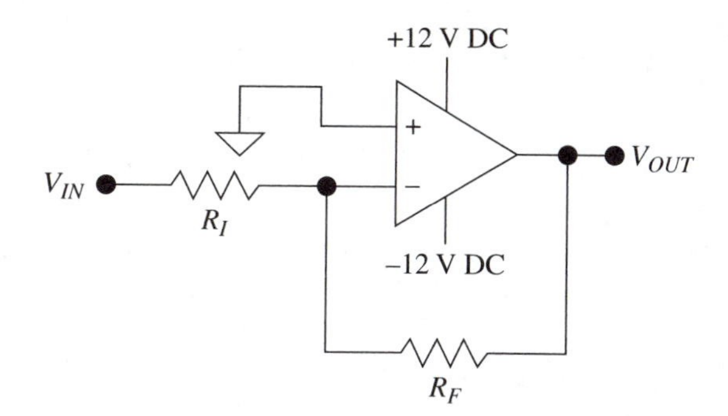

▶ FIGURE 7–7
Inverting amplifier

For R_F = 100 kΩ and R_I = 20 kΩ, the gain of the inverting amp is –4. For an input voltage range of 0 to 2 V, the output voltage is 0 to –8 V for the circuit shown in Figure 7–7.

The single-ended inverting amplifier can be converted into an inverting, summing amplifier by simply adding additional inputs through input resistors to the inverting input as shown in Figure 7–8. By varying the value of the input resistor, the gain for that particular input will vary when compared to the other inputs. The gain for each input is equal to $-R_F/R_I$.

Bias Current Values and Compensation

So far we have simplified the application of op amps slightly by ignoring bias currents, those currents that actually flow into the op amp input terminals because the input impedance is less than the ideal value of infinity. Bias currents must flow in each input terminal for the op amp's proper operation. Since the bias currents flow through the input resistors connected to a particular terminal, a voltage drop will occur across the input resistors due to the bias current that flows through them. The value of the voltage drop across the input resistors should be less than one

▶ FIGURE 7–8
Inverting summing amplifier

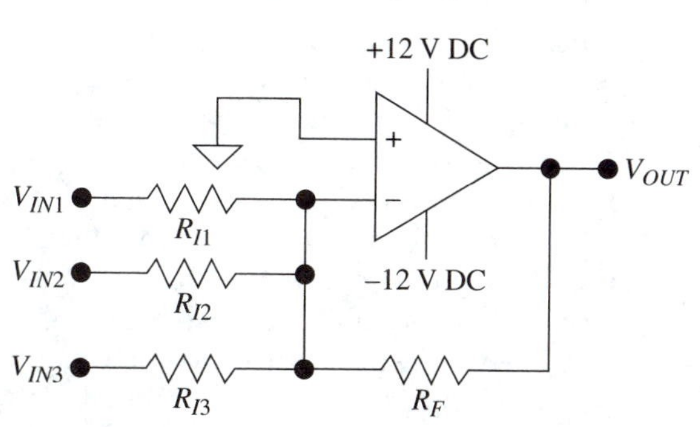

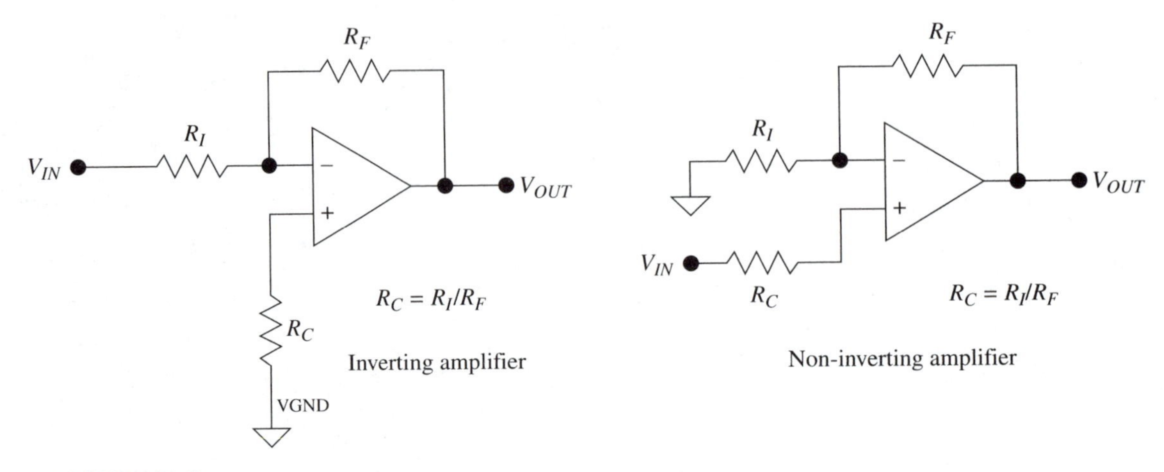

▲ **FIGURE 7–9**
Compensating resistor placement

tenth the value of the base emitter drop (usually $V_{BE} = .7$ V) across the op amp's input transistors. A particular op amp will also have a maximum value for bias current that is available from the data sheet. The maximum value for any input resistor should be calculated with the following formula:

$$R_{IN\text{-}MAX} = (V_{BE}/10)/I_{BIAS\text{-}MAX} \tag{7–4}$$

When the bias currents are not equal, there is a slight error that occurs at the output terminals. A good quality op amp design will try to minimize the difference between the bias currents flowing in the two input terminals. This can be accomplished by using a compensating resistor. Figure 7–9 shows the placement of the compensating resistor for the non-inverting, inverting, and summing amplifiers. The value for the compensation resistor is calculated by determining the parallel combination of all the resistors connected to the negative terminal. If a BIFET op amp is used, the input impedance is sufficiently higher than a BJT op amp and the error due to bias current is negligible.

Input Offset Voltage

Ideally the op amp should have an output of 0 V when the inputs to the op amp are 0. In an actual op amp, a small voltage is present at the output terminal when the inputs are 0. This value is called the *offset voltage.* The op amp circuit will amplify any offset voltage present at the output, so if the value of the offset voltage is significant when compared to the signal level, then some measure must be taken to minimize the value of the offset voltage. Many op amps provide terminals where a potentiometer can be connected to adjust the offset voltage to 0. Figure 7–10 shows a typical offset voltage adjustment circuit.

The value of the offset voltage also changes with the ambient temperature to which the op amp is exposed. So once the offset voltage is adjusted to 0 with the

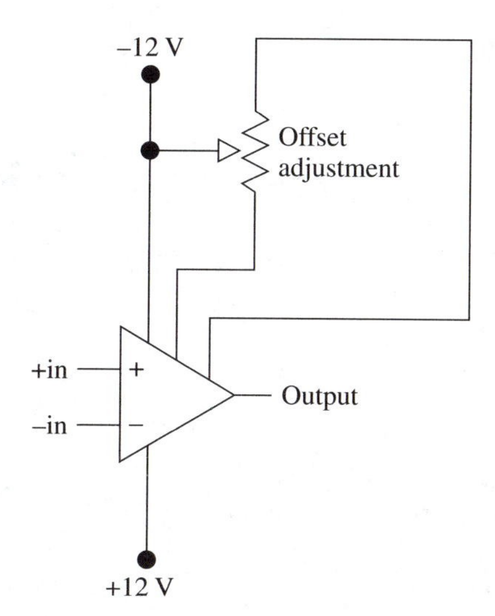

FIGURE 7–10
Offset adjustment circuit

potentiometer connection described previously, it will still vary with temperature. The degree to which an op amp's offset voltage varies with temperature is listed in its specifications. Higher-quality, precision op amps have circuitry that minimizes the amplitude of the offset voltage as well as the amount of temperature drift. Care should be taken to review this area of the specs when selecting op amps for use in a particular application, so the impact of the offset voltage and its drift do not affect the op amp's output appreciably.

Differential Amplifier

The differential amplifier is a special amp that amplifies the difference between the two input voltages, as compared to the single-ended amplifier that amplifies the difference between one input voltage and circuit common. The ideal differential amplifier will amplify only the difference between the two input signals. It will reject (completely attenuate) any voltage common to both inputs, called *common mode voltage.* The degree to which a real op amp succeeds in attenuating common mode voltage is given by the specification called *common mode rejection ratio (CMRR).* CMRR is usually given as a decibel value and the formula is:

CMRR = 20 Log (Common Output Voltage/Common Mode input Voltage) (7–5)

The CMRR can be measured by making the differential signal zero with the common mode voltage at some value. Any voltage measured at the output under this condition is common mode voltage that has not been attenuated by the differential amp. The circuit for a differential amplifier is shown in Figure 7–11. The CMRR is largely determined by the degree to which the resistors are matched: resistor $R_2 = R_4$ and resistor $R_1 = R_3$. The gain is determined by the ratio of R_4/R_2. Differential amplifiers can be constructed from standard op amps

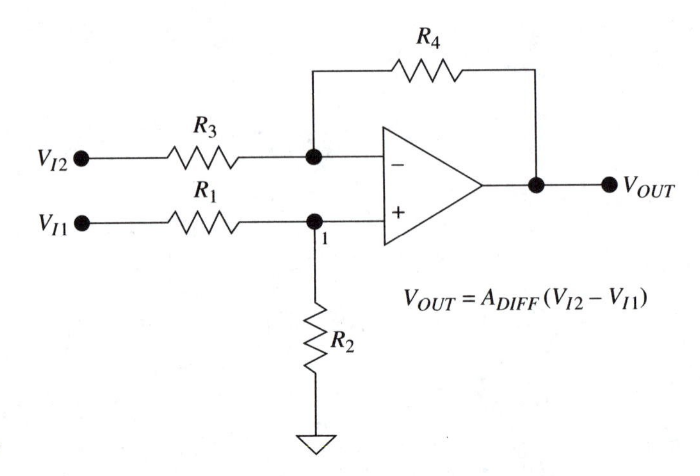

FIGURE 7–11
Differential amplifier

or are available as an integrated circuit. The differential amplifier is an improvement over the single-ended amplifier because of its ability to reject common mode signals. However, it does suffer from relatively low input impedance, as does the inverting amplifier. The instrumentation amplifier to be discussed next is an improved differential amplifier that enjoys the high input impedance of the non-inverting amplifier.

An example of an IC differential amplifier is the INA117 shown in Figure 7–12. The INA117 is a precision unity gain differential amplifier. Included on the IC are all of the necessary resistors, implemented on a thin film resistor network. The specifications state a minimum CMRR of 86 dB. More impressively, this amplifier accommodates common mode voltages on the order of ±200 volts. To put this in

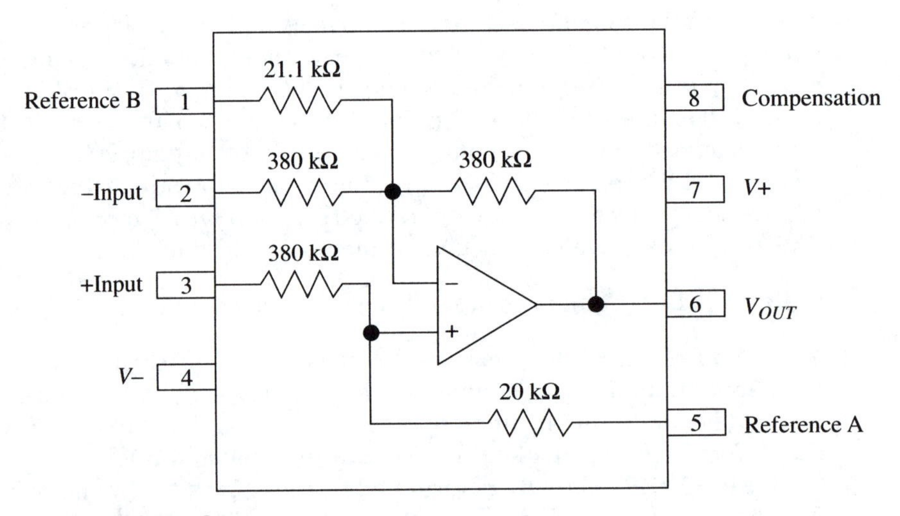

FIGURE 7–12
INA117 differential amplifier

perspective, let's say we have a 200-mV DC signal riding on a 120-V AC RMS common mode signal input to this unity gain differential amplifier. The output would be 200-mV DC signal riding on top of a 6 mV AC RMS signal.

Instrumentation Amplifier

The differential amplifier actually possesses a number of design deficiencies: the input impedance is less than desirable, the CMRR is highly dependent on matching the resistor values, and gain adjustment is accomplished by the adjustment of two resistors instead of just one. An amplifier circuit called an *instrumentation amplifier* improves on each of these problem areas. An instrumentation amplifier is shown in Figure 7–13. High input impedance is achieved by using two non-inverting amplifiers on the front end of each input to the instrumentation amplifier. The output stage of the instrumentation amplifier is simply a differential amplifier with a gain of one (all the resistor values equal R). The output of each non-inverting amplifier is connected in a creative scheme that results in the following equation for the instrumentation amplifier's output:

$$V_O = (1 + 2R/R_A)(V_1 - V_2) \qquad (7\text{–}6)$$

Instrumentation amplifiers can be constructed from discrete components or are available as an IC in many configurations. The INA101 is an example of a high-performance IC instrumentation amplifier. The internal schematic and external connections for the 14-pin DIP package are shown in Figure 7–14. The gain of this amplifier is determined by the value of resistor R_G.

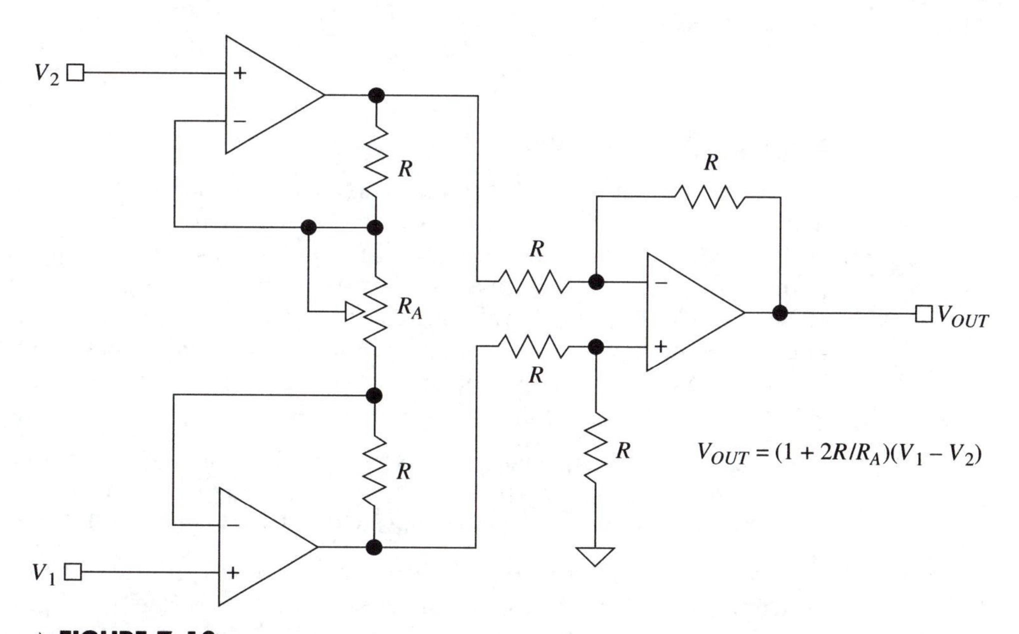

▲ **FIGURE 7–13**
Instrumentation amplifier

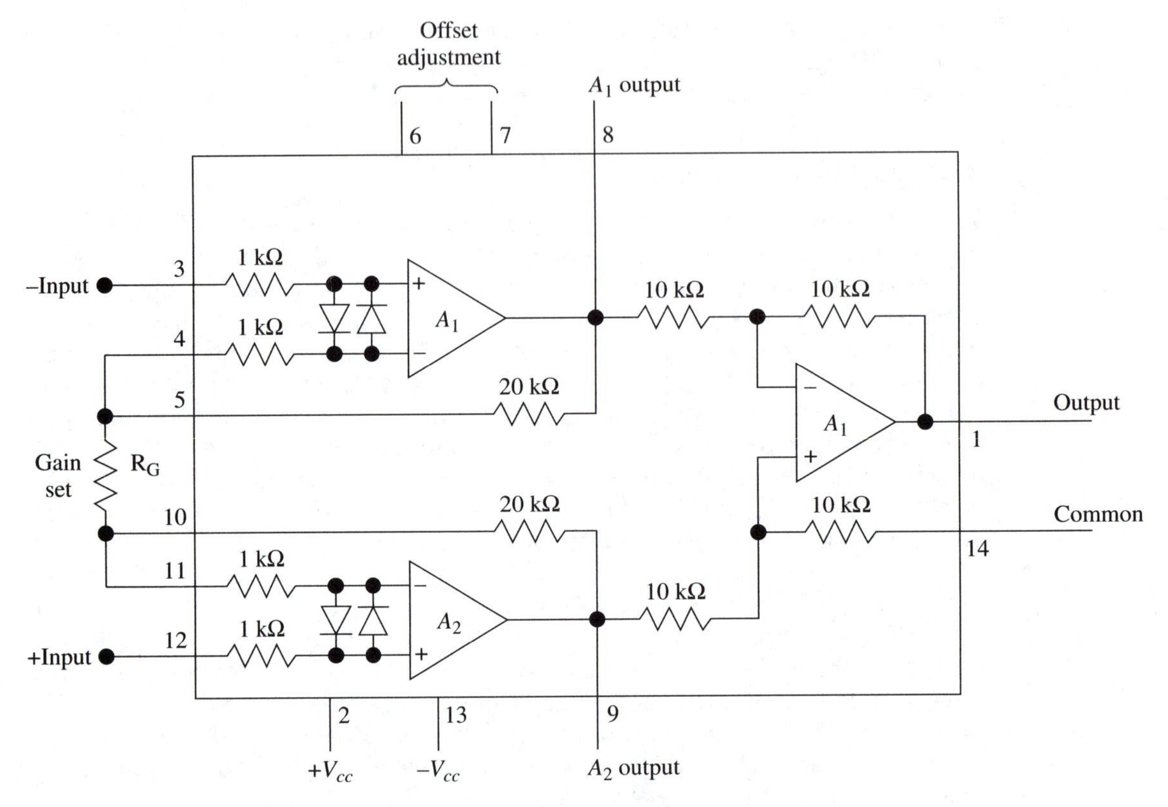

▲ **FIGURE 7–14**
INA101 Instrumentation amplifier

Single-supply Operation

In all of our discussions using the op amp as a DC amplifier, the power supply connected to the $V+$ and $V-$ connections has been equal in value and opposite in polarity. All of the circuit examples given thus far have used a ±12-V DC power supply. With a ±12-V DC power supply, the output of the op amp has a maximum range close to ±10 V DC. A dual symmetrical power supply is required when the op amp must output both plus and minus signals, or when the output must have a range that includes 0 under normal operating conditions. The symmetry of dual polarity power also promotes a 0-V output when the input is 0.

> *Note:* There are special function op amps that approach 0-V outputs with 0 input when operated with a single supply, but typical op amps can only drive the output voltage to values within $V+$ minus 2 V and V– plus 2 V.

In many applications op amps are not required to provide negative or zero output voltages. In these cases, the power supply circuitry can be simplified by using a single-power supply connected to $V+$ and $V-$. The voltage for the single supply should be greater than the minimum supply range for the op amp being used, yet

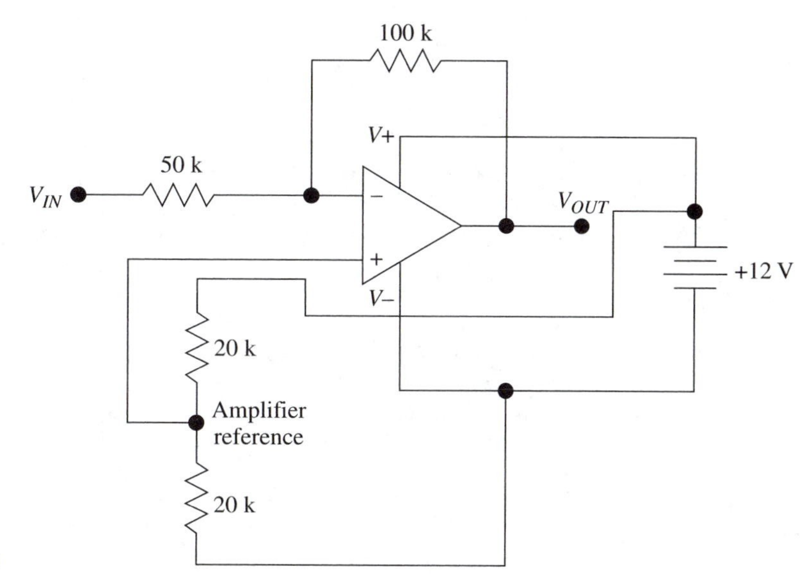

▶ **FIGURE 7–15**
Single-supply inverting op amp

less than its maximum voltage rating. The difficulty presented by this type of circuit is the establishment of circuit common. The dual symmetry power supply develops a circuit common exactly between the *V*+ and *V*– values. When a single supply is used, the circuit common must be developed, usually with resistor voltage dividers. Figure 7–15 shows an example single-ended op amp, inverting amplifier powered with a single 12-V power supply. The circuit common reference voltage in this circuit is actually *V*+/2, which is 6 V. When the input is equal to 6 V, the output also equals 6 V. If the input increases to 7 V, with the gain of –2 shown, the output will be reduced to 4 V. Likewise, if the input falls to 5 V, the output will increase to 7 V. The overall range of the op amp in Figure 7–15 is 2 to 10 V.

Power Op Amps

A severe limitation of most IC op amps is the low maximum output current specified for most op amps, usually on the order of 25 ma or 500 mW. Connecting an output transistor to the op amp output, as shown in Figure 7–16a, can readily increase the power and current output capabilities of any op amp. In this circuit, the output current flowing through the load resistor R_L is limited only by the current capabilities of the power supply, and the current rating for the transistor. The op amp will drive the collector voltage to a value determined by the op amp input voltage and the gain of the circuit according to the formula $V_O = (1 + R_F/R_1)\ V_1$ for positive input voltages only. If the op amp input voltage is negative, the NPN transistor in the circuit turns off. In order to drive the output in both polarities, a push-pull arrangement of NPN and PNP transistors could be connected.

There are many varieties of operational amplifiers that have been developed and are available as an IC that increase the op amp's power output. These can take

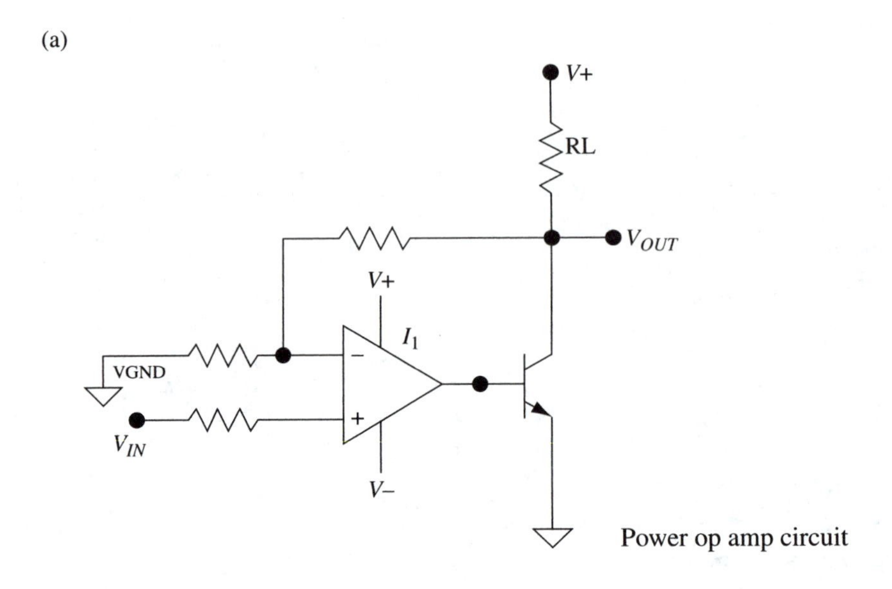

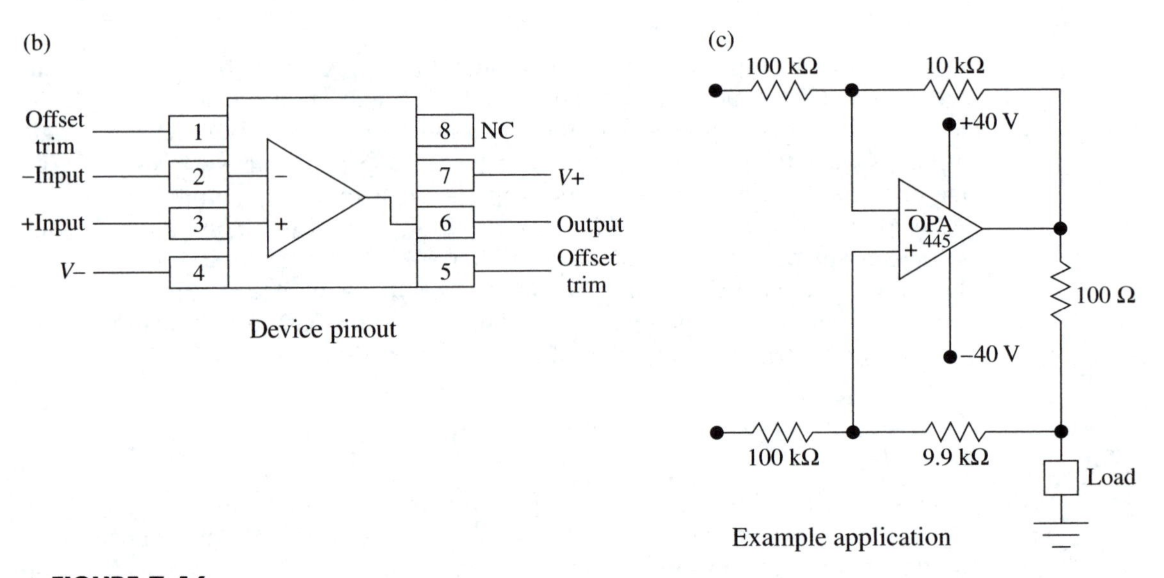

▲ **FIGURE 7–16**
Operational amplifier

the form of higher voltage and or current. Most common variety op amps can output about 10 to 20 ma at voltage levels up to ±18 V, or about 500 mW of power. The OPA445 is a good example of an IC that can provide much higher voltage outputs at current levels of 15 ma. The connection pinout for the OPA445 is shown in Figure 7–16b. In the diagram the device appears as a standard op amp but the OPA445 can handle power supply voltages and inputs up to ±45 volts. An example application of the OPA445 is shown in Figure 7–16c.

DC Amplifier Performance

This is a general summary of design and performance factors for DC amplifiers and should be used as a guide in their design and the analysis of their performance:

Input impedance (Z_{IN})

Output impedance (Z_{OUT})

Open loop gain (A_{OL})

Closed loop gain (A_{CL})

Power supply, single, dual, dual tracking

Power supply rejection ratio (PSRR)

Output voltage range

Input voltage range

Output current

Output power

Offset voltage

Offset voltage temperature coefficient

Bias current compensation

Input mode, single-ended or differential

Common mode rejection ratio for differential amplifiers (CMRR)

Type: voltage controlled/voltage source (VCVS)

voltage controlled/current source (VCIS)

current controlled/voltage source (ICVS)

current controlled/current source (ICIS)

7–3 ▶ AC Amplifiers

This section covers the use of the op amp as an AC amplifier over a wide range of frequencies. These circuits will be discussed in later sections where they will be applied to audio, video, and other specific application.

There are two frequency oriented limitations to the function of an op amp used as an AC amplifier. The first is the normal reduction in gain as the signal frequency increases, which is inherent in any amplifier. The other is the speed at which the op amp can change its output, the parameter called "slew rate". The open loop frequency response is available from the data sheet for any op amp. This is usually called the unity gain frequency or simply the bandwidth. The closed loop frequency response for any op amp circuit equals the open loop bandwidth divided by the closed loop gain of the circuit.

$$BW_{CL} = BW_{OL}/G_{CL} \tag{7–7}$$

where BW_{CL} = Bandwidth Closed Loop

BW_{OL} = Bandwidth Open Loop

G_{CL} = Gain Closed Loop

The frequency limitation due to an op amp's slew rate is approximated by the following formula:

$$BW_{SR} = S/(2\pi \times V_{PEAK}) \tag{7–8}$$

where BW_{SR} = Bandwidth due to slew rate

S = Slew rate from op amp data sheet

V_{PEAK} = maximum amplitude of the signal

Let's start with a simple op amp voltage follower as discussed in Section 7–2. Because the input signal is connected to the positive input, the output will be in phase with the input. To make this circuit an AC amplifier, it is desirable to capacitively couple the input and output of the op amp, as shown in Figure 7–17a. There is a problem with this circuit when compared to the DC voltage follower shown in Figure 7–6. A DC bias current flows into the non-inverting op amp input from circuit common in the DC voltage follower shown in Figure 7–6. The circuit shown in Figure 7–17a will not function because any DC bias current is blocked by the input coupling capacitor. Resistor R_{IN} is added to the circuit shown in Figure 7–17b, for the purpose of providing the bias current needed for the op amp to function.

It is important to note that in the AC circuits to be discussed, where the output of the op amp is capacitively coupled to the load, there is no attempt made to make the bias currents equal, as recommended for DC amplifiers. This is because any offset generated by unequal bias currents will be blocked from the output by the output coupling capacitor.

The circuit formed by C_{IN} and R_{IN} is a high pass circuit whose values should be determined, such that the lowest frequency in the signal range to be amplified is at the 3-dB cutoff point, f_1 for the high pass circuit. At the 3-dB cutoff point the capacitive reactance of C_{IN} should be equal to one-tenth the value of R_{IN}.

$$C_{IN} = 1/(2\pi f_1 (R_{IN}/10)) \tag{7–9}$$

R_{IN} becomes the effective input impedance of the circuit, significantly reducing the input impedance of the AC voltage follower from that seen by an input source to the DC voltage follower. It is therefore desirable to make R_{IN} as large as possible, but this is limited by the maximum value for any input bias resistor, as calculated using Equation 7–4. The maximum value for R_{IN} should be determined using Equation 7–4, and the minimum value should be determined for C_{IN} using

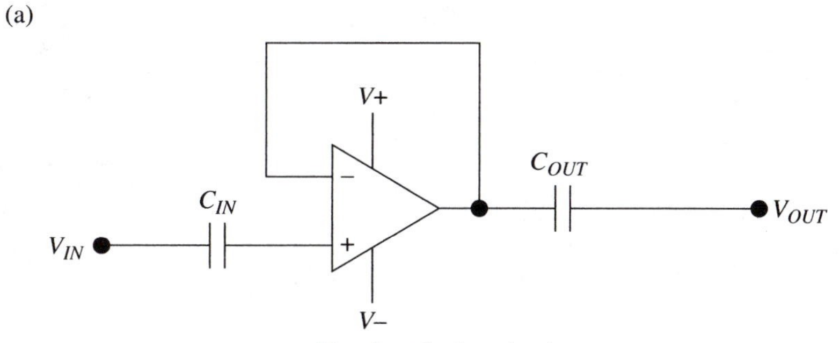

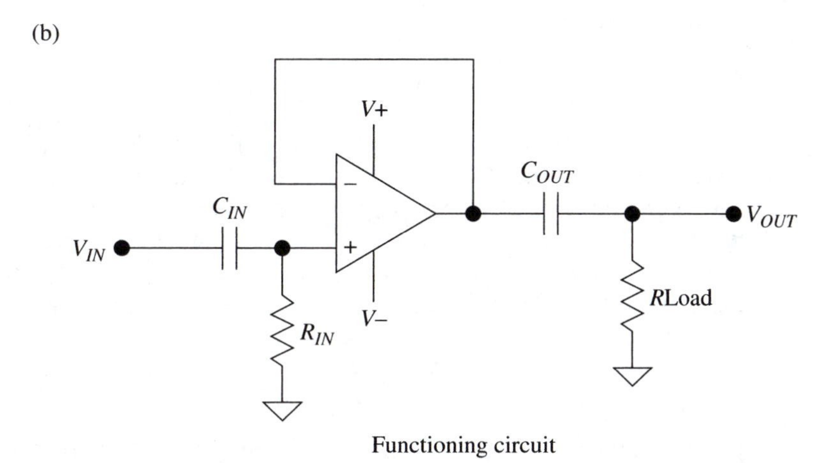

FIGURE 7–17
AC Voltage follower

Equation 7–8. The input impedance of this circuit can be increased dramatically by using a BIFET op amp, whose bias currents are much lower and can accommodate much larger input resistors, on the order of 1 MΩ. Figure 7–18 shows the AC voltage follower circuit in block diagram form. The output terminal of the op amp also drives a high pass circuit made up of C_{OUT} and R_{LOAD}. The voltage seen across the load resistor can be calculated as follows:

$$V_{LOAD} = V_{OUT}\,(R_{LOAD}/Z_{LOAD}) \qquad (7\text{–}10)$$

where Z_{LOAD} = the impedance of C_{OUT} and R_{LOAD}

The voltage across the load is at the –3 dB point when $C_{OUT} = R_{LOAD}$ C_{OUT} can be calculated by equating $X_{COUT} = R_{LOAD}$ at the lowest frequency of the range being amplified.

$X_{COUT} = R_{LOAD}$ at the lowest operating frequency

$$C_{OUT} = 1/(2\pi f_1 R_{LOAD}) \qquad (7\text{–}11)$$

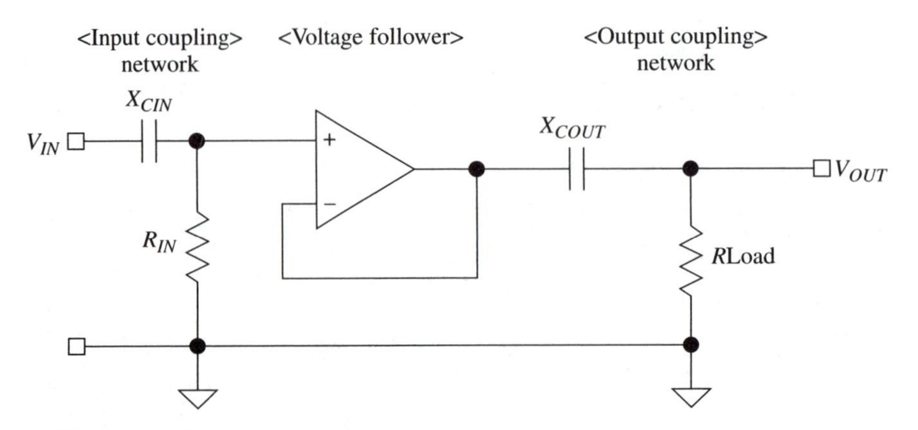

▲ **FIGURE 7–18**
AC voltage follower block program

Another method for increasing the input resistance of the AC voltage follower is the addition of another capacitor, C_Z, and a resistor, R_Z, to the circuit as shown in Figure 7–19. Capacitor C_Z connects the AC component of the output voltage to the junction between R_{IN} and R_Z. The current flowing through R_Z to ground creates a voltage drop across R_Z that opposes V_{IN} that effectively increases the input impedance. The theoretical input impedance for this circuit is given by the formula:

$$Z_{INPUT} = R_{IN}\,(1 + A_{OL}) \qquad (7\text{–}12)$$

However, the actual input impedance experienced with the circuit is significantly less than calculated with Equation 7–11, due to the typical stray capacitance that

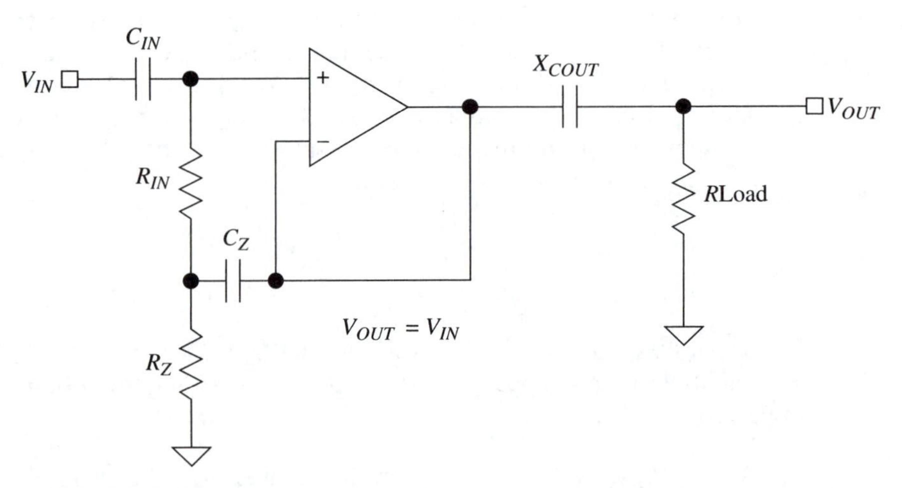

▲ **FIGURE 7–19**
AC voltage follower with high-input impedance

exists between the op amp terminal and ground. Nevertheless, normal values of stray capacitance will yield input impedances much higher than achieved with the circuit shown in Figure 7–18.

Example 7–1

Calculate all component values for the AC voltage follower circuit shown in Figure 7–17b, which must accommodate a signal with a low-frequency 3-dB cutoff point of 40 Hz. The op amp used in the circuit has a maximum bias current of 500 nA, and the load being driven is 2 kΩ. Also calculate the input impedance of the circuit.

Solution

1. Calculate R_{IN} by determining the largest input resistor value using Equation 7–4.

   $R_{IN\text{-}MAX} = (V_{BE}/10)/I_{BIAS\text{-}MAX} = .07\text{ V}/500\text{ nA} = 140{,}000$

   Use next lowest 1% value of $R_{IN} = 137\text{ k}\Omega$.

2. The effective input impedance of the circuit is simply the input resistor value $R_{IN} = 137\text{ k}\Omega$. Calculate C_{IN} using Equation 7.9 with 40 Hz for the lower 3-dB breakpoint and the R_{IN} value just calculated.

   $C_{IN} = 1/(2\pi f_1\ .1\ R_{IN}) = 1/(2\pi \times 40\text{ Hz} \times .1 \times 137\text{ k}\Omega) = .29\ \mu\text{F}$

   Use the next largest standard capacitance value, $C_{IN} = .33\ \mu\text{F}$.

3. Calculate C_{OUT} using Equation 7–11 with 40 Hz for the lower 3-dB breakpoint and the load resistance value stated.

   $C_{OUT} = 1/(2\pi f_1\ R_{LOAD}) = 1/(2\pi \times 40\text{ Hz} \times 2\text{ k}\Omega) = 1.99\ \mu\text{F}$

   Use the next largest standard capacitance value, $C_{IN} = 2.2\ \mu\text{F}$.

Example 7–2

Repeat Example 7–1 for the AC voltage follower circuit with high-input impedance shown in Figure 7–19. The typical open loop gain of the op amp being used is $A_{OL} = 200{,}000$.

Solution

1. Calculate the largest input resistor value as before using Equation 7–4.

   $R_{IN\text{-}MAX} = (V_{BE}/10)/I_{BIAS\text{-}MAX} = .07\text{ V}/500\text{ nA} = 140{,}000$

   Use next lowest 1% value of $R_{IN\text{-}MAX} = 137\text{ k}\Omega$.

   Split $R_{IN\text{-}MAX}$ into two equal resistors $R_{IN} = R_Z = 68.1\text{ k}\Omega$.

2. Calculate C_Z such that $X_{CZ} = R_Z$ at the 40-Hz breakpoint.

   $X_{CZ} = 6810 = 1/(2\pi \times 40 \times C_Z)$ $\qquad$ $C_Z = 1/(2\pi \times 40 \times 6810) = .584\ \mu F$

   Closest C_Z value = .56 μF

3. Calculate the input impedence of the circuit using Equation 7–12.

   $Z_{INPUT} = R_{IN}\ (1 + A_{OL})\ Z_{INPUT} = 68{,}100\ (200001) = 1362\ M\Omega$

   As stated earlier in the section, this is the theoretical value for the input impedance. The actual value will be determined by the amount of stray capacitance in parallel with the op amp inputs. Instead of calculating C_{IN} so that its capacitive reactance is one tenth the calculated Z_{IN}, it is better to use a value for C_{IN} that will be larger than any expected stray capacitance. A good design rule for high-impedance circuits is to use .001 μF.

   $C_{IN} = .001\ \mu F$

4. $C_{OUT} = 2.2\ \mu F$ as calculated in Example 7–1.

The next step is to turn the AC voltage follower shown in Figure 7–17b into an AC non-inverting amplifier. This can be accomplished with the addition of a feedback resistor, connected from the output to the negative input terminal, as shown in Figure 7–20a. The gain of this is circuit, $A_V = 1 + R_F/R_1$. All other component values can be calculated as in Example 7–1. It is obvious that this circuit suffers from the same low-input impedance, which is effectively equal to the value of R_{IN}, as the voltage follower shown in Figure 7–17b. As before, the input impedance can be increased significantly with the addition of capacitor C_Z and resistor R_Z as shown in Figure 7–20b.

Example 7–3

Calculate all component values for the AC voltage amplifier circuit shown in Figure 7–20b, which must accommodate a signal with a low-frequency 3-dB cutoff point of 60 Hz. The op amp used in the circuit has a maximum bias current of 500 nA, and the load being driven is 1 kΩ. The gain of the circuit should be 20.

Solution

1. Calculate the largest input resistor value as before using Equation 7–4.

   $R_{IN\text{-}MAX} = (V_{BE}/10)/\ I_{BIAS\text{-}MAX} = .07\ V/500\ nA = 140{,}000$

   Use the next lowest 1% value of $R_{IN\text{-}MAX} = 137\ k\Omega$.

   $R_{IN\text{-}MAX}$ will be split into two resistors ($R_{IN} + R_Z$) such that their total = 137 kΩ

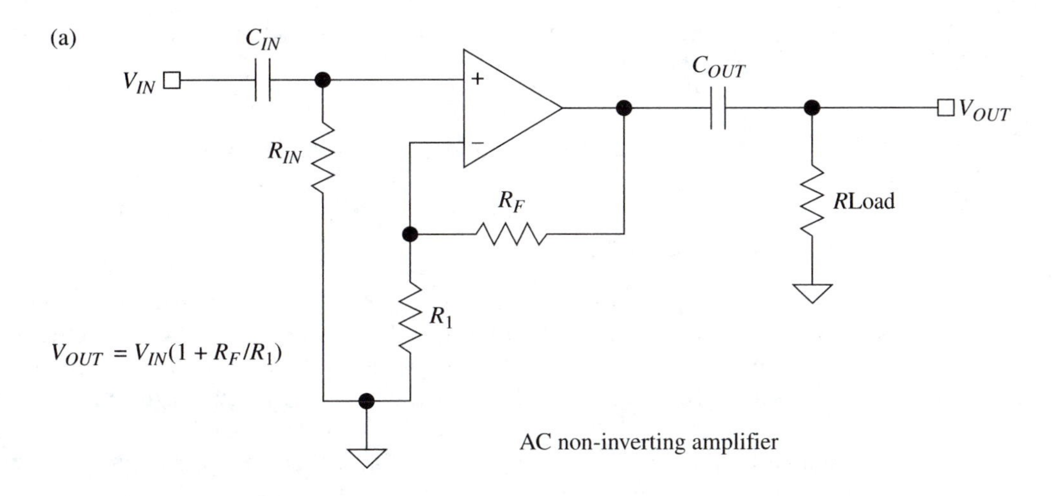

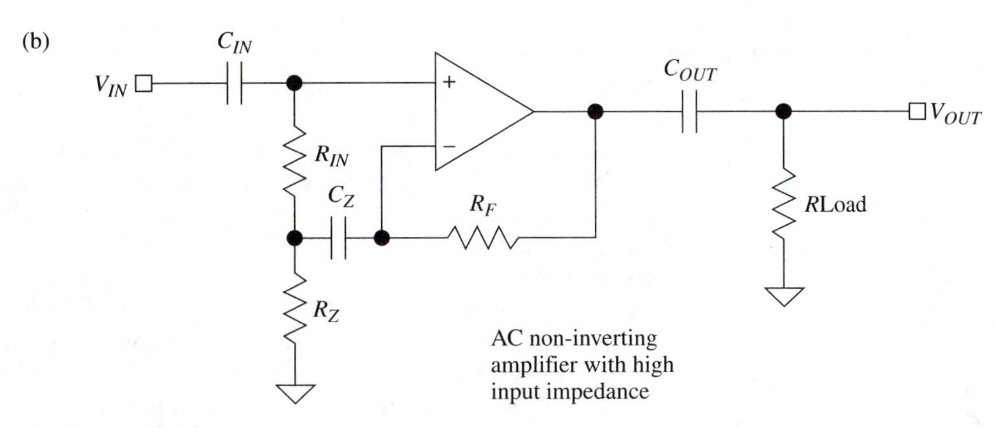

▲ **FIGURE 7–20**
Non-inverting amplifiers

2. Calculate R_Z and R_F to achieve the desired gain of 20.

   $A_V = 1 + R_F/R_Z = 20$
   $R_F/R_Z = 19$ Let $R_Z = 37.4$ kΩ, then $R_{IN} = 100$ kΩ
   $R_F = 710{,}600\ \Omega$
   Let $R_F = 715$ kΩ, the closest 1% value.

3. Calculate C_Z such that $X_{CZ} = R_Z$ 10 at the 60 Hz breakpoint

   $X_{CZ} = 3740 = 1/(2\pi \times 60 \times C_Z)$ $\qquad C_Z = 1/(2\pi \times 60 \times 3740) = .709\ \mu F$
   Closest C_Z value = .68 µF

4. Using the design rule for high impedance circuits, C_{IN} is selected = .001 µF

5. Calculate C_{OUT} using Equation 7–11 with 60 Hz for the lower 3-dB breakpoint and the load resistance value stated.

   $C_{OUT} = 1/(2\pi f_1 R_{LOAD}) = 1/(2\pi \times 60 \text{ Hz} \times 1 \text{ k}\Omega) = 2.65\ \mu\text{F}$
   Use the next largest standard capacitance value, $C_{IN} = 2.7\ \mu\text{F}$.

AC Inverting Amplifiers

An inverting op amp AC amplifier can be developed from a standard inverting op amp DC amplifier, with the addition of input and output coupling capacitors, as shown in Figure 7–21. As with the non-inverting amplifier, the input coupling capacitor C_{IN} stops the flow of DC bias current through R_{IN}; however, bias current does flow through the feedback resistor R_F. Because the output of this circuit is capacitively coupled to the output, there is no bias current compensation resistor placed between the plus terminal and circuit common. However, if bias current compensation is required, the value of the compensating resistor should be equal to just the feedback resistor R_F, not the parallel combination of R_F and R_{IN} as the case with the DC amplifier. This is because there is no bias current flowing through R_{IN} because of the coupling capacitor.

The value of capacitor C_{IN} is calculated such that its capacitive reactance is one-tenth the value of resistor R_{IN} at the low cutoff frequency. C_{OUT} is calculated so that its capacitive reactance is equal to the load resistance at the low cutoff frequency. The gain of this amplifier equals $-R_F/R_{IN}$.

In situations where an amplifier must process a limited range of frequencies, the low cutoff frequency is determined by the coupling capacitors that a high pass circuit comprises. An AC op amp circuit can create a low pass circuit that will limit the high frequency of operation to something less than the highest frequency the op amp can handle. This is accomplished by adding a feedback capacitor in parallel with the feedback resistor, as shown in Figure 7–22. The upper frequency is set by the capacitive reactance of C_F, relative to R_F, at the desired frequency. The capacitive reactance X_{CF} should equal the value of R_F at the desired upper cutoff frequency. The same approach can be used to limit the frequency of operation of the AC non-inverting amplifier.

▶ FIGURE 7–21
Inverting op amp AC amplifier

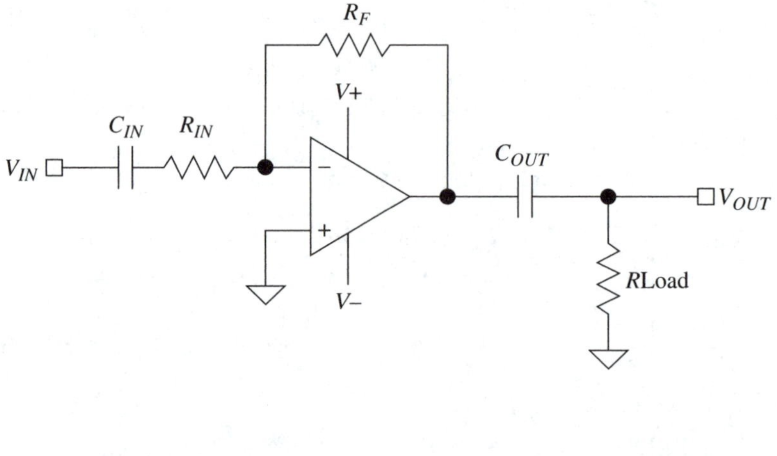

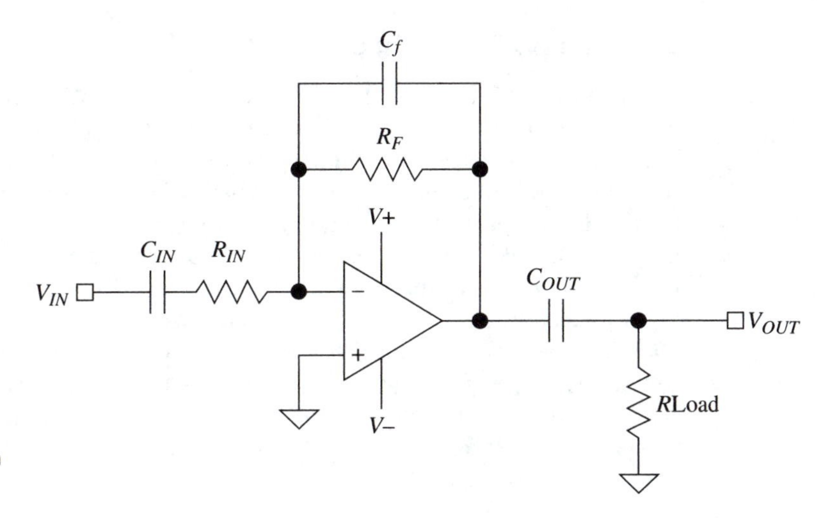

FIGURE 7–22
AC inverting amplifier with high-frequency limit

Example 7–4

Design an AC inverting amplifier circuit with a gain of –20 that will have an operating frequency range from 40 Hz to 15 kHz and will drive a load resistance of 60 Ω.

Solution

1. The circuit solution for this problem is the one shown in Figure 7–22. All of the resistor and capacitor values must be selected.
2. Determine the value of R_F and R_{IN}. $R_F/R_{IN} = 20$. Let $R_F = 100\ \text{K}\Omega$, and then $R_{IN} = 5\ \text{k}\Omega$.
3. Calculate C_{IN} and C_{OUT} to support the low cutoff frequency of 40 Hz.

   $X_{CIN} = .1 \times R_{IN}$ at the low cutoff frequency of 40 Hz
   $X_{CIN} = .1 \times 5000\ \Omega = 500\ \Omega$
   $C_{IN} = 1/(2\pi \times 40\ \text{Hz} \times 500) = 7.96\ \mu\text{F}$
   Use the next largest standard value, $C_{IN} = 8.2\ \mu\text{F}$

   $X_{COUT} = R_{LOAD}$ at the low cutoff frequency of 40 Hz = 60 Ω
   $C_{OUT} = 1/(2\pi \times 40\ \text{Hz} \times 60) = 66.3\ \mu\text{F}$
   Use the next largest standard value so $C_{OUT} = 68\ \mu\text{F}$.

4. Calculate C_F so that the circuit will support frequencies up to the upper cutoff frequency of 15 kHz.

   $X_{CF} = R_F$ at the high cutoff frequency of 15 kHz = 100 kΩ
   $C_F = 1/(2\pi \times 15\ \text{kHz} \times 100\ \text{k}\Omega) = 106\ \text{pF}$
   Use the closest standard value so $C_F = 100\ \text{pF}$.

Single-supply AC Op Amp Circuits

Many AC amplifier circuits are designed to operate off of a single power supply. This is because capacitive coupling eliminates concern about offset voltage errors, plus the fact that the actual value of the amplitude is usually not important, as the output signal should look like the input signal only larger. In Section 7–2, the operation of DC amplifiers with a single supply was discussed, as well as the need to create a circuit common reference in lieu of a ground. The single-supply AC amplifier creates some interesting circuit variations when considering bias current and the creation of a circuit common reference.

The circuit shown in Figure 7–23a shows a low impedance AC voltage follower circuit powered by a single power supply. The bias current flows from the positive supply through R_1 into the positive op amp input. The current flowing through R_2 to circuit common should be on the order of 100 times larger than the

▶ FIGURE 7–23
Z_{IN} single-supply AC voltage follower

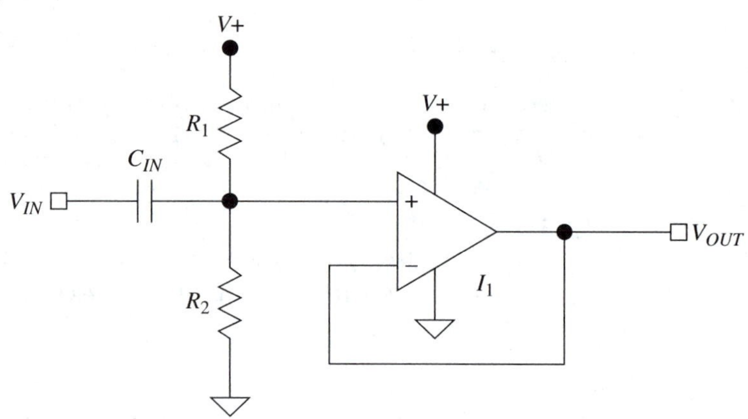

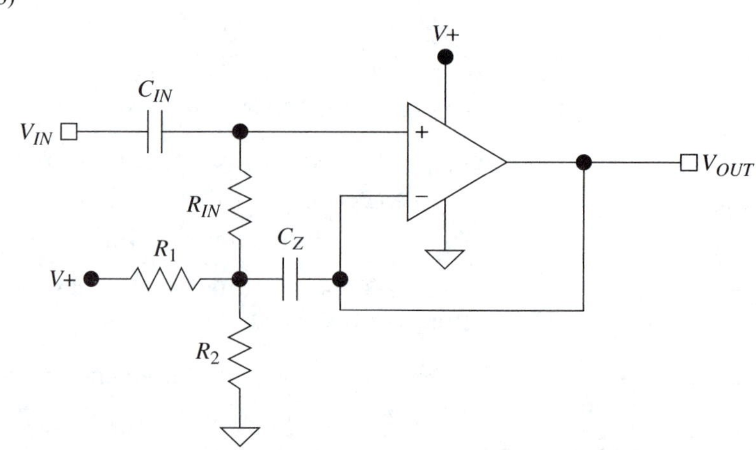

bias current in order to establish a common reference point that is exactly between $+V$ and circuit common. The input impedance of this circuit equals R_1 in parallel with R_2. Therefore, the value of C_{IN} should be one-tenth the value of R_1 in parallel with R_2 at the lower cutoff frequency of operation.

The corresponding single-supply high input impedance AC voltage follower circuit is shown in Figure 7–23b. This circuit is identical to that of Figure 7–19, except for the addition of the positive supply connection through R_1 to the junction of R_{IN} and R_Z. In this case, $R_1 = R_Z = R_{IN} = R$ where $2R = .07/I_{BMAX}$. Capacitor C_Z should equal one-tenth the value of the resistance in series with it (R_1/R_Z) at the lower cutoff frequency.

Single-supply AC Non-inverting Amplifier

Figure 7–24 shows an AC non-inverting amplifier operated from a single voltage supply. It is identical to the circuit shown in Figure 7–20a, except for the addition of capacitor C_1 and resistor R_B. Resistors R_B and R_{IN} are equal and create a common reference equidistant between $V+$ and circuit common. A bias current also flows into the plus terminal from $V+$ through R_B. The current flowing through R_{IN} should be 100 times the input bias current flowing into the plus input. The input impedance of this circuit equals R_B in parallel with R_{IN}. Therefore, the value of C_{IN} should be one-tenth the value of R_B in parallel with R_{IN} at the lower cutoff frequency of operation.

A unique situation occurs with this amplifier that is different from DC amplifiers and the AC voltage follower discussed thus far. The DC common reference generated by the voltage divider formed by R_B and R_{IN} equals $V+/2$. This voltage will be amplified, and with even low values of amplifier gain, the amplifier output will

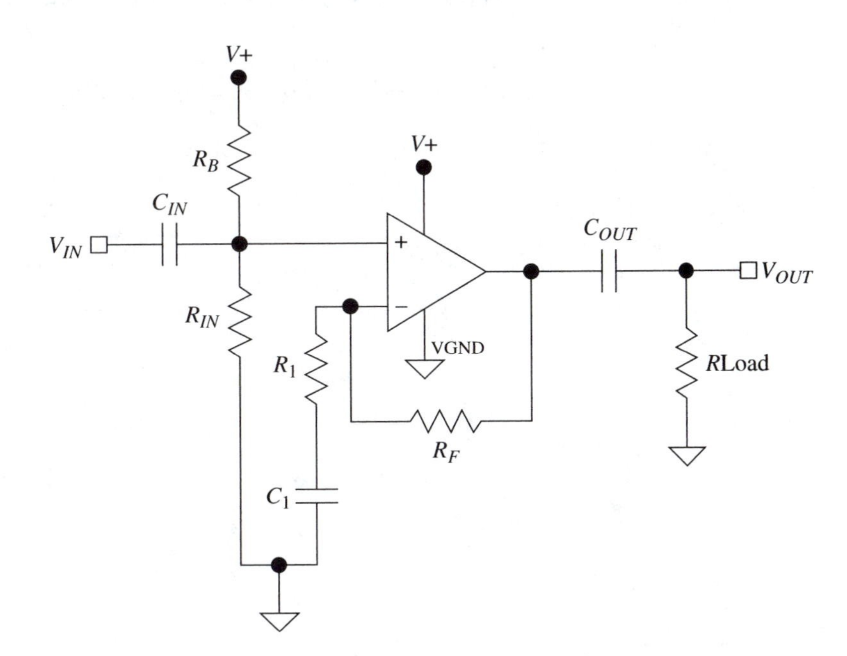

▶ **FIGURE 7–24**
Single-supply non-inverting amplifier

be driven into saturation. Capacitor C_1 causes the circuit to act like a voltage follower for DC voltages and a normal non-inverting amplifier for AC signals. Therefore, the DC output will equal the DC input, while the AC output will result in the AC input amplified by a gain that equals $1 + R_F/R_1$. The value of C_1 is selected so that its capacitive reactance is equal to one-tenth the value of R_1 at the low cutoff frequency.

Single-supply Non-inverting Amplifier with High-input Impedance

Figure 7–25 shows a high input impedance added to the single-supply non-inverting amplifier that is similar to the voltage follower circuit shown in Figure 7–23b, with the addition of resistor R_F that adds gain to the circuit. As with the voltage follower, $R_1 = R_Z = R_{IN} = R$ where $2R = .07/I_{BMAX}$. Capacitor C_Z should equal one-tenth the value of the resistance in series with its AC path to ground ($R_1//R_Z$) at the lower cutoff frequency. The gain of this circuit is also modified such that the gain to an AC signal equals $1 + R_F/(R_1//R_Z)$.

Example 7–5

Let's repeat Example 7–3 for single-supply operation by calculating all component values for the high input impedance AC voltage amplifier circuit shown in Figure 7–25. The desired low-frequency 3-dB cutoff point is 60 Hz. The op amp used in the circuit has a maximum bias current of 500 nA and the load being driven is 1 kΩ. The gain of the circuit should be 20.

Solution

1. Calculate the largest input resistor value as before using Equation 7–4.

$R_{IN\text{-}MAX} = (V_{BE}/10)/\ I_{BIAS\text{-}MAX} = .07\ \text{V}/500\ \text{nA} = 140{,}000\ \Omega$

$R_{IN\text{-}MAX}$ will be split into two resistors ($R_{IN} + R_Z$) such that their total = 137 kΩ

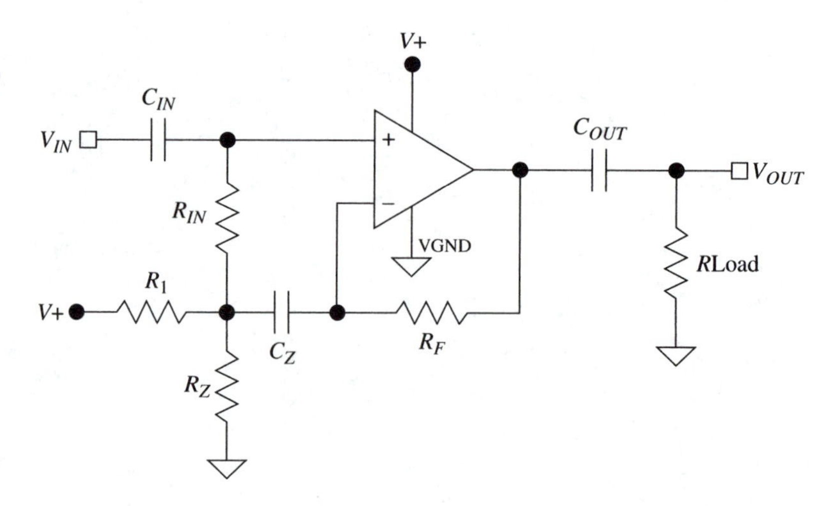

▶ FIGURE 7–25
Non-inverting amplifier with single supply

Let $R_{IN} = R_Z = R_1 = 68.7$ kΩ.

2. Calculate R_F that will achieve the desired gain of 20.

   $A_V = 1 + R_F/(R_1/R_Z) = 20$
   $R_F/(R_I//R_Z) = 19$
   $R_F = 652{,}650\ \Omega$
   Let $R_F = 649$ KΩ the closest 1% value.

3. Calculate C_Z such that $X_{CZ} = (R_I//R_Z)/10$ at the 60 Hz breakpoint

   $X_{CZ} = 3435 = 1/(2\pi \times 60 \times C_Z)$ $\qquad C_Z = 1/(2\pi \times 60 \times 3435) = .772\ \mu F$
   Closest C_Z value = .82 μF

4. Using the design rule for high impedance circuits, C_{IN} is selected = .001 μF.

5. Calculate C_{OUT} as before using Equation 7–11, with 60 Hz for the lower 3-dB breakpoint and the load resistance value stated.

   $C_{OUT} = 1/(2\pi f_1 R_{LOAD} = 1/(2\pi \times 60\ \text{Hz} \times 1\ \text{k}\Omega) = 2.65\ \mu F$
   Use the next largest standard capacitance value, $C_{IN} = 2.7$ μF.

Single-supply Inverting Amplifier

In order to operate an inverting amplifier from a single supply, a circuit common reference is connected to the positive input terminal as shown in Figure 7–26. The component value calculations involve selecting R_1 and R_2 such that the current flowing through R_2 is 100 times the maximum op amp bias current. R_1 must equal R_2 to create $V+/2$ at the positive op amp terminal. The gain of the circuit is $-R_F/R_{IN}$, and C_{IN} should be selected such that its capacitative reactance is one-tenth the value of R_{IN} at the low-frequency cutoff point.

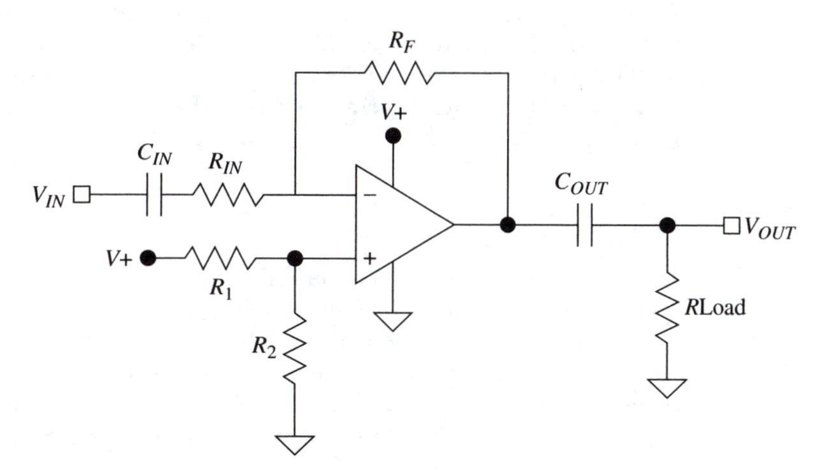

▶ **FIGURE 7–26**
Single-supply inverting AC amplifier

This section has discussed the design and operation of a variety of AC op amp circuits, including voltage followers and non-inverting and inverting amplifiers. Methods were provided to determine proper bias current and the calculation of component values to ensure circuit function over the specified frequency ranges. We discovered that in order to provide a bias current to the AC voltage follower, its input impedance is reduced significantly. The improved voltage follower with high input impedance resolved this shortcoming. The details relating to single power supply operation of AC op amps were also explored. All of these circuits and concepts will be applied to specific AC amplifier applications in the following sections.

7–4 ▶ Audio Amplifiers

Audio amplifiers must amplify signals in the audio frequency range, 20–20 kHz at either the pre-amp (less than 50 mW), medium (50 mW to 500 mW), or power amp (greater than 500 mW) level. As it is with most AC amplifiers, the actual amplitude of the amplified audio signal is usually not as critical as with DC amplifiers. Therefore, design considerations, such as offset voltage and offset voltage drift, are less significant. The goal of the audio amplifier is to provide an output signal that is the duplicate of the input signal except at a higher level.

An ideal audio amplifier will output a perfect duplicate of the input signal, over the entire frequency range of the signal, at the specified higher power level. The input to a practical audio amplifier will include a small noise signal and the amplifier itself will add in some small noise signals, along with the amplifier's power supply, that will also induce unwanted noise signals. The gain of the practical audio amp will vary with the frequency of the input signal. Therefore some frequencies will be amplified more than others.

The rate at which the output signal can change is important when the amplifier must duplicate an isolated fast changing signal. This is called the *slew rate*, and it is an important parameter for any AC amplifier. Stability is also an issue with a practical audio amplifier. It is desirable to eliminate the possibility of positive feedback in phase with the input of the amplifier that will send the amplifier into oscillation. It all comes down to gain, noise, frequency response, and stability. These are the areas where the performance of the audio amplifier is determined, measured either by an array of calibrated test equipment or a discriminating human ear.

Pre-amplifiers

Pre-amplifiers are a class of amplifiers that are designed to take a very small signal and amplify it to a level usually less than 50 mW. Because the pre-amp must work with such small signal levels, the primary design concern is to minimize distortion and provide a flat frequency response. Consequently, high power gain and efficiency are secondary concerns. The types of inputs that are possible for an audio pre-amplifier include a microphone, an instrument pickup, a magnetic tape

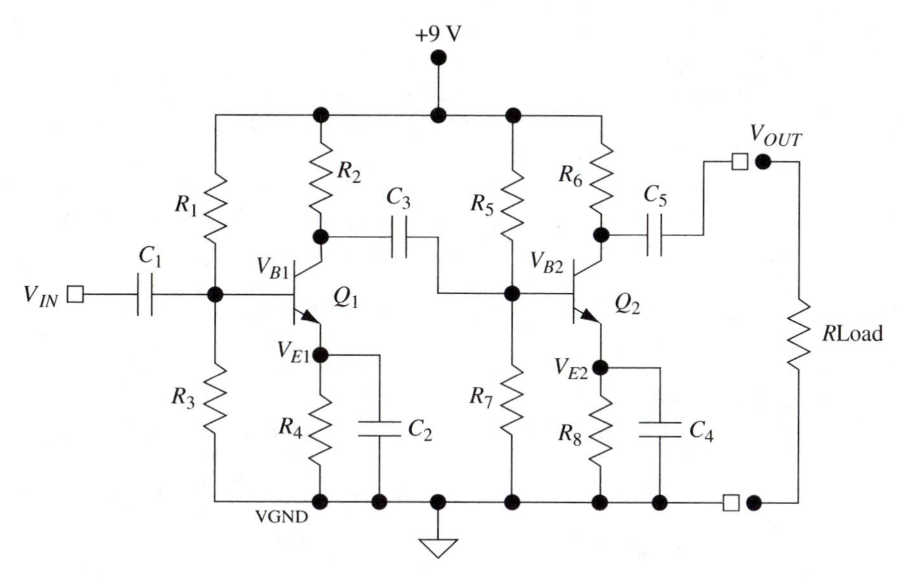

▲ **FIGURE 7–27**
Two-stage BJT audio pre-amp

playback head, a record cartridge and the audio output signal from a frequency selector-demodulator circuit that receives radio signals from an antenna. The gain required of pre-amps handling these many different types of inputs can range from 10 to 10,000, assuming input signal peak-to-peak amplitudes ranging from 100 µV to 100 mV and output signal requirements of 1 to 2 V at low current levels.

Using current technology, there are three possible ways to develop an audio pre-amplifier: complete a design using discrete transistors, develop a circuit that uses standard low-noise op amps, or utilize special purpose ICs designed to function as audio pre-amps. Let's begin our design discussion with an example of a typical two-stage bipolar junction transistor amplifier shown in Figure 7–27.

Example 7–6

This design problem is to complete a discrete transistor voltage pre-amplifier that will amplify AC signals with peak-to-peak amplitudes as small as 100 µV and frequencies over the audio range of 100 to 20,000 Hz. The peak-to-peak output of the amplifier should be around 1.25 V. DC power is to be supplied by a 9-V battery.

Solution

Since the overall gain of this pre-amplifier is on the order of 12,500, this design will be accomplished in two stages using a voltage divider biased common emitter pre-amplifier, shown in Figure 7–27. The transistors in the design will have $\beta_{DC} = \beta_{AC} = 150$.

Stage one of the amplifier will have a gain of roughly 62.5 times the gain of stage two, which will approximate 200 ($62.5 \times 200 = 12500$). The AC gain of each stage is equal to the effective collector resistance of the stage divided by the AC emitter resistance $r'e$ $A_V = R_C/r'e$ where $r'e = 25\ \text{mV}/I_B$ for the particular stage.

Stage two design calculations: If I_{E2} is selected to be 1 ma, then the desired $r'e = 25\ \Omega$ and $R_{C2} = A_V \times r'e = 200 \times 25\ \Omega = 5000\ \Omega$. R_{c2} for stage two is resistor R_6. Since the closest standard 5% resistor value is 5100 let $R_6 = 5100\ \Omega$.

R_5 is selected to be 10 times the value of R_6 or 51 kΩ. Let R_7 be approximately 20% of R_6. Therefore $R_7 = 10\ \text{k}\ \Omega$. The DC base voltage V_{B2} for stage two = $(R_7/R_7 + R_5)\ V_{CC,}$ if R_7 is less than $.1 \times \beta_{DC} \times R_E$. Substituting:

$V_{B2} = (10\ \text{k}\Omega/10\ \text{k}\Omega + \text{k}\Omega)\ 9\ \text{V} = 1.48\ \text{V}$
$V_{E2} = V_B - V_{BE} = 1.48\ \text{V} - .7\ \text{V} = .78\ \text{V}$
$I_{E2} = V_{E2}/R_E$ If I_{E2} is desired to be 1 ma, then
$R_E = .78\ \text{V}/1\ \text{ma} = 780\ \Omega$
The closest 5% standard resistor value is $R_E = 750\ \Omega$.

The design for stage two is complete with the following values:

$R_3 = 51\ \text{k}\Omega$, $R_6 = 5.1\ \text{k}\Omega$, $R_7 = 10\ \text{k}\Omega$ and $R_8 = 750\ \Omega$
The actual DC emitter current $I_{E2} = 1.03\ \text{mA}$.
$r'e = 25\ \text{mV}/I_{E2} = 24.2\ \Omega$
The actual AC gain for stage two = $R_6/r'e = 5100/24.2 = 210.7$.
This is slightly higher than the design goal of 200.

Stage one design calculations: Since the overall gain for the pre-amp should be 12,500, the new design goal for stage one should equal the overall gain divided by the actual gain of stage two, which is 12,500/210.7 or 59.32.

The emitter current for stage one, I_{E1}, is selected to be 1 ma, which makes stage one's design goal for $r'e = 25\ \Omega$, which is the same value used for stage two.

$A_{V1} = R_{C1}/r'e$, therefore $R_{C1} = A_{V1} \times r'e = 59.2 \times 25 = 1480\ \Omega$

However, the effective collector resistor for stage one, R_{C1}, is the parallel combination of the collector resistor R_2 and resistors R_5 and R_7 and the input base resistance of stage two (which = $r'e \times \beta_{AC}$). To achieve the gain of 59.2 the effective R_{C1} must equal 1480 Ωs. Therefore:

$1480 = R_2//R_5//R_7//(R_8 \times \beta_{AC}) = R_2//51\ \text{k}\Omega//10\ \text{k}\Omega//(24.2 \times 150)$
$1480 = R_2//51\ \text{k}\Omega//10\ \text{k}\Omega//(24.2 \times 150)$
$1480 = R_2//2531$ $R_2 = 3900\ \Omega$ is the closest standard value.
For $R_2 = 3900\ \Omega$ the effective $R_{C1} = 1535\ \Omega$.
Select $R_1 = 51\ \text{k}\Omega$, $R_3 = 10\ \text{k}\Omega$, and $R_4 = 750\ \Omega$

The design for stage one is complete with the following values:

$R_1 = 51\ \text{k}\Omega$, $R_2 = 3.9\ \text{k}\Omega$, $R_3 = 10\ \text{k}\Omega$, and $R_4 = 750\ \Omega$
The actual DC emitter current $I_{E1} = 1.03$ ma.
$r'e = 25\ \text{mV}/I_{E1} = 24.2\ \Omega$
The actual AC gain for stage one = $R_{C1}/r'e = 1535/24.2 = 63.4$.
This is slightly higher than the design goal of 59.2.

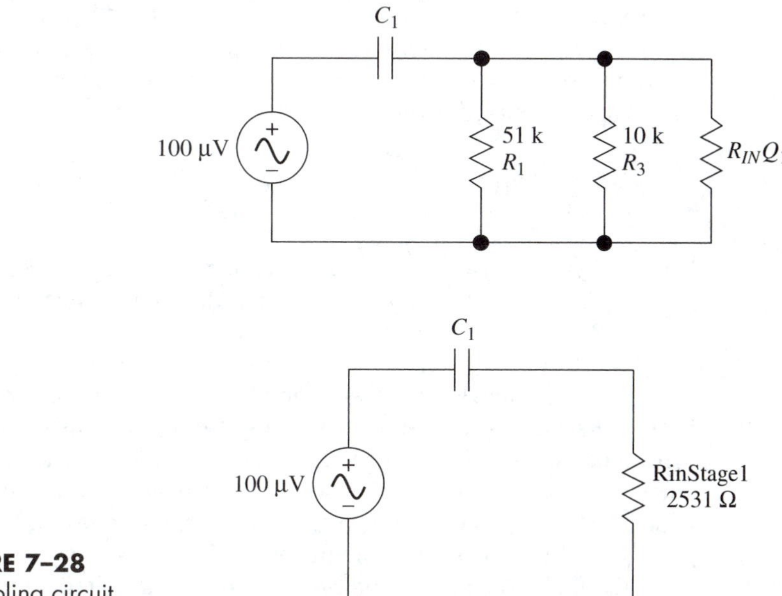

▶ **FIGURE 7–28**
Input coupling circuit

Coupling capacitor calculations: Capacitors C_1, C_5 and C_5 are coupling capacitors that remove any DC component voltage from the AC signal. C_1 and C_3 are input coupling capacitors while C_5 couples the amplifier output to the connected load. The optimum capacitance value for coupling capacitors depends upon the frequency range of the AC signal and the resistance in series with the capacitance. The target value often used is to make the capacitive reactance, X_C, of the coupling capacitor, calculated at the lowest frequency of the AC signal, equal to one-tenth of the resistance in series with the coupling capacitor.

Figure 7–28 shows the effective circuit for the input to stage one. The total input resistance of stage one equals the parallel combination of R_1, R_3, and R_{INQ1}.

$R_{INSTAGE1} = R_1//R_3//R_{INQ1} = 51\ \text{k}\Omega//10\ \text{k}\Omega//3630\ \Omega = 2531\ \Omega$
X_{C1} (at the lowest frequency) $= .1 \times R_{INSTAGE1}$
X_{C1} (at 100 Hz) $< .1 \times 2531 < 253\ \Omega$
X_{C1} (at 100 Hz) $= 1/(2\pi f C_1) = 1/(2\pi \times 100 \times C_1) = 253\ \Omega$
Solving for $C_1 > 6.3\ \mu\text{F}$

The input resistance seen by the AC signal going into stage two is essentially the same as the values used for stage one, so the optimum value for $C_2 = C_1$. Since there is no load specified for the amplifier, the value for C_5 will be made equal to C_2 and C_1.

C_5, C_2, and C_1 are all selected to be 10 µF.

Next, the capacitance values for the bypass capacitors C_2 and C_4 must be calculated. The function of the bypass capacitors is to allow the AC signal to bypass collector resistors R_4 and R_8, allowing these resistors to function in their DC bias role of stabilizing the circuit from changes in the transistor beta values. The only consideration in determining the value of the bypass capacitor is the value of the emitter resistor. The capacitive reactance of the bypass capacitor should be less than the one-tenth the value of the emitter resistance.

X_{C2} (at 100 Hz) $< .1 \times R_4 < .1 \times 750 < 75\ \Omega$
X_{C2} (at 100 Hz) $= 1/2\pi f C_2 = 1/2\pi \times 100 \times C_2 = 75\ \Omega$
Solving for $C_2 > 21.2\ \mu F$
C_4 bypasses the same value resistance so its value should equal that of C_2.
C_2 and C_4 are selected to be 22 µF. All of the values of the discrete transistor audio pre-amp have been determined.

A significant deficiency of the discrete BJT amplifier is the low value of the input impedance that requires the input coupling capacitance to be fairly large. This is partially a result of the high gain included in the amplifier that requires a smaller emitter resistor. This low emitter resistance translates into a small transistor base input resistance. A much higher input resistance can be obtained with a lower gain BJT amplifier or by using discrete FET transistors.

Example 7–6 shows the process for designing discrete transistor amplifiers as well as the expected performance and limitations. Pre-amplifier design is often accomplished today utilizing low noise op amps or op amps that are customized to perform as IC audio amplifiers.

Op Amp Audio Preamp

An equivalent audio pre-amp can be developed with op amps using the discussion of AC op amp circuits presented in Section 7–3. This particular design uses two op amps to accomplish a net gain of 12,500.

Example 7–7

This design problem is to complete an op amp voltage pre-amplifier utilizing an LM318 op amp that will amplify AC signals with peak-to-peak amplitudes as small as 100 µV and frequencies over the audio range of 20 to 20,000 Hz. The peak-to-peak output of the amplifier should be around 1.25 V. DC power is to be supplied by a single 12-V supply. It is desirable that the input impedance of the amplifier should be at least 1 MΩ and the load being driven has a resistance of 2 kΩ.

Solution

This design uses two op amps connected as shown in Figure 7–29. The first op amp, IC_1, is configured as a non-inverting amplifier with a gain of 50, configured such that it offers a very high input impedance to the audio input. Consequently, this circuit can accept audio inputs with very high source impedances. C_1 is the input coupling capacitor and R_1 supplies bias current to the non-inverting input. R_3 and R_2 serve as a voltage divider to generate a 6-V reference to allow operation

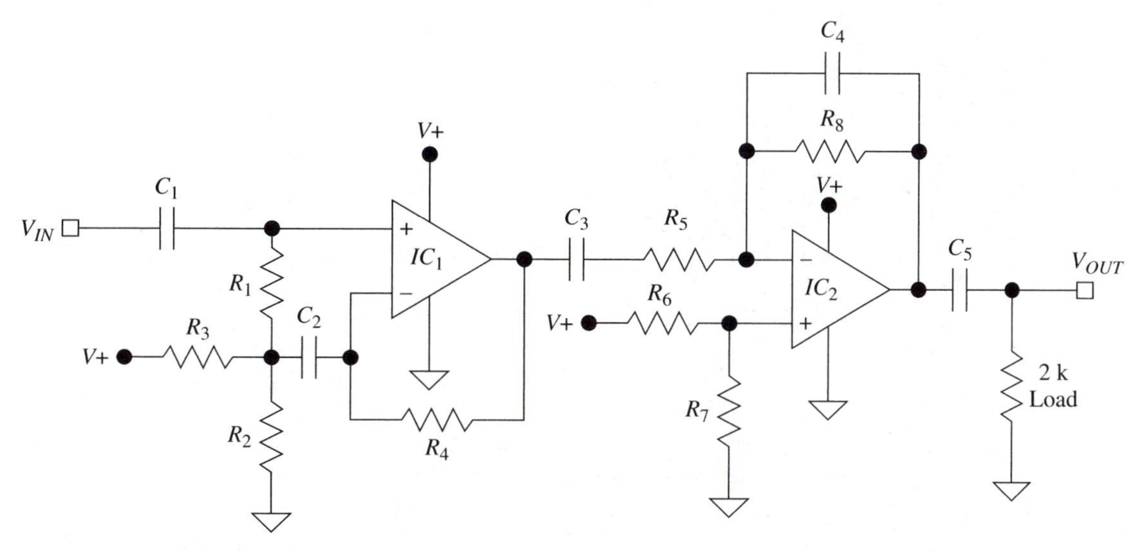

▲ **FIGURE 7–29**
Op amp audio pre-amp

from a single 12-V supply. IC_2 is an inverting amplifier with a gain of 250, whose input is coupled to the output of IC_1 by coupling capacitor C_3. Resistors R_6 and R_7 form a 6-V reference for C_2 and C_5 is a coupling capacitor for the audio output.

Component Calculations

1. IC_1 is a non-inverting amplifier with a gain of 50 that drives an inverting amplifier with a gain of 250. The net gain of the circuit is 12,500.

2. The maximum bias current for the LM318 op amp is 250 nA.

   $R_{IN\text{-}MAX} = (V_{BE}/10)/\ I_{BIAS\text{-}MAX} = .07\ \text{V}/250\ \text{nA} = 280{,}000\ \Omega$

   $R_{IN\text{-}MAX}$ will be split into two resistors ($R_1 + R_2$) such that their total = 280 kΩ

   Let $R_1 = R_2 = R_3 = 140\ \text{k}\Omega$

3. Calculate the value of coupling capacitor C_1.

   $X_{C1} = .1 \times \text{R}_1$ at 20 Hz

   $C_1 = 1/(2\pi \times 20\ \text{Hz} \times 14\ \text{k}\Omega)$

   $C_1 = .56\ \mu\text{F}$

4. Calculate the value of capacitor C_2.

   $X_{C2} = .1\ (R_3/R_2)$ at 20 Hz

   $C_2 = 1/(2\pi \times 20\ \text{Hz} \times 7\ \text{k}\Omega)$

   $C_2 = 1.14\ \mu\text{F}$ use the closest standard value = 1.2 μF

5. Calculating R_4 to yield a gain of 50 for IC_1.

   Gain = $1 + R_4/(R_2//R_3) = 50$
   $R_4/(70\ k\Omega) = 49$
   $R_4 = 3.43\ M\Omega$ use the closest standard value = $3.44\ M\Omega$

6. Calculate the values for the voltage divider resistors R_6 and R_7. The current passing through R_7 should be greater than 100 times the bias current flowing into the positive op amp terminal. The maximum bias current for the LM318 is 250 nA.

   $R_7 = V_{R7}/I_7 = 6\ V/25\ \mu A = 240\ K\Omega$
   $R_6 = R_7 = 240\ k\Omega$

7. Calculate the values of R_5 and R_8 to achieve a gain of 250.

   Gain = $250 = R_8/R_5$
   Let $R_8 = 1\ M\Omega$, therefore $R_5 = 4000\ \Omega$.
   Use closest 1% standard value, $R_5 = 4.02\ k\Omega$.

8. Calculate the value of coupling capacitor C_3.

   $X_{C3} = .1 \times R_5$ at 20 Hz
   $C_3 = 1/(2\pi \times 20\ Hz \times 4.02\ k\Omega) = 1.98\ \mu F$
   Use next highest standard value, $C_3 = 2.2\ \mu F$.

9. Calculate C_4 to provide a high-frequency cutoff of 20,000 Hz.

   $X_{C4} = R_8$ at 20,000 Hz
   $C_4 = 1/(2\pi \times 20{,}000\ Hz \times 1\ M\Omega) = 7.95\ pF$
   Use the next lowest standard value of $C_4 = 6.8\ pF$.

10. Calculate C_5 to meet the low-frequency cutoff of 20 Hz.

    $X_{C5} = R_{LOAD}$ at 20 Hz
    $C_4 = 1/(2\pi \times 20\ Hz \times 2\ k\Omega) = 3.98\ \mu F$
    Use the closest standard value of $C_5 = 3.9\ \mu F$.

IC Audio Amplifiers

There are a wide variety of IC audio amplifiers that comprise op amps enhanced for performance in audio applications. These ICs occupy the medium power level between pre-amps and power amps. They can drive speakers directly at power levels around .5 watts. The LM386 is an example of this class of IC. It is called a low voltage audio power amplifier. It is designed for single-supply operation over the voltage range of 4–18 V. The gain is set internally to 20, but with the addition of a

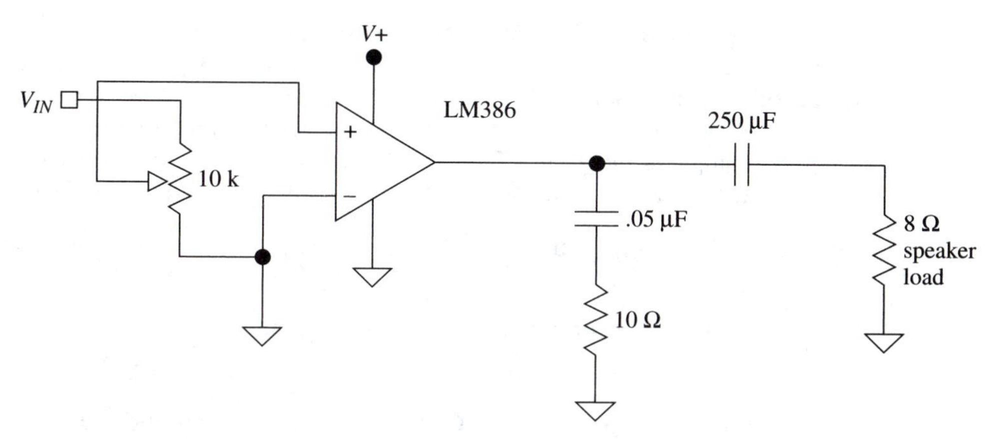

▲ **FIGURE 7–30**
LM386 audio amplifier

resistor and capacitor the gain can be varied between 20 and 200. Figure 7–30 shows the LM386 as an audio amplifier with a gain of 20. Note that to change the gain to 200, a 10 µF capacitor is connected between pin 1 and pin 8. Connecting a resistor in series with the 10 µF modifies the gain between 20 and 200 depending on the resistor value. The distortion rating of this amplifier is .2%.

The LM833 is a dual-audio amplifier with significantly improved distortion levels, about .002%, and a high slew rate, typically 7 V/µs. It functions more as a traditional pre-amp, capable of higher gains and is not intended to drive a speaker directly. Figure 7–31 shows one half of the LM833 connected as a tape deck pre-amp.

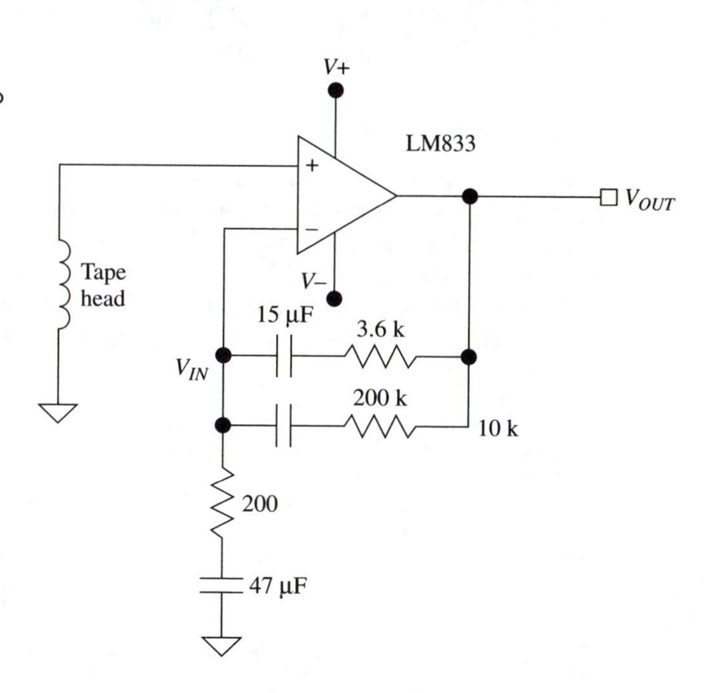

▶ **FIGURE 7–31**
LM833 dual audio amplifier

As audio amplifiers are included with personal computers and for operation with other digital circuits, there is pressure for them to improve their performance in these environments. The ability to operate from a noisy 5-V digital supply or low-voltage battery supplies, and an increased emphasis on efficiency and slew rate, are performance issues that relate specifically to desktop and laptop computer applications. Following are the amplifier performance parameters that are affected most:

Power Supply Rejection Ratio: The ability of the amplifier to reject the noise inherent on digital power supply lines

Power Efficiency: Improved efficiency at the .5-watt level

Slew rate: Improved slew rate at low voltage operation

Power Supply Voltage: Operation from low-amplitude, single-voltage power supplies

Output Voltage Swing: Because of the low voltage operation, the audio amp must be capable of driving the output very close to the positive supply value and ground. These are called *"rail-to-rail"* amplifiers and can drive the output to within approximately 55 mV of the positive or negative supply.

The MAX4490 audio op amp exemplifies this type of amplifier. It features operation from a single supply over a voltage range of 2.7 to 5.5 V. The PSRR is typically 100 dB, slew rate equals 10 V/µs, and the output voltage can swing to within 55 mV of either side of the power supply. The MAX4490 is available in a small five-pin surface mount package as shown in Figure 7–32.

The MAX4298 is called a *stereo driver* and it too features a high PSRR and can drive speakers directly to within 55 mV of either supply side. It also features a low noise level of .008%. Figure 7–33 shows the MAX4298 used as a typical headphone stereo driver in a computer application.

Audio Power Amplifiers

Audio power amplifiers have been in use for more than 60 years. Surprisingly, there are relatively few books available that describe the art as well as the science of power amplifier design. Most electronics courses discuss the basic classes of power amplifiers and their typical bias circuits, yet few courses or textbooks deal with the intricacies of their design. In this section we will only scratch the surface but will attempt to take the student a few steps further into this interesting and developing technology.

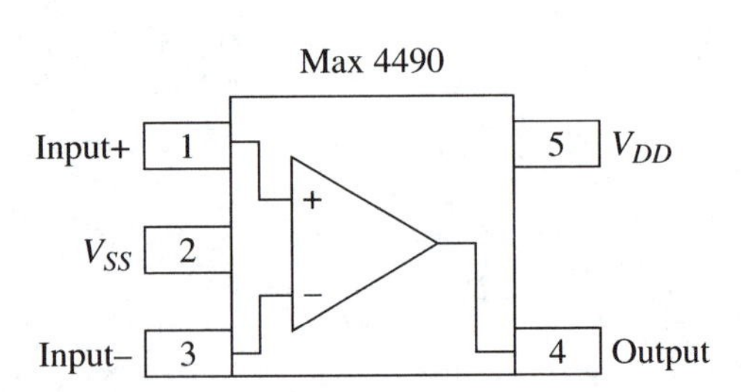

FIGURE 7–32
MAX4490 audio op amp

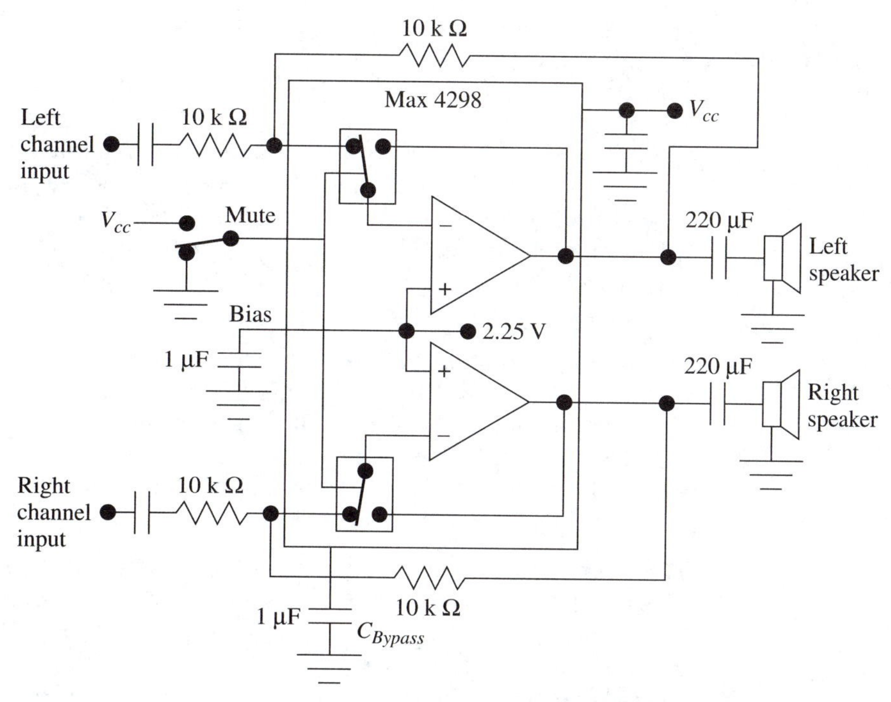

▲ FIGURE 7–33
MAX4298 stereodriver

Audio amplifier designs have progressed through vacuum tubes, bipolar junction and field effect transistors, and most recently the development of custom IC audio power amplifiers. During this process a number of facts as well as myths have been dispersed and widely accepted regarding the design, testing, and analysis of audio power amplifiers. At the center of the controversy is the belief by some that the human ear is such a complex device that it is impossible to qualify audio performance with measured parameters. The degree of disagreement has led to two schools of thought within the audio community: the audio scientists that believe in verified technical measurements as the way to determine audio performance, and subjectivists that accept only certain measured results combined with the sound as it is heard by the human ear.

The limit to which the human ear can detect audio differences is an important part in developing the criteria for amplifier performance. Following is a list of widely accepted limitations of the human ear:

1. Changes in amplitude can generally be detected as small as .5 to 1 dB. For a single frequency, level changes of .3 dB can be detected.
2. The ear is most sensitive to frequency changes in the range of 500 Hz to 2 kHz. Frequency changes can be detected for changes at least .2% in magnitude.
3. The smallest detectable THD level is approximately .2%.

Audio power amplifiers are classified by categories that characterize a particular design philosophy. Following are the various amplifier classes and their general definitions:

Class A: In Class A the transistors are biased in the middle of the load line; therefore, current is flowing at all times through the output connection. When biased in this fashion, the non-linearities associated with turning the transistor on and off are not experienced. Because current is always flowing to the output, these types of amplifiers are usually power inefficient.

Class B: Class B amplifiers use two transistors that are biased at the transistor's cutoff point. The transistors are said to be in a push-pull arrangement, where one transistor is controlling current flow in one half of the AC cycle, and the other transistor controls current flow in the second half of the cycle. Because the transistors are biased at the cutoff point, no current flows when the signal is 0.

This promotes the much-improved power efficiency experienced with Class B operation. However, Class B operation does suffer from something called *crossover distortion*, which is caused by the slightly different bias point achieved for the two transistors. This causes slight distortion when control of the signal is being passed from one transistor to the other.

Class AB: These are essentially Class B amplifiers that have a bias point that is above the transistor's cutoff point. This makes the amplifier operate somewhat like a Class A amplifier with two transistors connected in a Class B amplifier's push-pull configuration, which is the basis for the "AB" classification. This type of operation considerably reduces crossover distortion at the cost or decreased power efficiency.

Class C: The Class C amplifier locates the bias point below the cutoff region for the transistor, which means the transistor conducts less than half of the AC cycle. The Class C amplifier is used in radio signal amplification and is not really suited for audio applications.

Class D: The Class D amplifier is a pulse-width-modulating (PWM) type of amplifier. In other words, it switches a voltage on and off at a high frequency. The average value of the duty cycle (the on time vs. the off time) equals the amplitude of the amplified signal. The benefit of this approach is similar to switching power supplies where efficiency is improved significantly. However, as was also the case with switching power supplies, the PWM process is inherently noisy.

Class G: This is an innovative technique that attempts to improve amplifier efficiency by increasing the power supply voltages in steps as the signal amplitude is increased. This strategy utilizes the fact that less energy is wasted when the power supply voltage is closer to the signal level. Transistor circuits are used to sense and switch in typically three levels of increasing voltage supply. These amplifiers do offer increased power efficiency without significant increases in distortion at a cost of the added circuitry needed to perform the power supply sensing and switching.

Class H: This class of amplifier operates as a Class B amplifier with the added ability to boost, or not boost, the power supply voltage. The goal again is to improve efficiency. This is somewhat similar to the technique that was just described for Class G operation; however, in this case the power supply is simply increased instead of using the Class G approach, in which a different power supply is switched into operation.

Class S: Class S amplifiers are basically Class A amplifiers with a reduced output current drive capability. The Class A amplifier section of the Class S amplifier is backed up by a Class B amplifier section that makes a low resistive load appear to be higher than the minimum load the Class A amplifier section can drive.

Aside from the amplifier classes just discussed, there is also the question of whether an audio amplifier should be AC or DC coupled. AC-coupled amplifiers operate off of a single voltage supply, and with a 0 signal input the output is midway between the power supply rail and ground. A coupling capacitor is used to remove the DC component from the signal provided to the output. DC-coupled amplifiers use equal plus and minus power supplies that bias the input voltage at 0 V, eliminating the need for a coupling capacitor.

AC-coupled amplifiers offer the advantage of single-supply operation. They always provide a 0 DC offset voltage and do not require protection against DC output faults. Also, AC-coupled amplifiers seldom need inductors to provide stability. DC-coupled amplifiers require two power supplies but eliminate the need for large output coupling capacitors. The output coupling capacitors are expensive and they also contribute to distortion levels. By keeping the loads connected to the DC-coupled amplifier's plus and minus power supply voltages balanced, the power supply currents in the ground circuit will be essentially zero. This further reduces the amount of distortion and crosstalk that are inherent in AC coupled amplifiers, because the ground currents include power supply current flow combined with signal current flow.

Audio power amplifiers can be designed using discrete components or custom IC amplifiers. IC amplifiers offer a quick and easier approach but there are some issues that allow discrete amplifiers to still offer a higher performance solution. In IC amplifiers linear resistors are difficult to fabricate, and compensation capacitors are made as small as possible to conserve space and minimize costs. Also, many design techniques are based upon IC fabrication processes instead of focusing directly on amplifier performance. Whether to use IC audio amps or a discrete design will be dictated by the requirements for amplifier performance stated in the design specifications.

A typical audio power amplifier, IC or discrete, consists of three basic functional blocks: the input section, voltage amplifier, and output driver (see Figure 7–34). The input section is a differential type that takes the input signal and subtracts a portion of the output signal fed back to the negative input. The input section is usually a differential transconductance amplifier that takes the voltage difference between the

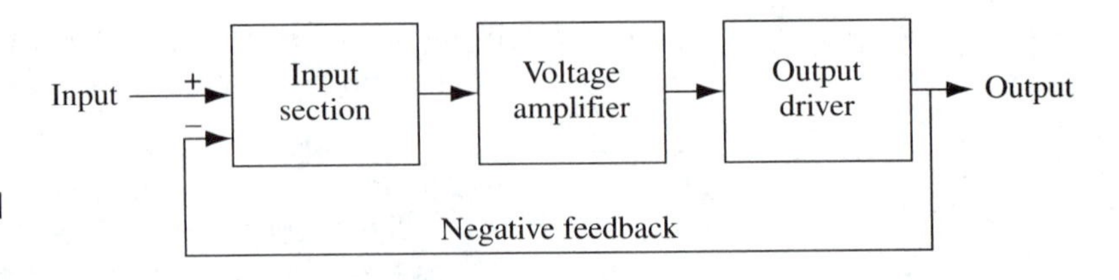

▶ **FIGURE 7–34**
Audio amplifier functional block program

input and the feedback signals and generates a current output. The voltage amplifier accepts the current signal from the input section and amplifies it to create a voltage output. This is called a *trans-resistance amplifier.* The output stage takes the amplified voltage signal from the voltage amplifier and supplies the capability to drive a low impedance load such as an 8- or 16-Ω speaker. Usually the output stage does not provide any further signal amplification but simply provides the proper voltage and current level to drive the load. A typical discrete audio power amplifier schematic including all three functional circuits is shown in Figure 7–35.

The LM1877 is a dual audio power amplifier available on a single IC. The maximum power output of this IC is low, just 2 watts per channel, but it exemplifies this type of IC power amplifier. This IC is designed to require the minimum number of external components. It includes internal compensation for all gains greater than ten and can be operated over a wide range of single or dual power supply voltages. Figure 7–36 shows examples of both single and dual power supply operation as a stereo power amplifier.

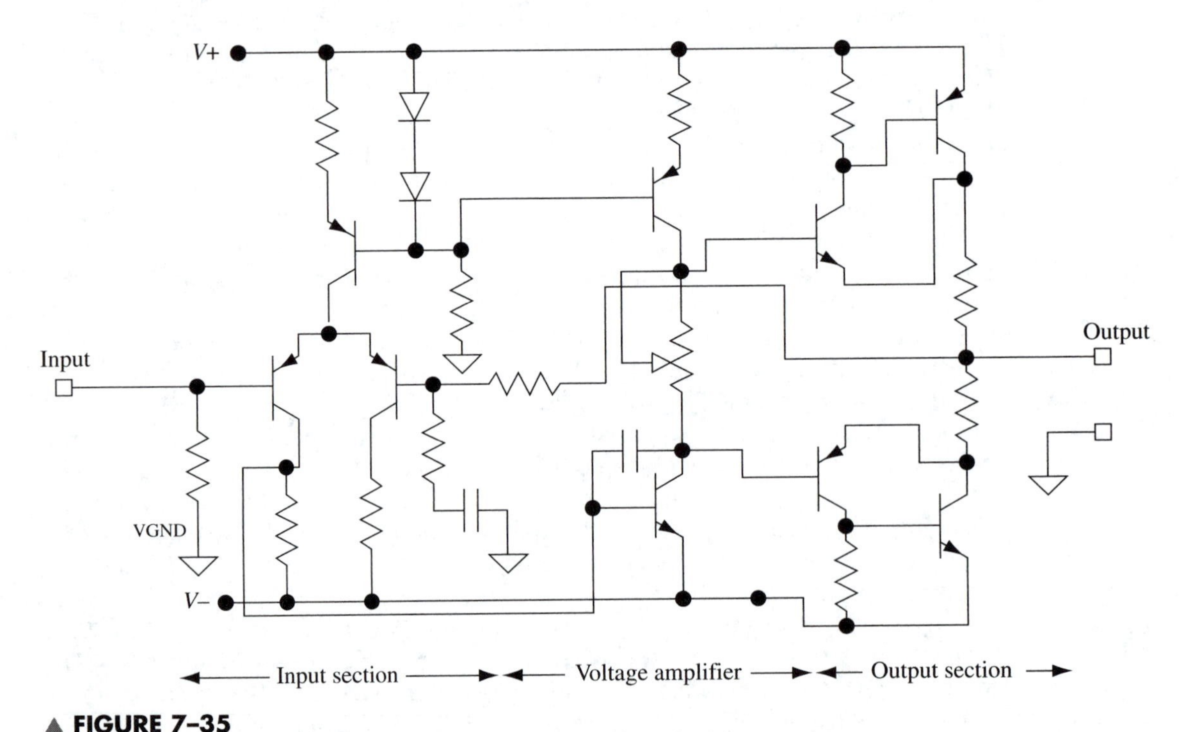

▲ **FIGURE 7–35**
Discrete audio amplifier schematic

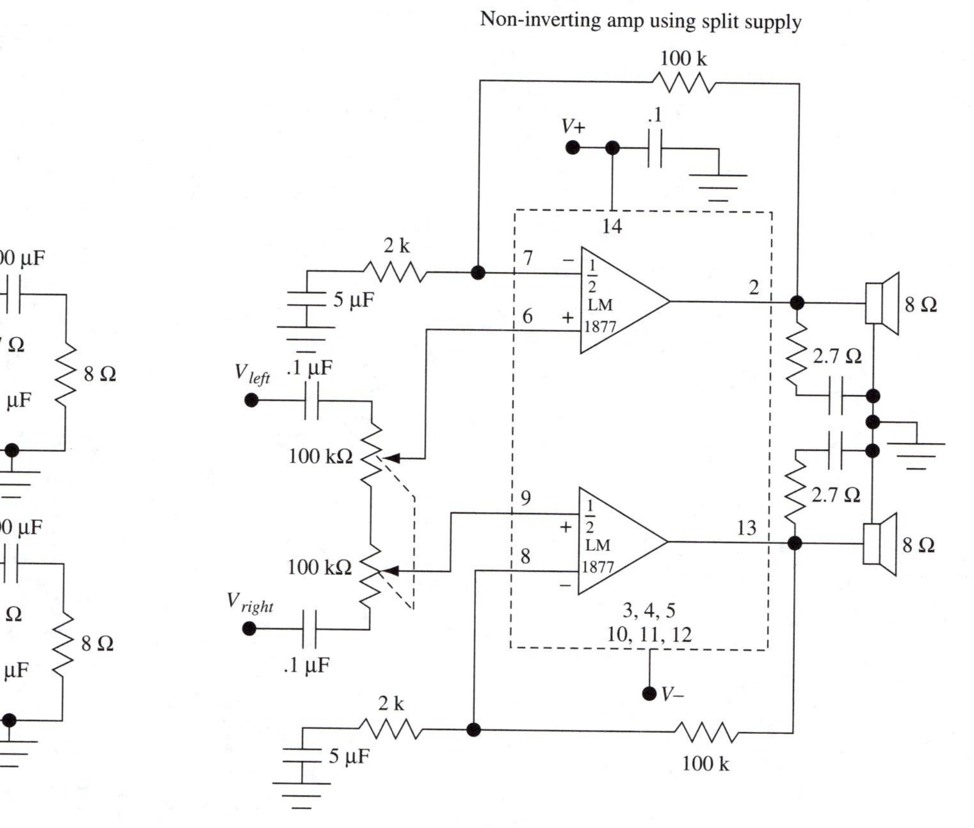

▲ FIGURE 7–36
LM1877 dual audio power amplifier

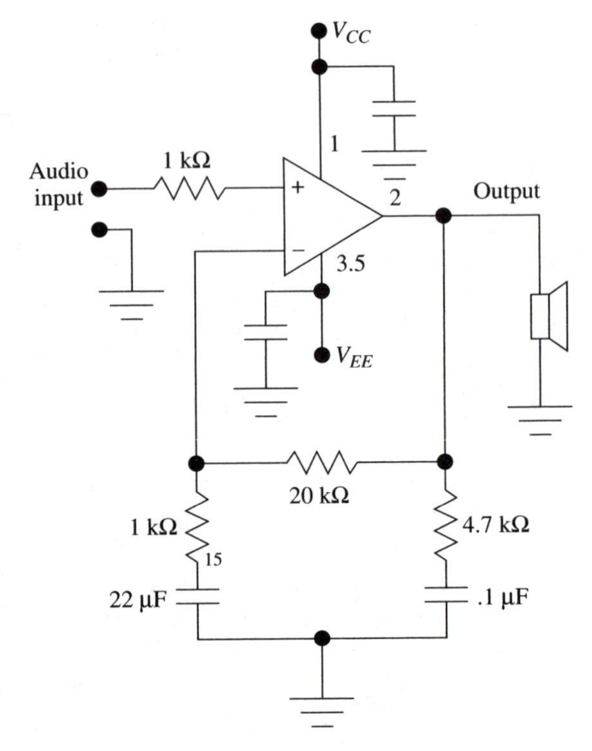

▶ **FIGURE 7–37**
LM4700 audio power amplifier example

The LM4700 is a single audio power amplifier capable of outputting 30 watts of power into an 8-Ω load with .1% distortion levels. The LM4700 offers a wide range of power supply options ranging from 20 to 66 V either as single or dual supplies. Figure 7–37 shows a typical power amplifier circuit using the LM4700.

7–5 ▶ Video Amplifiers

Video or wideband amplifiers are similar to audio amplifiers except that they must operate over a wider frequency range. The frequency requirements for transmitting video information can range from about 20 Hz up to 6 MHz and greater for other specialized applications. Specific television applications require a narrower operating bandwidth in the range from 4 to 6 MHz. Obviously, the operating frequency requirement for video amplifiers is much more demanding than those seen in audio applications and is why these types of amplifiers are often called *wideband amplifiers.*In order to meet this requirement, the gain level of video amplifiers must be reduced dramatically to provide a wider bandwidth of operation. Also, much effort must be made to consider the inherent capacitance of the transistor as well as source and load impedances.

The coupling and emitter bypass capacitors limit the lower frequency of an amplifier's operating bandwidth. These capacitors will be effective AC shorts at high frequencies and will have no effect on high-frequency operation so they will be ignored in this high-frequency analysis. It is the internal transistor capacitance

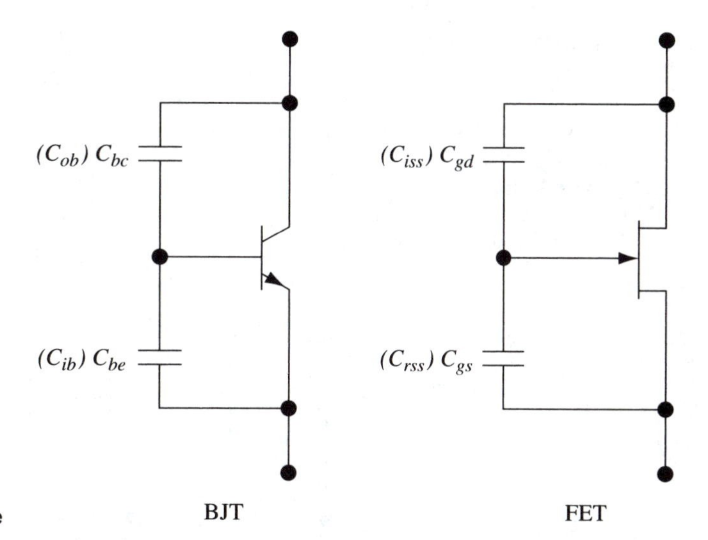

FIGURE 7–38
Transistor capacitance

that serves to limit high-frequency operation and is of prime concern to video amplifiers. These include the base-to-collector (C_{bc}) and base-to-emitter capacitance (C_{be}) for BJT's and the gate-to-drain (C_{gd}) and gate-to-source (C_{gs}) capacitance for FETs. Data sheets for BJTs often refer to C_{bc} as the output capacitance C_{ob} and to C_{be} as the input capacitance. FET data sheets list the input capacitance as C_{iss} and the reverse transfer capacitance as C_{rss}, from which C_{gd} and C_{gs} can be calculated (see Figure 7–38).

The base-to-collector capacitance is the most difficult to analyze. However, from Miller's Theorem we know that transistor capacitance from the base to the collector (gate to the drain for an FET) can be effectively reflected on the input side of the transistor as well as the output side, while the base-to-collector capacitance (gate to the source for an FET) is only seen on the input side. The amount of base-to-collector (gate-to-drain) capacitance reflected to the input and output side depends on the amplifier gain A_V and is defined by the following Miller's Theorem equations:

$$C_{IN} = C_{bc}\,(A_V + 1) \text{ or } C_{IN} = C_{gd}\,(A_V + 1) \qquad (7\text{–}13)$$

$$C_{out} = C_{bc}\,(A_V + 1)/A_V\ C_{out} = C_{gd}\,(A_V + 1)/\mathrm{Av} \qquad (7\text{–}14)$$

Thus C_{IN} is the most significant capacitance value as it results from the base-to-collector (gate-to-drain) capacitance multiplied by the amplifier gain plus 1. The net effect of transistor capacitance can be seen in the AC-equivalent schematic for a common emitter amplifier as shown in Figure 7–39. The impact of this circuit on the operating bandwidth is the generation of two high-frequency break points. The most significant break point is the one resulting from the input side of the circuit.

Looking at just the input side of the circuit we have the equivalent circuit shown in Figure 7–40. The critical frequency for this circuit can be determined by equating the capacitive reactance equal to the total resistance in parallel with it. Therefore:

R_s = Source resistance
R_1 and R_2 are voltage divider bias resistors
rb' = transistor base spreading resistance
$C_{in} = C_{bc}$ reflected at the input
$C_{out} = C_{bc}$ reflected at the output
R_c = Collector resistance
*C*Load = Load capacitance
*R*Load = Load resistance

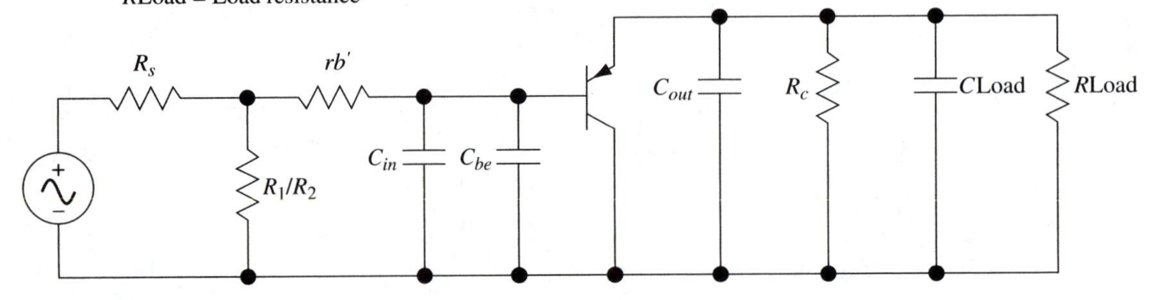

▲ **FIGURE 7–39**
High-frequency AC-equivalent common emitter amplifier

$$X_{CFI} = ((R_S // R_1 // R_2) + rb') \,//\, \beta_{AC}\, re' \quad (7\text{–}15)$$

$$f_1 = 1/(2\pi C_{F1}\, (((R_S // R_1 // R_2) + rb') \,//\, \beta AC\, re')) \quad (7\text{–}16)$$

where re' is the equivalent emitter resistance = 25 mV' I_E

The design problem for the video amplifier is to make the input circuit critical frequency equal to the highest operating frequency in the bandwidth of operation. To promote this, the following design considerations should be made for discrete transistor video amplifier designs:

1. Minimize the value of re' by making the emitter current as large as possible. This is accomplished by making the collector and load resistances as small as possible.
2. Minimize the net load capacitance.
3. Minimize the source resistance value R_S.

► **FIGURE 7–40**
High-frequency AC equivalent input circuit

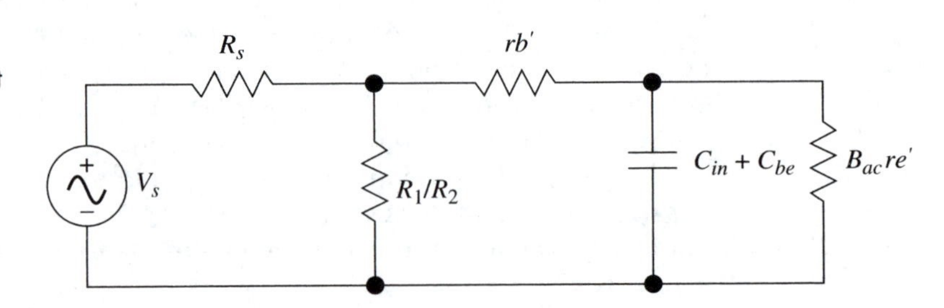

4. Select a high-frequency transistor with the following characteristics: low C_{bc} value, low C_{be} value and low *rb'* value.

5. Use negative feedback to reduce gain and widen the bandwidth.

A discrete video amplifier can be designed in stages using these design considerations at each stage.

The CLC410 is a specific IC video amplifier. It possesses the wide band required for video applications operating at low gain levels from ±1 to ±8. The CLC410 is available in an 8-pin DIP or surface-mount packages and can be treated as an operational amplifier with inverting, non-inverting output *V*+, *V*–, and offset connections. There is an additional input called a *quick disable* that allows the signal flow to be curtailed within 200 ns of the disable line becoming low. Figure 7–41 shows the recommended non-inverting circuit.

The CLC5602 is a dual video amplifier on one IC that can drive an output with current levels up to 130 mA. The CLC5602 can operate off of single or dual supplies as low as +5 V or ±5 V. The slew rate is 300 V/μs and it is available in an 8-pin DIP or surface-mount package. Figure 7–42 shows the CLC5602 being used as an AC-coupled non-inverting video amplifier operated from a single supply.

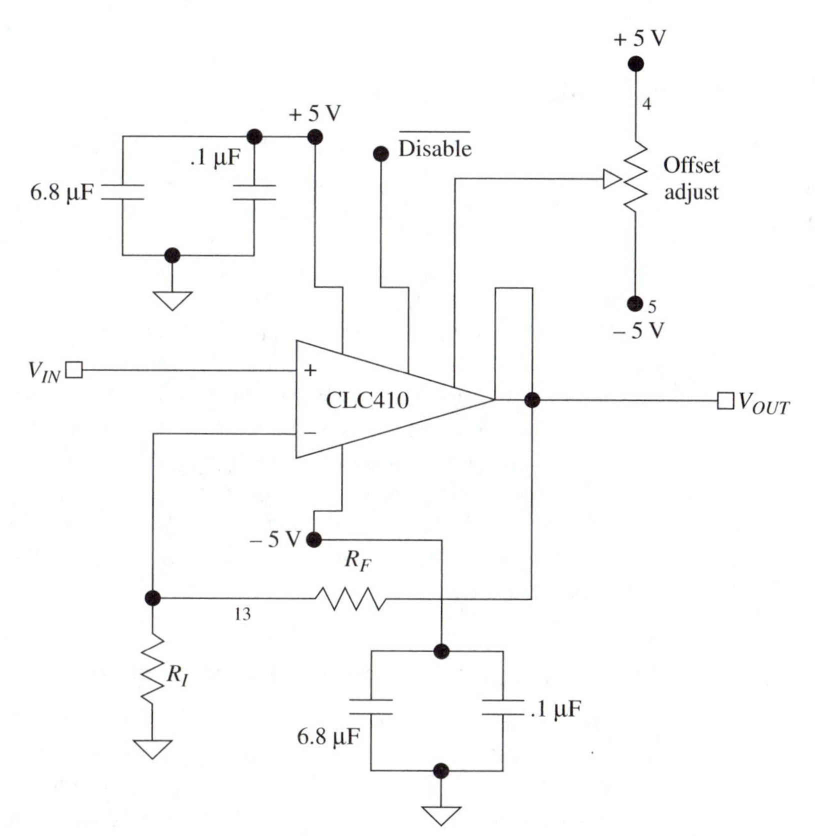

▶ FIGURE 7–41
CLC410 video non-inverting amplifier example

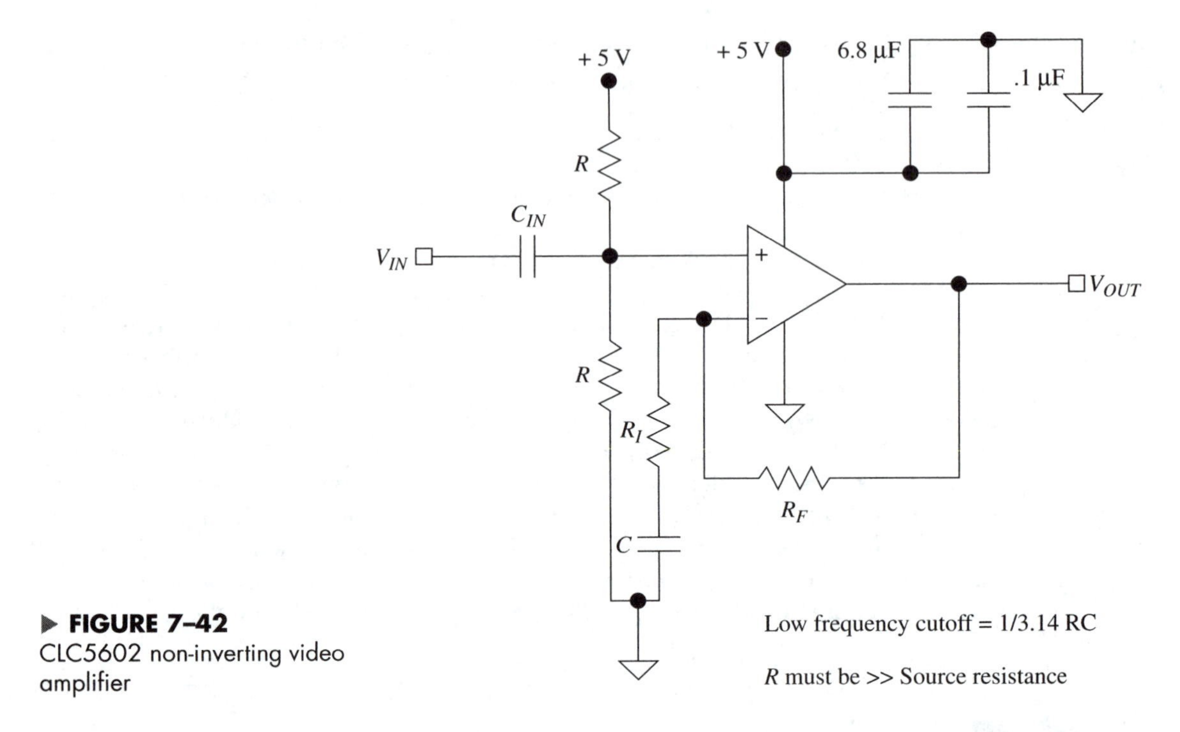

▶ **FIGURE 7–42**
CLC5602 non-inverting video amplifier

7–6 ▶ RF Amplifiers

RF amplifiers are increasingly applied in products and systems that involve telecommunications, global positioning systems, and wireless communications. Amplifiers designed for these applications utilize strategies developed for radio frequency (RF) and microwave (MW) circuits. It is important to realize that as transmitting signal frequencies increase, so does the information-carrying capacity of these circuit. This is easiest to see from the digital perspective. Let's say that one bit of information is transmitted for every low to high transition, and we compare the number of bits per second that can be transmitted at 1 MHz versus 1 GHz. At 1 GHz, 1000 times as much information can be transmitted in the same amount of time as a 1-MHz signal. This section will describe only the very basic elements of these strategies, as a more thorough coverage is well beyond the scope of this text. This topic requires a thorough re-discussion of many circuit elements due to the different ways that they react at these frequencies. Figure 7–43 shows the IEEE frequency spectrum. This section covers amplifiers that are designed to function generally at 1 GHz and above.

Passive Component Changes at High Frequencies

Conductors are typically viewed as simply a zero resistance, capacitance, and inductance at low-frequency operation. Actually, we know that conductors do have some resistance to all current flow and also some level of self-inductance for AC current flow. At high frequencies conductors experience significant increases in

Frequency Band	Frequency	Wavelength
ELF (Extreme Low Frequency)	30–300 Hz	10,000–1000 km
VF (Voice Frequency)	300–3000 Hz	1000–100 km
VLF (Very Low Frequency)	3–30 kHz	100–10 km
LF (Low Frequency)	30–300 kHz	10–1 km
MF (Medium Frequency)	300–3000 kHz	1–0.1 km
HF (High Frequency)	3–30 MHz	100–10 m
VHF (Very High Frequency)	30–300 MHz	10–1 m
UHF (Ultra High Frequency)	300–3000 MHz	100–10 cm
SHF (Super High Frequency)	3–30 GHz	10–1 cm
EHF (Extreme High Frequency)	30–300 GHz	1–0.1 cm
Decimillimeter	300–3000 GHz	1–0.1 mm
P Band	0.23–1 GHz	130–30 cm
L Band	1–2 GHz	30–15 cm
S Band	2–4 GHz	15–7.5 cm
C Band	4–8 GHz	7.5–3.75 cm
X Band	8–12.5 GHz	3.75–2.4 cm
Ku Band	12.5–18 GHz	2.4–1.67 cm
K Band	18–26.5 GHz	1.67–1.13 cm
Ka Band	26.5–40 GHz	1.13–0.75 cm
Millimeter wave	40–300 GHz	7.5–1 mm
Submillimeter wave	300–3000 GHz	1–0.1 mm

FIGURE 7–43
IEEE Frequency Spectrum

their resistance and inductance due to what is called the *skin effect* and a larger value of self-inductance. The skin effect is a phenomenon that describes the tendency for current to flow in the outer diameter of the wire as the frequency of the current flow increases. Because less and less current flows in the center region of the conductor, the resistance of the conductor increases with the frequency of the current flow. Also, the self-inductance that is always present when AC current flows through a wire is more pronounced at higher frequencies.

Circuits are also affected at high frequencies by the presence of stray capacitance that results from the existing capacitance between two conductors, two components, or between any conductor or component and ground. Stray inductance is also experienced by the parasitic inductance of conductors and components.

Resistors functioning in high-frequency circuits exhibit all three of the passive electrical impedance characteristics shown in Figure 7–44. The resistance is larger than the nominal resistance due to the skin effect, and the lead inductance and parasitic capacitance also are factors. Of course, the magnitude of each parameter value differs for resistors fabricated with different technologies—i.e., carbon composition, wire wound, and metal film. Carbon composition resistors experience a larger capacitance than other resistor types. Intuitively, wire wound

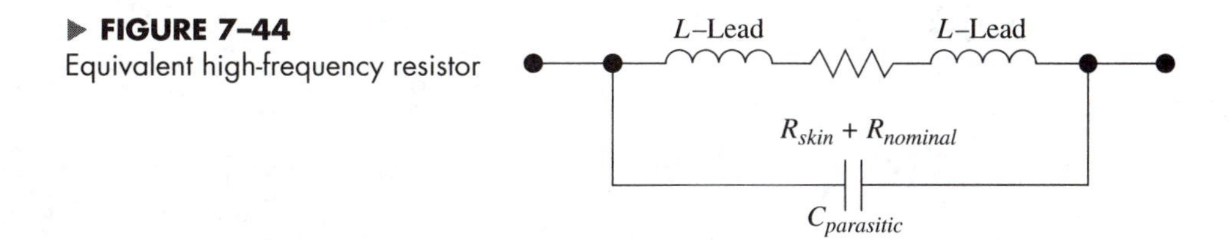

FIGURE 7–44
Equivalent high-frequency resistor

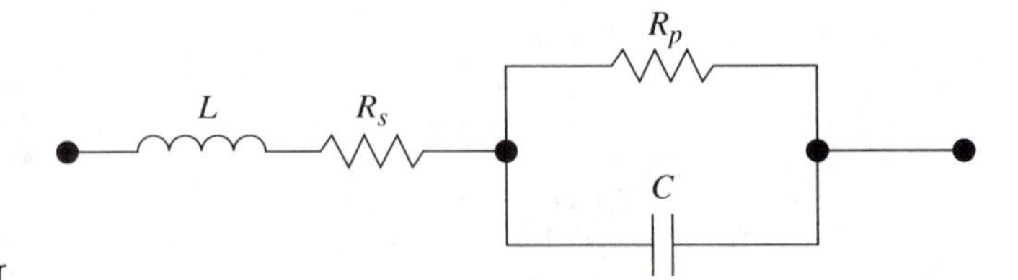

FIGURE 7–45
Equivalent high-frequency capacitor

resistors have a higher lead inductance. Thin film chip resistors offer the lowest values of inductance and capacitance. This results in a net decrease in resistance for all resistor types because the parasitic capacitance becomes the predominant factor.

The frequency effect of capacitors was discussed in Chapter 4; however, the high-frequency aspect was not discussed in detail. In this case the lead inductance becomes significant and at some point overtakes the capacitance at the resonant frequency. After this point, the capacitor begins to function like an RL circuit instead of the intended RC function (see Figure 7–45).

The equivalent high-frequency inductor is shown in Figure 7–46. As with the capacitor, there exists a resonant frequency after which the inductor begins to act more like a capacitor instead of the intended resistance to changing current.

It is easy to see that a simple RLC circuit becomes a complicated connection of resistive, capacitive, and inductive elements when analyzed at high frequencies.

Active Component Selection at High Frequencies

The typical pn semiconductor diode is not capable of operation in the RF region. Schottky diodes are a type of diode that can switch much faster than standard diodes and thus are capable of operation at high frequencies. Another style of diode that can operate in the RF region is the PIN diode, which is characterized by an intrinsic semiconductor layer that resides between the p and the n material. The PIN diode is used in RF applications for high-frequency switches and variable resistors. The inherent capacitance that exists between the pn junction is used to fabricate capacitors from diodes. These are called *varactor diodes* and are used in RF circuits as capacitors. Another very fast diode device, called the *tunnel diode*, has extremely high doping levels of the p and n materials that promote fast switching speeds.

Special BJT transistors are capable of operation at RF frequencies. These are called simply *RF transistors* and they feature very low base-to-collector and base-to-emitter capacitance. FET transistors require special changes to promote oper-

FIGURE 7–46
Equivalent high-frequency indicator

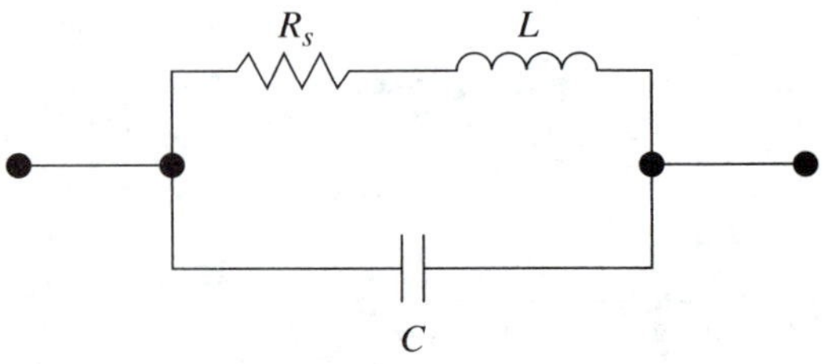

ation at high frequencies because of their inherent gate-to-source and gate-to-drain capacitance. Some of these special high-frequency FETs are called *Metal Insulator Semiconductor FET (MISFET)* and *Metal Semiconductor FET (MESFET)*.

RF Amplifier Design

The design of RF amplifiers is more complex than any other type of amplifier design because, as we have seen already, even a component as basic as a conductor now must be represented as a complex circuit element. A block diagram for an RF amplifier is shown in Figure 7–47.

The signal that is generated by the RF source is input through the input matching circuit because the impedance of the RF source must be matched to the input impedance of the amplifier. This is done to maximize power transfer and minimize signal reflections. After the signal is amplified, the output impedance of the amplifier must be matched to the load impedance again to maximize power transfer and minimize reflections.

The performance of the RF amplifier is characterized by the following parameters:

1. Power output and input
2. Efficiency
3. Operating frequency and bandwidth
4. Gain level over the operating bandwidth
5. Noise levels
6. Input and output power reflections

The design of RF amplifiers involves the development of a particular class of amplifier (A, B, AB, etc.), using a discrete transistor as completed for other transistor amplifiers. However, in this case the complex input impedance of the transistor amplifier is compared to the impedance of the RF source. Then an input matching circuit is developed. This process is repeated on the output side of the circuit.

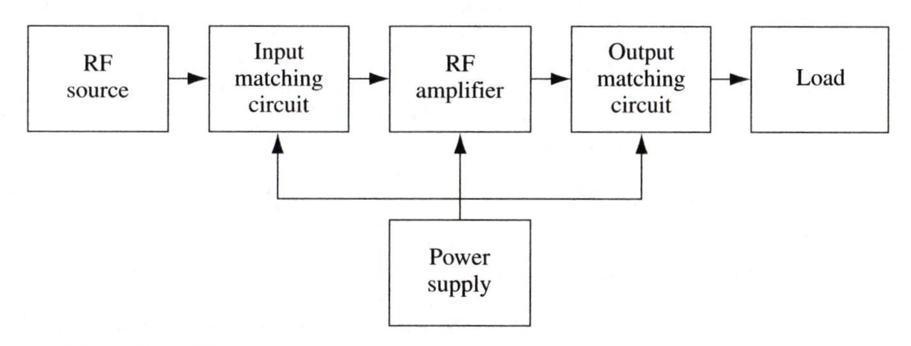

▲ **FIGURE 7–47**
RF amplifier block diagram

Summary

This concludes our discussion of amplifier design. The range of amplifier types reviewed in this chapter uncovers many different design criteria—from DC amplifiers, where DC offset and gain accuracy are most important, to RF amplifiers, where impedance matching and minimizing reflections are the most critical. Each amplifier application has its very own agenda, yet there are many aspects of amplifier design that are common to all. The applications for amplifiers range from industrial instrumentation, audio appliances, television, video networks, and telecommunications. The range of operating frequency most often drives the primary difference between the design criteria. Amplifier design concepts will continue development as new technology and applications pushes the frequency, performance, size, and efficiency requirements to even higher levels.

References

Bell, D. A. 1990. *Operational Amplifiers: Applications, Troubleshooting and Design.* Upper Saddle River, NJ: Prentice Hall.

Floyd, T. L. 1999. *Electronic Devices.* Upper Saddle River, NJ: Prentice Hall.

Ludwig, R., and Bretchko, P. 2000. *RF Circuit Design: Theory and Applications.* Upper Saddle River, NJ: Prentice Hall.

Malvino, A. P. 1999. *Electronic Principles.* Westerville, OH: Glencoe/McGraw Hill.

Radmanesh, M. M. 2001. *Radio Frequency and Microwave Electronics.* Upper Saddle River, NJ: Prentice Hall.

Self, D. 2000. *Audio Amplifier Design Handbook.* Woburn, MA: Newnes.

Soclof, S. 1991. *Design and Applications of Analog-Integrated Circuits.* Upper Saddle River, NJ: Prentice Hall.

Exercises

7–1 Explain the concept of "slew rate" and how it applies to amplifier design. Why is it an important amplifier design consideration? How can the slew rate for an amplifier be determined experimentally?

7–2 What is the relationship between an amplifier's closed loop gain and its bandwidth? In other words, if the closed loop gain of an amplifier increases, what happens to its bandwidth?

7–3 Explain the difference between single-ended and differential amplifiers.

7–4 In your own words, define the term *common mode rejection ratio (CMRR).* What type of amplifier does it most apply to?

7–5 Design an inverting DC amplifier circuit that will have a gain of –25 and an input impedance of approximately 100 kΩ. Specify the compensating resistor value as well as the feedback and input resistor values.

7–6 Design a non-inverting amplifier that will have a gain of +21. Show all resistor values.

7–7 A voltage follower circuit provides an output that equals the input. What function is provided by a such a circuit?

7–8 Why is it often necessary to have both + and –power supply voltages to power op amp circuits?

7–9 When designing DC amplifiers, why is it important to keep the input bias currents close to the same value?

7–10 Which two op amp characteristics limit their performance as the signal frequency increases?

7–11 Explain the benefits that an instrumentation amplifier offers over a standard differential amplifier.

7–12 Design an AC non-inverting amplifier with high input impedance and powered by a dual supply. See the circuit shown in Figure 7–20b. Calculate all component values that will accommodate a low frequency cutoff point of 40 Hz. The op amp being used has a maximum bias current of 500 nA. The gain of the circuit should be +25 and the load being driven is 1200 Ω.

7–13 Design an AC inverting amplifier with a gain of –30 powered by a dual supply. The operating frequency range is 60 Hz to 18 kHz. The amplifier must drive a load impedance of 32 Ω.

7–14 Design an AC non-inverting amplifier operated from a single power supply. See the circuit shown in Figure 7–25. Calculate all component values that will accommodate a low frequency cutoff point of 40 Hz. The op amp being used has a maximum bias current of 500 nA. The gain of the circuit should be +25 and the load being driven is 1200 Ω.

7–15 Explain the difference between Class A and B amplifiers. List at least one positive and one negative aspect of each.

7–16 Explain the concept of the Class AB amplifier. How does it resolve the negative aspects of both the Class A and B amplifiers?

7–17 What is the basis for disagreement between audio scientists and audio subjectivists, regarding the performance of audio amplifiers?

7–18 List the key design considerations when designing a discrete transistor video amplifier.

7–19 Which parameters most characterize the performance of RF amplifiers?

7–20 Why is RF amplifier design much more complicated than other types of amplifier design?

8 Oscillators and Function Generators

Introduction

There are two basic types of oscillators: relaxation and positive feedback oscillators. Relaxation oscillators are a class of oscillators that use the charge and discharge capacitors to induce the switching of circuit elements to create certain ongoing waveforms. The amplifier circuits discussed in Chapter 7 all use the concept of negative feedback to maintain a constant gain and provide good frequency response. Positive feedback results when an amplifier output signal is sensed at its input in phase with the original input signal. When positive feedback occurs, the amplified output signal is amplified further and further. This can happen in many amplifiers. The most common example is a public address system when the PA system speakers are placed too close and directed toward the microphone input. This is an example of positive feedback, as the amplified output is fed back into the input and then re-amplified. As the process is repeated over and over, the result is a high-pitched squeal most often referred to as, simply, *feedback*. The PA amplifier has become unstable and now functions as an oscillator. Positive feedback oscillators are a class of circuits using the concept of positive feedback to generate a particular periodic waveform, with a DC supply voltage as the only input signal.

It can be said that amplifiers and positive feedback oscillators perform opposite functions: amplifiers must amplify a signal and maintain stability (not go into oscillation) while positive feedback oscillators should generate a periodic waveform at a specific amplitude (without amplification). It can also be said that every amplifier has a tendency to oscillate while every positive feedback oscillator has some disposition toward amplification. In order for a positive feedback oscillator to sustain oscillation, the phase shift of the signal fed back to the input must be 0 and the closed loop gain of the fed-back signal must be 1. If the closed loop gain is less than 1, the signal will eventually attenuate to 0. For closed loop gains larger than 1 the signal increases continuously until limited by the DC power supply.

The simplest oscillator is a clock circuit similar to those used in most digital circuits. There are many other uses for other oscillator waveforms in what are classified as analog type timing circuits. Finally, oscillators are used to develop carrier frequencies in communications circuits. In this chapter we will discuss a variety of oscillator and function generator circuits. The specific areas covered are as follows:

- Oscillator and Function Generator Performance
- Clock Circuits
- Square-wave Generators
- Triangle-wave Generators
- Voltage-controlled Oscillators (VCOs)
- Sine-Wave Generators
- Pulse Generators
- Integrated Circuit Function Generators

8–1 ▶ Oscillator and Function Generator Performance

Oscillators and function generators produce an output waveform of a particular shape, amplitude, frequency, and duty cycle that is developed from a DC power supply input. The performance of an oscillator or function generator is determined by how well it meets these output signal requirements. In addition, oscillator performance is judged by how much current the output can provide as well as its efficiency relative to developing the output from the DC supply.

Amplitude: The amplitude determines the positive and negative peak values of the output waveform. The oscillator's amplitude accuracy is determined by how closely the oscillator can provide an output amplitude in the range required. In most cases the amplitude accuracy will be determined by the resolution provided for adjusting the oscillator output amplitude.

Frequency: The frequency performance is simply a measurement of how closely the actual oscillator output frequency equals the desired output frequency. This is usually determined by the tolerances and resolution of the circuit elements that the feedback mechanism comprises.

Stability: The amount of variation from the steady state frequency over time

Harmonic Content: The amount of frequencies other than the desired fundamental frequency that are present in the output signal

Waveform: This performance factor is a measurement of how well the output waveform approximates the desired output waveform. Square waves are relatively easy to generate. Triangle and sawtooth waves can be generated by charging a capacitor with a constant current source. Sine waves are developed almost perfectly below 1 MHz with RC components, while LC circuits are used above 1 MHz.

Duty Cycle: Duty cycle represents the amount of time the signal is positive vs. negative for analog signals and high vs. low for digital signals. In most oscillators and function generators, duty cycle is set at 50%. In more advanced function generators and pulse circuits, the duty cycle can be adjusted. In any event the duty cycle performance is determined by how well the oscillator provides the desired output duty cycle.

Output Power: This is usually specified by the output current that the oscillator can provide.

Efficiency: This is a measure of the efficiency of the oscillator—output power divided by the input power from the DC supply.

8–2 ▶ Clock Circuits

Clock circuits are most often used in digital circuits to synchronize their operation or to count real or system time. Digital clock circuits have a positive, peak amplitude around 3 to 5 V and a negative peak amplitude of approximately 0 V. The actual peak values will be defined by whatever type of logic (TTL, CMOS, etc.) is being driven by the clock circuit. The duty cycle will usually be 50%, meaning that the high and low time periods will be equal.

A very basic clock circuit, a type of relaxation oscillator, is shown in Figure 8–1. This type of clock can be constructed from many different types of logic, but we will focus on TTL. On power up, the input to the logic inverter is low; therefore, the initial inverter output is a high logic level. As soon as the inverter output goes high, it begins to charge up capacitor *C* from the output through the connection to resistor *R*. When the capacitor charges to a voltage equal to a logical high input, the inverter output will switch to a low logic level. The capacitor will then discharge through resistor *R* to ground until a logical low input voltage to the inverter is reached. The inverter output will then again become high.

The clock frequency is determined by the *RC* time constant. The formula is shown as Equation 8–1:

$$\text{Clock Frequency} = .8/RC \tag{8–1}$$

Because the charging and discharging *RC* values are the same, the duty cycle is very close to 50%. The other constraint placed on the circuit is the maximum values for

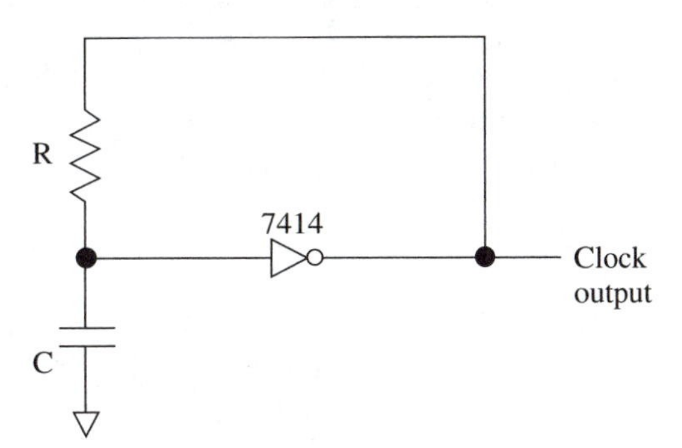

▶ **FIGURE 8–1**
Schmitt trigger oscillator

R. For a 74LS14 the maximum value for $R = 2\ k\Omega$. For a 7414 the maximum value for $R = 500\ \Omega$. Because the *R* and *C* values determine the clock frequency, its accuracy is determined by the tolerance of these components. While the *R* value could be adjusted with a potentiometer to compensate for any variations in the *C* value, this circuit would also be subject to temperature-induced variations in the *R* and *C* values and aging of these components over time. The overall frequency accuracy required, combined with the ambient temperature specs for the circuit, will determine the acceptability of this type of *RC* clock circuit.

A more accurate and temperature-independent clock circuit can be attained with the use of crystal oscillators. Crystal oscillators are constructed with a component called a *quartz crystal*. These crystals are two terminal components that are fabricated from a small piece of quartz crystal. When cut to a specific size and shape, the crystals can be made to resonate at a particular frequency. The frequency of resonance is very accurate, not sensitive to ambient temperature, and does not change over time. Crystals oscillate in either the fundamental or overtone mode. The fundamental mode is its normal listed resonant frequency. Because there is a limit to how small a crystal can be sliced, the maximum fundamental mode frequency for most crystals is 20 MHz. To oscillate at frequencies greater than 20 MHz, the crystal resonates in the overtone mode, which usually involves odd integer multiples of the fundamental frequency.

A variety of crystal-controlled clock circuits using different logic devices is shown in Figure 8–2. All of these circuits are based on the accurate resonant frequency established by the crystal as a feedback device from the output back to the circuit input. Figure 8–2a shows a crystal-controlled clock circuit using a 74LS04 TTL inverter. Figure 8–2b utilizes a CMOS inverter while Figure 8–2c shows a circuit that employs a CMOS Schmitt-trigger.

Up to this point all of the clock circuits discussed have been digitally oriented; their output amplitudes were compatible with digital logic circuits. Clocks, or astable multivibrators, are often needed in analog circuits where the output amplitude must be greater than digital logic levels. These can be developed with transistor or op amp circuits or with the use of 555 counter-timer ICs. A 555 counter-timer is a very versatile circuit that can function as a free-running astable multivibrator (a clock generator) or a monostable multivibrator (a one shot) or a time delay function. The 555 includes two comparators, an open collector transistor, a flip-flop, and an output buffer (see Figure 8–3). The upper comparator has its negative input connected to $V_{CC} \times .666$ and the lower comparator has its positive input connected to $V_{CC} \times .333$. Because the 555 is not designed around TTL or CMOS logic the V_{CC} power connection can be anywhere from approximately 4.5 to 18 V DC, which is the range of the possible output amplitudes from the circuit. Then again, if a digital logic circuit requires a relaxation oscillator, the 555 can be used in this application by making V_{CC} equal to 5 V. Let's look at an example.

Example 8–1

A free-running pulse circuit is needed for an analog circuit that will operate at 100 kHz. The output of the pulse circuit should be +12 V when high and 0 V when low. The duty cycle should be 60%.

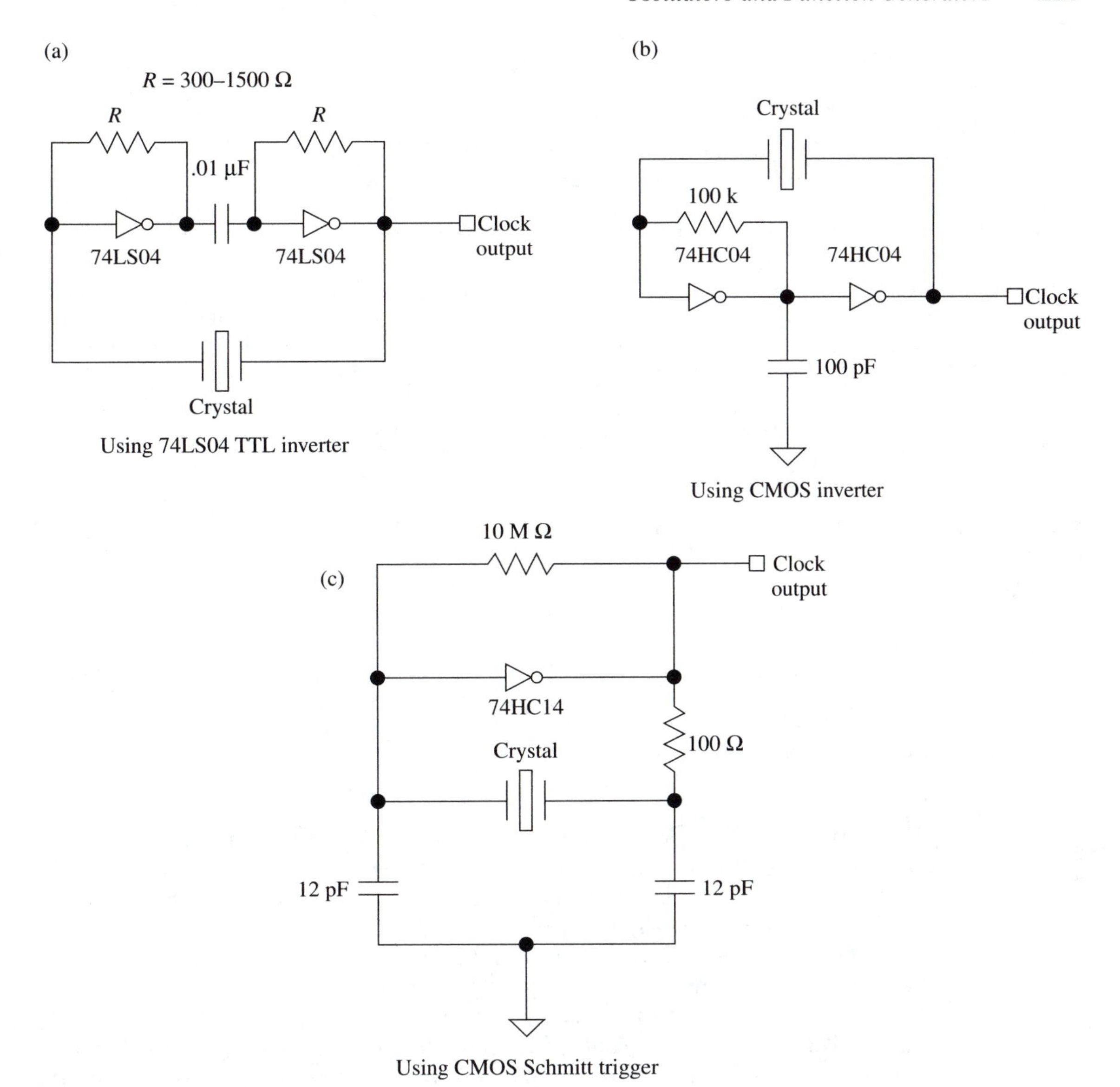

▲ **FIGURE 8–2**
Crystal-controlled clock circuits

Solution

Figure 8–4 shows the 555 counter-timer configured to operate in the free-running mode. Notice that both the upper and lower comparators have their available input connected to one side of capacitor *C*. The upper comparator has a $V_{CC} \times .66$ reference voltage of $12 \times .66 = 7.92$ V. The lower comparator is referenced at $V_{CC} \times .33$ $= 12 \times .33 = 3.96$ V. Notice that the 555's internal transistor collector is also connected to the same capacitor connection as the comparators.

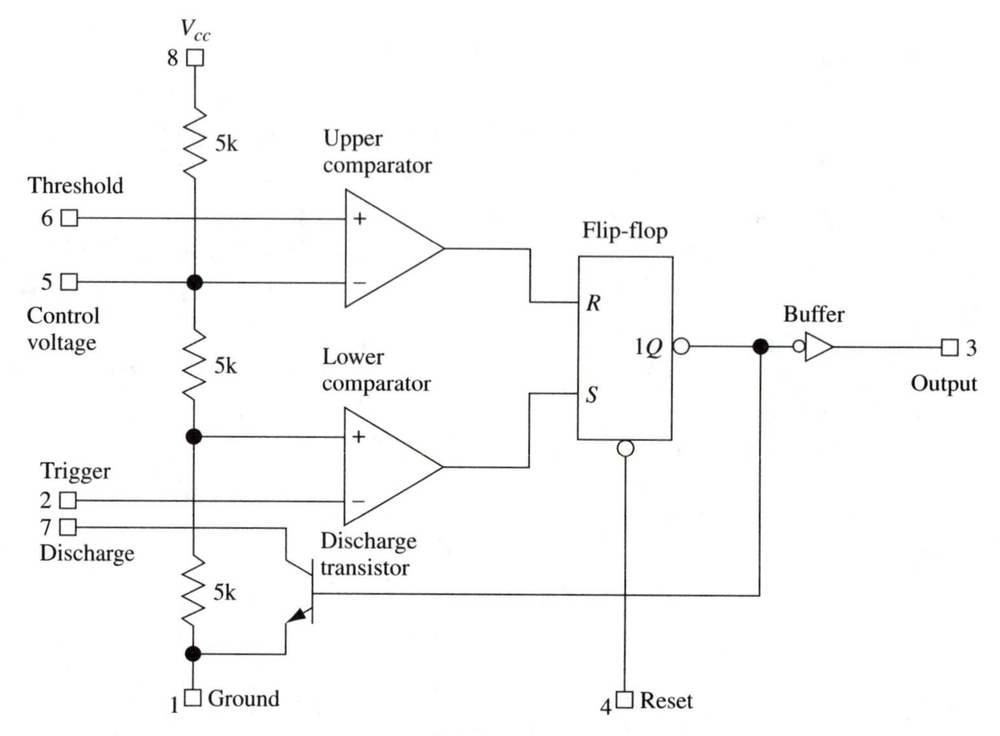

▲ **FIGURE 8–3**
555 counter-timer circuit

Initially the 555 powers up with Q = low and Q' = high. Since Q' is connected to the 555's output terminal, its output is initially high, or in this case, 12 V DC. The Q output from the flip-flop drives the transistor; therefore the transistor is initially off. The circuit begins operation by charging capacitor C through R_A and R_B. When the voltage at C reaches 7.92 V, the upper comparator will switch, setting the RS flip-flop and causing the 555 output to go to 0 V and the 555's internal transistor is turned on. When the transistor is on, it pulls the positive side of R_B to ground. This causes the capacitor to discharge through R_B to ground. When the capacitor voltage discharges down to 3.96 V, the lower comparator resets the RS flip-flop returning the 555 output to +12 V and turning the transistor off. This process repeats itself in time periods determined completely by R_A, R_B, and C.

It is important to note that because the capacitor charges through R_A and R_B, but discharges only through R_B, the 555 by itself can never produce a true 50% duty cycle output. When R_B is > R_A a 50% duty cycle is approached.

In order to provide a +12-V amplitude for the positive pulse, V_{CC} should be +12 V DC. To complete the design, all that is needed is the calculation of the values for R_A, R_B, and C. The formulas are as follows:

$$555 \text{ frequency } f = 1.44/(R_A + 2R_B)C = 100 \text{ kHz} \quad (8\text{–}2)$$

$$\text{Duty Cycle} = (R_A + R_B)/(R_A + 2R_B) = 60\% \quad (8\text{–}3)$$

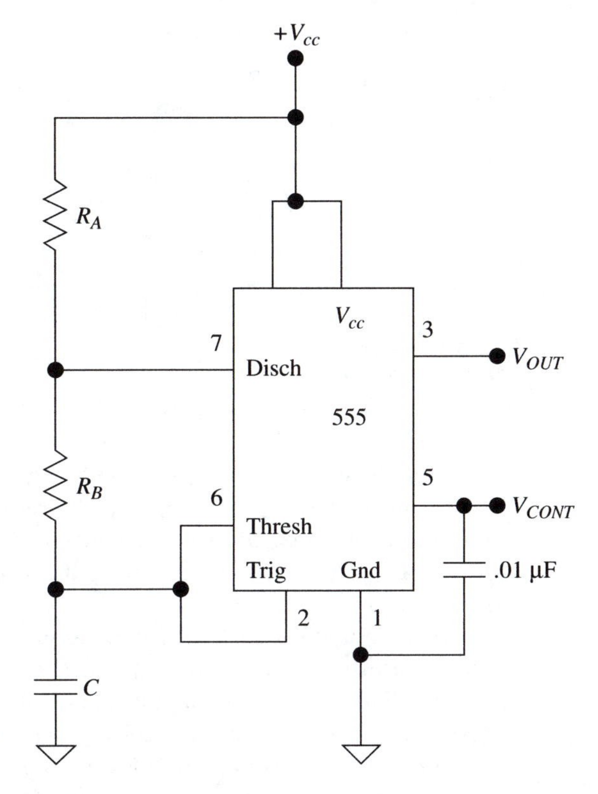

▶ **FIGURE 8–4**
555 Counter-timer in free running mode

There are two equations with three unknowns. First select a standard and available capacitor value and then solve the equations for R_A and R_B. Selecting the capacitors first is the best approach because there are more standard resistor values available, or a trimpot could be used for either resistor value.

Let $C = .001\ \mu F$

Solving the duty cycle equation (Equation 8–3):
$R_A = .5R_B$
Substituting in the frequency equation (Equation 8–2):
$R_B = 1.44/(.001\ \mu F \times 100\ kHz \times 2.5)$
$R_B = 5760\ \Omega$
$R_A = 2880\ \Omega$

8–3 ▶ Square-wave Generators

All of the circuits discussed in Section 8–2 developed waveforms that alternated between a positive high pulse and 0 V. A true square wave is a waveform that possesses positive and negative pulses with equal amplitudes and duration (see Figure 8–5).

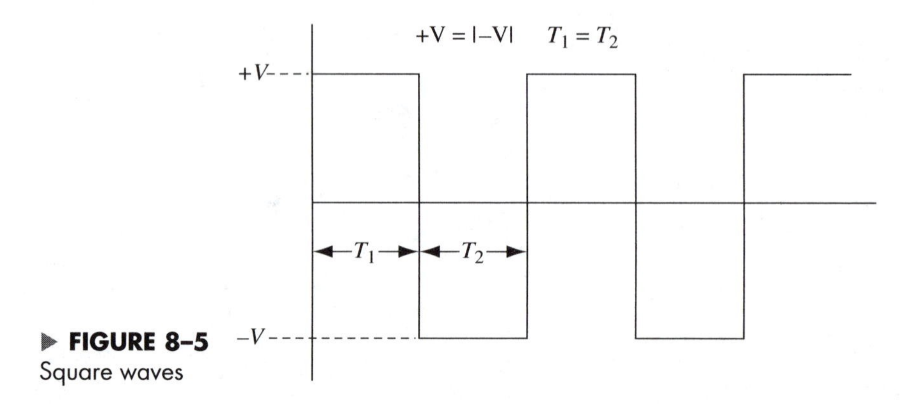

▶ **FIGURE 8–5**
Square waves

Square waves can be generated by a variety of op amp circuits where the op amp is configured as a comparator (with positive feedback), and a capacitor is charged and discharged in a manner similar to the clock circuits discussed in Section 8–2. Figure 8–6 shows an op amp relaxation oscillator that generates a square wave.

The circuit in Figure 8–6 is an op amp with positive feedback, which functions as a comparator. Initially, the capacitor C is uncharged making the negative op amp input equal to 0, so the op amp's output equals the positive saturation voltage, $+V_{SAT}$ for the op amp (usually V_{CC}–2 V). The voltage fed back to the positive input is determined by the voltage divider, $R/(R_{FB} + R)$. With the op amp output at $+V_{SAT}$ the voltage at the positive op amp input is $+V_{SAT}\ (R/(R_{FB} + R))$. The positive op amp output voltage begins to charge the capacitor through R_C. When the capacitor voltage exceeds the voltage present at the op amp's positive terminal, the op amp output will switch to $–V_{SAT}$. The voltage presented to the positive op amp terminal is now $–V_{SAT}\ (R/(R_{FB} + R)$. The capacitor now discharges through R_C until it becomes less than the voltage at the op amp's positive input, causing the op amp's

▶ **FIGURE 8–6**
Square-wave generator

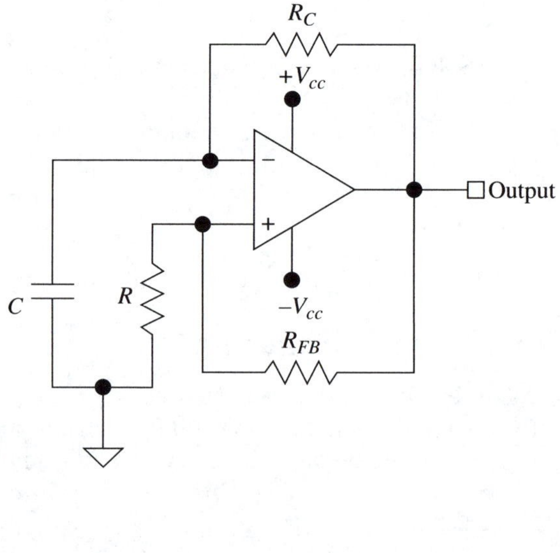

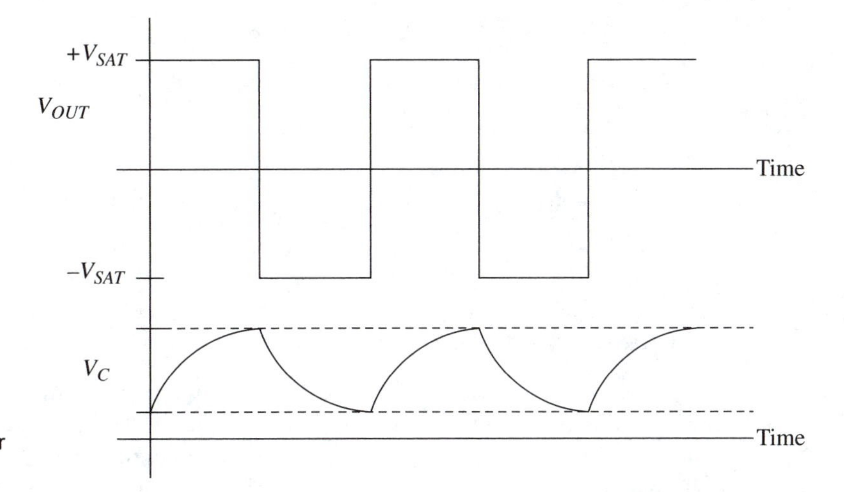

FIGURE 8–7
Square-wave generator waveforms

output to switch back to $+V_{SAT}$. This represents one complete cycle for this circuit. See Figure 8–7 for the resulting waveforms. This circuit develops a square wave with amplitudes equal to $+V_{SAT}$ and $-V_{SAT}$ at a frequency determined by R, R_{FB}, C, and R_C. The period for the waveform is given by the equation:

$$T = 2R_C C \ln (1 + 2R/R_{FB}) \qquad (8\text{–}4)$$

Example 8–2

Design a circuit that will generate a square wave with ±10 volt amplitudes at a frequency of 50 kHz.

Solution

The circuit shown in Figure 8–6 will perform this function if the proper values of V_{CC}, resistors, and capacitor are chosen.

1. V_{CC} is selected as ±12 V so that $\pm V_{SAT} = \pm 10$ V.
2. Equation 8–4 defines the relationship between the circuit component values and the period of the square waveform. If f must equal 50 kHz, then T = 1/50 kHz = 20 microseconds.
3. Equation 8–4 includes all four of the remaining circuit variables. In order to resolve this problem, some of the component values must be selected arbitrarily. If R and R_{FB} are selected such that $R = .86\ R_{FB}$, then the logarithmic function included in Equation 8–4 reduces to $\ln\ (1 + 2\ (.86\ R_{FB}/R_{FB})) = \ln 2.72 = 1$. Let $R_{FB} = 33$ kΩ, then $R = 28$ kΩ.
4. For R = .86 R_{FB}, Equation 8–4 reduces to T = 20 micro-seconds = 2 $R_C C$. Let C = .01 µF. Then R_C = 20 micro-seconds/.01 µF = 2000 Ω.

The complete design solution is:

V_{CC} is selected as ±12 V
R_{FB} = 33 kΩ
R = 28 kΩ
C = .01 µF
R_C = 2000 Ω

8–4 ▶ Triangle-wave Generators

A triangle waveform increases linearly to a maximum value and decreases at the same rate to a minimum negative voltage equal in magnitude to the maximum voltage. The duty cycle of a true triangle waveform is normally 50%. A common circuit that will generate both a square and triangle waveform is shown in Figure 8–8. This circuit includes two op amps; U_1 is connected as a comparator while U_2 is configured as an integrator. The circuit operation is as follows:

1. Initially assume that the output of U_1 is at $-V_{SAT}$ and the output of $U_2 = 0$.
2. The integrator circuit integrates the constant $-V_{SAT}$ presented to its inverted input causing the output of U_2 to increase linearly until $V_{TRIANGLE}/R_1 = V_{SQUARE}/R_F$. At this point the output of U_1, V_{SQUARE}, switches to equal $+V_{SAT}$.
3. The integrator U_2 integrates the constant $+V_{SAT}$ signal, decreasing linearly again until $V_{TRIANGLE}/R_I = V_{SQUARE}/R_F$, where one cycle of the circuit is complete and it is repeated again and again.

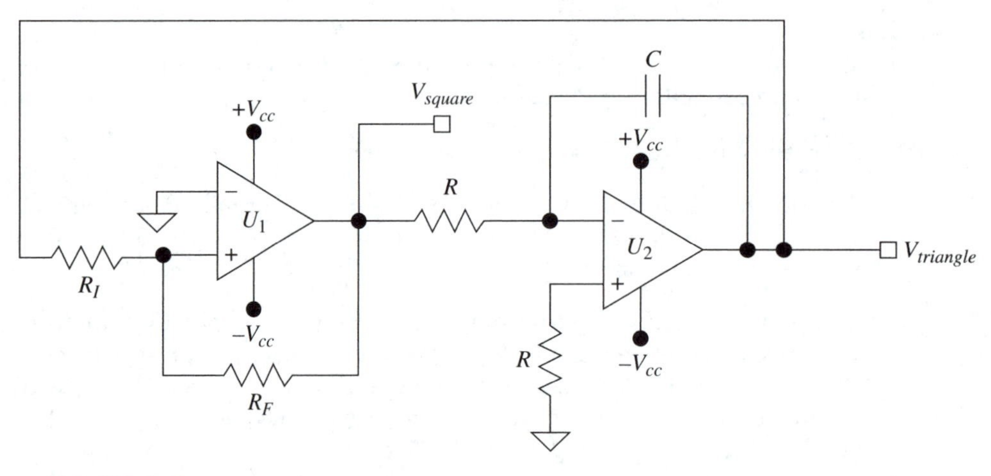

▲ FIGURE 8–8
Triangle/square wave generator

The equation that defines the relationship between the components and the resulting frequency for both the square and triangle waveforms is:

$$\text{frequency } f = (R_F/R_1)/4RC \tag{8–5}$$

The amplitude of the square wave equals $\pm V_{SAT}$, while the peak amplitude for the triangle output is determined by $V_{SAT} \times R_I/R_F$.

Example 8–3

Design a circuit that will generate both a square wave and a triangle wave at a frequency of 20 kHz. The amplitude of the square wave shall be ±8 V and the peak amplitude of the triangle wave should be ±5 V.

Solution

1. V_{CC} is selected as ±10 V so that the amplitude of the square wave will be approximately equal to ±8 V.
2. The peak amplitude of the triangle wave must equal 5 V, which equals $V_{SAT} \times R_I/R_F$. Therefore, $5 = 8 \times R_I/R_F$ or $5/8\ R_F = R_1$. Let $R_F = 20\ \text{k}\Omega$. Then $R_I = 12.5\ \text{k}\Omega$.
3. Equation 8-5 defines the relationship between the frequency of the circuit and the component values. Substituting in the desired frequency and the results of step 2 above, $5/8\ R_F = R_I$, we have:

   $20{,}000\ \text{Hz} = (R_F/.625\ R_F)/4RC$
   Let $C = C = .01\ \mu\text{F}$
   Solving for $R = 1.6/(4 \times 20{,}000\ \text{Hz} \times .01\ \mu\text{F}) = 2000\ \Omega$

   The complete design solution is:

   V_{CC} is selected as ±10 V
   $R_F = 20\ \text{k}\Omega$
   $R_I = 12.5\ \text{k}\Omega$
   $C = .01\ \mu\text{F}$
   $R = 2000\ \Omega$

8–5 ▶ Voltage-controlled Oscillators

Sawtooth waveforms are positive voltage waveforms that increase in amplitude at one rate and decrease in amplitude by a much greater rate. Sawtooth generators are a form of oscillator called *voltage controlled oscillators (VCOs)*, because the frequency of the sawtooth generator can be changed by a voltage input to the VCO circuit. An example sawtooth generator is shown in Figure 8–9.

The sawtooth generator shown in Figure 8–9 consists of an op amp integrator, U_1 and an op amp comparator, U_2. A negative voltage V_{IN} is input to the circuit and is the ultimate control voltage that determines the frequency of the oscillator

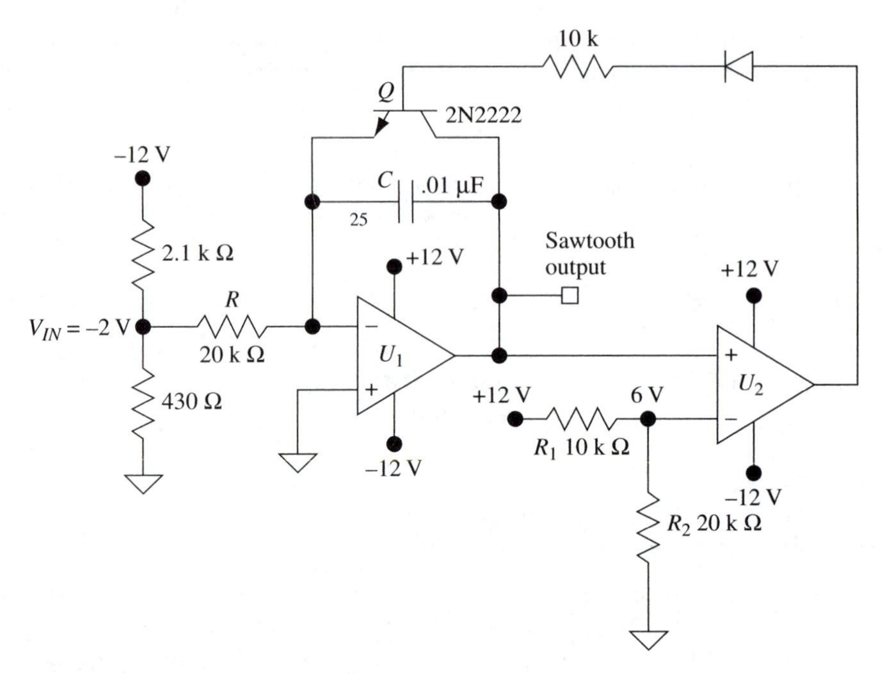

▲ **FIGURE 8–9**
Sawtooth generator

output. V_{IN} is connected to the integrator U_1 through R. Initially the comparator U_2 is at approximately negative rail and transistor Q is off. The integrator charges capacitor C at a rate determined using Equation 8–6:

$$\text{Rate of rise } V_{OUT} = V_{IN}/RC \quad (8\text{–}6)$$

Therefore the rate of rise for V_{OUT} is determined by the input voltage V_{IN}, R, and C. Since R and C are usually fixed in the circuit, V_{IN} is the primary voltage for determining the oscillator frequency. V_{OUT} increases until it is equal to the voltage labeled V_{PEAK}, at which point it causes the comparator op amp U_2 to switch bringing its output to approximately positive rail, switching transistor Q on. When transistor Q is turned on, it provides a relative short across the capacitor, forcing it to discharge very rapidly. Consequently, V_{OUT} decreases quickly to 0, which switches the comparator U_2 and transistor Q off. V_{PEAK} determines the peak amplitude of the sawtooth and also has an impact on the frequency. V_{PEAK} is fixed in most circuit applications, making the V_{IN} the exclusive oscillator control voltage. The frequency of the sawtooth waveform is defined by the following equation:

$$f = (1/RC)(|V_{IN}|/V_{PEAK}) \quad (8\text{–}7)$$

Example 8–4

Design a sawtooth generator circuit that will output a sawtooth waveform with a peak amplitude of 6 V, with an operating frequency of 10 kHz, when V_{IN} equals

–2 V. Calculate the new oscillation frequency for the circuit designed if the V_{IN} becomes equal to –6 V.

Solution

1. V_{PEAK} equals the peak amplitude of the circuit; therefore, $V_{PEAK} = 6$ V. R_1 and R_2 must be selected to develop a voltage divider that will divide the +12 V in half.

2. Using equation 8–7:

   $f = (1/RC)(|V_{IN}|/V_{PEAK})$
   $10\text{ kHz} = (1/RC)(|-2|/6)$
   $30\text{ kHz} = 1/RC$ Let $C = .01\ \mu\text{F}$
   $R = 1/(30\text{ kHz} \times .01\ \mu\text{F}) = 3.3\text{ k}\Omega$

   The complete design solution is:

   V_{CC} is selected as ±12 V
   $R = 3.3\text{ k}\Omega$
   $R_1 = 10\text{ k}\Omega$
   $R_2 = 20\text{ k}\Omega$ potentiometer
   $C = .01\ \mu\text{F}$

3. For the final circuit the oscillating frequency for $V_{IN} = -6$ V can be found as follows:

   $f = (1/RC)(V_{IN}/V_{PEAK})$
   $f = (1/3.3\text{ k}\Omega \times .01\ \mu\text{F})\ (|-6|/6) = 30{,}303\text{ Hz}$

It should be noted that the sawtooth generator discussed in this section generates a sawtooth waveform with a very steep decline in the output voltage after reaching the peak value. More complicated sawtooth generator circuits provide the capability to specify the rate of decline as well as that of the increasing ramp voltage.

The circuit shown in Figure 8–10 is a very versatile voltage-controlled oscillator circuit that can generate any range of triangle, sawtooth, and square waveforms where the frequency is determined by an input control voltage. The input control voltage V_C is used to generate the increasing ramp integrator input voltage V_{CR} and the falling ramp voltage C_{CF}. V_{CR} can be adjusted with trimpot TP_1, while TP_2 adjusts V_{CF}. The voltage at the trimpot wiper arms are input to voltage follower U_1 to develop V_{CR} and inverting amp U_2 to create $-V_{CF}$. These control voltages are alternately applied to integrator U_5 by the CMOS switches O_1 and O_2. The integrator rise and fall rates are as follows:

$$V_{OUT}\text{ Rise Rate} = +V_{CR}/RC \qquad (8\text{–}8)$$

$$V_{OUT}\text{ Fall Rate} = -V_{CF}/RC \qquad (8\text{–}9)$$

The square wave output voltage amplitude V_{SQUARE} is determined by the rating of zener diodes D_{Z1} and D_{Z2}:

$$V_{SQUARE} = (D_{Z1}\text{ or }D_{Z2}\text{ rating}) + .7\text{ V} \qquad (8\text{–}10)$$

(a)

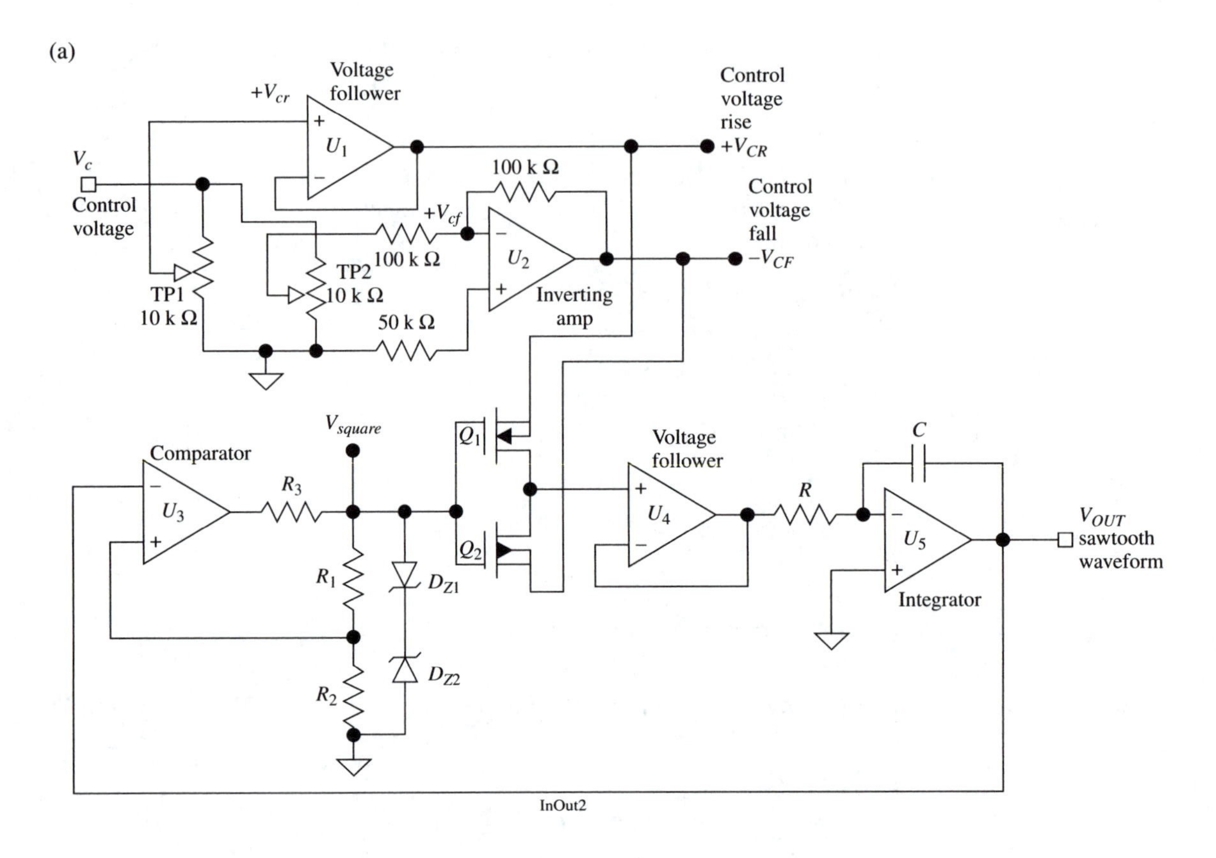

(b)

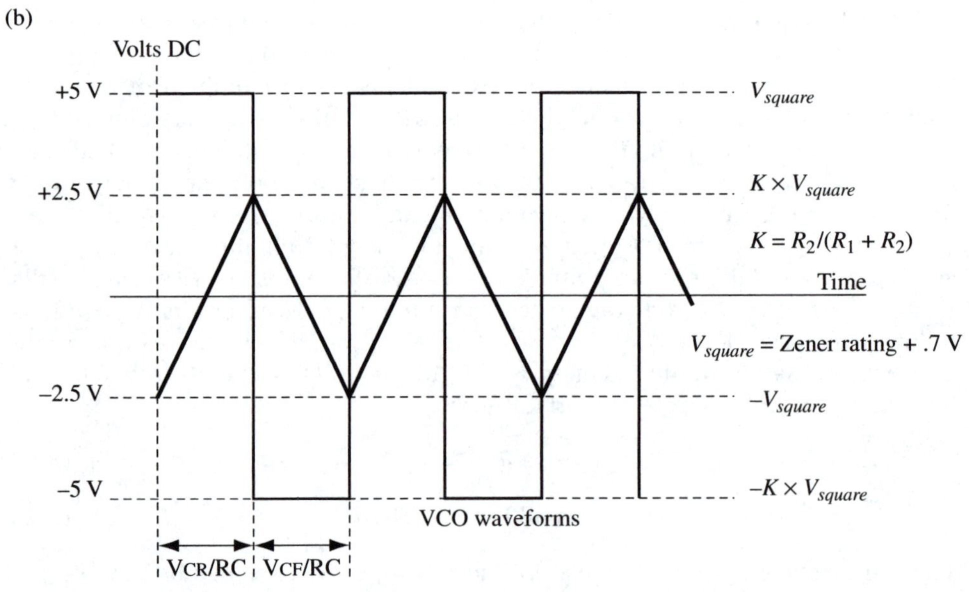

▲ FIGURE 8–10
Voltage-controlled oscillator circuit

The peak output voltage V_{OUT} is determined by comparator U_3 and the following equation:

$$V_{OUT}\ \text{Peak} = (R_2/(R_2 + R_1)) \times V_{SQUARE} \qquad (8\text{–}11)$$

The input control voltage will determine the frequency of V_{OUT} while the shape of V_{OUT} is set by V_{CR} and V_{CF}. If $V_{CR} = V_{CF}$, then the waveform is a triangle wave. If $V_{CR} \neq V_{CF}$, the waveform has more of a sawtooth shape. The peak of the output waveform is set by the comparator and is symmetrical if the zener diode ratings are the same. The waveforms shown in Figure 8–10b are for zener ratings of 4.3 V and $R_1 = R_2 = 10\ \text{k}\Omega$. Therefore, the peak voltages of V_{OUT} are ±2.5 V and V_{SQUARE} = ±5 V.

The 555 counter/timer discussed in Section 8–2 can also be configured to operate as a VCO. The circuit is almost identical to the circuit shown in Figure 8–4 except that the control pin labeled CONT is not connected through a capacitor to ground, but becomes a voltage control input that determines the frequency of the output pin's pulsed waveform (see Figure 8–11a). When a control voltage is applied to the CONT pin, it determines whether the upper or lower comparators are switched on (refer back to the internal 555 diagram shown in Figure 8–3). The upper comparator will switch off at the value of the control voltage, while the lower comparator switches off half of the control voltage. The waveforms for the circuit are shown in Figure 8–11b. The equation that determine the pulse widths and frequencies for the 555 VCO circuit are as follows:

$$V_{OUT}+ = -(R_A + R_B)C\ ln\ ((V_{CC} - V_{CONT})/(V_{CC} - .5V_{CONT})) \qquad (8\text{–}12)$$

(a)

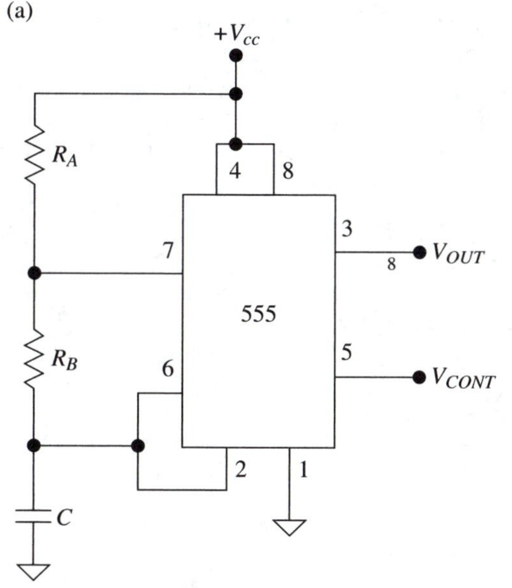

555 voltage-controlled oscillator with control voltage applied to CONT pin

(b)

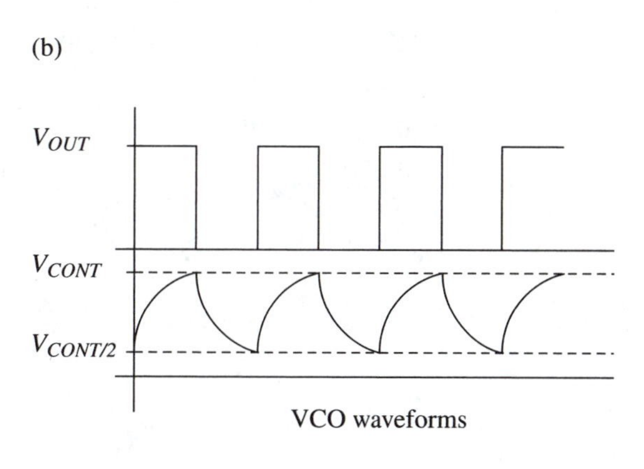

VCO waveforms

▲ FIGURE 8–11
555 voltage-controlled oscillator

$$555 \text{ frequency } f = 1/(V_{OUT+} + .693R_BC) \qquad (8\text{–}13)$$

where V_{OUT+} = the time the pulse is high.

Another variation of the VCO is a voltage-to-frequency converter. Actually, these two circuits perform the exact same function: both convert a voltage to a frequency. The difference between the two functions really lies in the waveform type and its function. VCOs are circuits that convert a voltage to a waveform operating at a particular frequency where the shape (sawtooth or pulse) of the waveform has some significance. A voltage-to-frequency converter usually converts an analog voltage to a stream of digital pulses that vary in frequency with the analog voltage.

Integrated circuits have been developed to perform voltage-to-frequency conversion on one IC. The pin out and block diagram for the VFC32 voltage to frequency converter is shown in Figure 8–12. The VFC32 is typically powered by ±15 V DC but can handle power supply voltages up to ±22 volts. The input signal is a voltage that is connected to an op amp. The feedback path for the op amp is external capacitor C_2, making it an integrator. The input current to the integrator $I_{IN} = V_{IN}/R_1$. The positive input produces a downward-ramping integrator output voltage. When the integrator ramps down to circuit common, the comparator switches, firing the one-shot. A 1-mA reference current is connected to the integrator while the one-shot is on, which drives the integrator back upscale. After the one-shot has completed one cycle, the integrator ramps downward again, beginning the next cycle. The net result is that the frequency of the output is directly proportional to the input voltage.

Figure 8–13 shows the VFC32 in a circuit that converts a 0–10 V input signal to frequencies ranging from 0–10,000 Hz. The VFC32 can output frequencies as high as 500 kHz. The range of the output frequency is set by capacitor C_1, so its selection (tolerance, temperature drift, and stability, as discussed in Chapter 4) is critical to the accuracy of this circuit. The tolerance, temperature drift and stability of R_1 are also critical. Therefore, metal film type resistors are recommended. The

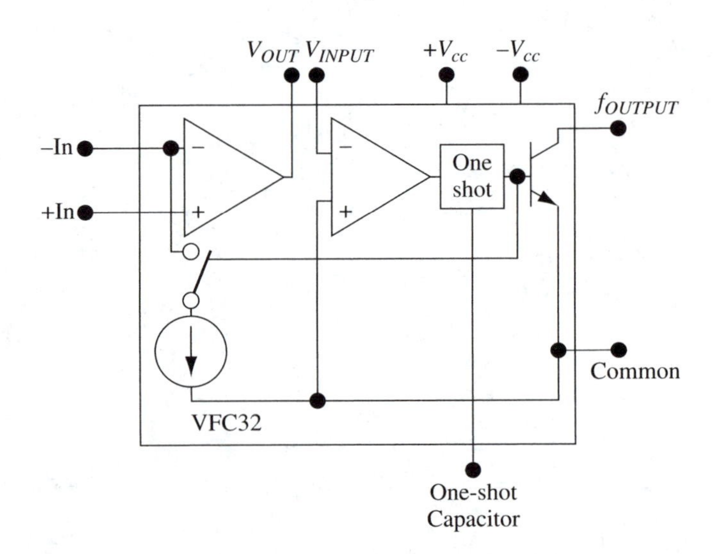

FIGURE 8–12 Voltage-to-frequency converter block program

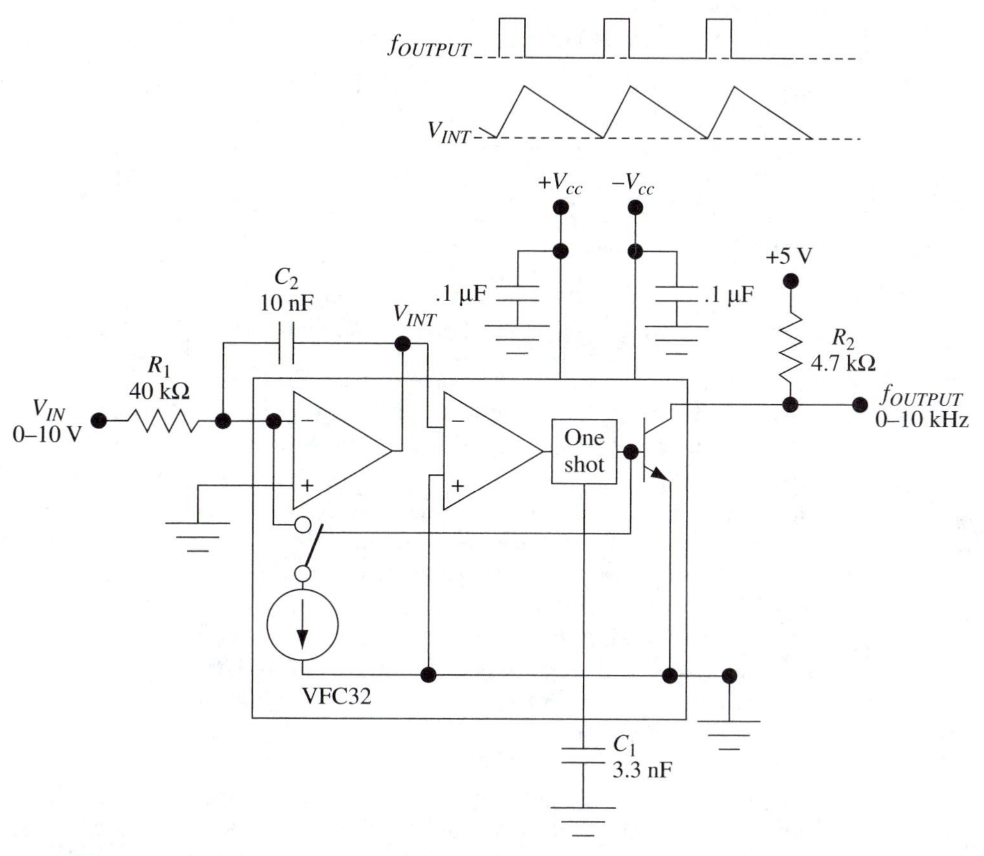

▲ **FIGURE 8–13**
Voltage-to-frequency converter circuit

data sheet for this chip provides graphs that help select R_1 and C_1 so that the duty cycle of the output pulse is generally 25% when the maximum output frequency is less than 200 kHz. For maximum output frequencies greater than 200 kHz, the recommended duty cycle is 50%.

The value of integrating capacitor C_2 is not extremely critical but must be within a range so that the up and down ramping occurs fast enough to accommodate the highest output frequency. The tolerance or temperature stability of C_2 is not important but it should have low leakage and low dielectric absorption (DA). Polycarbonate type capacitors are a good choice for C_2.

Notice that the output terminal is an open transistor collector, which can be connected to a separate power supply or V_{CC} of the VFC32. Figure 8–13 shows the output connected to a 5-V power supply through a 4.7 kΩ pull-up resistor. The pull-up resistor value is selected so that the maximum current passing through the transistor to circuit common, when on, is less than 8 mA. In Figure 8–13 the output is a 5-V pulsed waveform with a 25% duty cycle. The frequency of the output is directly proportional to the 0–10-V input voltage: 0 V in = 0 frequency out, 5 V in = 5,000 Hz out and 10 V in = 10,000 Hz out.

8–6 ▶ Sine-wave Generators

All of the oscillator circuits discussed thus far have been variations of relaxation oscillators; oscillators based on the charging and discharging of capacitors. In order to generate a good-quality sine wave, positive feedback oscillators are the best alternative. Positive feedback oscillators utilize an amplifier whose output is fed back to its input through a feedback circuit. The circuit will oscillate at a frequency determined by the circuit's component values if both of these conditions are met:

1. The net gain around the loop is equal to 1.
2. The phase angle between the feedback signal and the amplifier input is 0.

These condition are called the *Barkhausen criteria* (see Figure 8–14). The net gain around the circuit loop equals amplifier gain, A times feedback gain, A_{FB}. Because the feedback circuit usually consists of passive components, some signal attenuation occurs, which means the amplifier must have some gain to create a net loop gain equal to 1. If the feedback circuit is a resonant circuit, then the maximum feedback will exist at just one frequency, the resonant frequency of the circuit.

In some ways oscillator circuits seem like black magic. The concept of positive feedback is understandable once an input is presented to the amplifier, but where does the signal come from that starts the process? All it takes is some small source of noise voltage, likely thermally generated noise or power-up transients that will cause variations in the input to the amplifier and start oscillation. Once started, the circuit will automatically search out the frequency where the gain is a maximum and the phase shift is equal to zero. If the gain is 1 at this frequency, then oscillation will be sustained. In most sine wave oscillator circuits the initial loop gain is greater than 1 to promote oscillation in a short period of time. Once oscillation has been achieved, the loop gain is automatically adjusted to 1 so that steady oscillation will be maintained. There are two general classifications of sine-wave generators: RC and LC sine-wave generators.

RC Sine-wave Generators (Frequencies < 1 MHz)

Wien-Bridge Oscillators

RC sine-wave generators are used with op amps to generate sine waves at frequencies of less than 1 MHz. A common variety RC oscillator that has become almost an industry standard is the Wien-Bridge oscillator. The Wien-Bridge oscillator

▶ FIGURE 8–14
Positive feedback oscillators

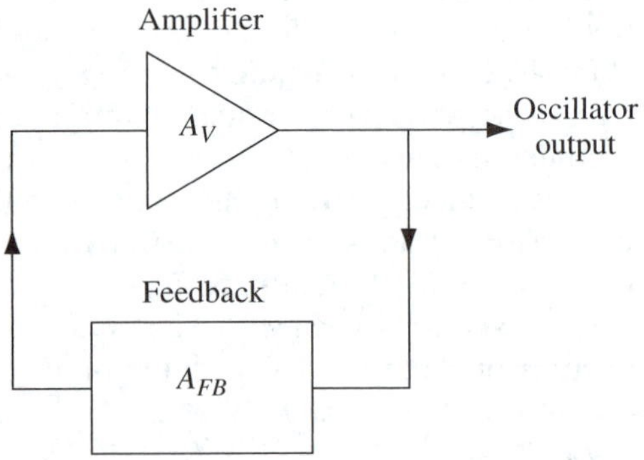

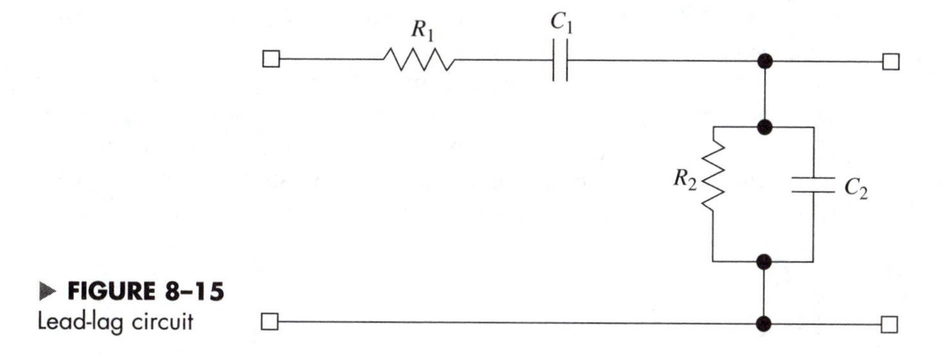

FIGURE 8–15
Lead-lag circuit

as shown in Figure 8–17 uses a lead-lag feedback circuit as shown in Figure 8–15. The concept of lead-lag simply means that this circuit element is both a low-pass filter and a high-pass filter at the same time, and that there is a resonant frequency that passes through this feedback circuit with minimal attenuation. At low frequencies both capacitors C_1 and C_2 are open and there is no signal path back from the output to the input. At high frequencies, C_1 and C_2 are both shorts and again there is no signal fed back to the input (see Figure 8–16).

Notice that the Wien-Bridge oscillator has both negative feedback to provide amplification, plus positive feedback to the op amp to initiate oscillation. The noninverting amplifier input provides 0 degrees of phase shift, as does the feedback circuit at the resonant frequency.

If $R_1 = R_2$ and $C_1 = C_2$, then the resonant frequency for the circuit is shown below:

$$f = 1/(2\pi RC) \qquad (8\text{–}14)$$

It can also be shown that the attenuation provided by the circuit at the resonant frequency is equal to 1/3. In order for the net loop gain to be unity, the gain provided

FIGURE 8–16
Lead-lag gain/phase angle plots

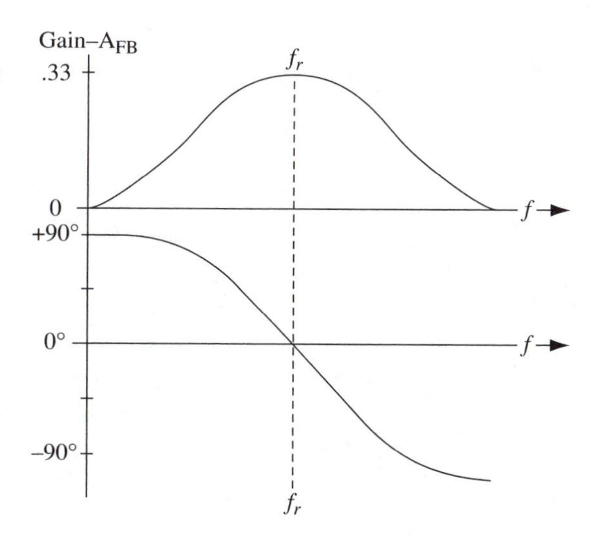

by the amplifier must be 3 to compensate for the attenuation provided by the lead-lag circuit. Once oscillation is achieved, the peak-to-peak amplitude of the sine wave will be close to the saturation voltage of the op amp. This is undesirable; with oscillation very close to the saturation level, some clipping and distortion is likely to occur. Also, when the circuit is powered up, the gain must be higher than 3 in order to cause oscillation to occur in a reasonable period of time. Let's ignore these details for a moment and complete a design example for the circuit shown in Figure 8–17.

Example 8–5

Design a Wien-Bridge oscillator that will generate a frequency of 200 kHz with an amplitude of approximately ±10 V.

Solution

1. Let $C_1 = C_2 = .001\ \mu F$.
2. Solving Equation 8–14 for $f = 200$ kHz:

   $f = 1/(2\pi RC) = 200\text{ kHz} = 1/(2\pi \times R \times .001\ \mu F)$
   $R = 1/(2\pi \times 200\text{ kHz} \times .001\ \mu F) = 796\ \Omega$
   Let $R_1 = R_2 = R_4 = 806\ \Omega$s 1% tolerance resistors.
   In order for the gain to be 3, then $R_3 = 2R_4 = 1612\ \Omega$.
   Use 1650 Ω 1% tolerance resistor for R_3 to make the gain slightly >3.

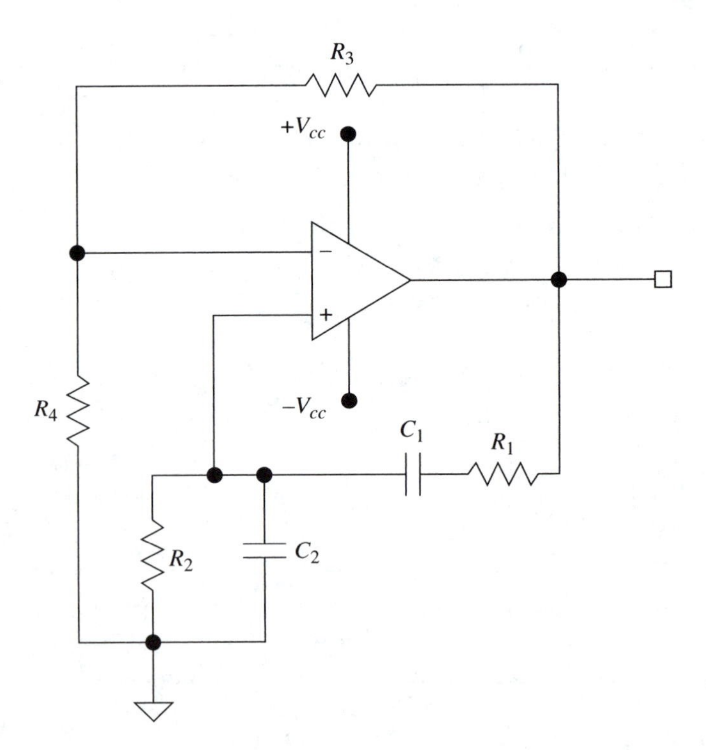

FIGURE 8–17
Wien-Bridge oscillator

The complete design solution is:

V_{CC} is selected as ±12 V
$R_1 = R_2 = R_4 = 806\ \Omega$ 1% tolerance resistors
$R_3 = 1650\ \Omega$ 1% tolerance resistor
C = .001 μF

The peak-to-peak amplitude of the sine wave will be about ±10 V with V_{CC} = ±12 V.

The circuit in Figure 8–18 shows one method of reducing the peak-to-peak amplitude to a value less than the saturation voltage, and it also provides a higher gain to get the circuit into oscillation. The circuit in Figure 8–14 has taken resistor R_3 and split its value into R_{3A} and R_{3B}. When this circuit is initially powered up, the amplitude of the output is low and the diodes have no effect on the circuit. The gain of this circuit on power up is given by the following equation:

$$A = 1 + (R_{3A} + R_{3B})/R_4) \qquad (8\text{–}15)$$

When the output amplitude reaches the desired amplitude (as determined by the zener diode voltage rating + .7 V) the diodes will effectively remove R_{3B} from

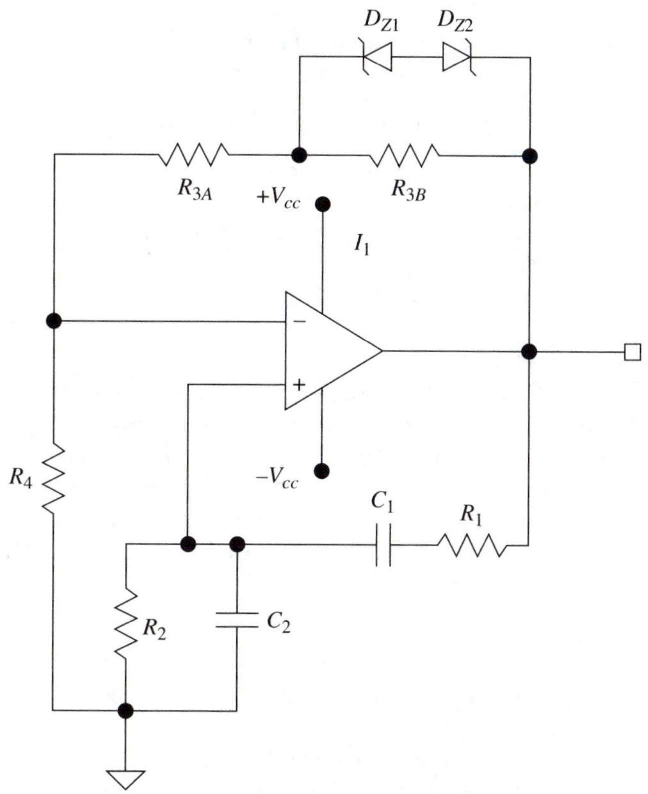

FIGURE 8–18
Wien-Bridge oscillator with gain adjustment

the circuit. The gain of the circuit once the zener diode voltages have been exceeded becomes:

$$A = 1 + (R_{3A}/R_4) \tag{8–16}$$

Example 8–6

Modify the circuit developed in Example 8–5 so that the circuit is self-starting and provides a peak-to-peak output amplitude of 6.3 V.

Solution

1. The circuit shown in Figure 8–17 will meet the requirements of this example with the proper component values.
2. Let $R_{3A} = 1650\ \Omega$ and $R_4 = 806\ \Omega$ as before then the gain after the zener voltages have been exceeded will equal:

   $A = 1 + (R_{3A}/R_4) = 1 + (1650/806) = 3.05$
3. If $R_{3B} = R_4$, then the gain on power up will be given by Equation 8–15:

   $\mathrm{A} = 1 + ((R_{3A} + R_{3B})/R_4) = 1 + (2456/806) = 4.05$
4. Zener diodes D_{Z1} and D_{Z2} are selected as 1N4734s with a zener voltage of 5.6 V. The peak-to-peak amplitude of the output will be 5.6 V + .7 V (the forward voltage drop of one zener diode) or 6.3 V.

The complete design solution is:

V_{CC} is selected as ±12 V
$R_1 = R_2 = R_4 = R_{3B} = 806\ \Omega$ 1% tolerance resistors
$R_{3A} = 1650\ \Omega$ 1% tolerance resistor
$\mathrm{C} = .001\ \mu\mathrm{F}$
$D_{Z1} = D_{Z2} =$ 1N4734

Phase-shift Oscillators

This type of oscillator creates a 360-degree phase shift to promote oscillation at a particular frequency. An inverting op amp circuit is used to generate 180 degrees of the needed phase shift, while *RC* networks that the feedback impedance comprise provide the remaining 180 degrees of phase shift. Each *RC* network can provide as much as 90 degrees of phase shift (see Figure 8–19).

If the two resistors are labeled $R = R_1$, then the equation for the resonant frequency of this circuit is as follows:

$$f = 1/(2\pi \times 2.45\ RC) \tag{8–17}$$

The attenuation provided by the *RC* feedback network at resonance = 1/29. Therefore, the gain of the op amp circuit must be slightly greater than 29. The design procedure for the phase-shift oscillator begins by using Equation 8–16 to calculate *R* and *C* values that will provide the proper oscillation frequency. Then R_1 is equat-

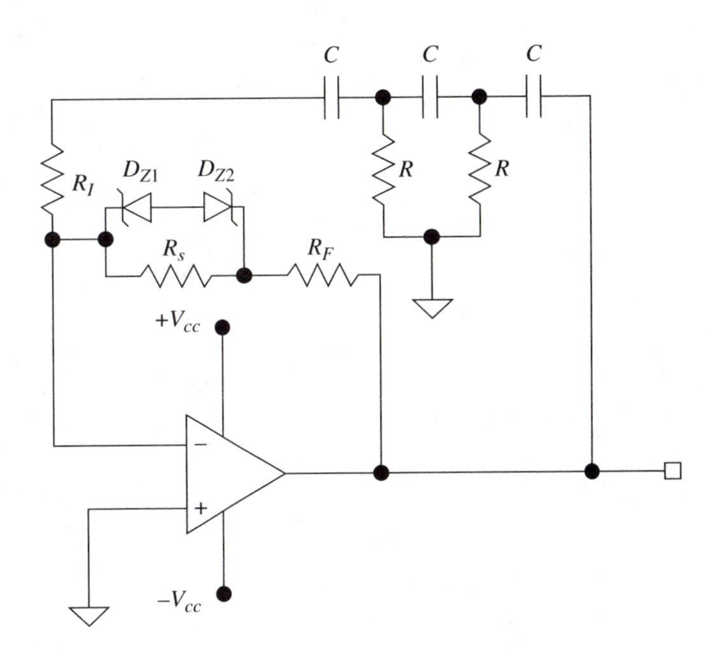

GURE 8–20
e-shift oscillator with gain adjustment

shown in Figure 8–18 function exactly the same as those discussed in Example 8–6. When the oscillator output exceeds the zener voltage rating of the diodes plus .7 V, resistor R_S is shorted out, leaving the gain of the circuit slightly higher than the required gain of 29 to support oscillation. When the oscillator output is low, R_S is added to R_F, increasing the gain of the op amp. Zener diodes D_{Z1} and D_{Z2} are selected to have a zener voltage of 4.3 V that will provide a ±5-V sine wave. To calculate R_S such that the gain is 40 on power up:

$-40 = -(R_F + R_S)/1300$
$R_F + R_S = 52{,}000$
$R_S = 52{,}000 - 37{,}900 = 14{,}000\ \Omega$
Let $R_S = 14{,}000\ \Omega$ 1% tolerance value.

The complete design solution is:

V_{CC} is selected as ±12 V
$R = R = R_I = 1300$ 1% tolerance resistors
$R_F = 37{,}900\ \Omega$ .5% tolerance resistor
$R_S = 14{,}000\ \Omega$ 1% tolerance value
$C = 100$ pF
$D_{Z1} = D_{Z2} =$ 1N4731 $V_Z = 4.3$ V

Twin T Oscillators

A third type of RC oscillator for use at frequencies up to 1 MHz is the Twin T oscillator; so named because of the dual T type filter that is the negative feedback leg of the circuit. Figure 8–21a a shows the Twin T oscillator circuit with the same

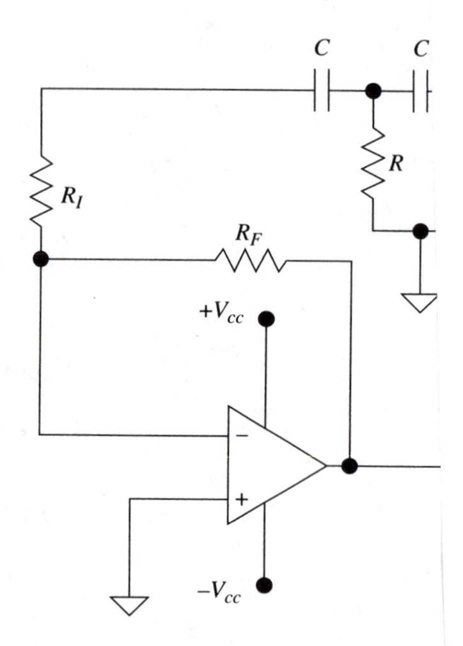

▶ **FIGURE 8–19**
Phase-shift oscillator circuit

▶
Ph

ed to the calculated R value and R_F is calculated to provide the i ($A = -R_F/R_1$) which must be slightly greater than 29. However, cillator has the same amplitude start-up and distortion problem for the Wien-Bridge oscillator. The example that follows will sh lem is rectified in the phase-shift oscillator.

Example 8–7

Design a phase-shift oscillator that will develop a 500kHz sine to-peak amplitude of ±5 V.

Solution

1. Utilizing the phase-shift oscillator circuit shown in Figure tion 8–9 for values of R and C that will promote oscillati

   $f = 1/(2\pi \times 2.45\ RC) = 500$ kHz
   Let $C = 100$ pF.
   $R = 1/(2\pi \times 2.45 \times 500\text{ kHz} \times 100\text{ pF}) = 1299\ \Omega$ use 1300 Ω

2. Let $R_1 = R$ and calculate R_F such that the gain = −29.

   $-29 = -R_F/ R_I = -R_F/1300\ \Omega$
   $R_F = 1300(29) = 37{,}700\ \Omega$ use .5% tolerance value of 37,90

3. The gain of the circuit on power up should be higher tha cillation to occur quickly. A gain value of 40 is reasonable.

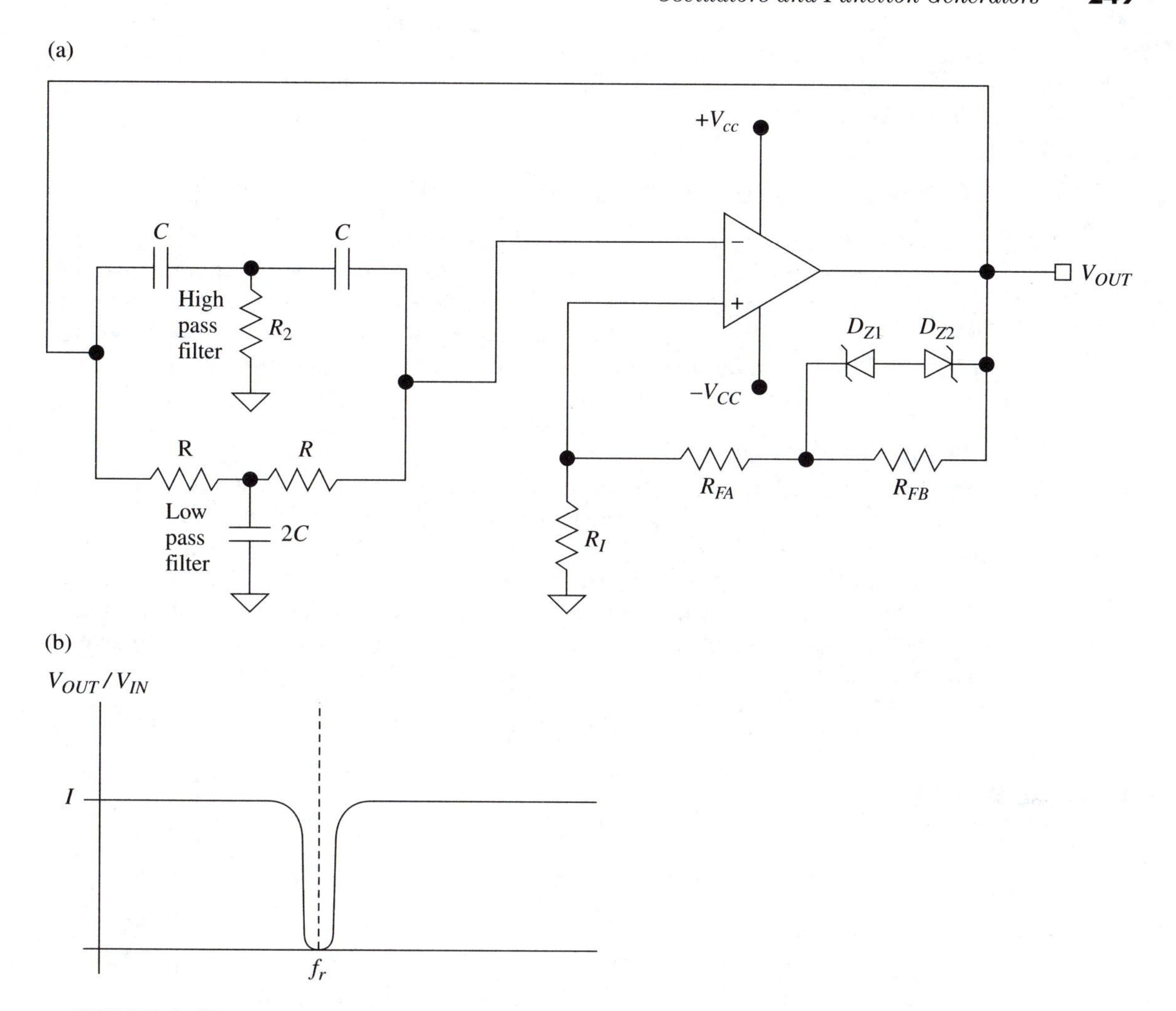

▲ **FIGURE 8–21**
Twin T oscillator with gain adjustment. (a) Using same gain adjustment as Wien-bridge and phase-shift oscillators. (b) Plot of transfer function.

gain adjustment scheme already discussed for the Wien-Bridge and phase-shift oscillators. The T filter with two resistors is a low-pass circuit while the dual capacitor T filter passes high frequencies. Figure 8–21b shows a plot of the transfer function for the Twin T filter, which combines the effects of the high- and low-pass filters to act as a notch filter.

The resonant frequency is given by the following equation:

$$f = 1/(2\pi RC) \qquad (8\text{–}18)$$

The amount of positive feedback is determined by the voltage divided $(R_{FA} + R_{FB})/R_I$ on startup, and after the zener diode ratings are exceeded, this reduces to simply R_{FA}/R_I. R_{FA}/R_I can range anywhere from 10 to 1000 and as before the peak amplitude of the generated sine wave will be equal to the zener rating of diodes D_{Z1} and D_{Z2}.

LC Sine-wave Generators (Frequencies > 1 MHz)

In order to build circuits that oscillate at frequencies higher than 1 MHz, LC sine-wave generators are most often used. These circuits are usually built out of discrete components because of the closed-loop frequency limitations of most op amps. The design of high-frequency oscillators is complicated by the effects discussed in Chapter 7 on RF amplifiers, such as stray capacitance and inductance and their increased effect on circuit behavior.

Colpitts Oscillator

The Colpitts oscillator is a relatively common LC oscillator whose feedback element, a resonant tank circuit, consists of two capacitors, C_A and C_B,and an inductor, L, as shown in Figure 8–22. The rest of the circuit is simply a voltage divider biased, bipolar AC amplifier. The gain of this circuit equals the effective collector AC resistance divided by the emitter resistance. The AC collector resistance is primarily the impedance of the tank circuit that has a maximum value at resonance. The resonant frequency of the circuit is given by the following equation:

$$f = 1/(2\pi\sqrt{LC}) \text{ where } C = C_A C_B/(C_A + C_B) \quad (8\text{–}19)$$

▶ FIGURE 8–22
Colpitts oscillator

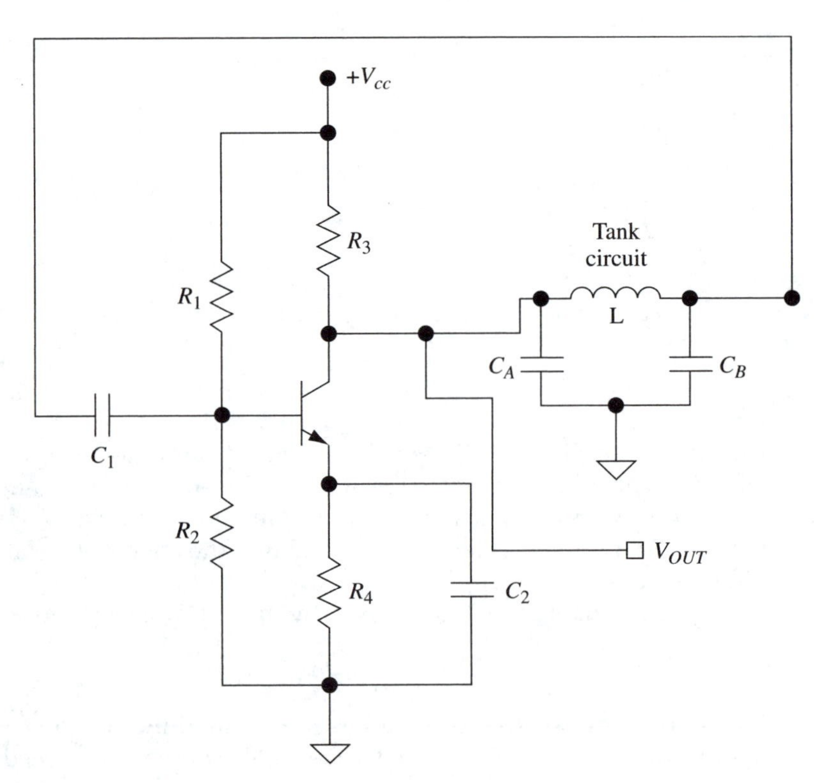

At resonance the amount of signal fed back to the input = C_A/C_B. In order for the circuit to start oscillation, the gain of the amplifier circuit must be > C_B/C_A.

Clapp Oscillator

A variation of the Colpitts oscillator is the Clapp oscillator, which employs an additional capacitor in series with the inductor in the Clapp oscillator's tank circuit (see Figure 8–23). The total capacitance of the tank feedback circuit equals the series equivalent capacitance of capacitors C_A, C_B, and C_C. The equation that approximates the resonant frequency for the circuit is the same as for the Colpitts oscillator:

$$f = 1/(2\pi\sqrt{LC}) \text{ where } C = 1/C_A + 1/C_B + 1/C_C \qquad (8\text{–}20)$$

The advantage of this circuit comes from making the value of C_C significantly smaller than the values of capacitors C_A and C_B. In this case, the value of C_C is the primary determinant of the resonant frequency that eliminates variations caused by stray capacitance between C_A, C_B, and ground.

At resonance the amount of signal fed back to the input = C_A/C_B. In order for the circuit to start oscillation, the gain of the amplifier circuit must be > C_B/C_A.

Hartley Oscillator

Another variation of the Colpitts oscillator is the Hartley oscillator, which interchanges the capacitor and inductor positions in the circuit, so that two inductors are connected to ground in parallel with one capacitor (see Figure 8–24). The

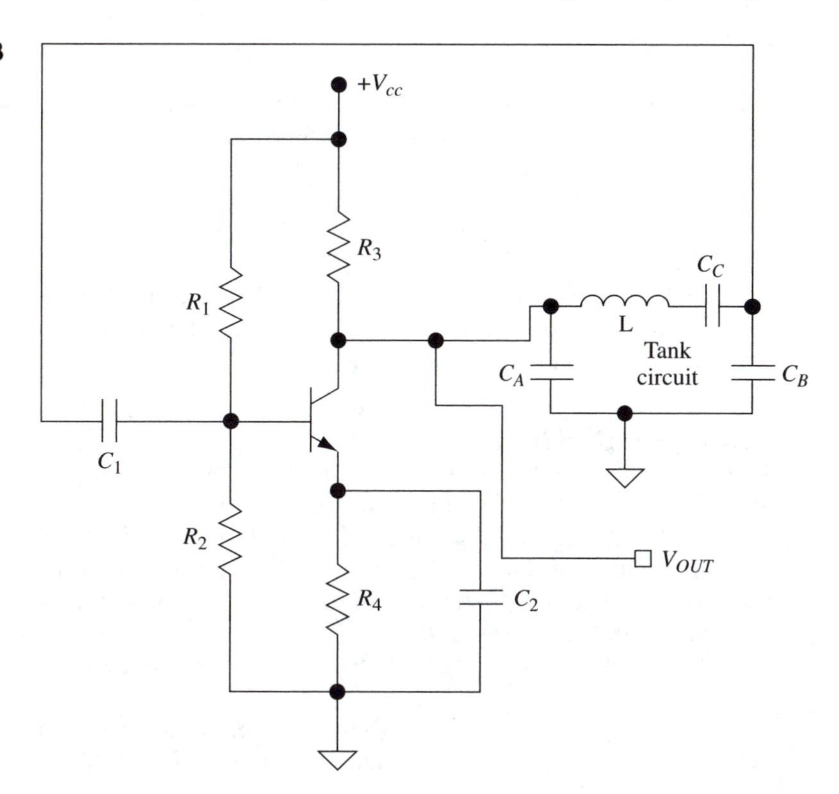

FIGURE 8–23 Clapp oscillator

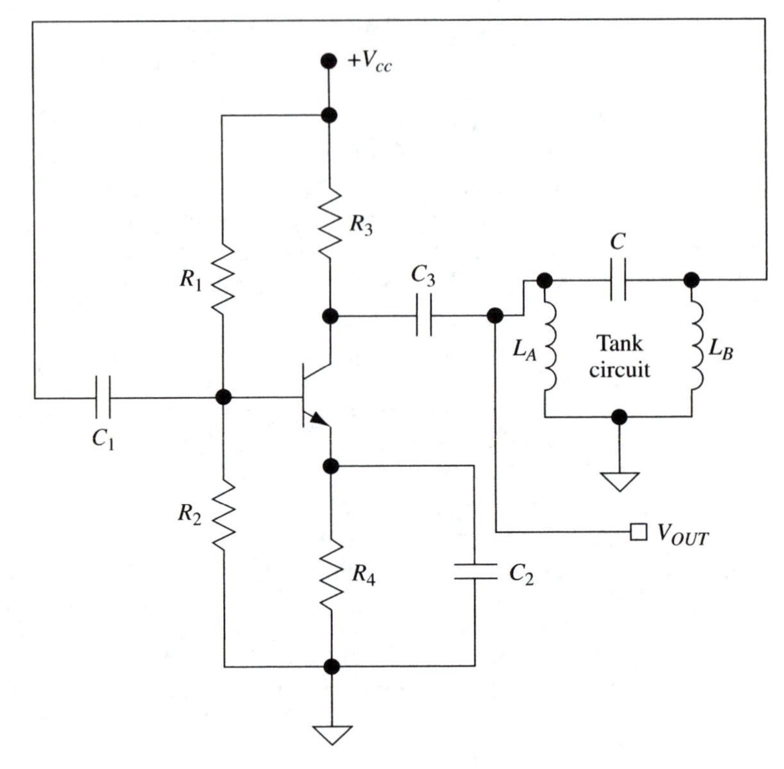

▶ **FIGURE 8–24**
Hartley oscillator

equation that approximates the resonant frequency remains the same as the Colpitts and Clapp oscillators, but inductance value of I is determined by the series equivalent of L_A and L_B:

$$f = 1/(2\pi\sqrt{LC}) \text{ where } L = L_A + L_B \qquad (8\text{–}21)$$

At the resonant frequency, the attenuation of the output signal fed back to the bipolar amplifier is L_B/L_A. Therefore, the minimum gain of the amplifier required to support oscillation equals L_A/L_B.

Armstrong Oscillator

The Armstrong oscillator uses the primary from a transformer, connected in parallel with a capacitor, as an LC tank circuit. The secondary of the transformer supplies the feedback signal of the bipolar amplifier (see Figure 8–25). The feedback network includes the LC tank circuit and the relationship between the transformer primary and secondary. The inductance L of the tank circuit is the inductance of the primary of the transformer.

The equation that approximates the resonant frequency for the circuit is:

$$f = 1/(2\pi\sqrt{LC}) \text{ where } C \text{ and } L \text{ are the capacitance and inductance of the tank circuit} \qquad (8\text{–}22)$$

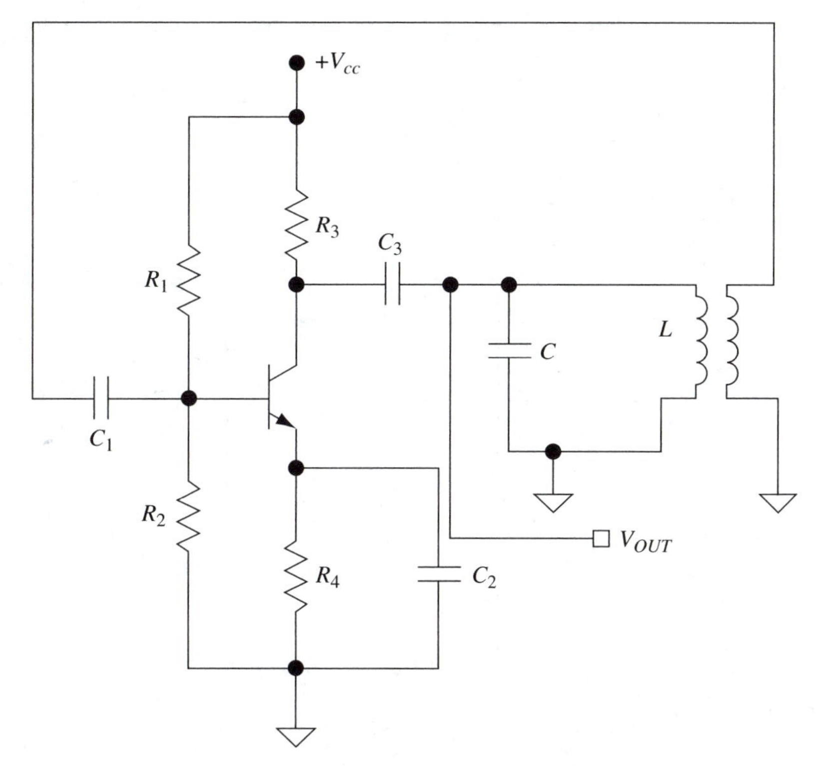

FIGURE 8–25
Armstrong oscillator

At the resonant frequency, the attenuation provided by the feedback network equals the mutual inductance between the primary and the secondary M, divided by the inductance L. The gain required to support oscillation equals L/M.

Crystal Oscillator

As discussed for clock circuit generators in Section 8–1, crystals can be used to generate sine waves with very accurate frequencies. A sine-wave crystal oscillator can be developed by replacing the inductor in a Colpitts oscillator with a crystal, rated at the desired oscillation frequency (see Figure 8–26). The resonant frequency of the circuit is the resonant frequency of the crystal. At resonance the amount of signal fed back to the input equals C_A/C_B. In order for the circuit to start oscillation, the gain of the amplifier circuit must be $> C_B/C_A$.

8–7 ▶ Pulse Generators

Pulse generators are very similar to the clock generators discussed in Section 8–2 except that pulse generators can develop a specific positive going pulse and repeat it at the selected frequency. In other words, pulse generators develop a pulse at a selected amplitude, duration, and frequency. Pulse generators can be developed with microprocessor-based circuits under software control or with analog/digital circuits. When using analog/digital circuits, a waveform of some type (usually square or triangle waveforms) is created at the desired frequency, and then a pulse

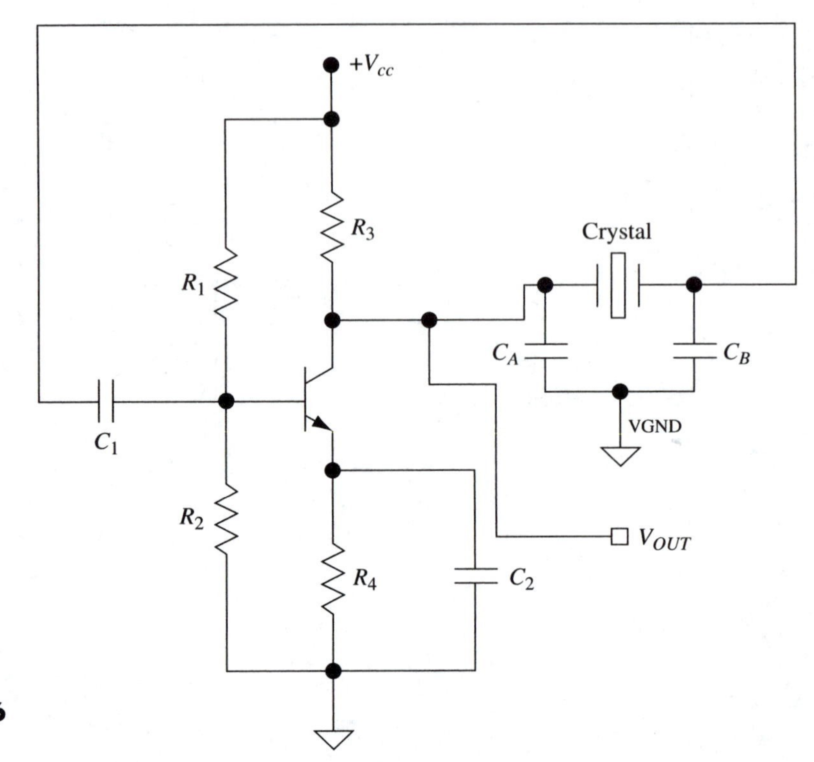

▶ **FIGURE 8–26**
Crystal oscillator

of the desired amplitude and duration is generated with either a "one shot" or a comparator circuit with an adjustable reference voltage. A "one shot" is another name for a monostable multivibrator, a device that when triggered puts out one pulse with a width that is determined by the RC network connected to it. Let's develop a pulse generator using a comparator circuit.

Example 8–8

Develop a pulse-generator circuit that will output a positive going pulse that is 10 µ seconds in duration with an amplitude of 6 V at a frequency of 20 kHz.

Solution

1. A triangle waveform can be used to develop a specific pulse waveform when it is input to a comparator circuit. Let's start with the triangle-wave generator circuit developed in Example 8–3. This circuit, shown in Figure 8–27, outputs a triangle waveform at 20 kHz with an amplitude of ±5 V.

2. The triangle wave is input to a comparator circuit with an adjustable voltage reference as shown in Figure 8–27. Trimpot TP_1, R_1, and R_2 form a voltage divider that is the reference voltage for the comparator. In order for TP_1 to develop any possible pulse duty cycle, it must be adjustable over the

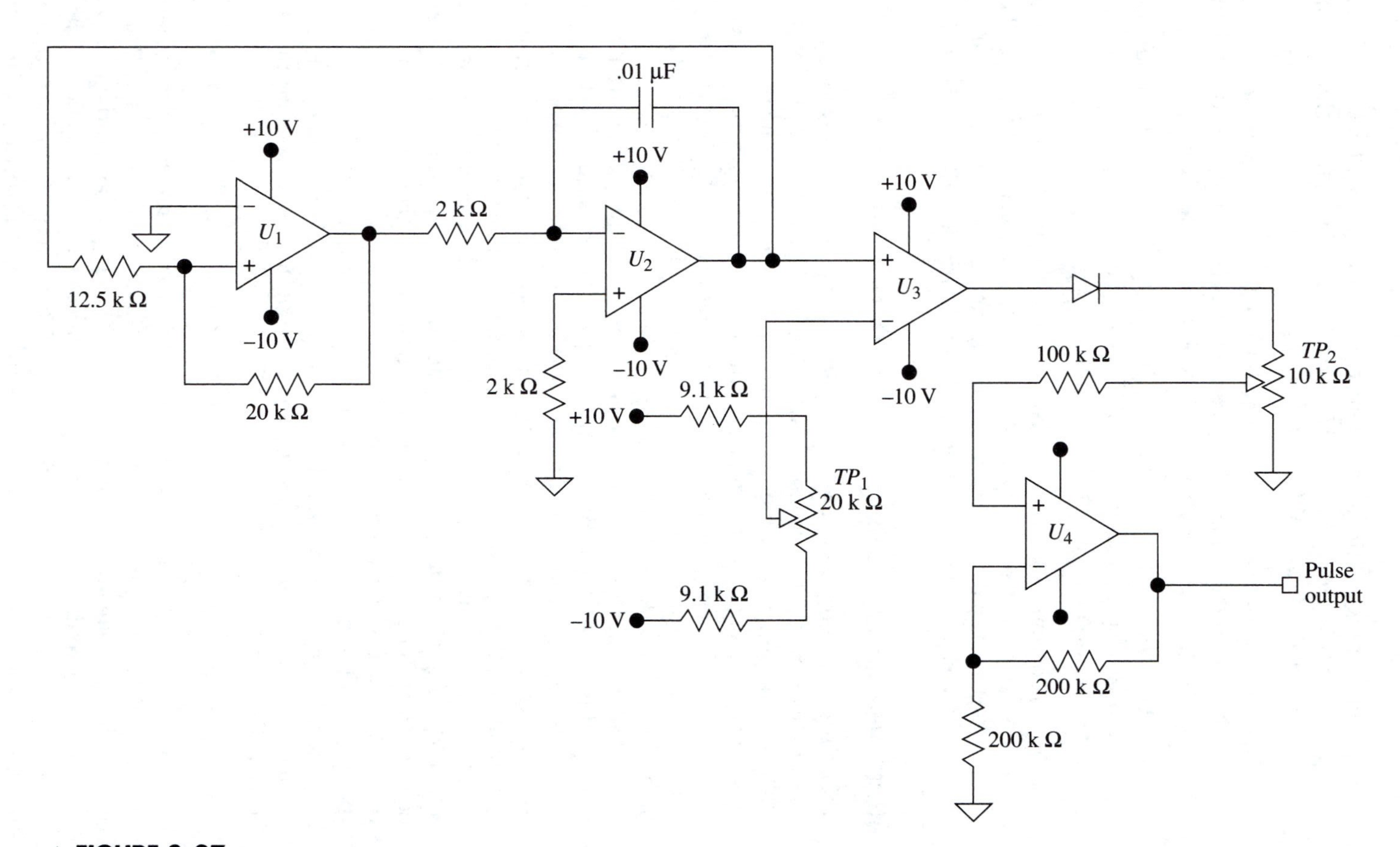

▲ **FIGURE 8–27**
Pulse-generator circuit

entire range of the input triangle waveform (±5 V). Let TP_1 = 20 kΩ while R_1 and R_2 = 9.1 kΩ. When TP_1 is adjusted full CW, the voltage at the wiper arm of TP_1 is +5.2 V. CCW TP_1 = –5.2 V.

3. In order for the output pulse to have a duration of 10 μ seconds, TP_1 must be adjusted to equal +3 V because this is the voltage that represents a 20% duty cycle (10 μ seconds/50 μ seconds = 20% duty cycle).

4. With TP_1 equalling 3 V, the output of comparator U_3 is approximately +8 V for 20 μ seconds and –8 V, for 40 μ seconds. ±8 V is the approximate output saturation voltage for the comparator.

5. The output of U_3 is connected to a diode to remove the negative half of the waveform, which is then input to an attenuation potentiometer TP_2 provided to adjust the amplitude of the peak amplitude down to the specified +6 V. U_4 is simply a non-inverting amplifier that provides the ultimate pulse output.

8–8 ▶ Integrated Circuit Function Generators

There are a number of integrated circuits that have been developed to generate many of the waveforms discussed in this chapter. The first and most common is the 8038 function generator, which is available from a number of manufacturers. The 8038 can create sine, square, triangle, sawtooth, and pulse waveforms with a minimum of external components over the frequency range of .001 Hz to 300 kHz. The sine, square, and triangle waves can be generated simultaneously and the chip can be powered from ±5 to ±15 V.

The block diagram for the 8038 is shown in Figure 8–28. Internally, there are two constant current generators, two comparators, a flip-flop, buffers for the triangle- and square-wave outputs and a triangle-to-sine-wave converter module. The external capacitor *C* is charged with current source *A* and discharged with current source *B*. External resistors *RA* and *RB* determine the value for each current source. The constant current sources generate a triangle wave across capacitor *C* when $R_A = R_B$. The triangle wave switches the flip-flop on and off to generate a square wave. Both the internal square wave and triangle wave are buffered and output to pins 9 and 3, respectively. The internal triangle wave is also sent to the triangle-to-sine-wave converter and then output to pin 2.

To provide 50% duty cycle let, $R = R_A = R_B$ by having just one resistor for both R_A and R_B, and then give the frequency for all of the output waveforms by the following formula:

$$f = .15/RC \qquad (8\text{–}23)$$

Figure 8–29 shows the 8038 connected to generate 50% duty cycle, square, triangle, and sine waves at frequencies determined by the values of *R* and *C* over the operating frequency range for the device. The amplitudes for the waveforms are approximately equal to the ± power supply values, with the exception of the square-wave output, which can be connected to a different power supply with load resistor R_L. Trimpot resistor TP_1 functions to minimize distortion of the sine-wave output. The optimum value is 82 kΩ, but it is best to make the value adjustable to compensate for potential circuit variations.

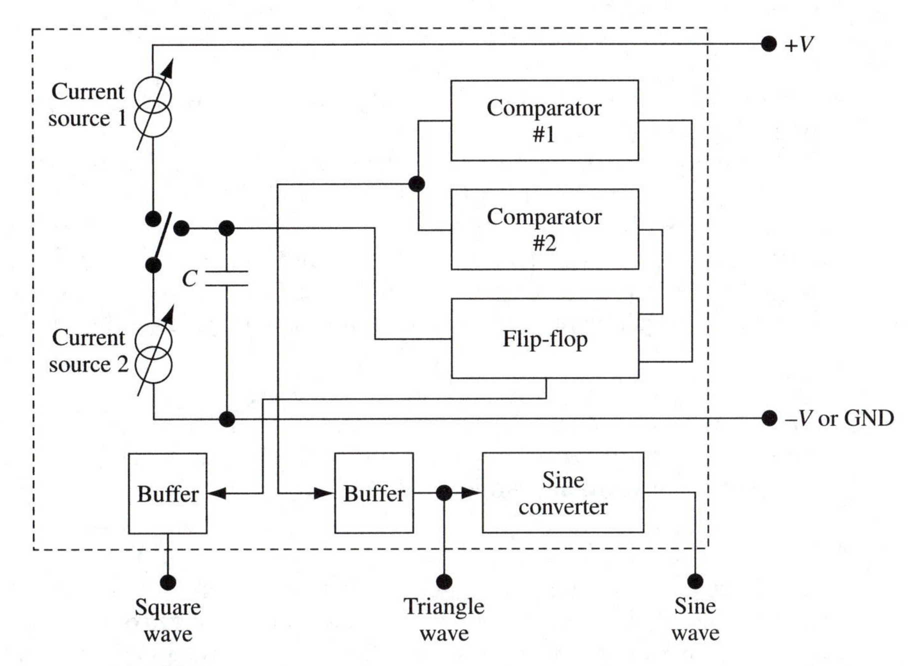

▲ **FIGURE 8–28**
8038 block diagram

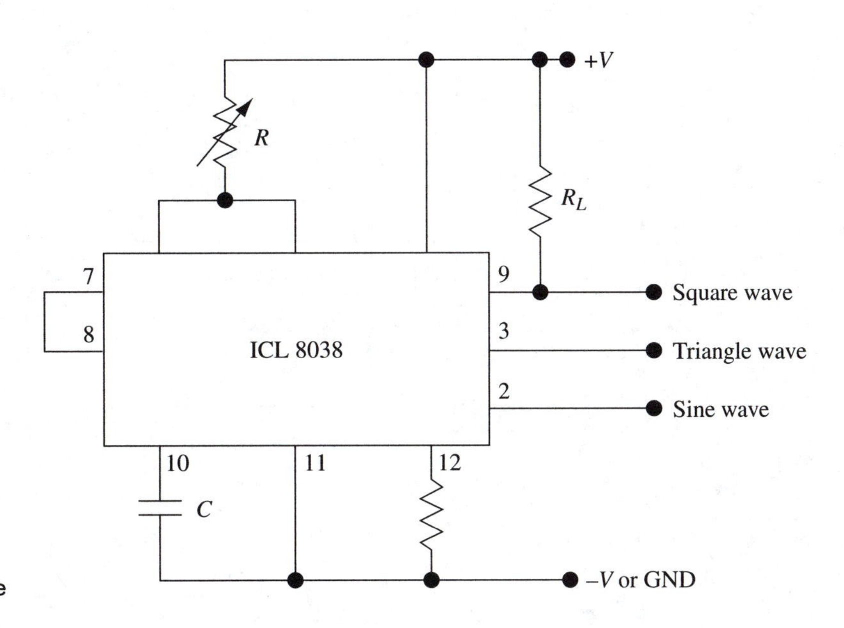

▶ **FIGURE 8–29**
8038 function generator connector, 50% duty cycle

To create non-symmetrical sawtooth and other waveforms, the values of R_A and R_B must not be equal. This creates a variation in the output frequency whenever either R_A or R_B is varied.

The MAX038 is a newer-function generator IC that operates at a much higher range of frequencies. The MAX038 is similar in operation to the 8038 chip. It uses constant current sources to charge capacitor C, which results in a triangle waveform (see Figure 8–30). The triangle waveform is used to generate both the square and sine waves.

Two logic inputs, A_0 and A_1, determine which of the waveforms are connected to the output through an internal analog multiplexer so the MAX038 can provide only one waveform output at a time. The output frequency is determined by a combination of the current injected into pin I_{IN} and capacitor C. I_{IN} can be an external current source or result from a voltage V_{IN} connected through resistor R. The formula for the fundamental frequency of the oscillator is as follows:

$$f = I_{IN}/C \text{ or } f = V_{IN}/RC \qquad (8\text{–}24)$$

The MAX038 includes a 2.5-V band-gap reference that can be used for V_{IN}. For example, the frequency of the MAX038 output if the 2.5-V reference signal is connected to a 20-kΩ resistor and $C = 100$ pF is:

$$f = V_{IN}/RC = 2.5\text{ V}/20\text{ k}\Omega \times 100\text{ pF} = 1.25\text{ MHz}$$

The MAX038 can generate frequencies from less than 1 Hz up to 20 MHz, significantly higher than the 8038 chip. It also possesses separate duty-cycle (DADJ) and frequency modulation (FADJ) inputs that operate independently: duty-cycle adjustments that don't affect the frequency and frequency modulation that does not

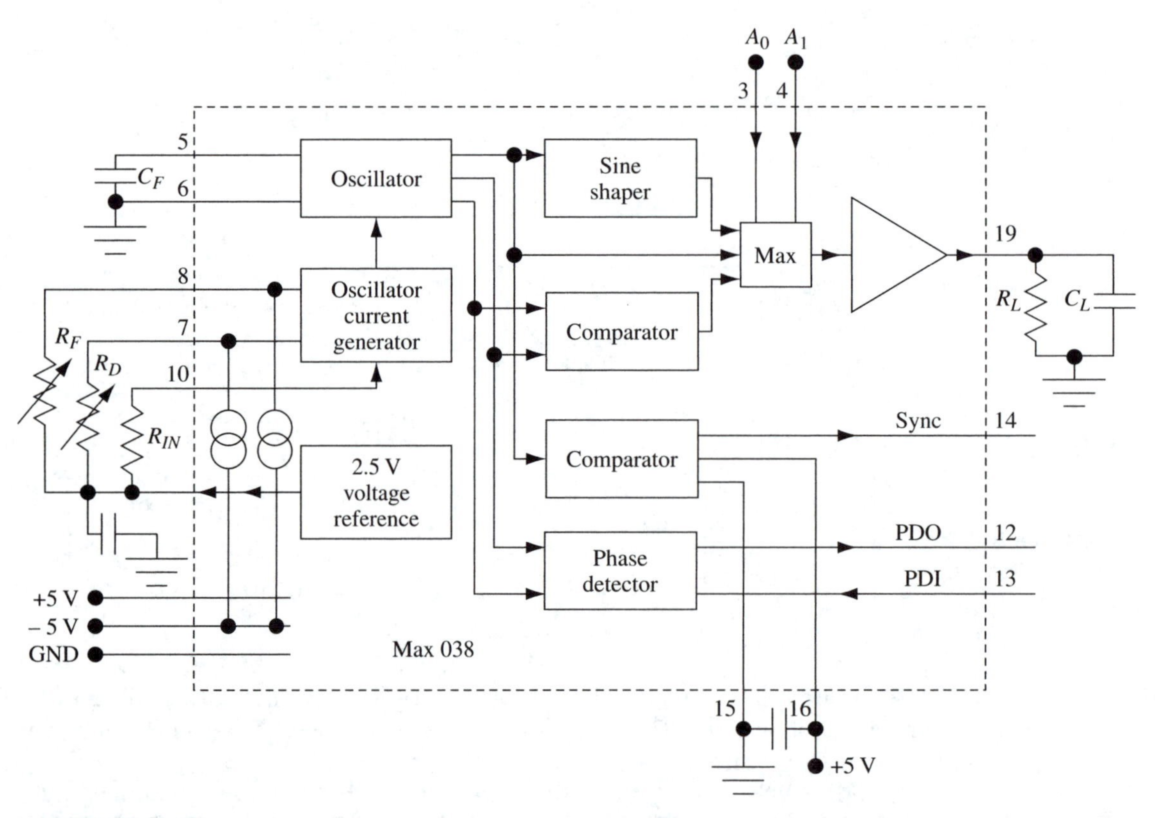

▲ FIGURE 8–30
MAX038 block diagram

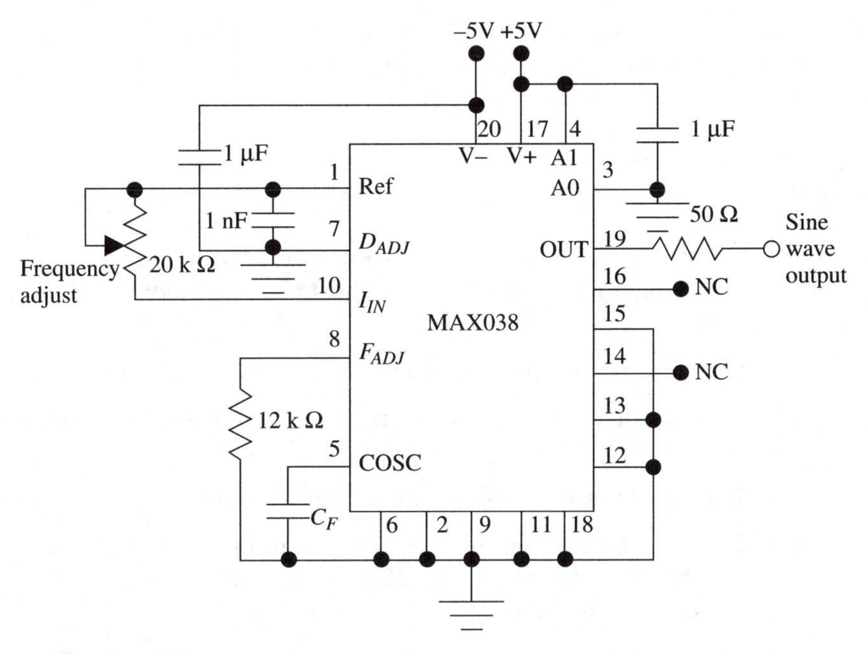

▲ **FIGURE 8–31**
MAX038 function generator circuit

impact the duty cycle. Unlike the 8038, the MAX038 can be powered only by ±5 V DC. Figure 8–31 shows a typical function generator circuit using the MAX038 without the duty cycle or frequency modulation functions.

Summary

In this chapter we have reviewed a wide variety of oscillator and clock circuits that are used in analog and digital circuits that range from computer clocks to RF communications. The two basic types of oscillators are the relaxation type and positive feedback oscillators. Relaxation oscillators function by the charging/discharging of capacitors. Positive feedback oscillators provide positive feedback to an amplifier through a resonant circuit that determines the oscillator frequency. The components of the feedback network are either RC networks, LC networks, or crystals. Voltage-controlled oscillators (VCOs) can be viewed as a subset of relaxation oscillators because they usually use the charging/discharging of capacitors to affect frequency. However, these oscillators all have a primary voltage input that proportionally derives the output frequency. In many ways the VCO is a voltage-to-frequency converter, but there are subtle differences between the two devices.

References

Bell, D. A. 1990. *Operational Amplifiers: Applications, Troubleshooting, and Design.* Upper Saddle River, NJ: Prentice Hall.

Floyd, T. L. 1999. *Electronic Devices.* Upper Saddle River, NJ: Prentice Hall.

Malvino, A. P. 1999. *Electronic Principles.* Westerville, OH: Glencoe-McGraw Hill.

Stanley, W. D. 1994. *Operational Amplifiers with Linear Integrated Circuits.* Upper Saddle River, NJ: Prentice Hall.

Exercises

8–1 Draw two cycles of a pulse waveform that has a frequency of 50 kHz, a peak positive amplitude of 5 V, a peak negative amplitude of 0 V, and a duty cycle of 65%.

8–2 Explain the basic principle of operation for relaxation oscillators.

8–3 What is the benefit of using a crystal as a resonant frequency source as compared to RC or LC circuits?

8–4 What are the three basic types of applications for a 555 timer-counter IC?

8–5 Use a 555 timer-counter to design a free-running pulse circuit that will operate at 50 kHz. The amplitude of the pulse should be +10 V to 0 V and the duty cycle should be 70%.

8–6 Explain why a 555 timer-counter can never achieve a 50% duty cycle without additional circuitry. What additional circuitry can be added to achieve 50% duty cycle?

8–7 Design a circuit that will generate a square wave with ±8 V amplitude at a frequency of 150 kHz.

8–8 Design a circuit that will generate both a triangle and a square wave at a frequency of 50 kHz. The amplitude of the square wave should be ±10 V and the peak amplitude of the triangle wave should be ±6 V.

8–9 Explain the principle of operation of a voltage-controlled oscillator. How can a VCO be used as a voltage-to-frequency converter?

8–10 Design a VCO that will output a sawtooth waveform with a peak amplitude of 5 V and an operating frequency of 20 kHz, when the input control voltage is equal to –2 V.

8–11 Explain the basic principle of operation for positive feedback oscillators. What are the requirements for sustained oscillation to occur?

8–12 Design a Wien-Bridge that will generate a sine wave at a frequency of 100 kHz, at an amplitude of approximately ±6 V.

8–13 Design a phase shift oscillator that will develop a 200 kHz sine wave with a peak-to-peak amplitude of ±8 V.

8–14 What is the benefit of the Clapp oscillator as compared to the Colpitts oscillator?

8–15 What are the major functional differences between the 8038 and the MAX038 function generator ICs?

Data Acquisition and Control Circuits

Introduction

Programmable logic controllers (PLCs), micro-based process controllers, and data acquisition systems are widely recognized as the heart of today's industrial control technology. Many advances have been made regarding the inputs to these controllers/systems, as well as the output devices they drive, such as integrated circuit temperature measurement ICs, high-performance A/Ds and D/As, high-resolution encoders and stepper motors, and high power drive electronics. Industrial control and data acquisition circuits utilize a wide variety of the available analog, digital, software and other technologies. Industrial control products are used by a very large group of companies that include manufacturers of all types of products.

Control circuits sense an event or measure a signal level and then generate an appropriate reaction. Sensing an event is usually a digital operation; the event is either occurring or it isn't, the switch is on or off, a photo-switch senses an object or it doesn't. Measuring a signal level is usually an analog operation, at least initially. A variety of parameters are measured with sensors designed to convert their variations into changing voltage/current levels and eventually digital data.

This chapter discusses the application of electronics to many of these control circuit applications. This chapter is subdivided into five major areas:

- Digital Input Devices: Switches, limit switches, electromechanical relays, proximity detectors, photo-sensors, encoders
- Analog Input Devices: Voltage/current inputs, temperature measurement, position and liquid level sensing, pressure sensors
- Data Acquisition and Control Circuits: Timers and time delay functions, voltage/current transmitters, data acquisition and recording, limit devices, ground fault circuit interrupters, on/off controllers, proportional, integral, derivative controllers (PID), PLCs

- Output Circuits: electromechanical relays, contactors, solenoids, solid state relays, current voltage outputs, proportional output devices, positioners, triac phase angle fire modules
- Data Acquisition and Control Systems: Process control systems, PLCs, SCADA systems

9–1 ▶ Digital Input Devices

Switches and Electromechanical Relays

Switches are very basic circuit components with a wide variety of configurations available. In Chapter 4 we discussed the general switch selection process and the difference between momentary and maintained switches, as well as the concept of poles and throws. Also, the varieties of mechanical configurations were demonstrated: toggle, slide, push, DIP, and so on. All switches are inherently digital in nature because they are either on or off.

Before going much further with the discussion on switches and relays, we must define a special type of schematic diagram commonly used in industrial control systems called ladder logic diagrams. Ladder logic diagrams consist of two vertical rails of the ladder that form the primary voltage that powers the circuit. The rungs of the ladder are drawn horizontally and include circuit elements that are in parallel with all of the other rungs and the rail power supply (see Figure 9–1). The rungs usually include circuit components such as switches, relays, indicator lights, or electronic modules of some kind. Switches or contacts that are considered inputs to the rung are located on the left side of the rung while the rung outputs are located to the right.

It is important to understand that all switches and contacts are shown on a ladder logic diagram in their disengaged, unenergized position. The meaning of the term *disengaged* depends on the type of device being considered. A mechanically engaged on-off switch is disengaged when the switch is in the mechanical OFF position. A float switch that senses the liquid level of fluid in a tank is disengaged when the liquid level is below the location of the float switch. An electro-

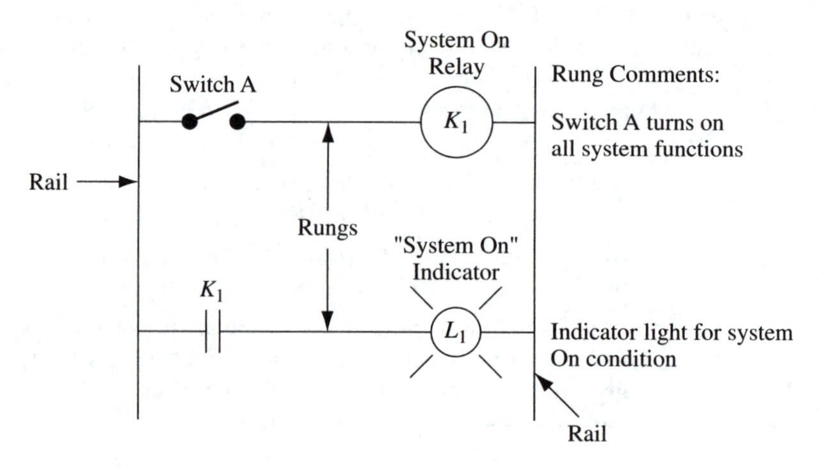

▶ FIGURE 9–1
Ladder logic diagram example

mechanical relay is disengaged when its coil is not powered. Each switch or device can have normally open and or normally closed contacts that are associated with the device. Normally open contacts are those that are open when the device is disengaged and closed when engaged. Normally closed contacts are closed when the device is disengaged and open when engaged.

Each rung is an independent circuit whose output is energized when the switches and contacts on the left side of the rung allow current to flow from the left rail through to the output connected to the right side rail. However, the operation of the rungs are linked together by parallel sets of contacts or contacts that are engaged by relays that are outputs from other rungs. Each rung must have at least one output, or the rung may create a dead short across the power supply. A rung can have more than one output, but they must be in parallel. If two outputs with nominal operating voltages equal to the rail voltage of the ladder circuit are placed on a rung in series, they will split the power supply according to their impedance and will not function.

While any rung theoretically can be placed in any location on the ladder diagram, it makes good sense to position the rungs in order of their sequence of operation and to group rungs that operate together in the same general area. Attaching comments, called "rung comments," is also a good idea to help anyone trying to understand circuit operation. This is analogous to using comments while writing software.

The simple ladder diagram in Figure 9–1 shows switch A as a maintained switch that, when engaged, energizes relay K_1. A normally open set of contacts associated with K_1 close when K_1 is energized, lighting the "system on" indicator light L_1. Figure 9–2 shows a summary of the rules for developing ladder logic diagrams.

1. Two vertically drawn rails represent each side of the power supply.
2. Circuit components are located on horizontally drawn rungs connecting the rails.
3. Inputs are drawn on the left-hand side and outputs are drawn on the right side.
4. A rung cannot have two outputs connected in series and must include at least one output.
5. All contacts are shown in their *disengaged* position.
6. Each rung is an independent circuit but should be located logically close to other interacting rungs and shown in the order of their operational flow.
7. Use rung comments liberally to explain circuit operation.

▲ **FIGURE 9–2**
Rules for ladder logic diagrams

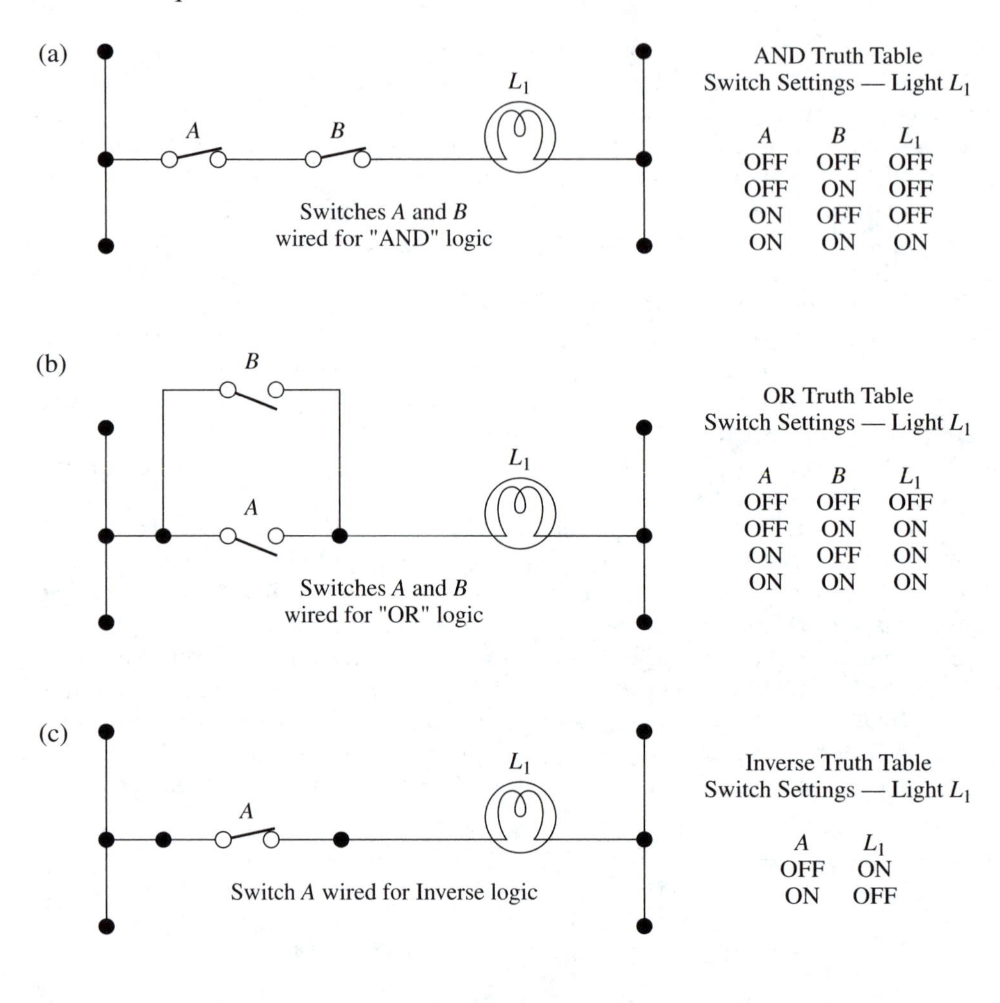

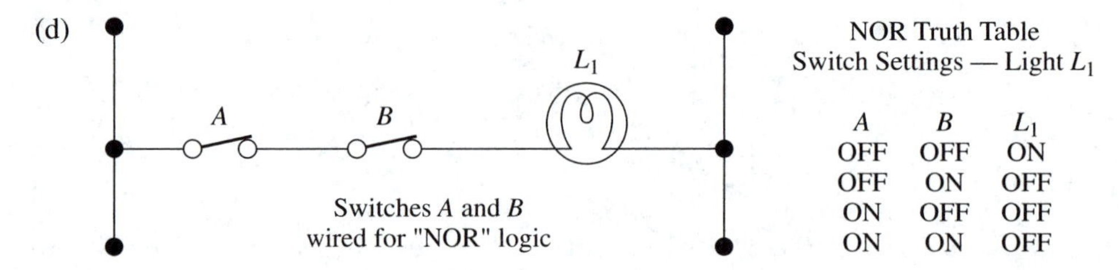

▲ **FIGURE 9–3**
Switch logic ladder diagrams

For industrial control applications it is important to understand how simple switches can be wired together to form the basic logical functions: AND, OR, NOR, and so on. The easiest to reason through are the AND and OR functions. If there are two switches, A and B, where an AND function is required, the normally open contacts for switches A and B should be wired in series. Figure 9–3 shows the lad-

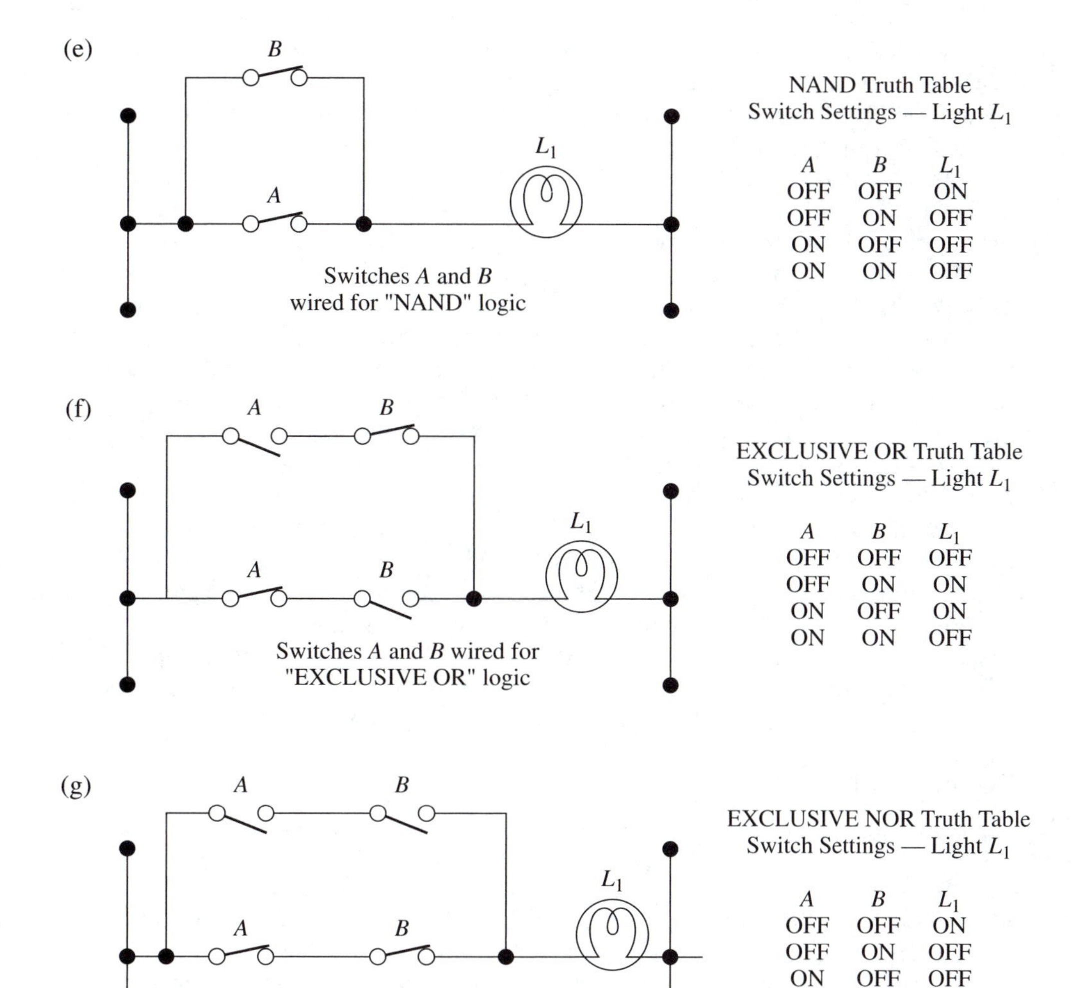

▲ **FIGURE 9–3 *(continued)***

der diagrams and truth tables for the switch logic functions being described here. If the OR function is required, the normally open contacts for switches A and B are wired in parallel. Invert functions can be achieved simply by using the normally closed set of contacts instead of the normally open ones. In Figure 9–3c the indicator light L_1 will light when switch A is not activated, providing the inverse function of using the normally open contacts.

Figure 9–3d shows the truth table for the NOR function and the ladder diagram. The NOR function is accomplished with switches by wiring the normally closed switch contacts for A and B in series, while the NAND function results from wiring the normally closed contacts from switches A and B in parallel (see Figure 9–3e). The remaining two combinational logic functions are the EXCLUSIVE-OR and the EXCLUSIVE-NOR. These are shown in Figures 9–3f and 9–3g, respectively. Both of these logical functions require using two sets of contacts from each switch; both normally closed and normally open contacts are required for each switch.

Electromechanical relays are contacts that are energized by electromagnetic coils. The contacts are often used as inputs to circuits in the same manner that mechanical switches are. The contacts arrangements and terminology for electromechanical relays are the same as for mechanical switches. (Refer back to Chapter 4 for the discussion of "poles" and "throws" and how they define the contact arrangement for mechanical switches and relays.) On schematic diagrams and ladder logic diagrams the coils for electromechanical relays are shown as a circle with a unique name or identifier shown within the circle. Very often the letters used for identifying relay coils are *C*, *K*, *R* or *CR* (stands for control relay) but any name scheme can be used. In addition to the basic name, numbers are attached to make the relays unique. For example, if K is chosen as the basic name scheme, the first relay used might be called K_1, the second K_2, and so on (see Figure 9–4). The contacts associated with any relay coil carry the same basic name as the coil. If more than one set of contacts are associated with a particular relay, such as a DPDT arrangement, then various schemes are used. It is most common to see a dash with a number added to differentiate between the physical contacts. For example, if relay K_1 has DPDT contacts, which means that there are two complete sets of contacts with common, normally open and normally closed connections, one set might be labeled K_{1-1} and the other K_{1-2} (see Figure 9–4).

Example 9–1

Review Figure 9–4 and analyze the circuit operation.

Solution

Figure 9–4 is a schematic shown in ladder logic format. On the first rung of the diagram, if momentary switch A is depressed before B, coil K_1 is energized. When K_1 is energized, normally open contacts K_{1-1} close, latching around the momentary

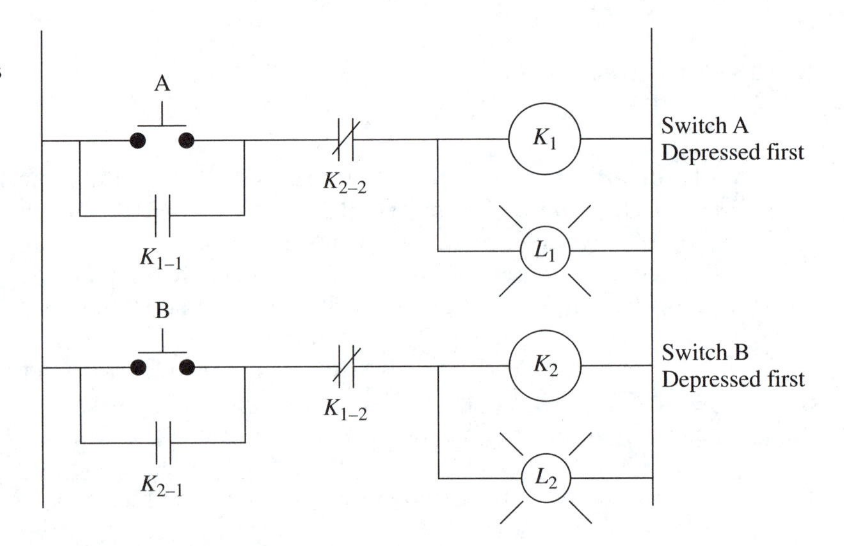

▶ **FIGURE 9–4**
Relay coils and contacts

switch A when it is released. (These are called hold-in or seal-in contacts.) At the same time normally closed contacts K_{1-2} open, breaking the connection from switch B to relay coil K_2. Indicator light L_1 is in parallel with K_1 and lights whenever K_1 is energized.

On the second rung of the diagram, if momentary switch B is depressed before switch A, coil K_2 is energized. When K_2 is energized, normally open contacts K_{2-1} close, latching around the momentary switch B when it is released. At the same time normally closed contacts K_{2-2} open, breaking the connection from switch A to relay coil K_1. Indicator light L_2 is in parallel with K_2 and lights whenever K_2 is energized.

The circuit operates to determine which switch, A or B, is depressed first and latches in to maintain this information for use in the system or by a system operator.

Limit Switches

Limit switches are special switches that are engaged mechanically by an object as opposed to being depressed by a person, as regular switches are. In other words, limit switches detect when the degree of mechanical movement has reached a particular level by the switching of the contacts. Limit switches can have a variety of contact arrangements and are identified in the same way as other switches. Figure 9–5 shows a limit switch and its schematic symbol.

Temperature, Pressure, and Float Switches

Temperature switches measure temperature and compare it with a preset or adjustable control point, engaging a set of contacts when the control point is exceeded. The temperature level can be measured mechanically (fluid/gas filled or bi-metallic sensors) or electrically (thermocouples, RTDs, thermistors, etc.). A thermostat is just another name for a temperature switch. The switches incorporated in temperature switches can have contact arrangements like those already discussed for standard mechanical switches. Figure 9–6 shows the schematic symbol for a SPDT temperature switch.

Pressure switches are similar to temperature switches. Pressure can be measured mechanically or electrically to engage a set of contacts when the control

FIGURE 9–5
Limit switch example

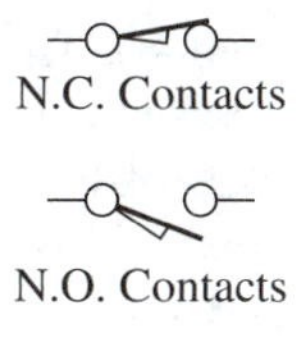

N.C. Contacts

N.O. Contacts

FIGURE 9–6
Temperature switch example

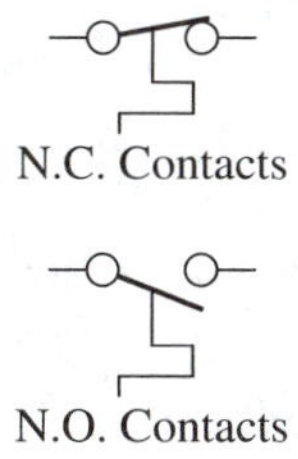

N.C. Contacts

N.O. Contacts

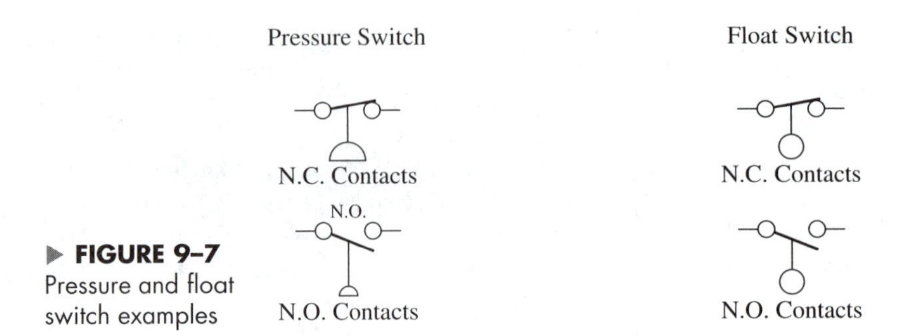

FIGURE 9–7
Pressure and float switch examples

point is exceeded. Figure 9–7 shows the schematic symbol for a pressure switch. Float switches usually include a mechanical arm with a float that indicates the level of a liquid. When the liquid level exceeds the mechanical trip point, the switch contacts are engaged.

Limit, temperature, pressure, and float switches are those most often used to measure and compare a parameter value to a control point and engage a set of contacts accordingly. However, there are many other varieties of switches that operate in a similar manner.

Proximity Detectors

Proximity detectors detect the presence of an object without making physical contact with the object. They perform the same function as mechanical limit switches except that actual contact between the sensor and the object is not required. There are three classes of proximity detectors: inductive, capacitive, or Hall-effect activated.

Inductive proximity detectors utilize LC oscillator circuits, which are similar to those discussed in Chapter 8, to develop an oscillator signal that is based upon the inductance value of the sensing inductor. The sensing inductor is a coil that is wrapped around a ferrite core. When the ferrite core is not in close proximity to any ferrous metallic materials, the inductance of the coil wound around the ferrite core has a value that equates to the resonant frequency of the LC tank circuit fed back from the amplifier output (see Figure 9–8). Therefore, when the sensor does not detect any ferrous materials, the oscillator will oscillate at the resonant frequency. Detection of ferrous materials causes the effective inductance of the coil to change, forcing the oscillator out of high-level oscillation. This change in the oscillator output is easily detected with a demodulator and comparator that switches on either a transistor or a relay output to indicate the presence of magnetic material.

Inductive proximity sensors are used in applications to detect metal objects, in poor environmental conditions (humidity, temperature, vibration and shock) without any physical contact or contact bounce, and with very quick response time.

Capacitive proximity sensors have the same basic structure as inductive proximity detectors. They include an oscillator, demodulator, comparator, and output stage. However, they overcome the most significant shortcoming of inductive proximity sensors: they can detect the presence of both conductive and nonconduc-

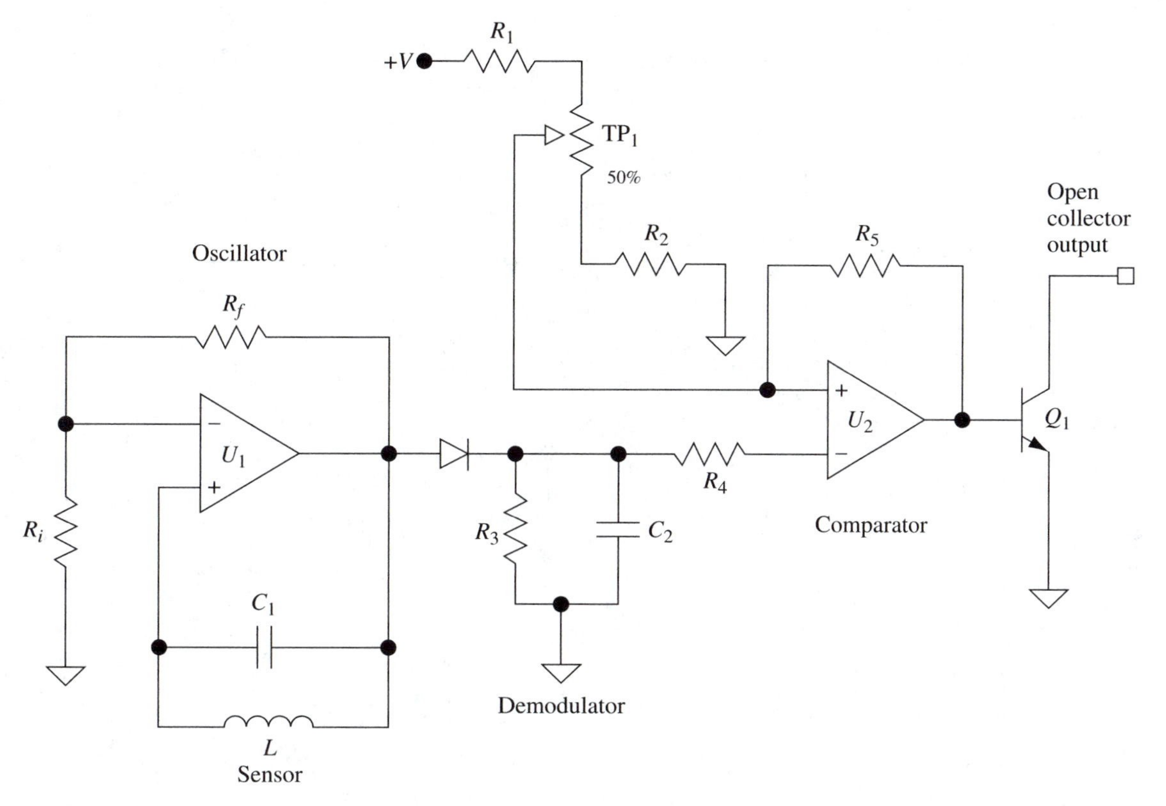

▲ **FIGURE 9–8**
Inductive proximity detector

tive materials. The sensor is a capacitor, the dielectric of which is either air or the object being sensed. When the object sensed is nonconductive, the capacitance increases because the dielectric constant is greater than one, the dielectric constant for air. When a conductive object is sensed, this effectively creates another electrode for the capacitor, reducing the distance between the plates and thereby increasing the capacitance.

The circuit for the capacitive proximity sensor is based upon an RC oscillator circuit (see Figure 9–9). The capacitance sensor becomes the capacitance value of the RC circuit. It either induces oscillation at the resonant frequency when an object is present or there is low amplitude or no oscillation when no object is detected. This is the opposite function of the inductive proximity detector. As with the inductive proximity detector, the demodulator and comparator detect oscillation and switch the output accordingly. However, the output function is reversed to compensate for the inverted oscillator properties (oscillation occurs when an object is sensed in the capacitive sensor, when no object is detected in the inductive sensor).

Capacitance proximity sensors can be used to detect the presence of solids, powdered materials, or liquids. They can sense both conductive and nonconductive materials as long as their dielectric constant is reasonably high.

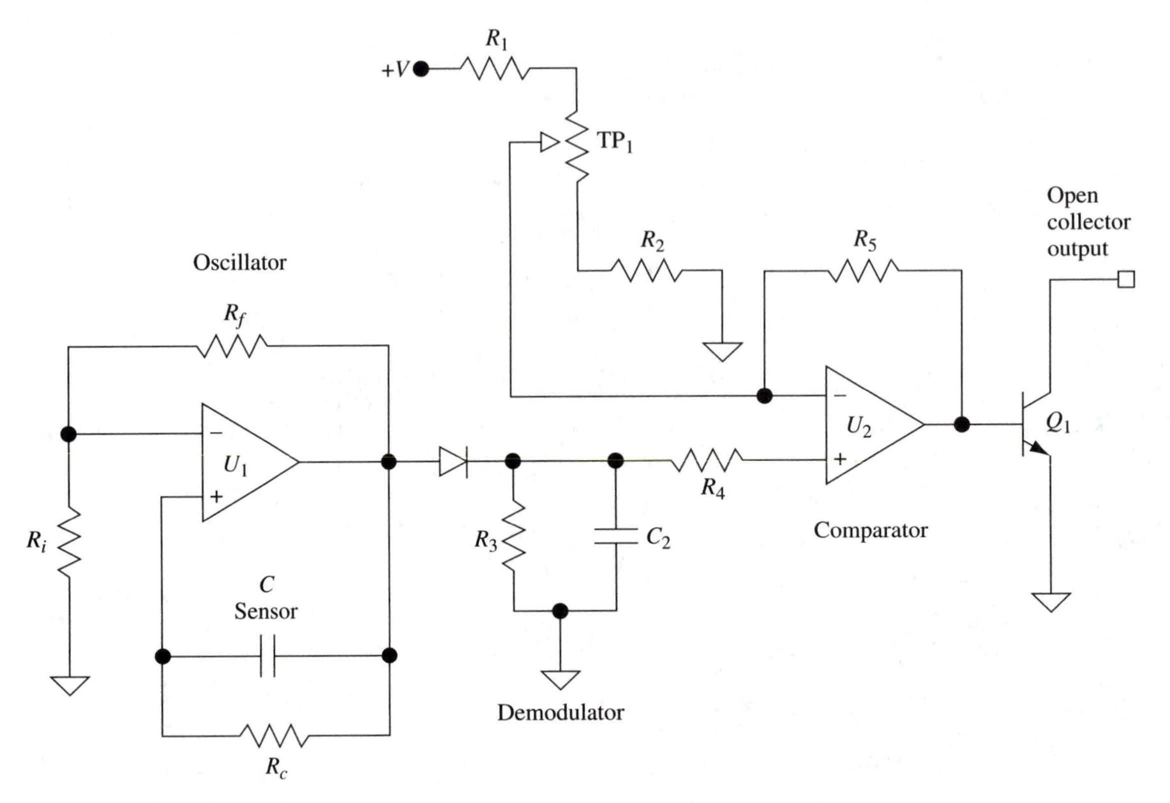

▲ **FIGURE 9–9**
Capacative proximity detector

Hall-effect sensors detect the presence of a magnetic field. The sensor itself is constructed from P-type semiconductor material. A power supply is connected to two sides of the P-type materials and the opposing sides are the detector connections for the "Hall" voltage. The Hall voltage is generated when a perpendicular magnetic field is in close proximity to the P-type material. A comparator circuit senses the Hall voltage and switches the output transistor on or off accordingly (see Figure 9–10). Hall-effect sensors can only be used to detect permanent magnet or electromagnetic materials. They can function in applications with poor environmental conditions that require fast response times and high frequencies.

Photo-sensors

Photo-sensors use the detection of light to determine the presence of an object. All photo-sensors include both a light source and a light detector. The types of light generated by the transmitters are red, green, or infrared. Red is used for general applications, green for the detection of color marks, and infrared is utilized to detect objects over long distances. There are three different methods for using light to detect objects: through-beam, retro-reflective, and diffuse (also called *reflective*). The performance of photo-sensors is determined by the following specification parameters:

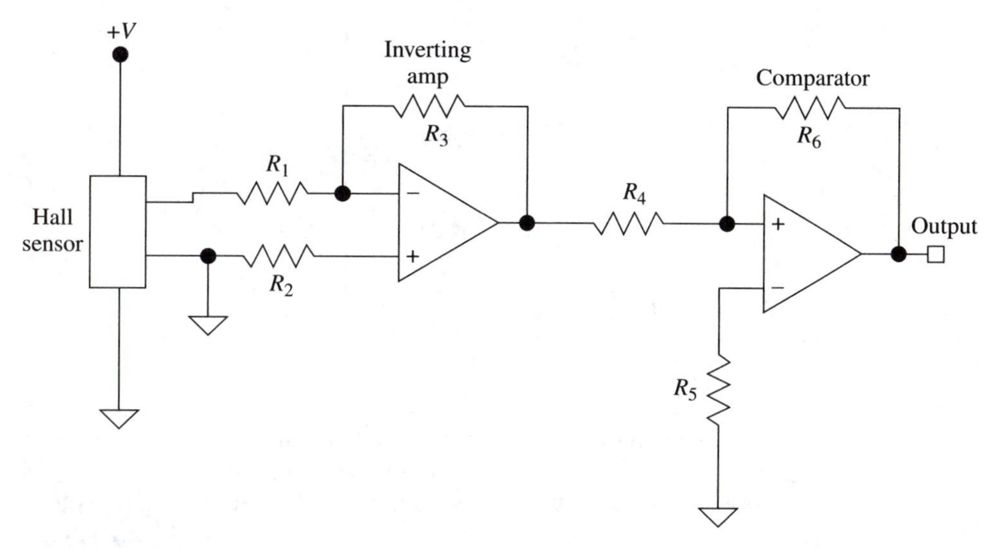

▲ **FIGURE 9–10**
Hall-effect detector

1. Minimum size object that can be detected
2. Maximum distance over which an object can be detected
3. Immunity to ambient light
4. Minimum amount of light needed for detection

Lenses can be used on either the light source, to reduce the size of the beam, or the detector, to reduce the sensed area.

Through-beam photo-sensors have a separate light source and detector that face each other, as shown in Figure 9–11. The light source and detector are aligned so that the light transmitted from the source is received continually by the detector when no object is blocking its path. Objects are detected when they break the beam, causing the detector to change the state of its output accordingly. Through-beam sensing can detect smaller objects over longer distances and is the most accurate

▶ **FIGURE 9–11**
Through-beam photo-sensor

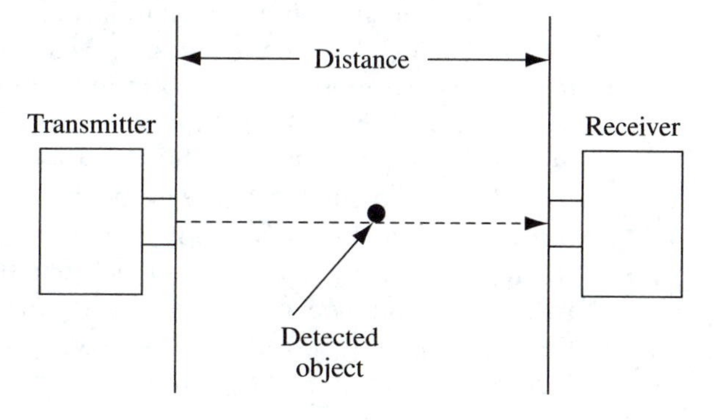

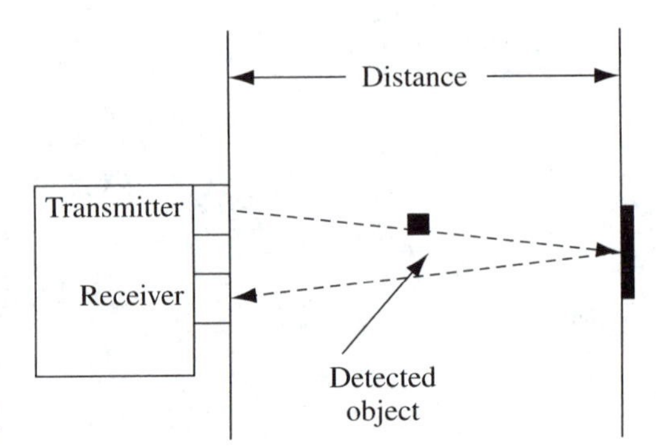

▶ **FIGURE 9–12**
Retro-reflective photo-sensor

photo-sensing method. However, through-beam sensors require two separate packages (light source and detector) that must be carefully aligned.

Retro-reflective sensors contain the light source and the detector in the same package, yet operate in a manner similar to that of the through-beam photo-sensor (see Figure 9–12). Retro-reflective sensors use a reflective disc (similar to a bicycle reflector) to reflect the light back to the detector that resides in the same package as the light source. Objects are detected when they break either the transmitted beam or the returned beam. The angle between the two beams depends upon the displacement between the light source and the detector as well as the distance between the sensor package and the reflector. As the angle increases, the accuracy of the location of the sensed object decreases. This is because the object can break either the transmitted or returned beam, and its location will differ slightly in each instance.

Retro-reflective photo-sensors include the electronics in a single package so they are easier to wire and install. However, the photo-sensor package must still be aligned with the reflector. The amount of light returned to the detector is significantly less than occurs with the through-beam sensor. The lower detected light level combined with the small angle between the transmitted and reflected light limit the size of the object and the range of detection. Retro-reflective sensors are limited to detecting larger objects over medium distances.

Diffuse photo-sensors also have the light source and the detector included in the same package. However, diffuse photo-sensors react to light that is reflected from the object being sensed. This is contrary to the operation of through-beam and retro-reflective photo-sensors, which detect no light when an object is being detected. The light source residing in the diffuse photo-sensor transmits light continuously. When no object is present in the path of the transmitted light, the detector sees only ambient light or the transmitted light reflected from other objects in the area of the sensor. When an object with enough reflective properties is placed in the path of the transmitted light, enough light is reflected back to the detector to cause the output to switch, indicating detection of the object. Diffuse photo-sensors offer the least accuracy and performance when compared to through-beam and retro-reflective photo-sensors, because they are more sensitive to ambient light conditions, and, like retro-reflective sensors, the amount of light reflected from the object is significantly less than the amount transmitted (see Figure 9–13).

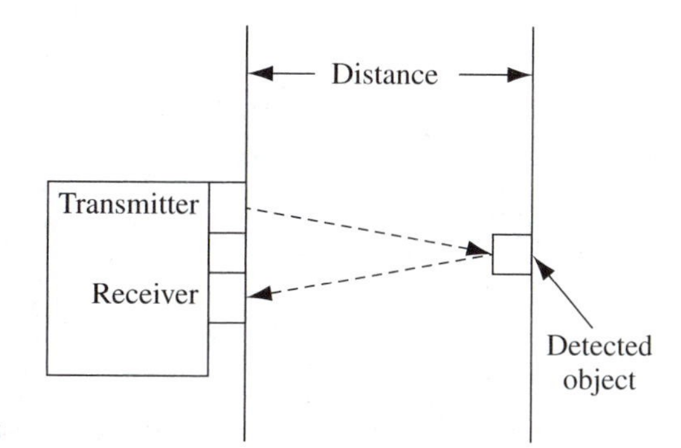

▶ **FIGURE 9–13**
Diffuse photo-sensor

There are a variety of schemes to minimize the effect of ambient light on photo-sensor detection. One method is to detect ambient light levels electronically, subtracting them from the total light detected by the sensor. Another method is to strobe the light source and enable the detector simultaneously at a frequency where the ambient light is a minimum. Fluorescent lights powered by 60 Hz AC power actually vary their light intensity at a frequency of 60 Hz, with the minimum light output occurring at the zero-voltage crossing point. A sensor can detect the zero crossing, trigger the light source, and enable the detector simultaneously, creating a situation in which the detector is "looking" for the light signal at a time when the ambient light is at a minimum.

Encoders

Encoders are sensors used to monitor the position of any device attached to the encoder. Encoders can be used to monitor either angular or linear motion, but rotary encoders measuring angular position are the most commonly used. Encoders are included in the digital input category because the output of the encoder is a digital pulse. However, when the device that receives all of the pulses counts them, the net result is a value for the position of the encoder shaft. There are also two other classifications of encoders: relative encoders and absolute encoders. Relative encoders put out a pulse for any relative movement. For example, a relative rotary encoder might output a pulse for every 3 degrees of angular rotation. The net angular position results from knowing the starting position and counting the pulses in either direction. If there is a power outage, the relative encoder loses all previous information. Relative encoders usually employ a zero position sensor for use as a reference for the start position. On system start-up the encoder shaft is positioned at the zero position and relative motion is determined from the zero point. Absolute encoders maintain and provide an output for the actual position of the encoder at all times.

Rotary encoders are constructed using a through-beam photo-sensor concept. Inside the housing there are two light sources, a disk, grid, and two photodetectors (see Figure 9–14). The disk and grid are attached to the shaft. The light sources and detectors are arranged in through-beam pairs, 90° out of phase with each other. The disk and the grid have opaque and transparent sections etched

▶ **FIGURE 9–14**
Rotary encoder *(Courtesy Omron Electronics, LLC. Used by permission.)*

into them to allow light to pass through to the detector after a particular amount of rotation. The transparent sections are spaced equidistant around the disk, providing an output pulse after a certain amount of angular rotation. Photographic etching processes applied to glass have greatly improved the performance and cost of encoders. These processes provide precision etching of the glass disk and grid, allowing very small transparent areas finely spaced around the disk assembly. This results in encoders with very fine resolution. Relative position, rotary, glass disk encoders offer resolutions in the range of 100–6000 segments per 360 angular degrees of rotation. A 100-segment encoder breaks up the 360° of rotation into 3.6°. A 6000-segment encoder provides one pulse for every .06° of angular movement.

Most encoders provide three output signals that are used by the device receiving the signals to determine the actual position of the shaft. Channels A and B are the signals generated for each transparent segment. The signals output to Channels A and B are out of phase by approximately 90°. The receiving device uses this out-of-phase relationship to determine the direction of angular rotation. If Channel A leads B, then the encoder shaft is moving in a positive (clockwise) direction. Channel B leading A means that the shaft is moving in a negative direction (counterclockwise). (see Figure 9–15). A third signal, called a *control signal*, generates an output pulse after every complete revolution in one direction.

9–2 ▶ Analog Input Devices

Analog input devices convert the value of a physical parameter, such as temperature or pressure, into a voltage or current so that its value can be monitored, displayed, controlled, or recorded. Section 9–3 discusses the devices that perform these operations. There are many types of analog input devices used to measure the following physical parameters:

Voltage/current

Temperature

Position/level

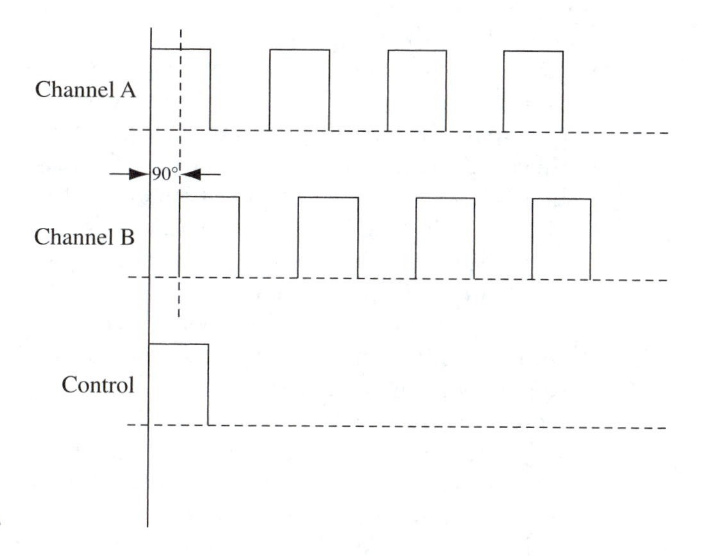

▶ FIGURE 9–15
Encoder output signals

Pressure

Flow

Force/weight

Speed/acceleration

Humidity, conductivity

PH, density, viscosity

Many of these signals are fairly slow in changing value, especially temperature. Pressure and flow are examples of fast-changing process variables. Each signal is viewed as a DC or low-frequency input as far as the speed of input changes that occur. This section will review the application of electronics to a few of the more popular types of analog input devices.

Voltage/Current Inputs

Voltage and current are the two most easily measured parameters and represent the starting point for our discussion of analog input devices. Voltage can be easily measured, amplified, digitized, and indicated. Voltage is the most desirable form for any measured parameter to be measured. Current inputs are handled almost as easily; they are simply passed through a resistor to generate a voltage. Currents can also be measured indirectly by measuring the electromagnetic field around a wire carrying current with current "donuts" or clamp-on ammeters.

Many times after a parameter has been converted to a voltage or current, it is desirable to amplify the signal further before connection to a controller or indicator. This is especially true if the controller or indicator is located some distance away from the sensor. A low-level signal transmitted over a long distance is much more likely to experience significant attenuation and pick up unwanted noise. Because the signal level is smaller, noise levels appear relatively larger and are harder to ignore or reject. To resolve this problem, devices called *transmitters* are used to amplify the input to standard voltage/current signal levels. The most common of these signals for industrial instrumentation are 4–20 mA and 1–5 V. When 4–20 mA or 1–5-V signals are used to transmit analog signal information, the low end of the signal (4 mA or 1 V) represents the minimum range of measured input while the high-end signal (20 mA or 5 V) represents the maximum-input signal value.

For example, if a temperature transmitter uses a thermistor to measure temperature over a range of 0° to 300°F and has a 4–20 mA output, then a 4 mA signal represents 0°F and 20 mA represents 300°F. Otherwise, the 4–20 mA is proportional to the input such that a 12 mA output equals 150°F, 16 mA equals 225°F, 8 mA equals 75°F, and so on. The output of this transmitter would be connected to a device, such as a digital indicator with a standard 4–20 mA input. The digital indicator would need to know the type of signal represented by the 4–20 mA (temperature in °F, pressure, etc.) and the input range vs. the 4–20 mA range. This information is keyed into the digital indicator in the following form: Units, °F; Minimum range, 0; Maximum range, 300. The indicator will then scale the display to show the actual temperature measured by the thermistor.

4–20 mA signals are constant current signals. This means that when the low-end input signal is provided, 4 mA will be the output to any load device as long as its load impedance is below the maximum impedance the constant current source can drive. The 4–20 mA signals are easily converted to 1–5 V by passing the current through a 250-Ω resistor.

Temperature Measurements

Temperature can be converted to an electrical medium by using sensors that vary their voltage or resistance as temperature changes. The most commonly used temperature sensors are thermistors, RTDs, thermocouples, and IC temperature sensors.

Thermistors

Thermistors vary their resistance as temperature changes. There are many varieties and types of thermistors. Most provide a large change in resistance per degree of temperature change and exhibit a negative temperature coefficient (the resistance decreases as the temperature increases). Thermistors are categorized by their resistance value at 32°F (0°C), which is relatively large at 1000 Ω or greater. The relationship between the temperature and the resistance is nonlinear (see Figure 9–16).

In order to convert the resistance change to a voltage, a resistor is placed in series with the thermistor and a voltage applied to the circuit (see Figure 9–17a). Because of the high thermistor resistance value, the relative resistance of the connecting leads is not high enough to cause significant lead length error, as experienced with other resistance sensors. Either an analog or digital circuit that indicates, monitors, records, or controls temperature receives the voltage signal across the thermistor. Another common thermistor circuit is the bridge circuit shown in Figure 9–17b. Thermistors are an inexpensive and simple method for

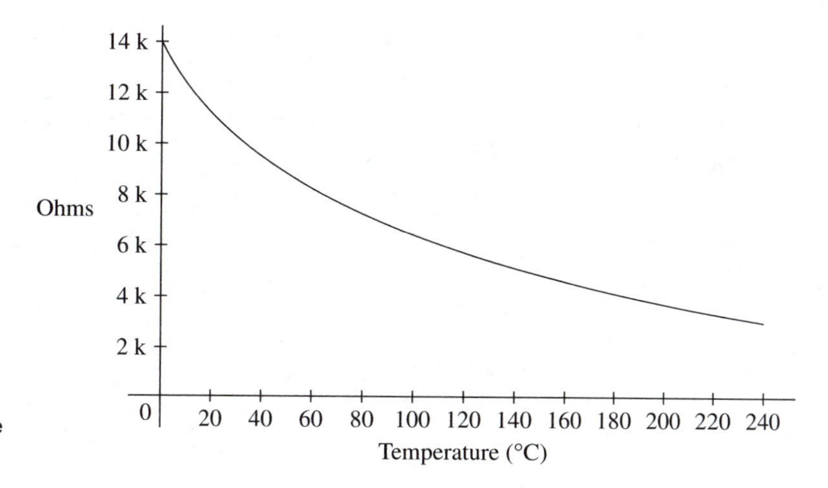

FIGURE 9–16
Thermistor Resistance vs Temperature

measuring temperature over ranges of approximately 0° to 300°F. Their negative temperature coefficient, large nonlinear change in resistance, and limited accuracy limit their use to the least critical applications.

Resistance Temperature Detectors (RTDs)

RTDs also vary their resistance with temperature. However, the nature of the resistance change is much different than that experienced with thermistors. RTDs offer a very linear response and positive temperature coefficient. The amount of resistance change is small, only about .2 Ω/°F. The nominal resistance value (the resistance at 32°F) for an RTD is usually 100 Ω. They are fabricated most often with a nickel or platinum material. Their construction takes the form of wire, wrapped around a bobbin or thick fim substrates. The resistance values are controlled with great accuracy and RTDs can be used over a temperature range of

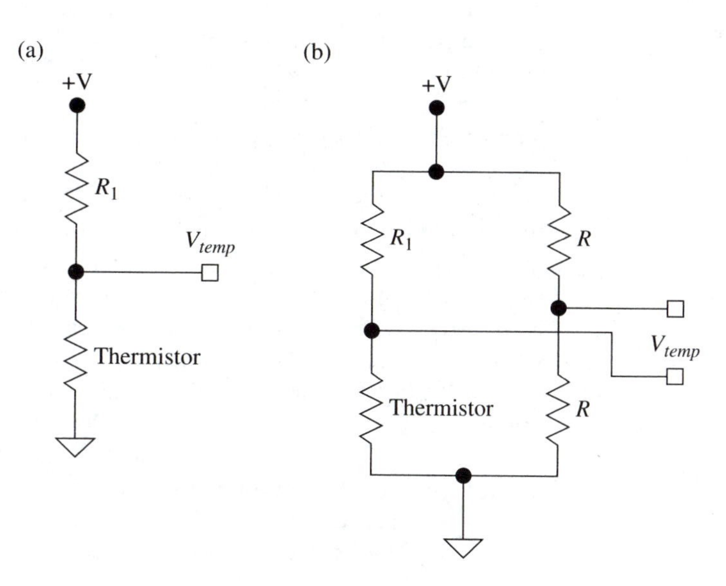

FIGURE 9–17
Thermistor temperature measurement circuits

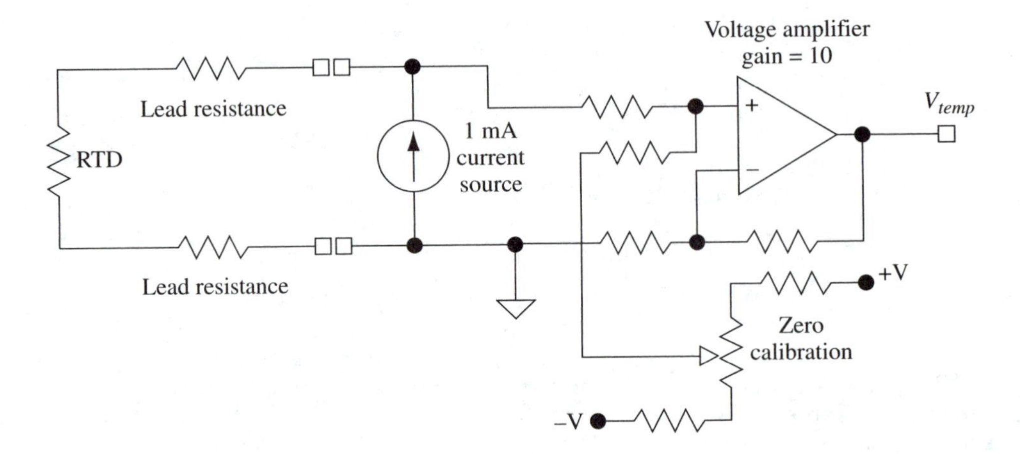

▲ **FIGURE 9–18**
RTD temperature measurement circuit

−100° to 800°F. RTDs are the most accurate temperature sensors available and are commended for use in critical applications. RTDs are linear enough so that further linearization is not required for many applications when the RTD signal is converted to a digital value.

Because both the nominal resistance value and the change in resistance of RTDs are small, the near linear resistance vs. temperature relationship can be preserved by using a constant current source to generate a voltage signal across the RTD. To prevent self-heating of the RTD, the current is usually kept below 1 mA. (see Figure 9–18). Because the resistance value of the RTD is relatively small, lead resistance is more significant. The resistance of the leads adds to the resistance of the RTD, making the voltage seen by the sensing circuit larger than it would be without any lead resistance. The size and length of the connecting leads, as well as the required measurement accuracy, determine whether the amount of lead length error must be compensated for.

The simplest method for minimizing lead length error is to use the shortest lead lengths and the largest conductor size possible, thereby reducing the net lead resistance. A calibration circuit can be added to compensate the circuit for any lead length error. This is shown in Figure 9–18. To calibrate the circuit, the sensor should be placed in a medium for which the temperature value is known. The voltage output from the voltage amplifier is adjusted with the zero adjustment to read the value expected for the temperature being sensed. For example, let's say that the sensor is placed in an ice bath maintained at exactly 0°C. Ideally, the RTD will have a resistance of 100 Ω at 0°C. If the current source is exactly 1 mA and the lead resistance is 0, the voltage across the RTD would be 100 mV. If the voltage gain of the voltage amplifier is +10, then its output is ideally 1 V. Any lead length resistance present in the circuit would add voltage to the output of the voltage amplifier. The zero calibration potentiometer can be used to eliminate the effect of the lead length error.

In some applications calibration is undesirable and a more permanent and automatic form of lead length compensation is needed. This is accomplished by using

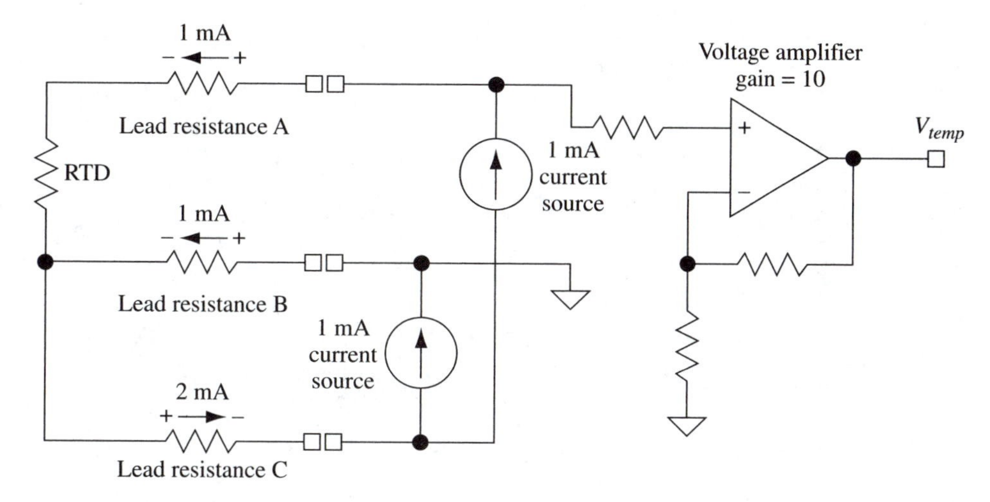

▲ **FIGURE 9–19**
Three-wire RTD lead length compensation

an RTD with three leads. Two leads are attached to one side of the RTD and the third lead to the other side. A special circuit is developed to drive the three RTD leads. The circuit uses two constant current sources connected to the RTD leads, as shown in Figure 9–19. The trick of this circuit is to connect the voltage amplifier sensing leads to the positive side of each constant current source. By sensing these two points, the voltage drop caused by the lead resistance of lead A and lead B is equal in magnitude and opposite in polarity. The two requirements for this are that the lead resistance must be the same and the constant current source values must be identical. Because neither of these situations is ever completely true, there will always be some amount of lead length error. In actual practice the error is minimal and the system works very well. Three-wire RTDs dictate a higher cost due to the additional lead wire and constant current source required, but the concept is used often to resolve lead length error problems without calibration in industrial appilications. Most industrial instrumentation manufacturers offer both two- and three-wire RTD inputs as a standard feature on their equipment.

An even better method of lead length compensation for RTDs is the utilization of four lead wires. This method employs Kelvin sensing, a concept used on many high-accuracy ohmmeter systems. Current is passed through the resistance being measured and the voltage across the resistor is sensed on two separate lines connected to a high-input impedance amplifier. Because of the very high input impedance, there is very little current flow in the sensing leads and therefore only minute levels of lead length error (see Figure 9–20).

Thermocouples

Thermocouples have been used to measure temperature for a long time. While they might represent old technology, they are still commonly used today because of their accuracy, overall versatility, and the wide range of temperature ranges they cover. Thermocouples are constructed by combining two dissimilar metals together at a

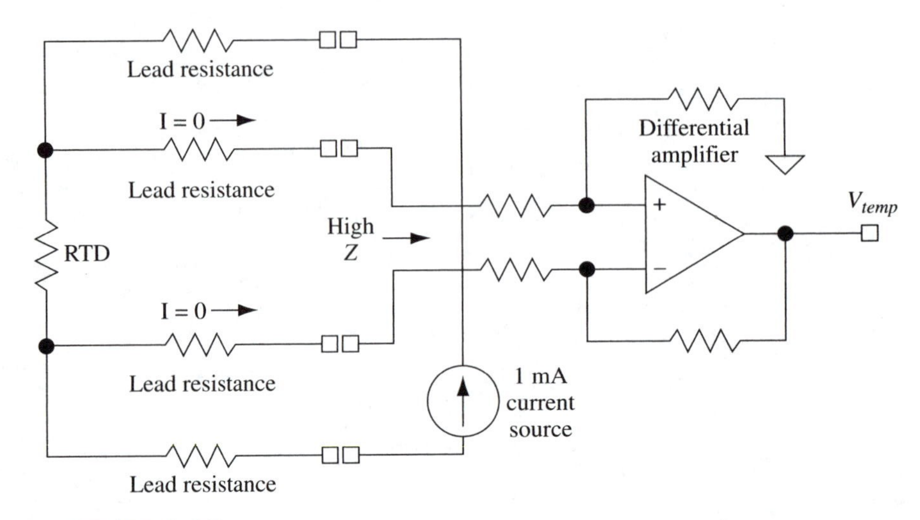

▲ **FIGURE 9–20**
Four-wire RTD lead length compensation

junction point to create a temperature sensor. Whenever dissimilar metals are connected, a small voltage is generated across the junction of the two materials. The value of the voltage is in the millivolt range and varies in a predictable way as the temperature of the junction varies. This is called the *Seebeck effect* and occurs whenever dissimilar metals are connected. When certain dissimilar metals are combined, they create functional thermocouples that have become standard temperature sensors. For example, type J thermocouples are fabricated with iron and constantan and function over a range of 0° to 1400°F. Type K thermocouples are made from chromel and alumel and are used in the range from 0° to 2500°F. There are many other types of thermocouples that have been developed for particular temperature ranges and applications.

Thermocouples are excellent temperature sensors but their use in these applications is complicated by one simple fact: the Seebeck effect is also experienced when sensing leads are connected to the thermocouple to measure the voltage across the junction. Because of this, the wire leads connecting to the thermocouple junction must consist of the same dissimilar metals that make up the thermocouple junction. A type J thermocouple has an iron-constantan junction as a temperature sensor with the iron side of the junction connected to an iron wire lead and the constantan side of the junction connected to a constantan lead. Constantan is a copper-nickel alloy. The sensing iron-constantan junction is often called the *hot junction.* The hot junction is the point where the temperature measurement is being taken (see Figure 9–21).

In order to use the thermocouple as a sensor, a connection to an electrical circuit must be made. A voltage will then be induced by the Seebeck effect at the point of connection, and this must be compensated for. Let's look at what happens when we connect the thermocouple to a DVM. The connection points from the iron and constantan leads to the DVM leads actually create two additional thermocouple junctions (see Figure 9–22a). If the DVM leads are made out of aluminum

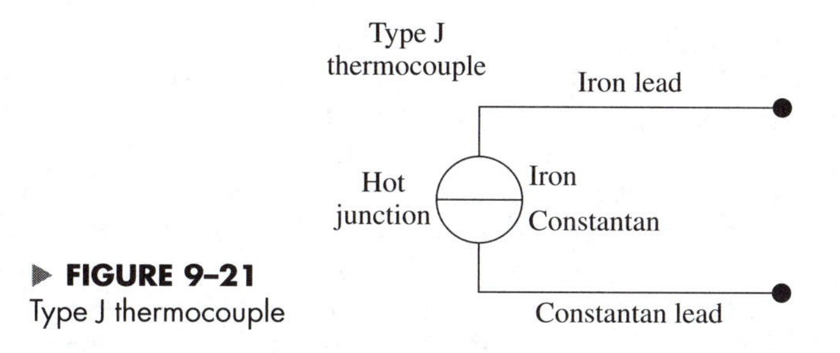

FIGURE 9–21
Type J thermocouple

conductors, the created thermocouple junctions are iron-aluminum and constantan-aluminum thermocouples. The connection points to the thermocouple wires are called the *cold junction.* Note that the voltages generated by the cold junction thermocouples subtract from the hot junction voltage. This phenomenon is called *cold junction error* and results from the Seebeck effect that occurs when connections are made to the thermocouple wire. If the connection points to the thermocouple wires are maintained at the same temperature and are made out of the same material, the net effect of the two junctions is one type J cold junction that opposes the hot junction (see Figure 9–22b). The net voltage seen by the DVM is the hot junction voltage minus the cold junction voltage. Let's calculate the DVM reading if the hot junction is placed in a temperature medium of 200°F and the cold junction is at room temperature, 70°F. The table for type J thermocouples shows that at 200°F the type J junction generates 10.777 mV, and at 70°F the voltage is 3.649 mV. The voltage read on the DVM would be 10.777 – 3.649 = 7.128 mV. Converting back to °F

FIGURE 9–22
Thermocouple cold junction

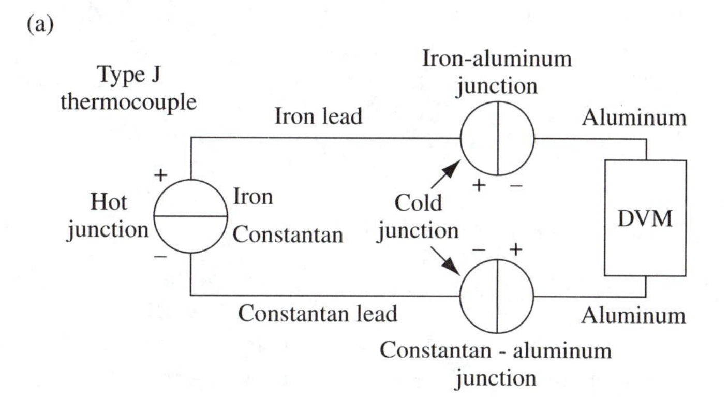

using the thermocouple tables, 7.128 mV is equivalent to 134°F, significantly less than the actual 200°F seen at the hot junction. The cold junction error is not a steady error but varies with the temperature of the cold junction.

In order to use thermocouples to measure temperature, we need some method to compensate for the error induced by the cold junction. Cold junction error can be eliminated by maintaining the cold junction at 32°F, because the voltage generated by all thermocouple junctions at 32°F is 0 mV. However, maintaining 32°F as a cold junction temperature is usually impractical. The preferred method employed on most industrial instrumentation is to measure the cold junction temperature and subtract the equivalent value from the voltage measured for the hot junction. This is done with a small cold junction sensor, usually a thermistor, diode, or IC temperature sensor. Figure 9–23 shows a circuit that measures the thermocouple hot junction value and subtracts the cold junction value.

Because very little current flows in the thermocouple leads, small leads can be used and very little lead length error occurs. Thermocouple types are available to measure temperature ranges that go from approximately –300°F to 4200°F. Some thermocouples are more linear than others but all have areas of significant nonlinearity. When thermocouple signals are converted to digital, linearization is required. This is usually accomplished by software linearization tables.

IC Temperature Sensors

Most IC temperature sensors are based upon the variation in voltage drop across the PN junction as temperature changes. Most IC temperature sensors are available in two- or three-pin transistor packages or IC mini DIP packages. The LM335 IC temperature sensor is a very popular IC sensor. The LM335 uses the breakdown voltage across a zener diode as a measurement of the temperature. The LM335 has been configured to provide 10 mV/°K output and operates over the range of –40° to 100°C. The output is very linear and is accurate within 1°C over the central 100°C span. A third terminal is available for calibration of the output. At 0°C the output of the LM335 is ideally 2.73 V (0°C = 273°K, 10 mV/K × 273 K = 2.73 V). Figure 9–24 shows a typical application of the LM335.

Position and Liquid Level Sensors

Analog position sensors output a voltage or current that indicates the absolute position of an object. The most common position sensors are rotary or linear potentiometers. Rotary potentiometers indicate angular position in terms of resistance, while linear potentiometers output a resistance that relates to linear position. Both types of potentiometers offer a multitude of resistance ranges with very linear resistance vs. position relationships. Resolution is an important concern for this type of position sensor. In potentiometers constructed with fine conductive wire, resolution is determined as the number of turns of the conductor used to develop their resistance. Conductive plastic types of potentiometers offer negligible resolution. Potentiometers can be used to develop simple voltage dividers or they can be part of a bridge circuit, as shown in Figure 9–25.

Linear Variable Differential Transformers (LVDTs) are another variety of position sensor. They use basic transformer theory to develop an output that indicates

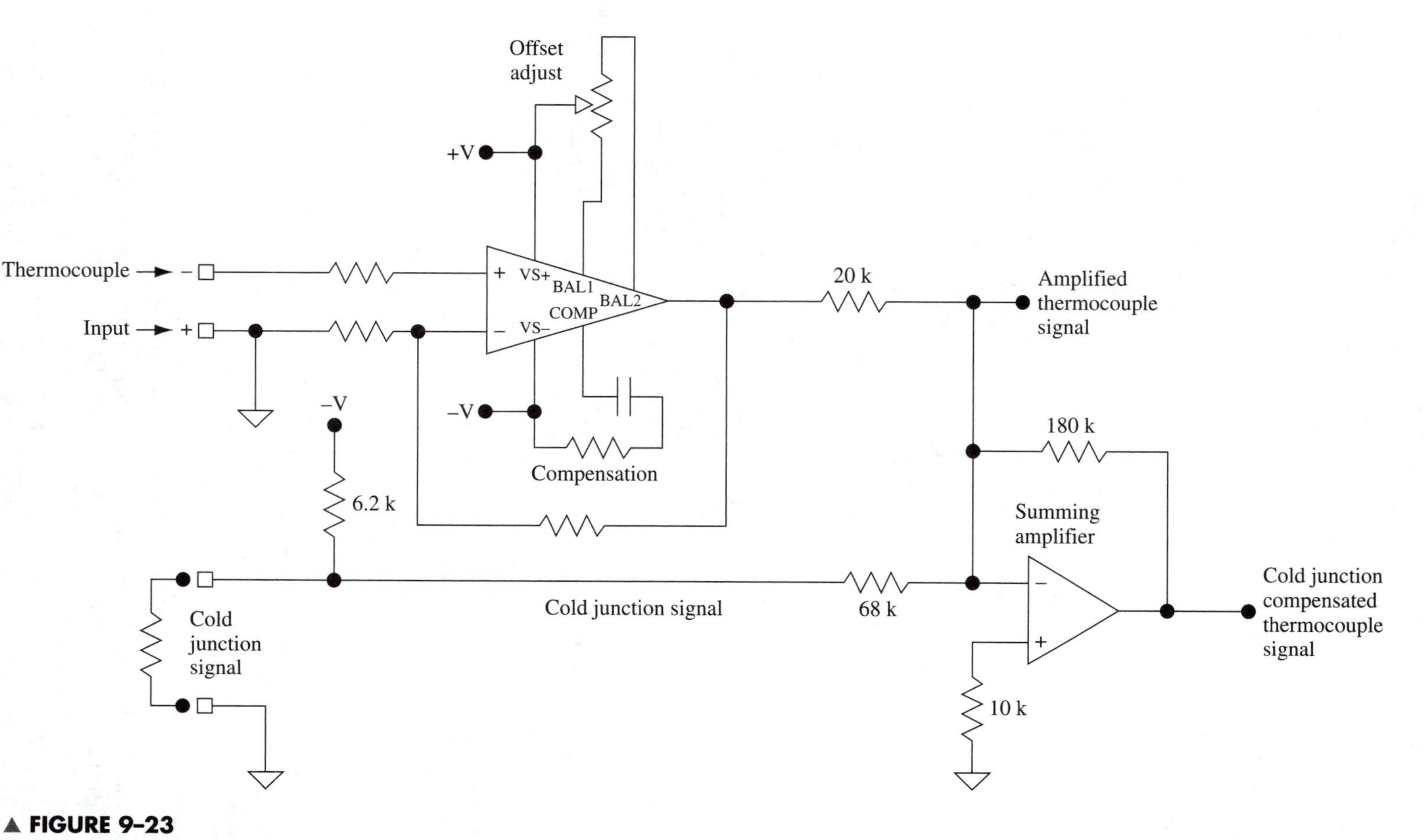

▲ **FIGURE 9–23**
Thermocouple cold junction condensation

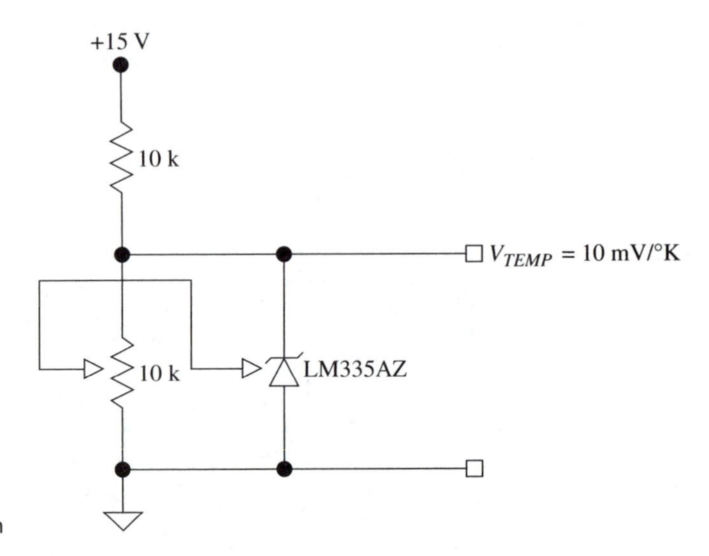

▶ **FIGURE 9–24**
LM335 IC temperature sensor application

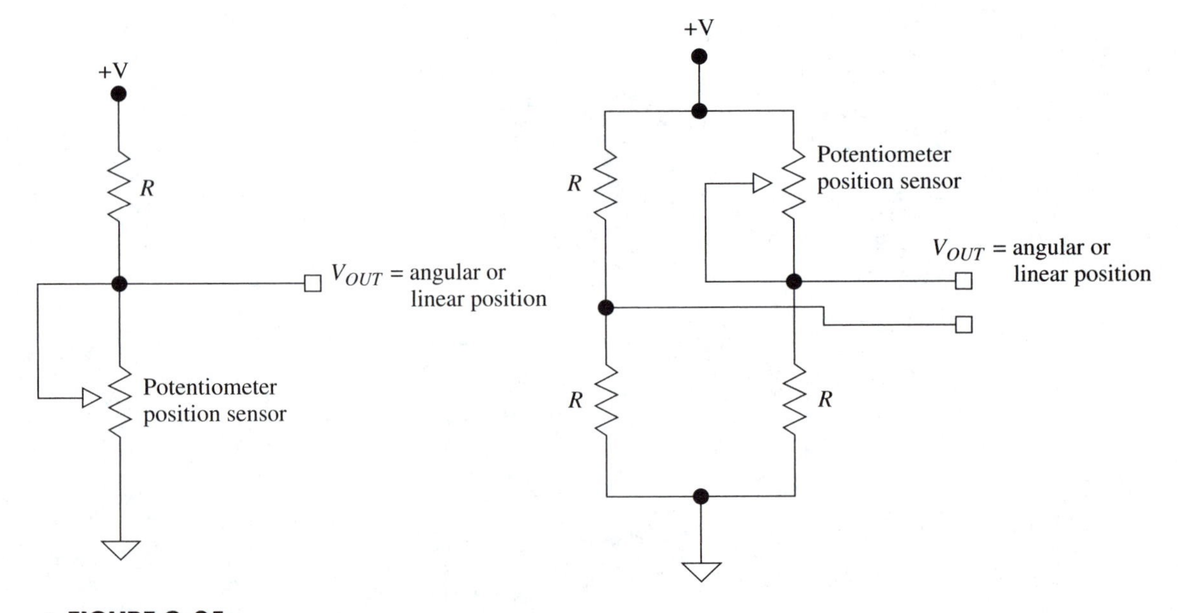

▲ **FIGURE 9–25**
Potentiometer position sensors

position. LVDTs consist of a transformer whose primary and two identical secondaries are wrapped around a tube. A movable core is positioned inside the tube. As the core is moved in and out of the tube, the AC voltage presented to the secondary output is proportional to the position of the moving core (see Figure 9–26).

When a float arm is attached to a position sensor, the result is a liquid level indicator. Sensing the fuel level in a gasoline tank is a very common application of potentiometer position sensors as liquid level indicators. Both potentiometers and

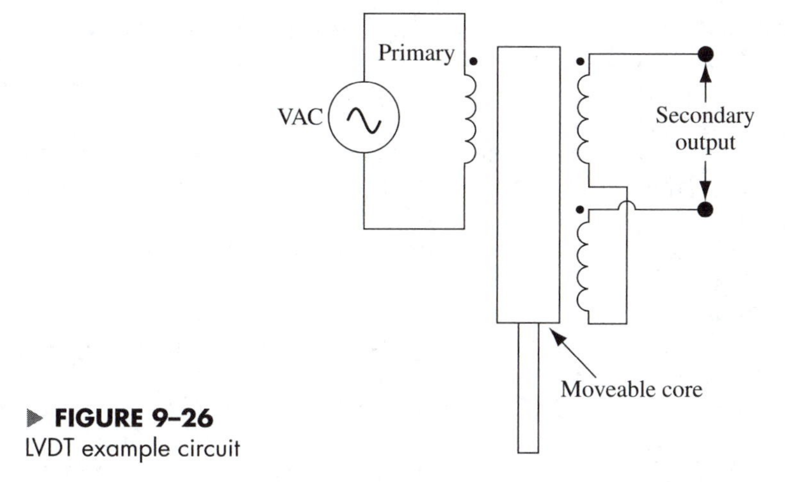

FIGURE 9–26
LVDT example circuit

LVDTs can be used as liquid level indicators but potentiometers are the sensors used most often in these applications.

Capacitance sensors can also be used for liquid level detection. Capacitance sensors are constructed by placing two plastic tubes one inside the other. The liquid level being sensed becomes the dielectric in this case, and as the level rises between the two tubes, the capacitance of the sensor changes. If the capacitance of the sensor is used to create a relaxation oscillator as discussed in Chapter 8, the frequency of the relaxation oscillator will be proportional to the capacitance of the sensor and the liquid level being sensed.

The liquid level measurement methods described thus far work well for smaller tanks where relatively small mechanical movement is experienced. However, in large tanks either weight-and-line, differential pressure, or sonic liquid level sensors are used. Weight-and-line sensors are mounted in the top of a tank. They extend a floating weight on a line until the level is detected by the release in tension of the line. The weight is then retracted as an encoder counts the distance of the retraction. Differential pressure level indicators measure the pressure at the top of the tank and the pressure at the bottom of the tank. The differential pressure between these two measurements is proportional to the liquid level. Sonic liquid level detectors use sound wave transmitter/receivers mounted on the top of the tank. The transmitters send out sound waves. The time for the sound waves to be reflected back to the receiver is measured. This time value is proportional to the liquid level in the tank.

Pressure

Pressure can be measured electronically with sensors that vary in resistance, capacitance, or inductance. Piezoelectric sensors output a small voltage when pressure is applied to certain crystal materials. One of the most popular methods for measuring pressure employs strain gages as the principal measuring element. A strain gage is a continuous back-and-forth pattern of fine wire affixed to a flexible surface. As pressure is applied to the surface, the resistance of the wire changes proportionally as the deflection of the surface material. This is readily converted to pressure in units of pounds per square inch (psi). A common nominal strain gage resistance value (with only atmospheric pressure applied) is 120 Ω with resistance changes of 2 Ω/psi.

Capacitance pressure sensors exist where pressure is applied to the plate of a capacitor, thereby changing the capacitance of the sensor. One capacitor plate is made stationary while a moveable plate is attached to a pressure-activated diaphragm. The capacitance of the sensing capacitor varies as the pressure applied to the diaphragm.

Inductive properties can also be used to measure pressure by taking an LVDT, as discussed in this section, and attaching a pressure diaphragm to the moveable coil. These are called *variable reluctance/inductance pressure sensors.*

Because pressure measurement generally requires a mechanical assembly around the sensing mechanism, the sensor is usually converted to what is called a *transducer* or a *transmitter* within the assembly. A sensor is the bare sensing element by itself, such as a strain gage. If the strain gage is connected in a bridge circuit powered by a DC voltage, the result is called a *pressure transducer.* If the output of the transducer is further amplified to develop a 1–5-V output, this device is called a *pressure transmitter* (see Figure 9–27). The same definitions for sensors, transducers, and transmitters are applied to temperature, flow, and other parameter measurement. For example, a thermocouple temperature transmitter includes all the electronics to accept a thermocouple input and develop a 1–5-V cold junction compensated output.

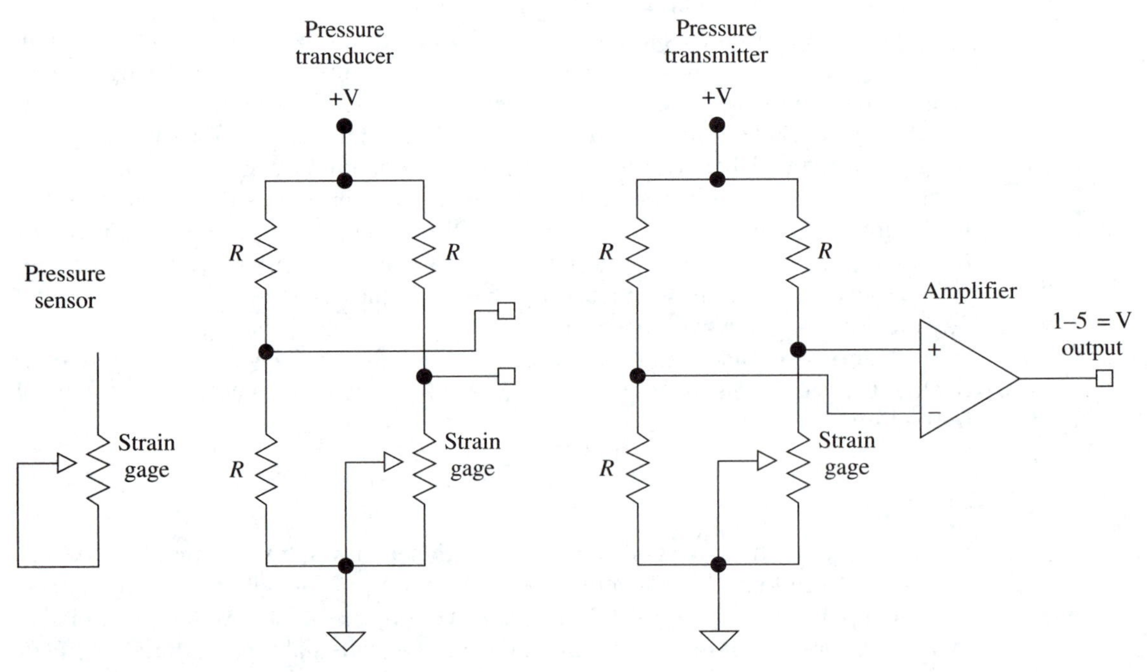

▲ **FIGURE 9–27**
Sensor-transducer-transmitter

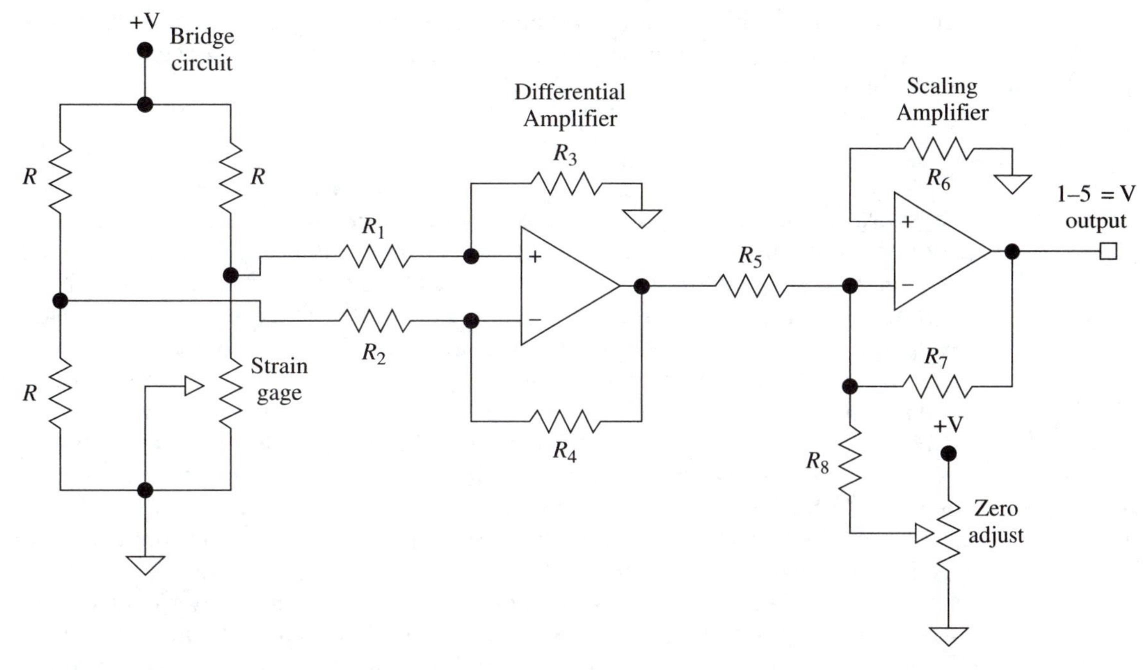

▲ **FIGURE 9–28**
Pressure transmitter circuit

Pressure sensors are usually converted into pressure transducers or transmitters within the mechanical housing required for mounting the mechanical diaphragm. This is done for two reasons:

1. There is usually room for the electronics in the assembly.
2. The output of a strain gage is a relatively low resistance signal and is subject to the lead length error problem discussed for RTDs.

By including the transducer or transmitter electronics within pressure sensing housing, the lead length problem is resolved with the 4–20 mA or 1–5-V pressure output signal from the pressure transmitter. Figure 9–28 shows the circuit for a very basic strain gage pressure transmitter with a 1–5-V output.

9–3 ▶ Data Acquisition and Control Circuits

Data acquisition and control circuits receive the signals from the input devices described in Sections 9–1 and 9–2. They can either indicate or record the value of the parameter measured, or they can provide outputs that can control its value or the value of some other parameter. Most often these circuits are microprocessor-based and include advanced math functions; timing/counting functions; programmable logic; and proportional, integral, and derivative operations. This section discusses these functions and how they are incorporated in a variety of data acquisition and control products.

Timers and Time Delay Functions

Timers are an important requirement for most control systems. Timers are used to either count the time between events or to engage outputs at an appropriate time. Many times in control circuits, time delays must be incurred to ensure the smooth operation of control systems. There are generally two timing devices that are used to implement timing functions: on-delay timers and off-delay timers.

On-delay timers are available separately as electromechanical devices or as software functions on a variety of microprocessor-based controllers. When powered, each on-delay timer counts the passage of time. As with any other switch or electromechanical relay, there can be a variety of contacts associated with each timer (i.e., SPST, DPDT, etc.). Each timer also has a preset or time set value. When the accumulated time count equals the preset time value, the timer becomes engaged. When engaged, all of the contacts associated with the timer change state. Previous to this, all contacts were in their disengaged condition; normally closed contacts were closed, normally open contacts were open. Once the timer becomes engaged, it remains engaged, still counting time while the outputs are maintained in their energized state. Whenever power is taken from the on-delay timer, it is reset immediately and all outputs revert to their de-energized condition. When energized again, the cycle repeats itself. The on-delay timer is so named because the contacts are delayed from activation after the relay is turned on. They are straightforward in their operation and are the most commonly used time delay function. See Figure 9–29a for a block chart representation of on-delay timer operation.

Off-delay timers are also available as separate electromechanical or software devices. In many ways they operate opposite to the function of on-delay timers. Off-delay timers also can have many different contact arrangements. When the off-

(a)

On-delay timer operating cycle

Coil	Not energized	Energized	Energized
Time	Not being counted	Accumulated time less than preset time	Accumulated time greater than or equal to preset time
N.O. contacts	Open	Open	Closed
N.C. contacts	Closed	Closed	Open

(b)

Off-delay timer operating cycle

Coil	Energized	Not energized	Not energized
Time	Not being counted	Accumulated time less than preset time	Accumulated time greater than or equal to preset time
N.O. contacts	Closed	Closed	Open
N.C. contacts	Open	Open	Closed

▲ **FIGURE 9–29**
On- and off-delay timer function

delay timer is energized, all of the contacts become energized immediately, but time is not counted. The off-delay timer also has an adjustable preset time value. The off-delay counter begins counting time when power is removed from it or it is disabled. At this point, time is being counted and the contacts are maintained in their energized condition. This condition remains until the accumulated time from the point at which power was removed from the timer equals the off-delay timer's preset value. When this occurs, the contacts revert to their disengaged condition. Whenever the off-delay timer is powered up, the cycle repeats from the beginning. The off-delay timer gets its name from the fact that it delays contacts from being turned off for a period of time after the off-delay timer is de-energized. Figure 9–29b shows a block chart of the off-delay timer's operation.

Voltage/Current Transmitters

Transmitters are DC analog devices used in industrial applications to amplify a process parameter such as temperature or pressure and convert it to one of the standard voltage or current ranges used by most data acquisition or control systems. Most common are the 4–20 mA and 1–5-V ranges, but many others are possible. Transmitters are called by the name of the parameter sensor with which they are designed to work. Thermocouple transmitters accept a thermocouple input and develop a standard range signal output that corresponds to a specified range of temperature. There are RTD transmitters, pressure and flow transmitters, and, literally, transmitters for any process parameter typically measured. These transmitters accept the relatively low signal input from a sensor and convert it to a high-level signal that can be sent some distance to a data acquisition or control system. The transmitter will have an input signal range and an output signal range. Sometimes the transmitter will also include a display to indicate the output signal value, either in the units of the parameter (°F for a temperature transmitter) or the output signal (volts, if 1–5 V is the output signal range).

Example 9–2

Determine the temperature value indicated by a thermocouple transmitter that accepts the input of a type J thermocouple over the range of 0° to 800°F and outputs 4–20 mA if the output of the transmitter is 15.6 mA.

Solution

1. Determine the percent of range the output represents.

   The range of a 4–20 mA signal is 16 mA, starting at 4 mA and ending at 20 mA. A 15.6 mA signal is 15.6 mA – 4 mA = 11.6 mA above the 4 mA zero range point.

   11.6 mA/16 mA = 72.5 % of range

2. The temperature value equals the percent of output range times the input temperature range value.

   72.5% × 800°F = 580°F

Data Acquisition and Recording

In many industrial process applications it is necessary to maintain a record of process variables for efficiency and quality control or to meet government regulations. The federal Food and Drug Administration, for example, requires that critical variables be maintained for all processes relating to foods and drugs. An operator of a boiler system desires that the boiler maintain optimum levels of steam pressure. A metal heat-treater strives to maintain accurate and uniform temperatures in their heat-treating processes. Data acquisition systems and recorders provide a means of analyzing and comparing system performance on an ongoing basis.

Before the advent of microprocessor technology, all data acquisition and storage was done on paper. Strip chart or circle chart recorders performed all data storage. Strip chart recorders are available in paper widths from 20 mm to 100 mm and can monitor and record up to 32 channels of data. Circle chart recorders can typically process up to 4 channels of data on circular charts ranging from 6 to 12 in. in diameter. Data loggers and data acquisition systems are available that store all process data in memory and display old or current data on a CRT or LCD screen. The number of channels that can be processed is limited only by the size of the system.

The input capabilities of most of these systems are similar. Most data acquisition systems can accommodate thermocouple, RTD, 1–5-V, and 4–20 mA inputs. The 1–5-V and 4–20 mA inputs can be transmitters for any number of process parameters (temperature, pressure, flow, humidity, PH, etc.). When transmitters are used, the variable must be scaled and the units identified. For example, let's take a pressure transmitter that is connected to channel 3 of a 16-channel strip chart recorder. The pressure transmitter outputs 1–5 V over an input range of 0–300 psi. When configuring the strip chart recorder, channel 3 would be programmed for a zero value of 0, a span (maximum) value of 300 and units selected as psi.

Limit Devices

Limit devices are safety devices that monitor a process parameter and compare its value with a set point. When the set point is exceeded, an output is engaged that either sounds an alarm or shuts down the system. In most cases it is desirable to have the limit device shut down the system. Most limit devices are configured to shut down the system when the set point is exceeded. They maintain the shutdown condition even after the process value goes below the set point. System operation can only commence when the process value is below the set point value and after someone has reviewed the situation and decided that it is all right to continue. That person then depresses a reset button to restart the system operation.

A good example of a limit device application is an environmental test chamber. If the test chamber is designed to function over a temperature range of 0° to 400°F, a safety limit device is added with a set point of 450°F. If a temperature greater than 450°F is sensed, the limit device shuts down the environmental test chamber until the temperature has cooled to less than 450°F and the reset button has been depressed.

Ground Fault Circuit Interrupters (GFCIs)

A very common limit device used in industry and the home is the GFCI. GFCIs work by measuring the current through both the hot and common leg of an AC circuit. If these currents are approximately the same, the operation of the outlet is allowed to continue. However, if a difference of at least 4 mA is detected between the two legs of the circuit, the GFCI trips, which shuts down the outlet. This condition, called a *ground fault*, indicates that current is going from the hot leg back to ground by some other path. This is symptomatic of a human in contact with the hot conductor while grounding some other part of the body. The GFCI breaks the circuit and will maintain this condition until both the ground fault is removed and the reset button has been depressed. Figure 9–30 shows a basic GFCI circuit.

On-Off Control

On-off controllers are similar in operation to limit devices except that their purpose is to control the value of the input parameter so that it equals the value of the set point. A home thermostat is a very simple on-off controller that turns the heat source or air conditioner on or off whenever the temperature goes outside an acceptable temperature band. The inputs to most industrial on-off controls accommodate thermocouple, RTD, 1–5-V and 4–20 mA inputs. When the input is a thermocouple, the on-off control automatically compensates for cold junction error and linearizes the resulting temperature value. For RTD inputs the option of two- or three-wire connections is provided to address lead length error. Soon after the controller receives the signal, its value is digitized and compared to the set point value.

Many outputs can be provided as an option with on-off controls that share either the same or an independent set point. Each output can be selected to operate in either direction: to be activated when the process value is less than the set point (called *reverse action*) or when the process value is greater than the set point

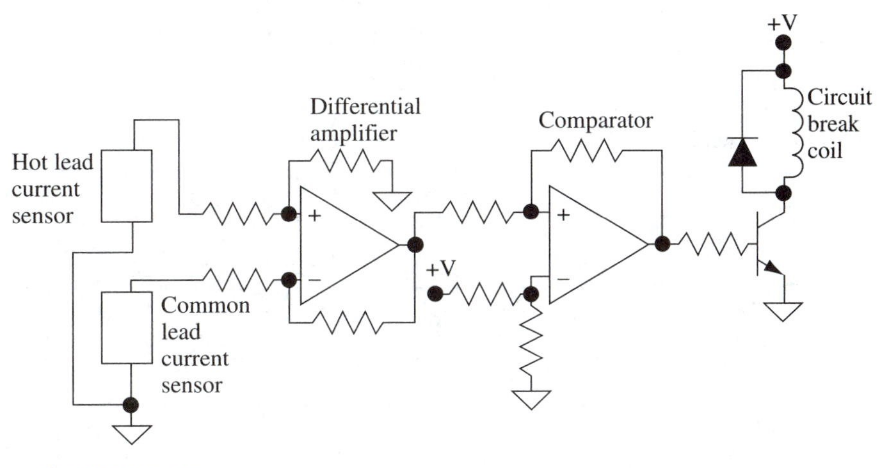

▲ **FIGURE 9–30**
GFCI circuit

(called *direct action*). Each output also has hysterisis, otherwise called *switch differential*, associated with its operation. Hysterisis means that the output becomes engaged and disengaged at different points. Let's take the example of a home thermostat with a set point of 70°F and just one output that operates in the reverse action mode (the output is engaged when the temperature is less than the set point as desired to control heating devices). If the hysterisis value was 0, the thermostat would constantly switch the output back and forth when the temperature was very close to 70°F. If the hysterisis value was set at 2°F, the heat would switch on when the temperature fell to 69°F and switch off when the temperature reached 71°F. The larger the hysterisis, the larger the band of temperature variation allowed by the controller. At the same time, the heating system is cycled on and off less frequently. This is the design tradeoff made when adjusting the hysterisis value of an on-off control: control error vs. on-off cycling of the output. The performance of on-off controls can be seen when recorded on a strip chart recorder as a sawtooth-appearing waveform that shows the hysterisis value (plus system time responses) as the difference between the maximum and minimum control point values as they vary around the set point (see Figure 9–31).

Current technology on-off controls digitize the input signal, linearize and scale the input signal, indicate the process value, and determine all control outputs under microprocessor control. The output devices themselves are either electromechanical or solid state relays driven by analog or digital electronics. Figure 9–32 shows some sample on-off and PID controls.

Proportional-Integral-Derivative (PID) Controllers

PID contollers differ from on-off controllers because they have proportional outputs. Proportional outputs are effectively analog outputs; they can have any value over a defined range of voltage or current. On-off controllers have control outputs that are either on or off and result in either 0% or 100% output to the load. Proportional outputs can have values anywhere from 0% to 100%. Electronic proportional outputs are usually in the standard form of 4–20 mA or 1–5 V, where 4 mA

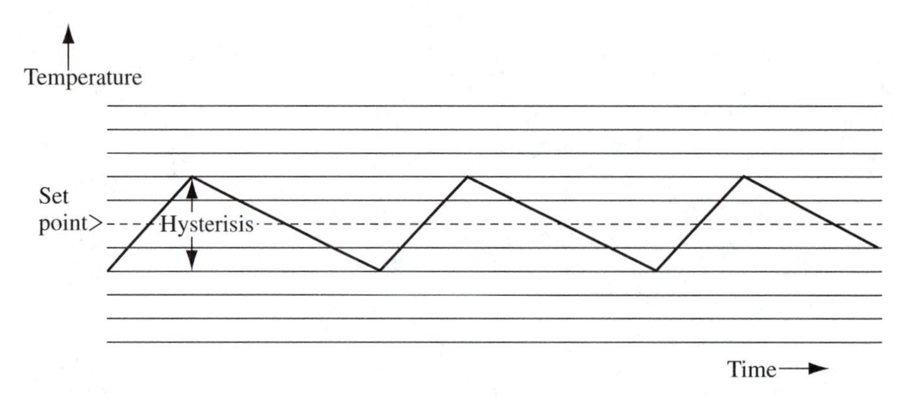

▲ FIGURE 9–31
On-off control performance

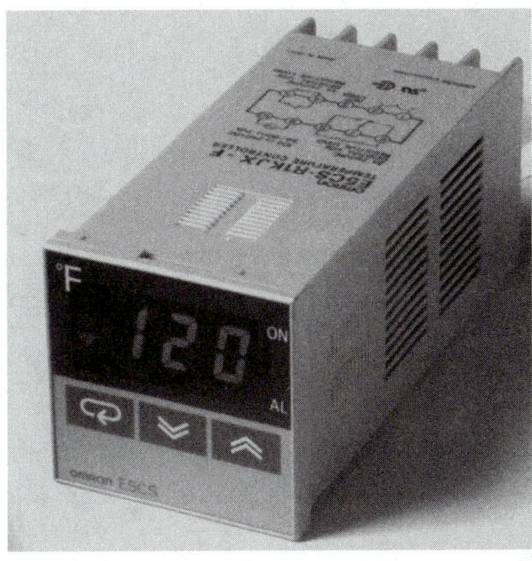

▲ **FIGURE 9–32**
On-off and PID controllers *(Courtesy Omron Electronics, LLC. Used by permission.)*

or 1 V corresponds to 0% output and 20 mA or 5 V equates to 100% output. A 12-mA signal results in 50% output.

To understand the proportional concept a little better, let's compare the operation of an on-off and proportional controller as it performs in a typical industrial heating application. In a gas-fired industrial oven the temperature is controlled by how much heat is applied to the oven. An on-off control applies either 100% or 0% of the heat capacity by turning the gas valve and heating system on or off. The performance of the system would appear like the jagged sawtooth shown in Figure 9–33a. If the temperature variation shown in the figure is unacceptable, the hysterisis of the controller could be decreased, but the increased on-off cycling of the equipment would also be unacceptable.

A proportional control can be used to control the oven temperature by replacing the on-off gas control valve with a motorized valve positioner. This valve positioner accepts a 4–20 mA signal from the controller and positions the gas valve completely open when the control output is 20 mA, completely closed with 4 mA, and any percentage open between 0–100% for 4–20 mA.

Proportional controls offer smoother and tighter performance because they gradually reduce the heat output as the temperature approaches the set point. Proportional controls are called PID controls because they have three parameters that are adjusted to provide optimum system performance: proportional band, integral action and derivative action, or P, I, and D. The proportional band is the band of process values over which the output is proportioned. It is centered at the set point to provide 50% output. It extends equally in both directions around set point. So if the set point of the oven was 200°F and the proportional band was adjusted to be 100°F, the proportional band would extend from 150° to 250°F. The reverse acting output of the proportional control would begin to reduce the heat into the oven when the temperature just exceeded 150°F by starting to close the gas valve. If the

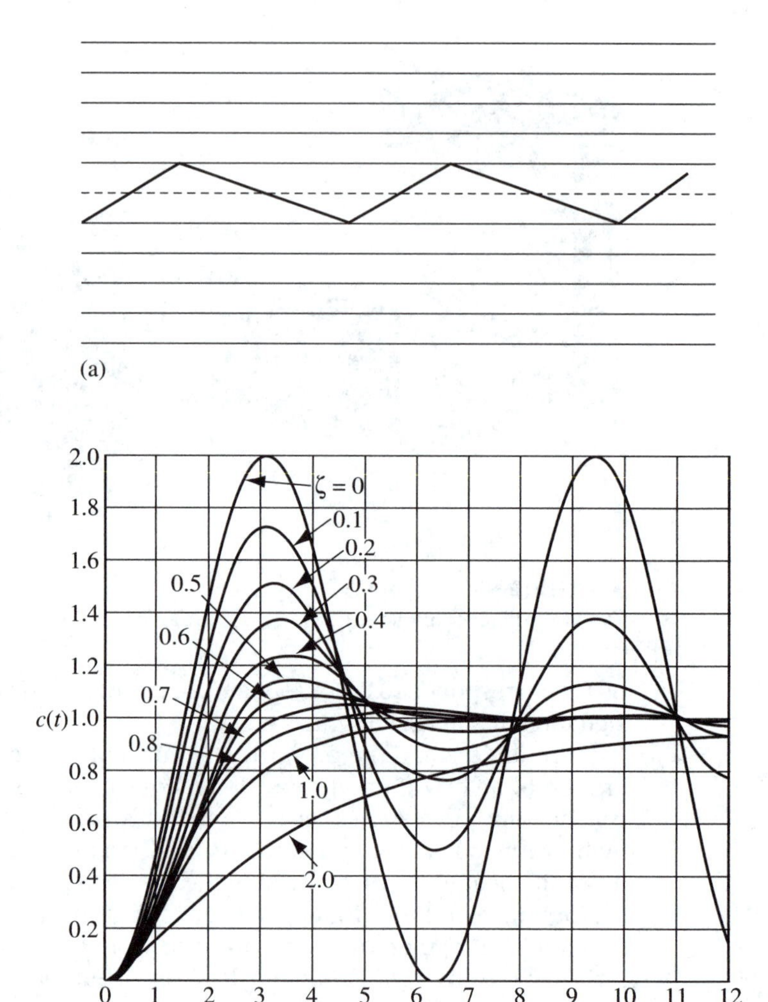

▶ **FIGURE 9–33**
On-off and PID response curves

temperature exceeded the set point and reached 250°F, the gas valve would be completely closed. You can imagine that as the proportional band is made small, the system will reach the set point faster, but it will have a tendency to overshoot and possibly oscillate, just as discussed in Chapter 7. The larger the proportional band, the slower the system response, but it will have less overshoot and be more stable. The proportional band is the *P* adjustment provided on PID controls and sometimes is called *gain*, which equals *1/P*. The proportional band adjustment creates a whole family of response curves shown in Figure 9–33b.

Because the proportional band is centered and provides 50% output at set point, the oven will be controlled at the set point only if the load happens to require exactly 50% output. The amount of load on any heating system is determined by the set point, the amount of material in the system, and the ambient temperature conditions around the system. As the load varies, the amount of output required to maintain the set point varies. All proportional controllers need a method for sensing the error from the set point to allow repositioning the proportional band to provide the output value needed to maintain the load at the set point. Integral action performs this function by integrating the set point error. It uses the area under the curve to add to or subtract from the nominal 50% provided by the proportional band at set point. An op amp integrator performs integral action in analog type controls. Software emulates the integrator function in microprocessor-based controls. The amount of integral action is adjustable by changing the gain of the integrator. The integrator gain, the *I* setting, changes the speed at which the proportional band is shifted by the integrator output. If the integrator output changes too fast, the system may overshoot or become unstable. Too low an integrator gain setting causes the system to be slow and unresponsive.

In many applications, system changes can occur rapidly, too rapid for the integrator to respond to them in the short term. This is because the integrator is always working from historical information, continually totaling the area under the error curve over time. Derivative action is used in these situations to provide immediate response to quick changes in the error. Derivative action calculates the rate of change of the error and provides a corrective signal to either increase or decrease the percent output immediately. An op amp differentiator performs derivative action in analog controls while software emulates the function in microprocessor-based controls. The gain of the differentiator, the *D* setting, determines the magnitude of the change induced by a certain rate of change of the error signal.

Programmable Logic Controllers (PLCs)

PLCs evolved after the introduction of the microprocessor in the late 1970s. They are essentially industrial computers that can be programmed to perform logic circuit functions. This is done with a graphical programming language based upon the ladder logic diagram discussed in Section 9–1. The initial PLCs included only digital inputs and outputs. The PLC examines the status of each digital input and stores this in a buffer memory. The PLCs processor then reviews the ladder logic program currently being executed and determines the status of each output as specified by the combination of the input status and the ladder logic program. Once determined, the new output status is loaded in a separate output buffer. Next, the contents of the output buffer are sent to the PLC's outputs. This process represents basically one scan of the PLC, the length of which is determined by the number of inputs, outputs, and the complexity of the ladder logic program.

Let's take an example of a small PLC that has eight inputs and six outputs. The 8 inputs are given address names of I0-I7; the output addresses are O0-O5. The PLC is to be programmed to energize output #1 whenever input 0 is on, and either input #1 or input #2 are on, and input #3 is not on. Figure 9–34 shows the ladder logic program that would be drawn and loaded into the PLC. Note that each input is shown graphically as either an open or closed set of contacts. In each case the graphical symbol shows the status of the unenergized input: open contacts represent an input that is open when not engaged, and closed contacts represent an input that is closed when not engaged. Also, the outputs are shown as an open parenthesis rather than a circle as was done to symbolize a relay.

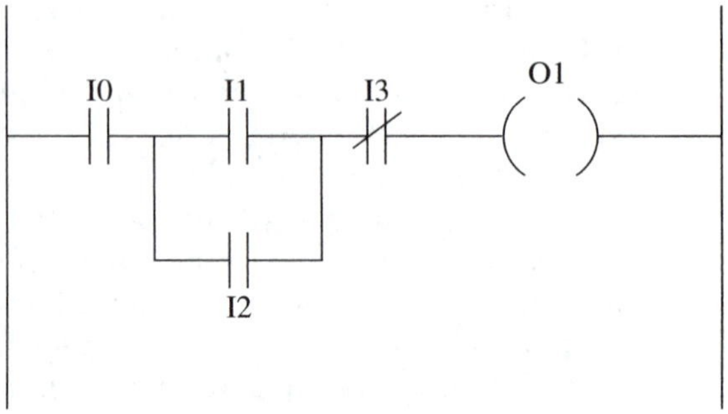

FIGURE 9–34
PLC ladder logic diagram

Today's PLCs have been expanded to include analog inputs and outputs, advanced mathematical functions, and PID control as well. They also possess extended communications capabilities and can be configured to function in a variety of networks. An example PLC is shown in Figure 9–35.

9–4 Output Circuits

Output circuits are used to implement the operational changes determined by the control system. Outputs can be either digital on-off outputs or analog outputs. Analog outputs indicate either the value of a parameter or the relative percent output that should be provided to the controlled process.

Electromechanical Relays, Contactors, and Solenoids

On-off outputs are provided by electromechanical or solid-state relays, contactors, or solenoid valves. All of these devices are either energized or not energized. The schematic symbols and contacts for electromechanical relays were discussed in

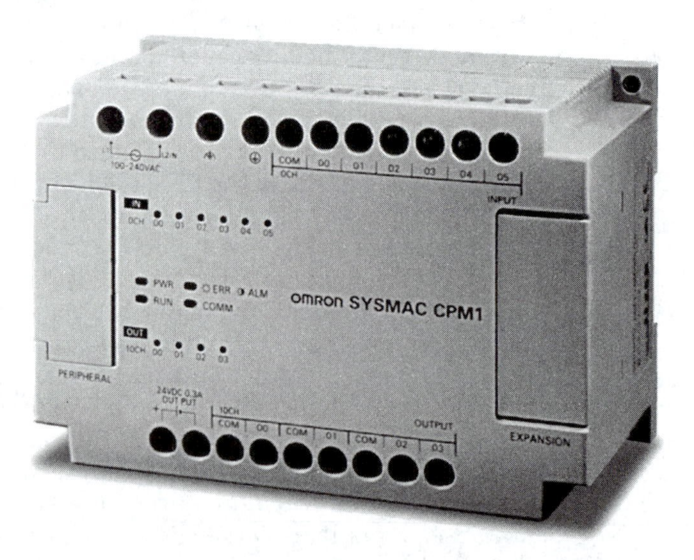

FIGURE 9–35
PLC example
(Courtesy Omron Electronics, LLC. Used by permission.)

▶ FIGURE 9–36
Contactor and solenoid example *(Courtesy Omron Electronics, LLC. Used by permission.)*

Section 9–1, which covered control inputs. Electromechanical relay contacts are configured the same as switches (SPDT, DPDT, etc.) and can also represent the input to a circuit. Contactors are simply relays that have contact ratings in excess of 15 amps. Contactors and relays are usually shown by the same schematic symbol, so from an analysis point of view, relays and contactors are viewed the same. Solenoids are simply electromechanical valves that are energized open or closed. Figure 9–36 shows an example electromechanical relay.

Electromechanical relays are available for AC or DC operation and with a variety of coil voltages. A transistor circuit is usually used to energize a DC relay coil. It is common practice to use what is called a *free-wheelin diode* to provide a path for current flow initially after de-energizing the inductive coil of the DC relay (see Figure 9–37). When the transistor turns off, the current flowing in the DC coil will cause a large voltage drop across the collector-emitter of the transistor unless there is a path for this current to flow. The normally reverse-biased free-wheelin diode provides this path for current flow.

▶ FIGURE 9–37
DC relay free-wheelin diode

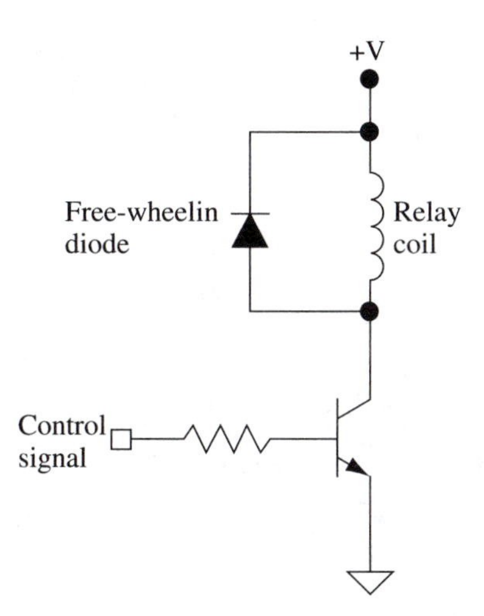

Whenever selecting a device that includes switching contacts, the voltage and current ratings must be considered. The voltage rating indicates the highest voltage that can exist across the contacts without the potential for arcing. The current rating specifies the maximum current that can flow through the contacts. It is also important to protect electromechanical contacts from the wear that occurs with repeated switching. Every time contacts are opened or closed, there is potential for contact damage when the contacts are in close proximity to each other. Damage occurs when there is current flow across the gap between the contacts when they are either just opening or just closing. The EMI generated at this time can also cause electronic systems to fail. Lower-voltage DC circuits (20 to 30 VDC) have the arcing capability of higher voltage (115 V) AC circuits.

The exact calculations for contact protection are rigorous and dependent on the actual load impedances. To protect contacts, RC snubber networks are most often placed across all output contacts from process controls and PLCs. Snubber networks should be applied across contacts of any device when the currents being disconnected are large and there are either capacitive or inductive load impedances. RC networks are commonly placed across contacts for protection, as shown in Figure 9–38. The *R* value should be equal or less than the load resistance and the resistor should have a power rating of at least ½ watt. The *C* value can be calculated as follows:

$$C = \sqrt{I_L / 300} \times L \qquad (9\text{–}1)$$

where I_L = the load current
L = load inductance

Make sure the voltage rating of the capacitor is sufficiently high. A 1000-V capacitor rating is often used for 115/230 V AC applications. Figure 9–38 shows RC values commonly used for 5-amp contacts rated at 230 V AC.

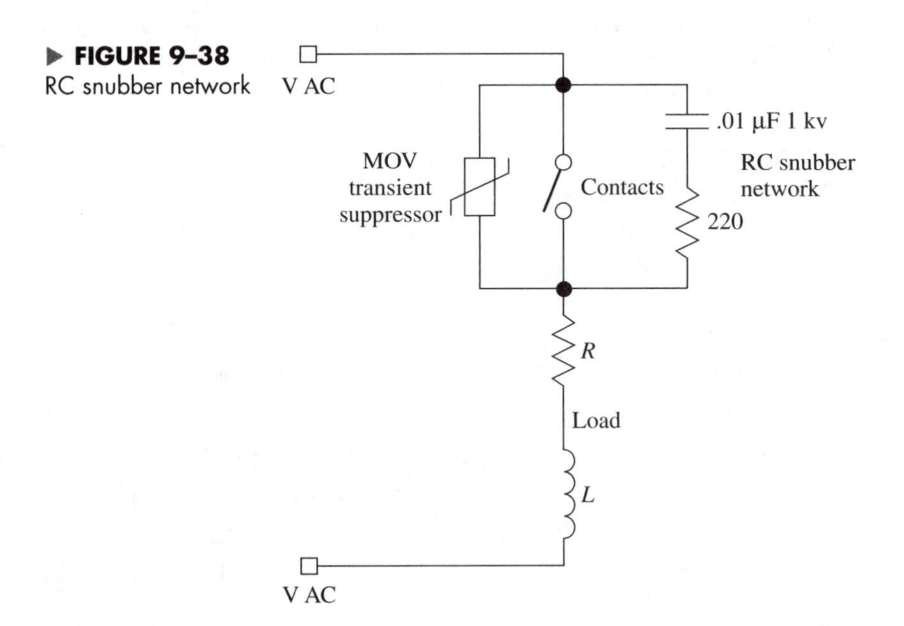

▶ FIGURE 9–38
RC snubber network

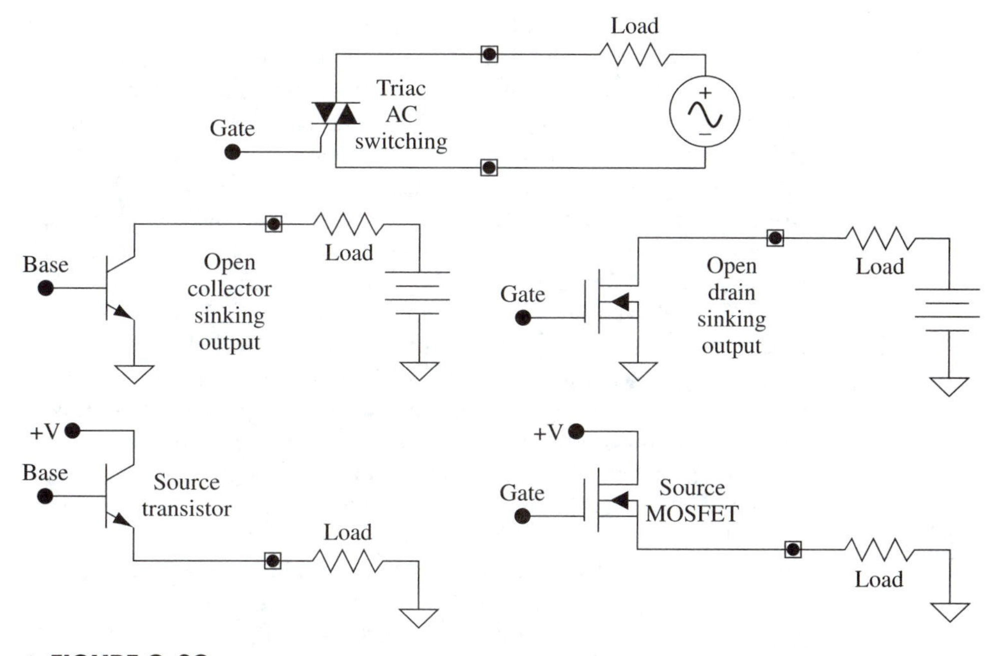

▲ **FIGURE 9–39**
AC solid-state relay, DC-solid-state relay, open collector transistor, and source transistor

For added protection with inductive loads, metal oxide varistors (MOVs) are also placed in parallel with the snubber network (see Figure 9–39). MOVs are semiconductor devices that reduce their resistance to absorb transient energy when their clamping voltage is exceeded. MOVs have a clamp voltage rating and an energy rating in joules. The clamp voltage rating indicates the voltage where the MOV will begin to absorb transient energy. The energy rating indicates how much energy the MOV can absorb before being destroyed. The key to selecting MOVs is finding a clamping voltage that is close to the highest operating voltage seen by the circuit but is never equal or below it. A MOV can be operative for only short periods of time, or even the highest energy rating will be exceeded.

Solid-state Outputs

Solid-state outputs are simply solid-state equivalents of electromechanical relays. Solid-state outputs exceed the performance of electromechanical relays in the areas of contact life and switching speed. However, electromechanical relays are much more versatile and robust and are easier to apply and troubleshoot.

The devices that make up the switching contacts in solid-state outputs are either triacs for AC circuits or transistors for DC circuits. Triac outputs for AC circuits are often called *AC solid-state relays.* Transistor switches for DC circuits are referred to as *DC solid-state relays*, *open collector outputs*, or *source transistor outputs* (see Figure 9–39).

Solid-state relays that employ triacs are commonly looked at as replacements for electromechanical relays in 115/230 V AC line voltage applications. Recall that triacs are excellent switching devices that are turned on by a gate voltage. When the gate voltage goes to 0, triacs rely on current flow in the reverse direction to turn off. Because DC circuits usually don't provide any reverse current flow, triacs used alone in DC applications, once energized on, never turn off. There exist commutation circuits that promote the use of triacs for DC switching applications, but they are not commonly used. Consequently, solid-state relays with triac outputs should be generally considered for only AC applications.

Most solid-state relays have inputs, either 3-32 V DC or 9-280 V AC, which, when present, turn the triac on. This results in current flow through the device. These devices use opto-isolators to provide electrical isolation from the input circuit to the output circuit. Figure 9–40 shows an example DC input solid-state relay circuit.

When compared to the electromechanical relay, the solid-state relay is capable of much faster switching speeds and many more contact operations. However, the contacts for solid-state relays are limited to SPST—one set of normally open contacts. Also, much care must be taken with the application of solid-state relays in high ambient temperatures or applications where high voltage spikes are prevalent.

When transistors are used as DC switching outputs, either the DC solid-state relay, open collector, or source transistor arrangements shown in Figure 9–39 are used. The DC solid-state relay is a stand-alone device that employs a MOSFET transistor to achieve low switch resistance. The inputs to the DC solid-state relay are similar to AC solid-state relay, 3-32 V DC. The output switch ratings are usually 0–100 V DC with load currents ranging from 3 to 10 amps.

Open collector or source transistor outputs are often provided as outputs on many discrete sensors or control outputs. For example, the output from a diffuse photo-sensor as described in this chapter is usually an open collector. Also, the outputs from many process controllers and PLCs are optionally electromechanical relays, solid-state relays, open collector, or source transistor outputs. When an open collector is provided, DC power is provided externally (see Figure 9–39). The DC power supply connects to the controlled load that in turn is connected to the open collector output. The transistor completes the circuit to a common circuit con-

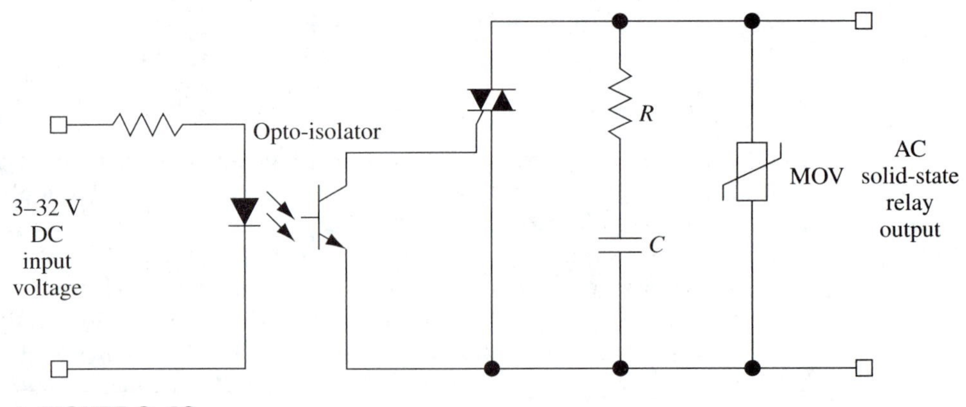

▲ **FIGURE 9–40**

nection when energized. The open collector is viewed as a sinking output. In the source transistor arrangement, the controlling device provides DC power. The load is connected to the transistor emitter and a circuit common connection. When the source transistor is energized, current flows through the load. In this case, the transistor is seen as a sourcing output.

Zero-voltage Switching

Zero-voltage switching is a simple concept that is very useful in eliminating transients that occur when devices are switched on or off in AC circuits. Through the study of transient analysis, we understand that the level of transient voltage spikes that occur when a circuit is powered up depends on the initial conditions present in the circuit. In most cases, the minimum transient spikes occur when the initial voltage is 0.

Because AC circuits offer two points in time, 0° and 180°, when the power voltage is 0, it makes sense that these are the best times for AC devices to be turned on or off. This is the concept of zero-voltage switching. To implement zero-voltage switching, all that is needed is a device that monitors the AC line to determine the zero crossing point. This circuit is called a *zero crossing detector* and can be a comparator connected to a stepped-down AC line voltage. The comparator output triggers a one-shot to output a zero crossing pulse. A microprocessor or other circuit device can use the zero crossing pulse to determine when to turn I/O devices on or off.

Proportional Outputs

Proportional outputs accept the control signals that are output from PID process controllers. They provide the means for controlling the value of some physical parameter that will result in control of the process. In an industrial oven a PID control outputs a 4–20 mA signal to a valve positioner that controls the flow of gas into the burner. This in turn determines how much heat enters the oven and the temperature within the oven. In a gasoline pipeline the flow rate of the gasoline is sensed by a flow meter and compared to the desired flow rate set point. The PID flow controller outputs a 1–5-V signal to a valve positioner, opening a control valve just the right amount to provide the desired flow rate.

The 4–20 mA and 1–5-V signals are the standard electrical signals used by process controllers to communicate to proportional output devices. However, there are many others, such as 0–20 mA, 0–10 V. If you are using an output module that requires a 0–20 mA signal, you must find a process controller with this type of output or develop some circuitry that will accomplish the conversion. In many industrial applications pressurized air signals called *pneumatic signals* are used as an input signal to many types of valve positioners. The standard pneumatic process control signal is 3–15 psi, where 3 psi = 0% output and 15 psi = 100% output. Many times a process controller with a 4–20 mA output must send a control signal to a pneumatic valve positioner. In this case a current to pressure (I to P) converter is used to convert the 4–20 mA electrical control signal over to a 3–15 psi pneumatic signal.

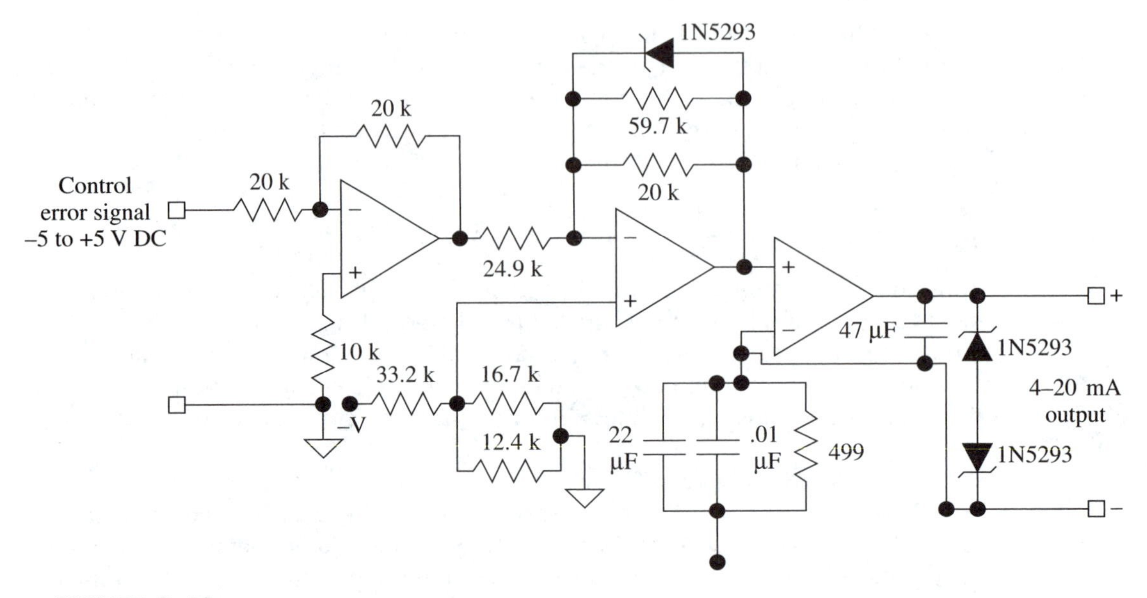

▲ **FIGURE 9–41**
4–20 mA output circuit

The circuit shown in Figure 9–41 uses an op amp to convert a ±5-V error signal over to a 4–20 mA signal capable of supplying a constant current to a load resistance up to 750 Ω. The 4–20 mA signals are easily converted to 1–5 V by passing a 4–20 mA current signal through a 250-Ω resistor. The voltage drop across the resistor is a 1–5-V signal.

Motorized Positioners

Motorized positioners are used to proportionally position all types of devices: valves, dampers, doors, hydraulic cylinders, and so on. These devices are actually a control system within a control system. The output from a PID process control requests that a certain percentage of some variable be added to the system under control. This control signal becomes a set point to the motorized positioner that moves the motor while sensing its position with a position sensor. The types of motors used to perform the positioning are servomotors, stepper motors, or any common fixed-speed electric motor. For each type of motor, the position of the motor shaft is fed back to the motor control circuit by a potentiometer, an LVDT, an encoder, or possibly photo-sensors.

Stepper motors are the opposite of encoders; they move the shaft of the motor a certain number of degrees for each input pulse to the motor and can be stepped in either direction. The key parameters for stepper motors are the step angle and maximum stepping rate. The step angle determines the amount of angular rotation that occurs for each pulse. The maximum stepping rates indicates the highest speed at which the stepper motors can be stepped. In a stepper motor positioner system, the motor is stepped while sensing its position with a potentiometer or encoder. The actual position is compared to the desired position and the motor is pulsed to correct its location if necessary.

Servomotors accept a voltage signal and position their output shaft according to this voltage, moving in either positive (clockwise direction) or negative (counter clockwise direction). Servomotors can be viewed as motorized analog indicators where the position of the meter pointer represents the position of the motor shaft. A potentiometer or encoder usually senses the shaft position, and a control circuit provides the correct output voltage signal to position the shaft properly.

Common variety fixed-speed electric motors are often used as positioners for slower changing outputs, such as dampers in temperature control systems. In these systems, the output of the controller is the input to the motor positioner control circuit. The control circuit senses the position of the motor shaft and compares it to the desired position. If the shaft position is in error, it energizes a relay to turn the shaft in either a clockwise or counterclockwise direction.

Triac Phase-angle Fire Outputs

Triac phase-angle fire devices are used to provide proportional amounts of heat, generated electrically with resistance, to a temperature-controlled system. These are complete electronic proportional output devices. They proportion the amount of power dissipated across resistance heating devices by switching a triac on and off at various times in an AC voltage waveform. Two SCRs, connected in parallel in reverse polarity, with a common gate are used to construct a triac. When the gate receives a pulse during the AC waveform, whichever SCR is currently forward biased turns on. The SCR stays on until the AC current flow reverses direction where it turns off. Another pulse, applied during the second half of the waveform, causes the other SCR to turn on and off in the same manner. The triac phase-angle fire output device is so named because it controls the amount of power applied to a resistive load by varying the phase angle at which the triac is fired or triggered.

If the signal sent to a triac phase-angle fire output calls for 50% output, the triac will be fired at 90° and 270°, as shown in Figure 9–42. The triac will conduct current only during the portions of the AC waveform between 90° to 180° and 270° to 360°, exactly half of the area under the sine-wave curve. A 75% output requirement requires the triac to be fired at 45° and 225°. Again, the signals used to communicate from the process controller to the triac output module are either 4–20 mA or 1–5 V, where 4 mA and 1 V represent 0% output and 20 mA and 5 V call for 100% output. Figure 9–43 shows a basic circuit that implements the phase-angle firing concept. As the voltage across the capacitor *C* becomes large enough, the gate of the triac is triggered. The resistance value adjusted for *R* determines the phase angle at which the triac will be triggered.

9–5 ▶ Data Acquisition and Control Systems

Data acquisition and control systems use all of the components discussed thus far to provide acquisition and control of one or more system variables. We have discussed a variety of digital and analog input and output devices as well as the concepts

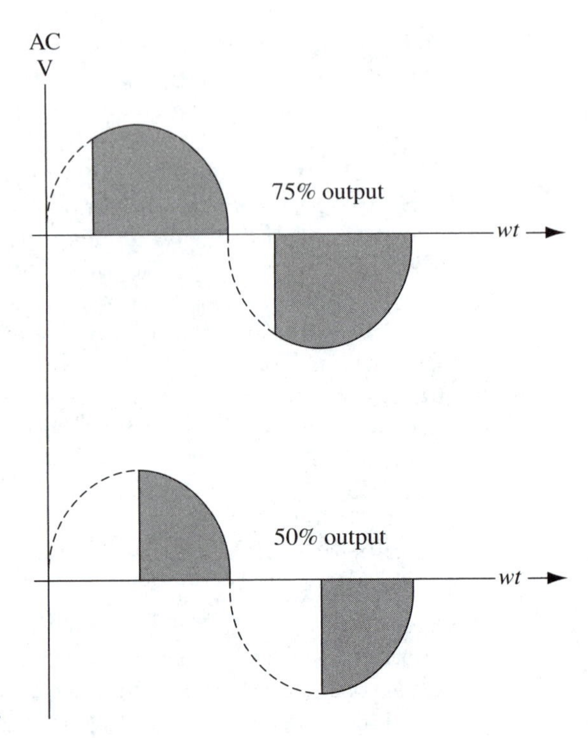

▶ **FIGURE 9–42**
Triac phase-angle fire waveforms

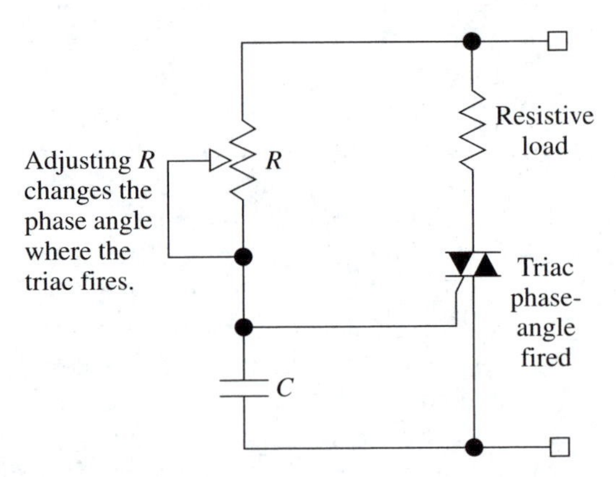

▶ **FIGURE 9–43**
Triac phase-angle fire circuit

of the process (PID) controllers and PLCs. All of these devices can now be connected to a personal computer in networks that result in very powerful data acquisition and control systems.

Process Controller Systems

A basic process control system includes just one control loop. A control loop consists of one process controller, an input measuring device, and a control output. There is just one set point and therefore one parameter under control. In many industrial applications it is necessary to control many parameters or take many different measurements of the same parameter at different locations in the system. These applications are called *multi-loop systems* because there are many control loops that monitor and control the system. Multi-loop systems can be developed in two ways: stand-alone controllers can be connected together in a network, or a multi-loop process controller or data acquisition system may be used. Figure 9–44 shows the block diagram for stand-alone controllers configured into a multi-loop network. A multi-loop block diagram is shown in Figure 9–45.

Process controllers used to accept only one analog input and provide up to three control outputs. These three outputs provided control of two output devices per controlled parameter and an alarm output. As more microprocessor power and surface-mount packaging technologies have been applied to these products, many additional inputs and outputs, both analog and digital, have been added to these controllers. There have been many added features as well: advanced math functions to manipulate and combine measured variables, many digital I/Os to implement ladder logic commands, and fuzzy logic algorithms to self-adjust PID values.

PLC Systems

PLCs are available as fixed or as rack-mounted units (see Figure 9–46). The fixed PLCs include a standard number of inputs, outputs, and control capabilities that cannot be expanded upon later. Rack-mounted units feature a card cage that provides almost limitless expansion. While the early PLCs offered only digital I/O, their capabilities have now been expanded to include a variety of analog I/O, which means that the PLC can actually measure an analog variable and output a proportional control output. In order to use the added analog I/O, PLCs have software

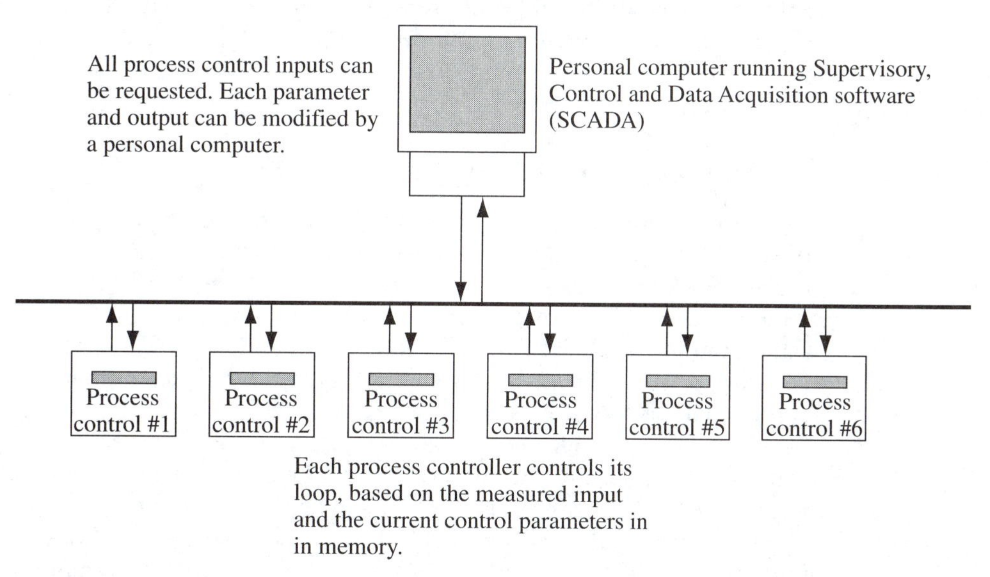

▲ **FIGURE 9–44**
Single-loop process control network

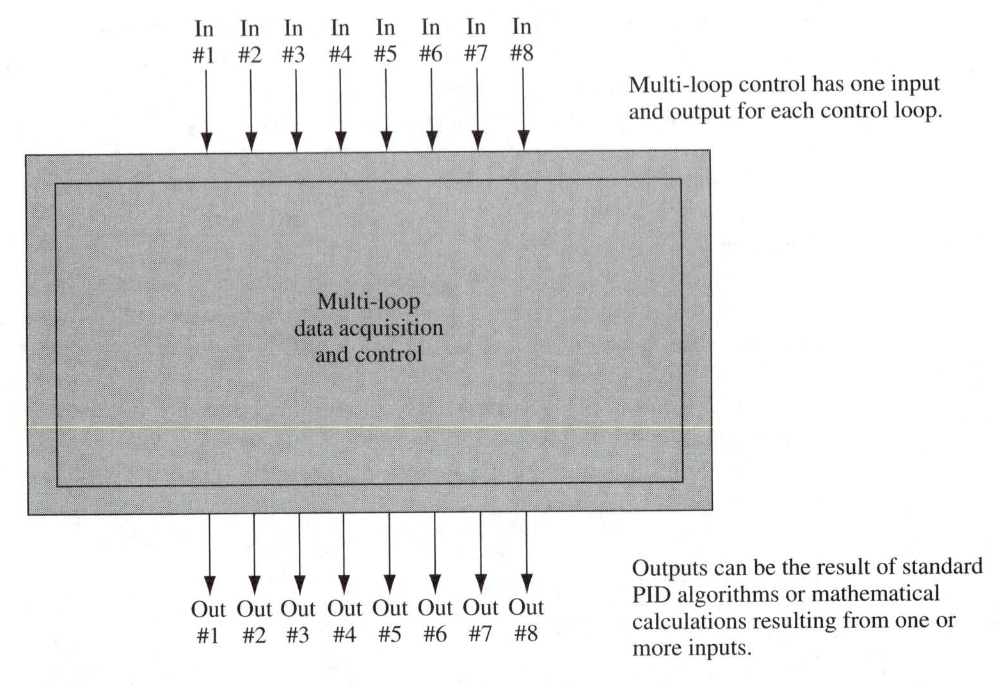

▲ **FIGURE 9–45**
Multi-loop process control network

PID control modules with full proportional control capabilities. Today's PLC also includes advanced digital communications capabilities and can be connected in a variety of network configurations.

Process Control and Data Acquisition Software

As process controllers and PLCs became larger and acquired additional capabilities, the personal computer became a standard device with the capability to tie all these devices together as a process control and data acquisition network. Beginning in the mid-1980s, personal computer software was developed with protocols that communicated with process controllers and PLCs. There are many high-level software packages available today that run on a personal computer and provide advanced data acquisition and control capabilities. Most of this software allows the user to create custom display screens that are pictorial representations of the process being monitored and controlled. The measured variables are displayed within the process screen at appropriate locations. For example, if the temperature of a chemical product processed in a tank is measured at various locations in the tank, this would be shown as a picture of the tank with the temperature values located at the appropriate locations on the tank.

(a)

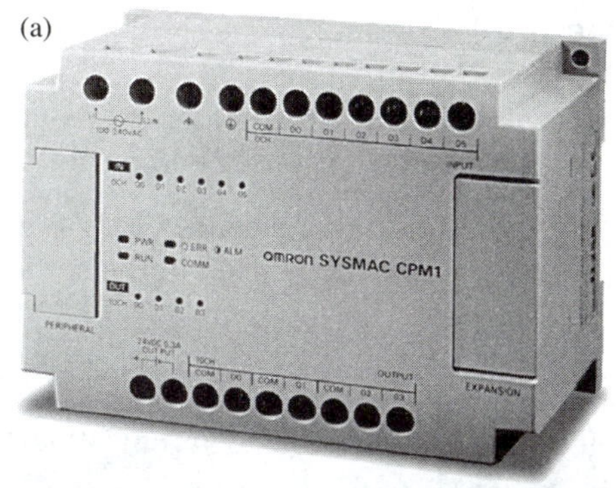

(b)

▶ **FIGURE 9–46**
PLC fixed (a) and rack-mounted (b) units. *(Courtesy Omron Electronics, LLC. Used by permission.)*

▶ Summary

In this chapter we have discussed a wide variety of the technology associated with industrial control and data acquisition systems. We have also reviewed much of the history and witnessed how the old world of industrial control has evolved from electromechanical relays, timers, and controllers to advanced process controllers, PLCs, and process control software. As we look back on all these advances, we can now see that the process controller has evolved to include many PLC (digital I/O and logic function) features while the PLC has acquired many process control (analog I/O and PID) capabilities. In many ways, when viewed strictly from an operational point of view, these two classes of products are difficult to tell apart.

We have also seen examples of old technology that continue to be used in spite of many efforts to replace them with faster, smaller electronic alternatives. At this point, we are unable to develop electronic alternatives to devices that are as rugged, versatile, and as easy to use as the electromechanical relay and the thermocouple. This chapter is just a snapshot of the industrial data acquisition and control world. It represents a broad combination of old technology and state-of-the-art embedded systems, software, and networks.

▶ References

Bartelt, L. M. *Industrial Control Electronics: Devices, Systems and Applications.* Albany, NY: Delmar, 2002.

Kissel, T. E. *Industrial Electronics.* Upper Saddle River, NJ: Prentice Hall, 1997.

Exercises

9–1 When reviewing ladder logic diagrams, how can you determine whether a set of contacts is energized or not?

9–2 When using one set of SPDT contacts (C, NO, and NC), can both the NO and NC contacts be used in the same control circuit?

9–3 How can you tell if a rung on a ladder logic diagram operates completely independent of all other rungs in the diagram?

9–4 Does the location of a rung affect the operation of the rung in any way? Explain.

9–5 Why does each rung require at least one output device? Why can't two output devices be connected in series? Can two output devices be connected in parallel?

9–6 Draw a ladder logic diagram that will operate as follows:

Output relay K_1 will be energized when either maintained switch A is on or momentary switch B is depressed. If switch B is depressed, a set of contacts from K_1 will seal-in momentary switch B until switch C, a normally closed momentary switch, is depressed, shutting off K_1. Whenever switch A is on, relay K_1 is energized.

9–7 What is the primary difference between a limit switch and a proximity detector?

9–8 What is the primary advantage of using capacitive proximity detectors instead of inductive proximity detectors?

9–9 List the three primary types of photo-sensors. List and discuss one advantage of each type.

9–10 List and discuss one disadvantage of each primary type of photo-sensor.

9–11 What technological developments have led to the development of high-resolution optical encoders?

9–12 Explain the difference between relative and absolute encoders.

9–13 A pressure transmitter converts a pressure signal from 0 to 400 psi to a 4–20 mA signal. If the transmitter outputs 6.5 mA, what is the pressure input to the transmitter?

9–14 When does an analog process signal require linearization?

9–15 What is lead length error and which sensor is most susceptible to it?

9–16 Discuss the three methods for resolving lead length error associated with RTDs.

9–17 What is cold junction error and what type of sensor is most affected by it?

9–18 What is the most important advantage of IC temperature sensors when compared to thermistors, thermocouples, and RTDs?

9–19 A thermocouple sensor is placed in a medium of 220°F. The leads are connected to a cold junction compensated temperature indicator. The ambient temperature in the area where the thermocouple leads are connected to the indicator is 66°F. What temperature does the indicator read?

9–20 An LM335 IC temperature sensor measures the temperature in an environmental test chamber. The output of the sensor is 3.23 V. What temperature is the LM335 measuring?

9–21 What determines the resolution of wire-wound potentiometers used as position sensors?

9–22 Explain the difference between a sensor, transducer, and a transmitter.

9–23 Why are pressure sensors usually combined in one assembly, with all the electronics necessary to make them a complete pressure transmitter?

9–24 Explain the difference between the on-delay and off-delay time delay functions.

9–25 How do GFCIs protect people from electrical shock?

9–26 If someone touches the hot side of a GFCI-protected AC circuit with one hand, and the common side with the other hand, will the GFCI shut down the circuit? Will the person receive an electrical shock?

9–27 Define the term *hysterisis* as it is applied to on-off control output devices.

9–28 Define the difference between proportional and on-off outputs.

9–29 Explain the main concepts behind the proportional, integral, and derivative functions that make up PID control.

9–30 Explain the basic function of a PLC with digital I/O only.

9–31 What is the difference between an electromechanical relay and a contactor?

9–32 Explain the function of a device called a *solenoid.*

9–33 List and explain the two methods for protecting the life of electromechanical contacts.

10 Discrete Digital Design

Introduction

Digital circuit design has also experienced rapid growth and change since the initial development of digital ICs. Overall speed, chip densities, and power efficiency continue to increase, providing a basis for increasingly complex functions in smaller packages. TTL, which was developed in the 1970s, represents mature technology that has seen great expansion and improvement over the years. Mature TTL devices are still used in many modern applications. CMOS technology has provided the foundation for significant improvements in the area of power efficiency and has also experienced much development. BiCMOS technology, which combines the speed capabilities of TTL with the power efficiency of CMOS, is providing the path for the future with further performance enhancements. There also is a push for lower logic voltage levels to help attain further speed and efficiency improvements. Lower voltage technology (LVT) reduces the nominal power for digital logic circuits from 5 V to 3.3 V, 2.5 V and even as low as 1.8 V. These lower power supply voltages reduce power consumption while increasing speed. In other application areas, the need for user programmable logic has increased. The variety and complexity of programmable logic devices (PLDs) has expanded to accommodate this need.

This chapter summarizes the circuit design concepts and considerations covered in most digital electronic courses. These concepts are then reviewed in light of the current digital devices available. Following are the specific topics covered in this chapter:

- Discrete Design Considerations
- Logic Families
- Package Considerations
- Programmable Logic Devices

10–1 ▶ Discrete Design Considerations

Digital components are often classified by their level of sophistication. Small-scale integration (SSI), medium-scale integration (MSI), large-scale integration (LSI), and so on, are terms used to classify ICs by their complexity. Basic digital electronic devices can also be classified as combinational, sequential, or bus-oriented. Combinational logic includes logic gates that are connected together to achieve a particular Boolean logic function. The outputs of combinational logic are determined by the inputs at any point in time. Combinational logic circuits are a combination of one or more of the following logic gates: AND, OR, NAND, NOR, INVERTER, EX-OR, EX-NOR. Figure 10–1 shows all of the logic symbols and truth tables as a reference.

Sequential logic utilizes digital components that can have three kinds of inputs: a clock, synchronous, and asynchronous. The clock input initiates changes to the outputs dictated by the synchronous inputs. Asynchronous inputs affect the

AND (A, B → O)

AND Truth Table

Inputs		Output
A	*B*	*O*
0	0	0
0	1	0
1	0	0
1	1	1

OR (A, B → O)

OR Truth Table

Inputs		Output
A	*B*	*O*
0	0	0
0	1	1
1	0	1
1	1	1

NAND (A, B → O)

NAND Truth Table

Inputs		Output
A	*B*	*O*
0	0	1
0	1	1
1	0	1
1	1	0

NOR (A, B → O)

NOR Truth Table

Inputs		Output
A	*B*	*O*
0	0	1
0	1	0
1	0	0
1	1	0

Inverter (A → O)

Inverter Truth Table

Inputs	Output
A	*O*
0	1
1	0

Exclusive-OR (A, B → O)

EX-OR Truth Table

Inputs		Output
A	*B*	*O*
0	0	0
0	1	1
1	0	1
1	1	0

Exclusive-NOR (A, B → O)

EX-NOR Truth Table

Inputs		Output
A	*B*	*O*
0	0	1
0	1	0
1	0	0
1	1	1

▲ **FIGURE 10–1**
Logic symbols and truth tables

output independent of the clock. The outputs from sequential circuits often depend on the state of the outputs prior to the clock. Sequential circuits form the basis for electronic memory and include devices such as flip-flops, monostable multivibrators (one-shots), and counters. Figure 10–2 shows the symbol and truth tables for two basic sequential circuits: the D flip-flop and the J-K flip-flop.

In many modern circuit applications, bus configurations are used to communicate signals to different areas of the circuit. A bus allows two or more devices to communicate with other devices over a common connection. Access to the bus is controlled so that only one device can output data at any one time. This is coordinated with the receiving device so that it receives only the data destined for it. The outputs from both standard combinational and sequential circuits are not designed to function with more than one output connected together. This precludes the use of these circuits in a bus environment. Tri-state logic was developed to resolve this problem. Logic devices with tri-state outputs can have their outputs connected together. Each tri-state device includes an enable input that, when active, connects the output of the device to the bus. When disabled, the tri-state device appears as a high input impedance to the bus, making it, essentially, electrically invisible to all devices on the bus. Figure 10–3 shows two commonly used tri-state logic devices: the 74LS373 and the 74LS374. Each of these ICs also has a clock input, so they are considered sequential, tri-state devices. As integrated circuits progress from SSI to MSI, LSI and so on, their classification and overall function usually becomes a mixture of combinational, sequential, and bus-oriented devices.

FIGURE 10–2
Sequential circuit logic symbols and truth tables

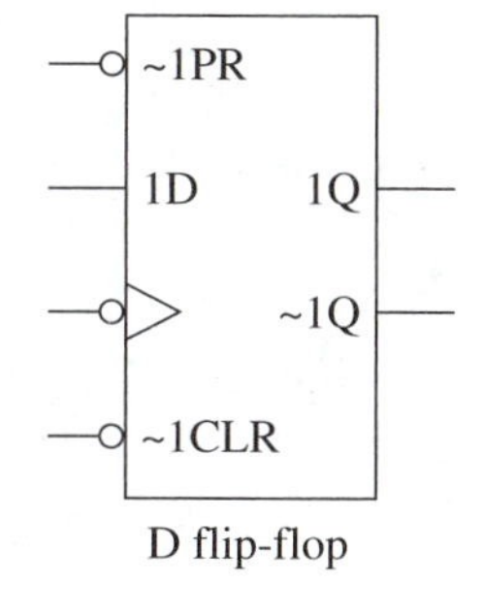

D flip-flop

Truth Table

D	Clk	Pre	Clr	Q
0	Active	1	1	0
1	Active	1	1	1
×	×	0	1	0
×	×	1	0	1

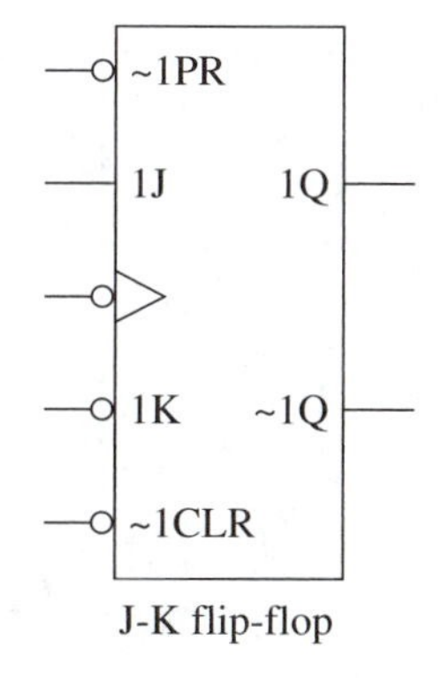

J-K flip-flop

Truth Table

J	K	Clk	Pre	Clr	Q
0	0	Active	1	1	0
0	1	Active	1	1	1
1	0	Active	1	1	1
0	1	Active	1	1	0
×	×	×	0	1	0
×	×	×	1	0	1

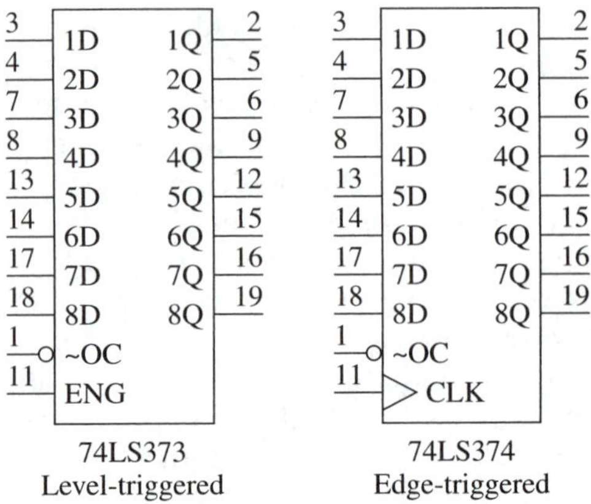

▶ **FIGURE 10–3**
74LS373 and 74LS374 tri-state logic devices

Design specifications vary significantly between the various logic families that are included within the standard TTL and CMOS categories. All of the logic families and subfamilies are listed in Figure 10–4 to serve as a reference for the discussion of each specification category.

The design considerations for digital devices can be simplified by focusing on their inputs and outputs. Let's consider a digital device as simply having logic inputs and outputs, with power inputs and power outputs as well, as shown in Figure 10–5. The basic design considerations for all digital devices can be summarized as follows:

1. The voltage levels of all inputs, including both the logic and the power supply inputs, must be within the range of values defined in the subject specifications.

2. The rise time, width, and fall time of all input signals must be slower and wider than the specification values. For the case of sequential circuits, the relative time relationship between synchronous and clock inputs must also be met.

TTL	CMOS	Low-voltage technology/ BiCMOS	ECL
74	4000/14000		10K/10H
74S	74HC		100k
74LS	74HCT	LVT	
74AS	74AC	ALVT	
74ALS	74ACT	BCT	
74F	74AHC	ABT	
	74AHCT		

▲ **FIGURE 10–4**
Digital logic families

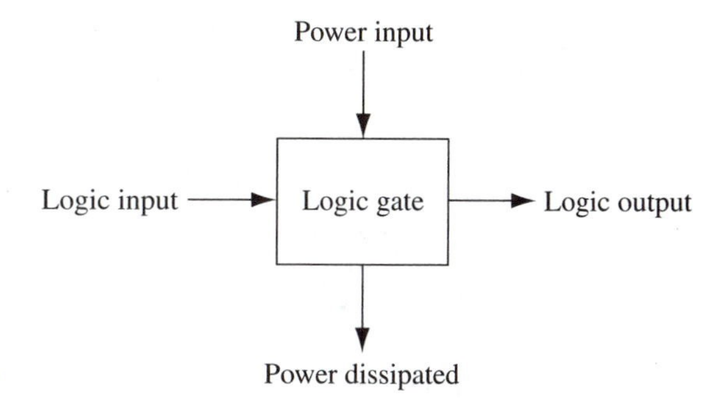

FIGURE 10–5
Digital device functional diagram

3. The voltage and current levels of all logic outputs must be less than the specification values that are noted as "Absolute Maximum Values."
4. The power dissipated by the device must be less than the maximum power dissipation allowed.
5. General considerations such as ambient temperature, noise immunity, shock, and vibration must also be considered (see the discussion included in Chapters 4 and 5).

Power Supply Voltage Requirements

The most important input voltage level is that of the power supply. Any logic device ceases to function properly if the power supply voltage is not maintained within an acceptable range. Digital ICs are unforgiving of power supply voltage levels higher than the *absolute maximum values* specified. The result is usually permanent damage to the IC. For standard TTL, the required operating range for the power supply input is 4.75 to 5.25 V. The absolute maximum power supply input must be less than 7.0 V. For CMOS the 4000/14000 and 74C devices will operate on power supply voltages that range from 3 to 15 V with an absolute maximum value of about 18 V. The 74HC/HCT, 74 AC/ACT, and 74AHC/AHCT will operate over a power supply range of 2 to 6 V with an absolute maximum value of 7 V.

Input Logic Levels

The input requirements for logic circuits are presented in data books and textbooks in the form of V_{IL} and V_{IH}. V_{IL} is the maximum input voltage that is considered a logical low or 0. V_{IH} represents the minimum input voltage that is a 1 or logical high. There is also the entire range of inputs that will be recognized as either logical highs or lows. To ensure reliable operation, the input logic levels provided to a digital circuit must be within these levels. The acceptable range of inputs, V_{IL} and V_{IH} for TTL and CMOS families, are shown in Figure 10–6.

When one logic device drives another, it serves as a current source to the input of the device being driven when its output is high; low outputs are a current sink for the input being driven. The value for the current flowing to an input when the input is a logical high is denoted as I_{IH}; I_{IL} is the current that flows back to the

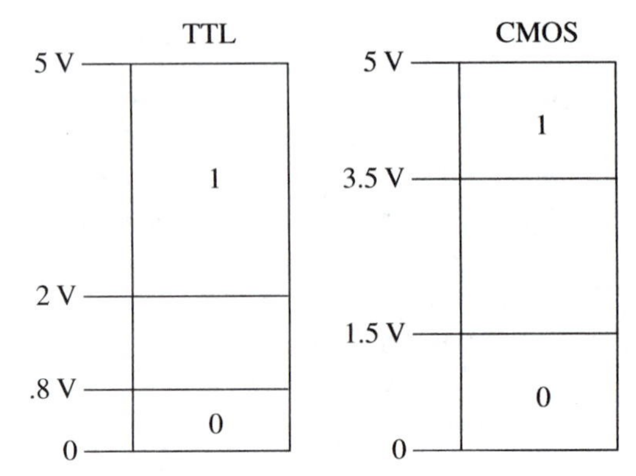

▶ **FIGURE 10–6**
Input and Output logic requirements for TTL and CMOS

output when it is low. Both of these values can be obtained from the data sheet for any logic device.

It is often necessary to bias a logic device to a logical high or logical low voltage when connecting switches and other devices as inputs. In these situations the input is either pulled high or low when the switch is open; when the switch is closed, the opposite logic level results (see Figure 10–7).

To calculate the proper pull-up resistor, determine the largest resistance value that will develop an input voltage larger than the minimum value for high logic ($V_{IH(MIN)}$) while supplying the input current for a logical high (I_{IH}). If the resistor value is too large, the voltage seen at the logic input will not be above the required minimum logic high input; too small a resistance results in excessive current flow and wasted power.

Example 10–1

Calculate the pull-up resistor value for the circuit shown in Figure 10–7b.

▶ **FIGURE 10–7**
Switch inputs

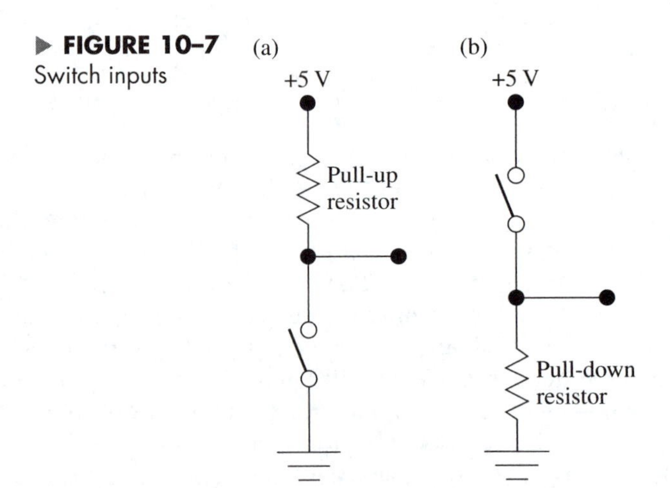

Solution

1. $V_{IH(MIN)} = 2.0$ V and $I_{IH} = 40$ µA for the logic device shown.
2. Let's select an input voltage of 4.6 to the logic device.
3. The voltage drop across the resistor when the switch is not depressed is:

   $V_R = 5.0\text{ V} - 4.6\text{ V} = .4\text{ V}$

4. Calculate the resistor that will develop a .4-V drop while 40 µA is flowing through it.

   $R = .4\text{ V}/40\ \mu\text{A} = 10\text{ k}\Omega$

5. 10 kΩ is commonly used as a pull-up resistor value.

To calculate the pull-down resistor values, determine the largest resistance value that will develop an input voltage less than the minimum value for low logic ($V_{IL(MIN)}$) while supplying the input current for a logical low (I_{IL}). If the resistor value is too large, the voltage seen at the logic input will be above the maximum logic high input; too small a resistance results in excessive current flow and wasted power.

Example 10–2

Calculate the pull-down resistor value for the circuit shown in Figure 10–7.

Solution

1. $V_{IL(MAX)} = .8$ V and $I_{IH} = -1.6$ µA for the logic device shown.
2. Let's select an input voltage of .16 to the logic device.
3. Calculate the resistor that will develop a .16-V drop while 1.6 µA is flowing through it.

   $R = .16\text{ V}/1.6\ \mu\text{A} = 100\ \Omega$

4. 100 Ω is commonly used as a pull-down resistor value.

Timing Requirements

Sequential circuits have a clock input with time requirements for synchronous inputs relative to the clock. Input signals present at the synchronous inputs must be at a logic level for a period of time prior to the active clock transition, called the *setup time* or t_S, in order for the output to respond to the input. The input signal must also remain at that logic level for a period of time, called the hold time or t_H, after the active clock transition.

Propagation delay defines how quickly the output responds to input conditions that require the output to change state. Propagation delay is present in all logic devices. It is specified as propagation delay low-to-high, or t_{pLH}, and propagation delay

t_S – setup time
t_H – hold time
t_{PHL} – propagation delay high to low
t_{PLH} – propagation delay low to high
t_W(L) – clock low time
t_W(H) – clock high time
t_W(L) – low pulse width required at set and clear inputs

FIGURE 10–8
Logic device timing specifications

high-to-low, or t_{pHL}. Sequential devices can have different propagation delays for synchronous inputs and asynchronous inputs (e.g., SET and CLEAR).

Sequential devices also have timing requirements for the clock input signal. The clock signal must remain low for a time denoted as t_W(L) and high for a time t_W(H). There is also a maximum frequency specification for the clock, called f_{MAX}. Asynchronous inputs such as SET and CLEAR have minimum pulse widths for these input signals for their active state. Active low asynchronous inputs have a specification, t_W(L), while active high inputs specify t_W(H). Figure 10–8 summarizes all of the timing requirements for logic devices.

Output Levels and Loading

The output from any logic device is in the form of a voltage that signifies either a logical 1 or 0. Of course, each logic family has different specifications for these voltages, which are designated $V_{OH(MIN)}$, the minimum logical high output voltage, and $V_{OL(MAX)}$, the maximum logical low output voltage. The values for these output voltages must be within the range of acceptable input voltage for the logic device being driven.

The logic output either sources current to or sinks current from the device being driven. It is important to make sure that the current in either case does not exceed the maximum ratings for the device. These specifications are listed with the other *recommended operating conditions* and are denoted as output current logical high, I_{OH}, and output current logical low, I_{OL}. The specification values for these two parameters will differ by a large amount.

The outputs of two or more conventional TTL and CMOS devices should never be connected together. Open collector outputs can be connected together to implement what is called a *wired-and* configuration. Otherwise, tri-state outputs should be utilized to allow the connection of multiple logic device outputs for the purpose of sharing a connection.

The output current specifications can easily be met when driving logic devices of the same family by verifying that the fan-out specification has not been exceeded. Fan-out is an indication of how many standard logic devices of the same logic family can be driven by the logic device. Standard TTL has a fan-out specification of 10. This means it can drive 10 standard TTL devices without exceeding the output current specifications. The fan-out specification varies for each logic subfamily and is discussed in the next section.

Noise Margin

One major advantage of digital devices is their relative immunity to input signal noise. The degree to which a device is immune to noise is known as the *noise margin.* The noise margin is defined for high and low outputs as follows:

$$\text{Noise Margin High Outputs} = V_{NH} = V_{OH(MIN)} - V_{IH(MIN)}$$
$$\text{Noise Margin Low Outputs} = V_{NL} = V_{IL(MAX)} - V_{OL(MAX)}$$

Noise margin varies between the different logic families and is usually higher in CMOS devices when compared to TTL.

Power Requirements

The total power dissipated by a logic device is the sum of the output power and the power consumed to bias all of the internal transistors. In both cases the amount of power dissipated varies for logical high- and low-output conditions. I_{CCH} is the amount of current drawn from V_{CC} for a high output and I_{CCL} is the value for a logical low output. Both of these parameter values are listed in the spec sheets for digital devices. In order to estimate the average power dissipated over time, an estimate of the percentage of time the output is high and low must be made.

It is important to estimate the amount of power dissipated by each device in a circuit because the net power of the entire circuit must be calculated. This is done for two reasons:

1. To provide guidelines for designing the power supply
2. To ensure that proper heat dissipation is provided for in the design of the enclosure

A summary of the design considerations for digital logic circuits is shown in Figure 10–9. The actual specification numbers for the parameters discussed in this section, as they apply to a specific logic family, are reviewed in the next section.

10–2 ▶ Logic Families

The two primary types of logic families are TTL and CMOS. TTL has always been noted for high speed and poor power efficiency. The small bias currents and inherent capacitance associated with CMOS logic devices have given them better power efficiency and slower speeds. Many different subsets of these two logic families have been developed over the years and new initiatives continue that include combining

1. Verify that power supply voltages are within acceptable operating ranges.
2. Verify that all input/output logic voltages are within the required levels for the respective high/low conditions of the logic device.
3. Verify that all timing requirements have been met — specifically, setup and hold times for all sequential circuits.
4. Verify that all clock signals meet minimum width and maximum frequency requirements.
5. Consider the effects of propagation delays using worst-case scenarios in all areas of the logic circuit.
6. Verify that all output currents are within acceptable ranges for all logic devices.
7. Estimate the power dissipation for each device and compare that with its maximum power rating.
8. Estimate noise levels at key circuit inputs and compare to noise margins to determine circuit noise immunity.

▲ **FIGURE 10–9**
Summary of digital design considerations

the best of TTL and CMOS to lowering logic voltage levels. These strategies serve to improve both speed and power efficiency at the expense of noise margin. These new design innovations have resulted in new logic families, such as BiCMOS and low voltage technology and many variations, combinations, and extensions to each of these. The part numbers for the digital devices currently available include an array of letters that indicate the type of design philosophy utilized. The variations seem endless as the older, mature technologies still available are flanked with newer high-performance devices.

TTL Logic Family

TTL logic devices were the first widely accepted logic ICs and they still enjoy broad use today. These devices are recognized by their 74xx designation. The 54xx devices function the same as equivalent 74xx devices except that they meet extended military specifications. It is important to note that the power consumed by TTL devices is relatively constant as the switching frequency changes. We will see that this is not the case with CMOS devices. Following is a summary for each currently available TTL logic subfamily that includes the typical performance specifications for 74xx00 devices:

74xx Series—Standard TTL:

	Propagation Delay	Power	Max Frequency	Fan-out
Standard TTL	9 ns	10 mW	35 MHz	10

74Sxxx Series—Schottky TTL: This series includes a Schottky barrier diode that is connected between the base and collector of all transistors in the circuit. This serves to reduce the propagation delay to 3 ns, which is one-third the propagation delay of standard TTL.

	Propagation Delay	Power	Max Frequency	Fan-out
S Series	3 ns	19 mW	125 MHz	20

74LSxxx Series—Low-power Schottky TTL: The LS series includes the Schottky barrier diodes but has larger resistance values that serve to reduce power consumption at the expense of reduced speed.

	Propagation Delay	Power	Max Frequency	Fan-out
LS Series	9.5 ns	2 mW	45 MHz	20

74ASxxx Series—Advanced Shottky TTL: As devices represent a major improvement in the design if the TTL circuit that still uses the Schottky barrier diodes on all transistors. These design improvements make the AS series the fastest TTL family.

	Propagation Delay	Power	Max Frequency	Fan-out
AS Series	1.7 ns	8 mW	200 MHz	40

74ALSxxx Series—Advanced low-power Shottky TTL: These devices are identical to the AS series but include larger resistors that reduce power consumption and increase propagation delays.

	Propagation Delay	Power	Max Frequency	Fan-out
ALS Series	4 ns	1.2 mW	70 MHz	20

74Fxxx Series—Fast TTL: F series devices feature new fabrication technologies that reduce the inherent coupling capacitance between circuit devices. These innovations yield performance that make the F TTL series as fast as the TTL S series, while consuming about one-third of the power.

	Propagation Delay	Power	Max Frequency	Fan-out
F Series	3 ns	6 mW	100 MHz	33

Good design practice calls for connecting all unused TTL inputs to a logical high or low voltage that will allow proper function of the device. All unused TTL devices should have their inputs connected to yield a logical low output all the time.

CMOS Logic Family

The CMOS logic family was actually developed before TTL. However, because of fabrication problems it was not commercially developed until well after bipolar TTL devices had become standard. CMOS logic devices typically feature better power efficiency at lower speeds. However, design enhancements have led to many CMOS subfamilies that have improved on the inherent weaknesses of CMOS technology. One of the inherent advantages of CMOS has always been its better power efficiency. It is important to note that the power consumed by CMOS devices increases with the switching frequency. This is because the internal capacitance must be continually recharged; the faster this occurs, the more power is lost.

Following is a summary for each currently available CMOS logic subfamily that includes the average performance specifications:

4000/14000 Series—CMOS: Series 4000/14000 devices were the initial CMOS families marketed. The 4000 series was developed by RCA while Motorola designed the 14000. For the most part, the 4000 and 14000 devices are equivalent to each other. However, they are not electrically or pin compatible with TTL devices. Series 4000/14000 devices are not usually used in new designs but can be found in many older products.

	Propagation Delay	Power (Static)	Max Frequency	Fan-out
4000/14000	100 ns	.3 mw	4 MHz	2

74Cxxx Series—CMOS, pin/function compatible with TTL: These are second-generation CMOS devices that featured pin compatibility with TTL devices. The C designation indicates that an IC is a CMOS device that is pin-for-pin and functionally compatible with TTL devices. This means that a 74C00 has the same pin-out and logic devices, which perform the same function as a 7400 or any other TTL series device (74LS00, 74ALS00, etc.). However, 74C devices are not electrically compatible with TTL, which means that their definitions of logical high and low voltages are different. The 74C series was discontinued after the 74HC/HCT series, which is discussed next, was introduced.

	Propagation Delay	Power (Static)	Max Frequency	Fan-out
74 C Series	100 ns	.3 mw	4 MHz	2

74HCxxx Series—High-speed, pin/function-compatible CMOS: This series improved upon the performance of the C series, which was discontinued after the HC series was introduced. The H stands for *high-speed.* HC-series propagation delays were reduced by a factor of ten when compared to the C series. Power consumption is also reduced. The HC series is pin and functionally compatible with TTL devices, but the two families are not electrically compatible. The HC series uses the standard CMOS logic voltage levels, as shown in Figure 10–6.

	Propagation Delay	Power at 100 kHz	Max Frequency	Fan-out
74HC Series	9 ns	.068 mw	55 MHz	11

74HCTxxx Series—High-speed, TTL-compatible CMOS: This family was introduced with the HC series. The primary difference is that the HCT series is also electrically compatible with TTL. In fact, the CT designation signifies that a device features CMOS technology that is functionally, pin-for-pin, and electrically compatible with TTL. The performance specifications for the HCT series differ slightly from the HC devices.

	Propagation Delay	Power at 100 kHz	Max Frequency	Fan-out
74HCT Series	10 ns	.050 mw	50 MHz	11

74 AHC Series—Advanced high-speed CMOS: This series features design enhancements made to the 74HC series that improve both speed and efficiency. The AHC and the TTL-compatible AHCT series discussed next represent the latest in CMOS technology and the highest performance levels. The AHC series also can operate from 3.3- or 5-V power.

	Propagation Delay	Power at 100 kHz	Max Frequency	Fan-out
74 AHC Series	5.2 ns	.073 mw	100 MHz	11

74AHCTxxx Series—Advanced high-speed, TTL-compatible CMOS: The TTL compatibility requirement affects the performance slightly, but for all practical purposes the AHCT and HCT series perform the same except for the logic voltage levels accepted and output by each.

	Propagation Delay	Power at 100 kHz	Max Frequency	Fan-out
74 AHCT Series	5.5 ns	.075 mw	100 MHz	11

IC manufacturers continue to seek performance improvements in the form of higher functionality, smaller packages, and quicker switching speeds with better power efficiency. To accomplish this, both power supply and logic level voltages

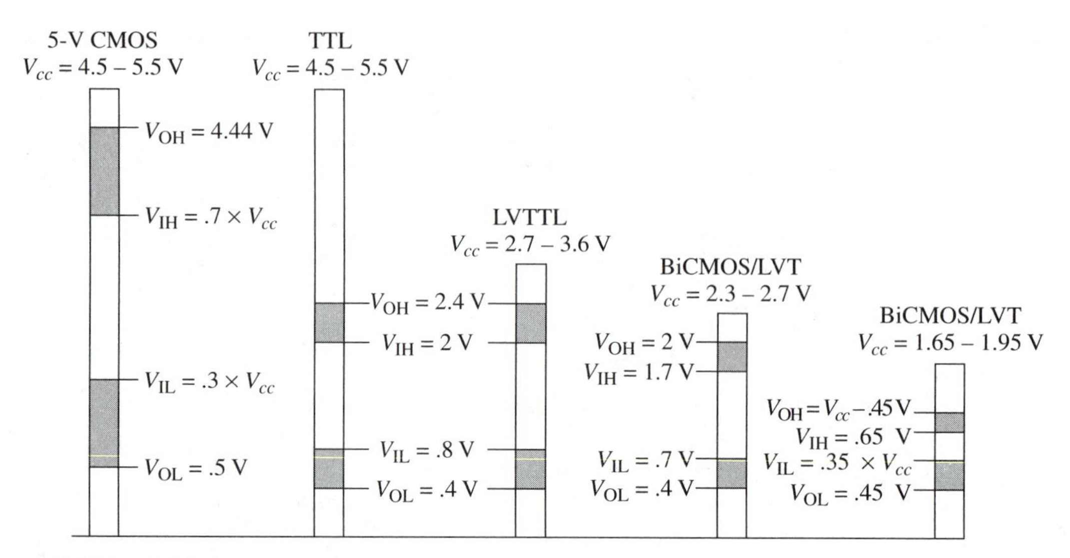

▲ **FIGURE 10–10**
Logic level definitions for standard and low-voltage logic families

have been decreased on new logic families. As transistor geometries become smaller, they can no longer withstand voltages in the 5-V range. Decreasing power supply voltages helps support performance improvements by allowing circuits to be placed closer together. Lower power supply voltages mean less current and therefore less power consumption. Propagation delays are reduced slightly because logic voltage levels are lower; the output can go from a logical zero to one quicker because the voltage margin is smaller.

Future plans call for three low-voltage standards powered by these power supply voltages: 3.3 V, 2.5 V and 1.8 V. New devices will be developed to operate at these logic levels and released in stages. Logic level standards for these low-voltage families are shown in Figure 10–10.

The following logic families represent the low-voltage families that have been developed most recently.

74LVxxx—Low-voltage series: This is a TTL low-voltage series and includes devices that are powered by 3.3 V. It features many of the more common SSI and MSI logic devices.

	Propagation Delay	Power at 100 kHz	Max Frequency
74 LV Series	5.4 ns	.11 mw	60 MHz

74LVCxxx—Low-voltage CMOS: This series is the CMOS version of the 74LV series; however, there is a broader range of devices available. Also, the 74 LVC series can accept regular 5-V logic inputs and it can drive 5-V logic inputs as long as the maximum output current specification is not exceeded.

	Propagation Delay	Power at 100 kHz	Max Frequency
74 LVC Series	5.4 ns	.05 mw	60 MHz

74ALVCxxx—Advanced low-voltage CMOS: This series features higher performance in the area of propagation delay and can be used with 3.3-V logic only. It is intended for tri-state bus-oriented applications.

	Propagation Delay	Power at 100 kHz	Max Frequency
74 ALVC Series	2 ns	.22 mw	180 MHz

74AVCxxx—Advanced very-low-voltage CMOS: these are the latest in pure CMOS devices and are designed to operate from power supplies over a range of 1.8 to 3.3 V.

	Propagation Delay	Power at 100 kHz	Max Frequency
74 AVC Series	1.9 ns	.22 mw	200 MHz

Special CMOS Considerations

When designing or working with CMOS circuits, there are special limitations that must be considered. These limitations include the sensitivity of CMOS to static electricity, requirements for unused CMOS inputs and devices, the effect of switching frequency on power consumption, and the impact of the load output on propagation delay.

CMOS inputs must always be connected to some voltage or device. If there is no need for an input, tie it to a logic voltage (low or high) that will allow the device to function properly. If a device is not being used at all, connect the inputs to fixed logic voltages that will yield a logical low output all the time.

The silicon dioxide layer that insulates the inputs of all CMOS devices is the primary mechanism for the high-input impedance and low-bias currents that are associated with CMOS technology. At the same time CMOS devices are known for their sensitivity to static electricity capable of generating thousands of volts at very low currents. These high-voltage shocks can destroy or seriously degrade the silicon dioxide layer to the point of failure or marginal operation. Consequently, the following considerations should be made when utilizing CMOS devices:

1. Handle CMOS devices only when using grounded wrist straps on conductive bench pads, and make sure that all test equipment and powered tools are properly grounded.
2. CMOS devices and circuit boards should be stored in anti-static tubes, anti-static bags, or other appropriate (conductive) containers. Styrofoam is not an acceptable storage medium.

3. Power up CMOS devices before signals are applied to their inputs and remove signal sources before taking power away.
4. Power down CMOS devices before inserting or removing them from a circuit.

CMOS Power Consumption

While it is true that CMOS generally exhibits better power efficiency when compared to TTL, it is important to note that increases in switching frequency also increase the amount of power consumed by a CMOS device. TTL devices consume relatively constant levels of power as the switching frequency increases up to frequencies around 3 MHz. Above 3 MHz, TTL power consumption also increases with the frequency. Therefore, it is critical that the digital designer review the specifications completely in regard to power consumption and switching frequency. It is conceivable that at a particular frequency, a TTL device may offer equal or better power efficiency when compared to a CMOS device operating at that same frequency.

Output Load Effect of Propagation Delay

The fan-out stated in the specifications for CMOS and TTL devices indicates the number of gate inputs that can be driven safely by the device's output. The number of CMOS devices driven can affect the propagation delay of the driving device. This is caused by the capacitance associated with the input of a CMOS circuit. The designer must review the specifications to determine the actual propagation delay expected for a given fan-out situation.

Latch Up

In the early development stages of CMOS technology, CMOS devices were highly susceptible to a condition called "latch-up." Latch-up occurred as a result of transistors mistakenly fabricated on the CMOS substrate as part of the fabrication process. Transient voltage spikes and noise can permanently switch on these parasitic transistors. Current CMOS devices have internal circuitry that minimizes the possibility of latch-up, but it is good design practice to use power supplies that are well regulated and filtered and feature current limiting.

BiCMOS Logic

BiCMOS is the result of combining the best aspects of bipolar TTL and CMOS technologies to develop devices that are fast and power-efficient. It was developed for microprocessor and bus-oriented applications in which speed and efficiency are most critical. BiCMOS devices are available in standard 5-V and 3.3-V power configurations.

74 BCTxxx—BiCMOS bus interface technology: This operates with standard 5-V logic and generally offers significant reduction in power consumption while maintaining low propagation delays. The BCT series operates off of 5-V power and is pin compatible with TTL devices and includes primarily bus interface devices.

74 LVTxxx—Low-voltage BiCMOS technology: This series includes mostly bus-oriented devices that can accept either 3- or 5-V logic level inputs where the outputs are compatible with TTL.

	Propagation Delay	Power at 100 kHz	Max Frequency
74 LVT Series	5.4 ns	.6 w	95 MHz

74ABTxxx—Advanced BiCMOS technology: This is the second generation of BiCMOS devices that operates off of 5 V and is also bus-oriented.

The 74ABT245 is a good example of the ABT series of devices. It is an Octal Bus Transceiver with tri-state outputs. This means that it can interface asynchronously with an 8-bit data bus, bi-directionally. The 74ABT245 is powered by 5 V, V_{IH} = 2 V and V_{IL} = .8 with propagation delays in the area of 4 nanoseconds. Power consumption is a maximum of .6 watts for all 8 transceivers. The pin-out and logic diagram for the 74ABT 245 is shown in Figure 10–11.

74 ALVTxxx—Advanced low-voltage BiCMOS technology: This series features high-performance BiCMOS devices capable of operation with power inputs of 3.3 or 2.5 V. ALVT devices offer low propagation delay, low static power consumption, and 64 mA current drive capability. They can interface with 5-V logic devices on a mixed-mode basis. This means that the input can be 5 V while providing a 3.3-volt output.

The 74ALVTH16244 exemplifies the ALVT series as it provides fast and efficient communications with 16-bit data buses. Propagation delay for this device is about

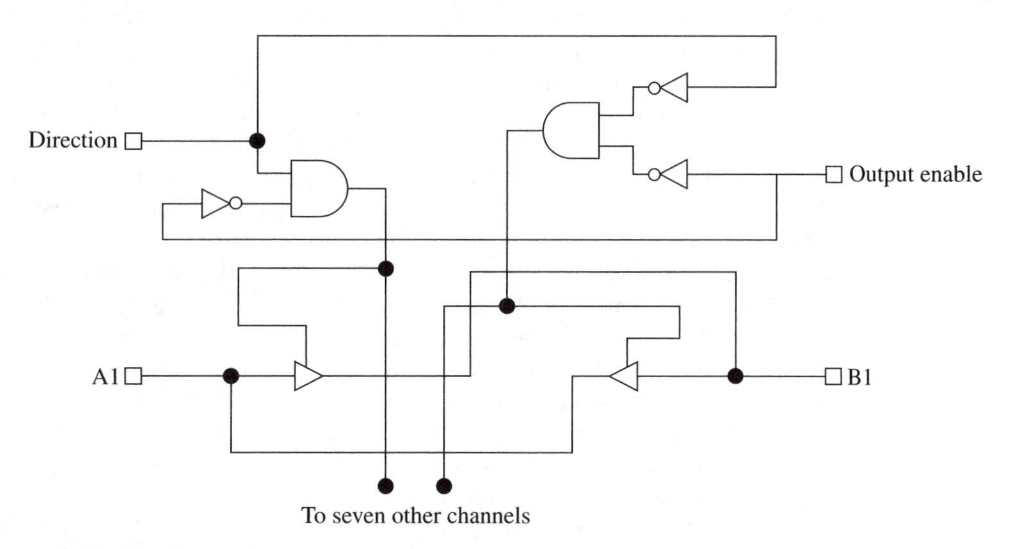

▲ **FIGURE 10–11**
74ABT 245 pin-out and logic diagram

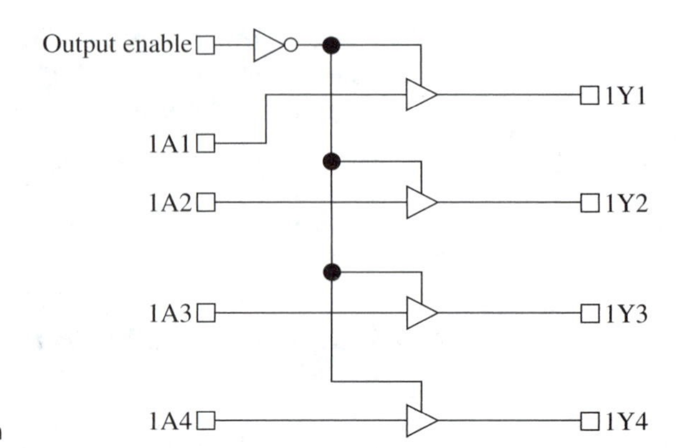

FIGURE 10–12
74ALVTH16244 logic diagram

4 ns and it is available in a Widebus (a Texas Instruments trademark) package that allows all 16-line connections to be made in line, one-on-one through the device. The pin-out and logic diagram for the 74ALVTH16244 are shown in Figure 10–12.

74ALBxxx—Advanced low-voltage BiCMOS: This series features bus-oriented devices that operate with 3.3-V logic levels.

Emitter Controlled Logic (ECL) Family

The ECL family was developed to provide very high switching frequencies that surpass the switching capabilities of all other logic families developed thus far. The ECL family utilizes bipolar transistors that are configured to select between one of two paths of current flow, depending on whether the input is high or low. This configuration, called *current mode logic* (*CML*), is a drastic change from the totem pole arrangement of the transistors used in both TTL and CMOS logic families. ECL can be viewed as a differential amplifier, where the current levels that define ECL high and low logic levels do not require the saturation of the transistors. This is the key to fast switching times that also results in a smaller difference between the logic high and low levels and high power consumption. Consequently, ECL is used only in applications where extremely high data rates are experienced and power efficiency is a much lower priority. These applications include large mainframe super computers and communications equipment that interfaces with fiber optic transceivers for gigabit Ethernet and ATM networks.

The difference between a logical high and low is generally less than 1 volt for the ECL family, so the noise margin provided is smaller than TTL or CMOS. To improve noise performance, ECL was designed to use a negative power supply. This was done because ground lines are usually more noise free than the positive supply side. Since ECL logic is referenced to the most positive voltage, it uses the more noise-free ground line as the most positive reference, with a negative power supply.

The CML structure of the ECL family provides another unique result: ECL devices have outputs for both the true logical output and the inverted output. In other words, ECL devices provide complementary outputs.

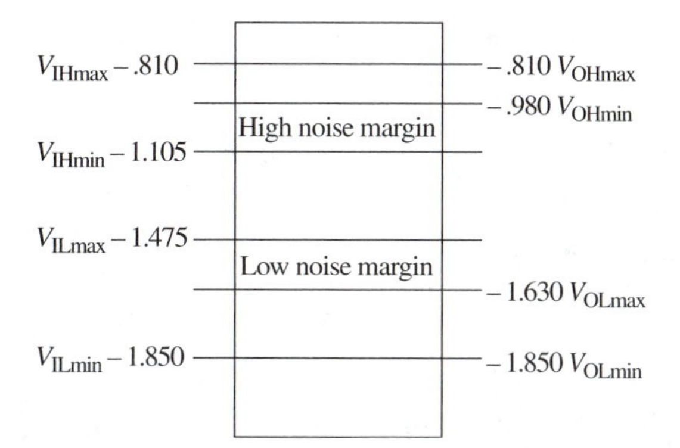

FIGURE 10–13
ECL 10K logic levels

ECL 10K/10H Series—Emitter coupled logic with 10xxx designations: This is the most popular ECL family that is identified with a five-digit number with the first two digits fixed at 1 and 0, respectively. This series is powered with V_{CC} = 0 V and V_{EE} = –5.2 V. The logic voltage levels for ECL are shown in Figure 10–13. The 10H designation means that these devices are compensated to operate from voltages other than –5.2 volts.

Internally, both TTL and CMOS logic devices incorporate a pull-up resistor within their totem-pole circuit arrangement. Because the switching times for ECL are so fast, circuit connections longer than a few inches must be viewed as a transmission line. In ECL devices a pull-down resistor is necessary because the outputs are emitter coupled. The value of the pull-down resistor should match the effective impedance of the connection and minimize power consumption. For this reason the selection of the pull-down resistor value (270 to 2 kΩ) is left up to the circuit designer. These pull-down resistors consume a lot of power, which is why the power consumed by ECL devices can vary so greatly.

	Propagation Delay	Power
ECL 10K/10H Series	2 ns	35 to 175 mw per gate

ECL 100K Series—Emitter coupled logic with 100xxx designations: The 100K series is identified with six digits, with the first three digits fixed as 1, 0, and 0. The numbers have no significance relative to the 10K/10H series. The 100K series offers improved speed performance at the price of lower power efficiency. The propagation delay for ECL 100K is in the area of .7 ns with approximately 40 mw/gate of additional power consumption. This series uses V_{CC} = 0 and V_{EE} = –4.5 V, which changes the logic level voltages also.

Positive emitter coupled logic: These are ECL devices designed to operate from positive power supplies. This promotes their interfacing with TTL and CMOS logic families.

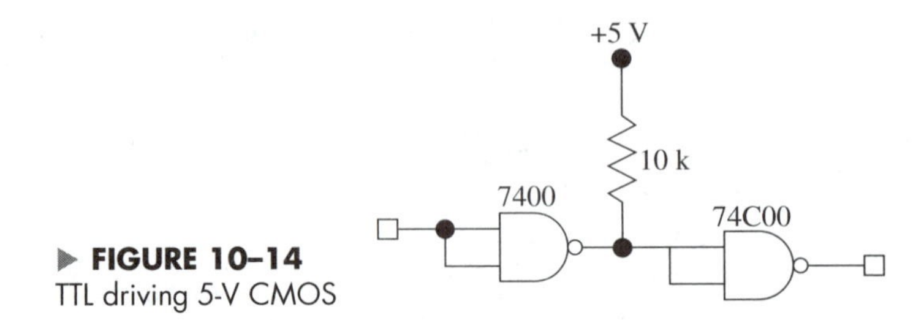

▶ **FIGURE 10–14**
TTL driving 5-V CMOS

Interfacing Logic Families

TTL to 5-V CMOS

TTL is very capable of supplying enough current to drive CMOS devices. When interfacing any TTL family device to a 5-V CMOS device, care must be taken to insure that the high output voltage is high enough to register as a logical high with the CMOS device. The minimum high output voltage, $V_{OH(MIN)}$ for all TTL devices is too low when compared to the minimum high input voltage, $V_{IH(MIN)}$ for most CMOS devices. The proper method for TTL driving CMOS is to connect a 10-kΩ pull-up resistor on the output of the TTL device. This will pull the high output voltage of the TTL device well within the acceptable range for CMOS devices (see Figure 10–14).

TTL to High-voltage CMOS

A TTL device with an open collector output is often used to drive CMOS logic powered by voltages greater than 5 V. The voltage rating for open collector outputs is usually around 30 V DC, greater than the maximum power supply voltage for CMOS logic. The open collector output is pulled up to the power supply voltage being used, as shown in Figure 10–15.

5-volt CMOS to TTL

Most 5-V CMOS families can drive TTL directly. It is important to review the specific data sheet in question to determine how many TTL devices can be driven safely. However, the early 4000 and 4000B series of CMOS devices cannot sink enough current when driving a TTL device to a logical zero. In order to cover this situation, the 4000 series includes special buffers. Two devices, the 4050 buffer and the 4049 inverting buffer, have higher current sourcing and sinking capabilities that resolve this interface problem. Figure 10–16 shows a 4000 series logic device driving TTL with a 4050B device acting as a buffer/driver.

▶ **FIGURE 10–15**
TTL driving high-voltage CMOS

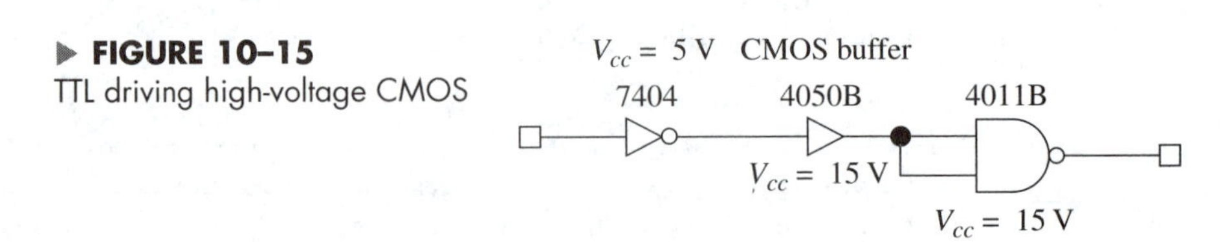

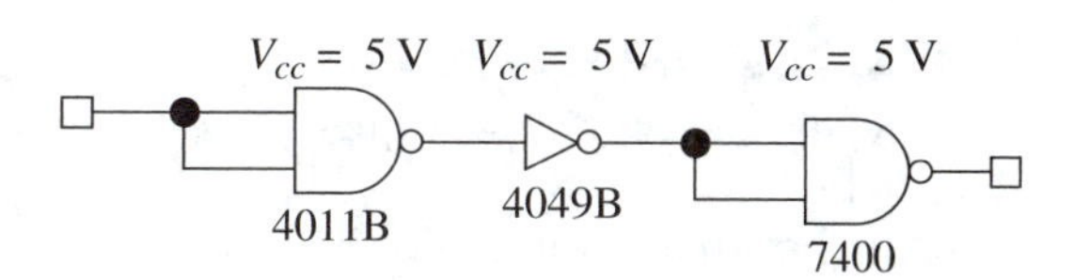

▶ **FIGURE 10–16**
4000 CMOS driving TTL

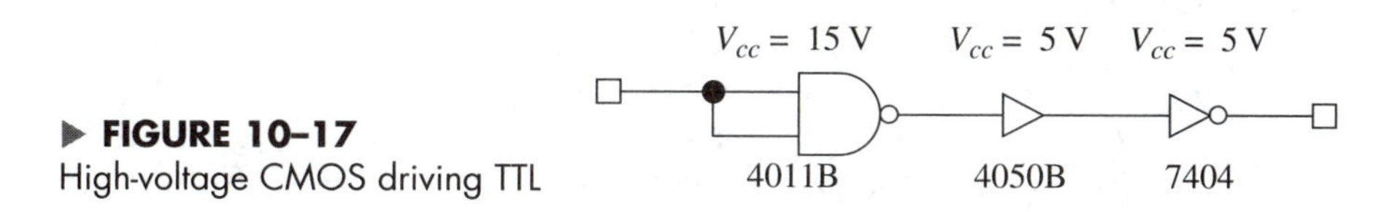

▶ **FIGURE 10–17**
High-voltage CMOS driving TTL

High-voltage CMOS to TTL

When high-voltage CMOS circuits must drive TTL, a CMOS buffer can be used to translate high-voltage CMOS levels down to 5-V CMOS. In this case either the 4049B or 4050B can be used, as was discussed previously for CMOS driving TTL devices. Figure 10–17 shows an example circuit in which high-voltage CMOS drives a TTL logic gate.

Interfacing with ECL

IC manufacturers have addressed the difficult problem of converting ECL to TTL and vice versa. ECL 10k series devices have been developed especially to accomplish this task. The 10125 interfaces ECL to TTL while the 10124 accepts TTL inputs and drives ECL outputs (see Figure 10–18).

▶ **FIGURE 10–18**
ECL and TTL interfacing

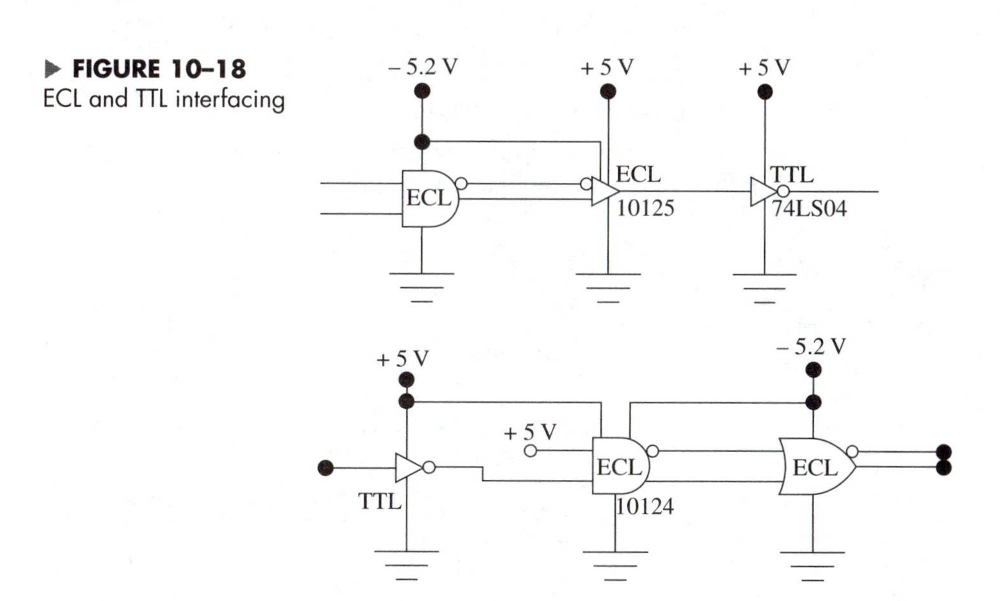

5-V to Low-voltage Logic

The development of low-voltage logic devices has brought about the need to translate between 5-V and 3.3-V logic systems. Most devices cannot tolerate input voltages higher than their power V_{CC}. Special circuitry must be added to 3.3-V devices to make them tolerant of 5-V inputs. Even though many 5-V and 3.3-V devices provide TTL-compatible interface levels and are 5-V input fault tolerant, some devices lack these capabilities. For example, there are certain CMOS devices that can drive to 5 V but cannot tolerate a 5-V input. These devices require 5-V to 3.3-V translation when the two logic levels must be interfaced.

A common design requirement is the interfacing of low-voltage microprocessor and memory circuits with 5-V I/O devices and memory (see Figure 10–19). The 74CBTD3384 is a special device that has been developed to provide translation between these two logic levels. The bus switches included for each bit of the 74CBTD3384 consist of an N-channel MOSFET. The inherent voltage drop across the drain and the source of the MOSFET serves to drop the 5-V levels that are input to the 3.3-V devices down to a maximum of 3.3 V (see Figure 10–20).

10–3 ▶ Package Considerations

The original package configuration for digital ICs was the dual-in-line package (DIP). DIPs are consistent with older through-hole package technology, where all electronic components are mounted to printed circuit boards by inserting their connecting leads into holes in the PCB. Copper pads present on the surface of the PCB surround these holes. The component lead is soldered to the copper pad, making the electrical connection and securing the component to the circuit board. The spacing between DIP leads, called the *lead pitch*, is .1 in. There are two rows of leads with an equal number in each row. The .1-in. lead spacing is the minimum dimension practical, when considering minimum hole and pad sizes for fabricating PCBs.

Early TTL and CMOS IC families were founded on the DIP. DIP packages are available in two package widths, .3 in. or .6 in. between lead rows. Following are the standard DIP configurations:

Mini-DIP—8 pins, .3-in. width

14-pin DIP—14 pins, .3-in. width

16-pin DIP—16 pins, .3-in. width

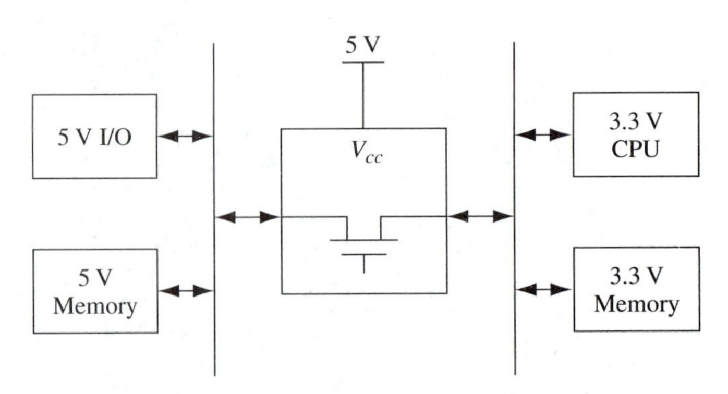

▶ **FIGURE 10–19**
5-V to 3.3-V interfacing (TI data sheet 74 CBTD3384)

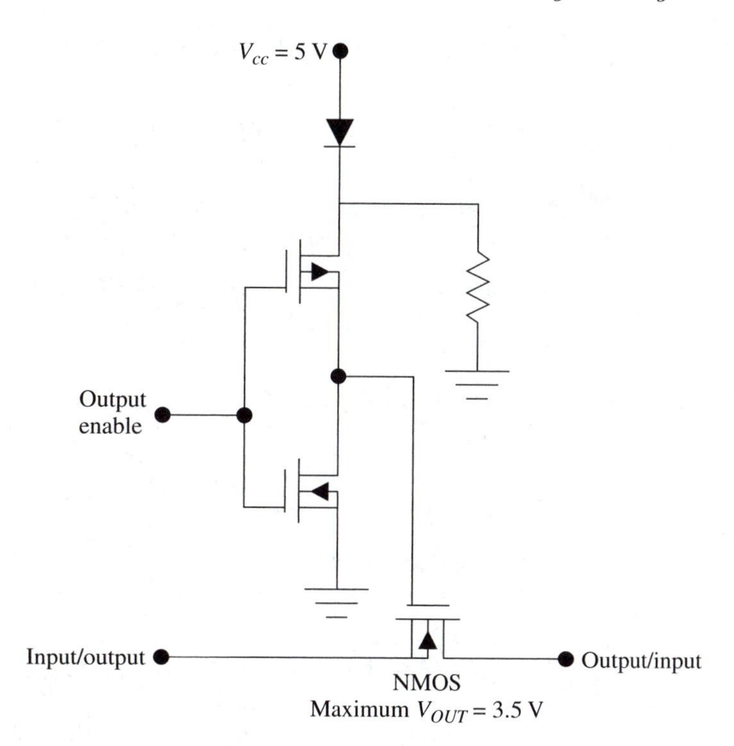

▶ **FIGURE 10–20**
NMOS Switch of 74CBTD3384
(TI data sheet 74 BTD3384)

20-pin DIP—20 pins, .3-in. width
24-pin DIP—24 pins, .3-in. width
24-pin DIP—24 pins, .6-in. width
28-pin DIP—28 pins, .6-in. width
40-pin DIP—40 pins, .6-in. width

As integrated circuit densities increased and package sizes decreased, there was a need to decrease lead pitch as well. A new method was needed to make component electrical connections and affix the component to the circuit board. Surface-mount technology (SMT) was the result. The first surface-mount package style to be developed was the Small Outline (SO) package developed for the watch industry. SMT quickly became the standard for circuit connections made on the surface of PCBs. SMT connections are made with small pads etched onto the surface of the PCB. Surface-mount devices (SMDs) have leads that fit circuit board pads. SMDs are placed on the circuit board pads and soldered into place, making electrical connections and affixing the component to the board. SMT does not require holes in the PCB and supports lead pitch as small as .4 mm.

Surface-mount devices incorporate different lead styles dependent on the overall package configuration. SMD lead styles include gull-wing, J-lead, and contact ball and pin. These lead styles are combined with the following package styles:

Small Outline Integrated Circuit (SOIC): gull-wing leads arranged in equal rows similar to DIP with lead pitch of 1.27 mm

Shrink Small Outline Package (SSOP): a small outline package with more dense gull-wing leads arranged in equal rows with a lead pitch of .65 mm

Thin Shrink Small Outline Package (TSSOP): a shorter height SSOP package with gull-wing leads arranged in equal rows and a lead pitch of .65 mm

Thin Very Small Outline Package (TVSOP): a more dense TSSOP package featuring gull-wing leads with a lead pitch of .4 mm

Plastic Leaded Chip Carrier (PLCC): square packages with J-leads on each side; lead pitch is 1.27 mm and standard package sizes are 28, 44, and 68 pins

Quad Flat Pack (QFP): square packages with gull-wing leads on each side; lead pitch is .635 mm with standard packages of 48 and 96 pins

Thin Quad Flat Pack (TQFP): more dense QFPs with gull-wing leads and a lead pitch of .5 mm

Low-profile Fine Pitch Ball Grid Array (LFBGA): feature contact ball leads placed on a grid along the bottom surface of the IC; lead pitch is .8 mm

Pin Grid Array (PGA): identical to the LFBGA except the ball leads are replaced with pins that are designed to mate with a surface-mount socket that is soldered into the circuit board; PGA ICs are easily removed and replaced. Most of the currently available personal computer motherboards feature PGAs for connecting the processor to the board.

Whichever package technology is used, it is important to review all specifications for the particular device package being utilized. For example, when using a 74 AHC triple, three-input Nand gate IC in a DIP package, its performance specifications when packaged as a SOIC surface-mount device may differ in some key areas. There are two areas where the differences in performance between through-hole and surface-mount technology are significant: high-frequency signal operation and maximum power dissipation levels. SMDs offer better high-frequency operation over DIPs because the IC is physically smaller, containing less inductance and capacitance. On the other hand, the small size of SMDs means there is less surface area to dissipate heat. Consequently, DIPs usually have a greater power dissipation capacity.

10–4 ▶ Programmable Logic Devices

In digital electronics courses, the various methods of circuit minimization are studied at length. Boolean algebraic manipulation and Karnaugh maps are the two methods discussed most often. Software tools are also available that simplify this process by allowing the circuit designer to enter a Boolean logic function. The software then determines the minimum combinational logic circuit that will perform that logic function. One well-known example of this type of software is the logic converter included with Electronics Workbench and MultiSim programs. Once the minimum circuit is known, the designer can implement the circuit with discrete logic devices.

Another approach to digital circuit design is to employ the use of programmable logic devices (PLDs). PLDs are a collection of logic devices that are connected as general logic circuits configurable to generate any possible logic function. PLDs have seen rapid development over the last decade as the requirements for complex programmable logic have increased dramatically. While PLDs usually require using more than the minimum logic circuit, they can often be the most versatile and economical design solution that consumes the least space. *PLD* is a general term that is used when conversing about combinational logic devices that are programmable with fuseable links or memory cells. The field of PLDs seems confusing and overwhelming because of all the different terms that are used and the complexity of these devices. SPLDs, PALs, PLAs, GALs, CPLDs, and FPGAs are some of the more common terms that define various PLDs.

PLDs are derived from sum-of-the-product logic circuits where the various combinations of input logic variables are ANDed together, then ORed to form the final output. There is one AND gate for each possible input combination. The standard logic circuit schematic for a sum of the product circuit is shown in Figure 10–21a. Figure 10–21b shows the simpler PLD method for representing the same circuit. The output from each AND gate is called a "product-line" because it represents the Boolean product of the inputs to the AND gate.

Generally, PLDs have programmable links located either at the inputs to the AND gates called Programmable Array Logic (PALs), or at the AND gate's inputs

▶ FIGURE 10–21
(a) Standard sum of the product logic circuit. (b) PLD format.

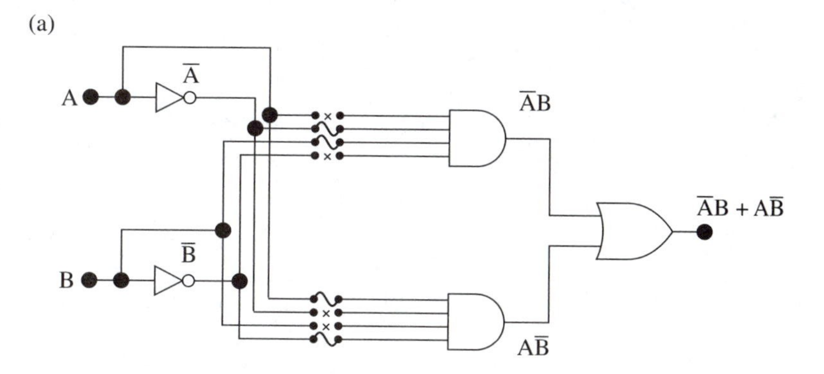

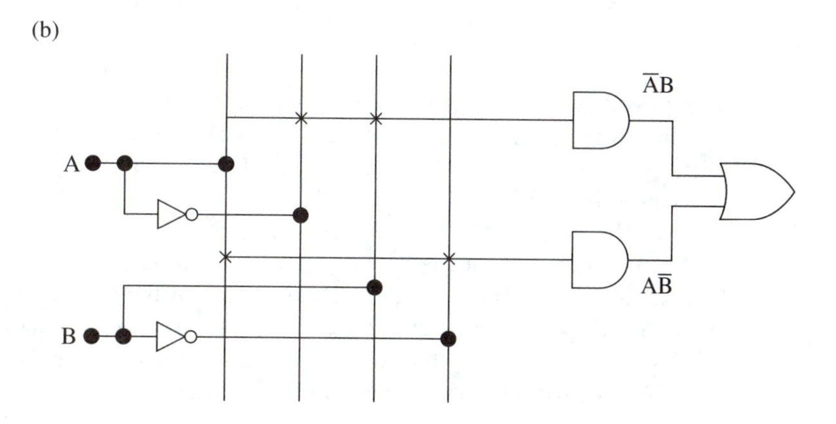

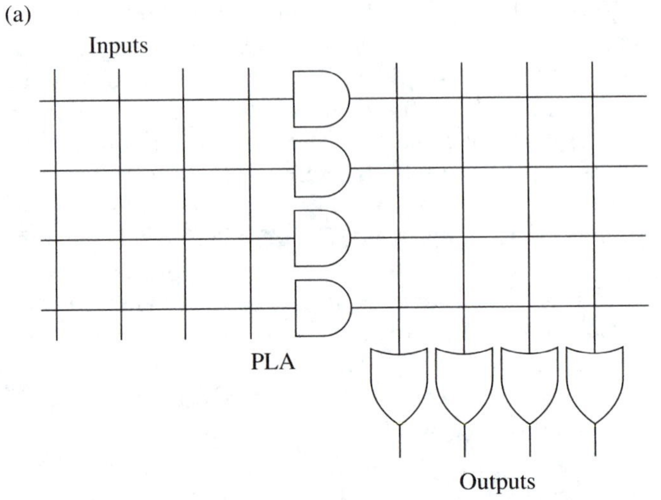

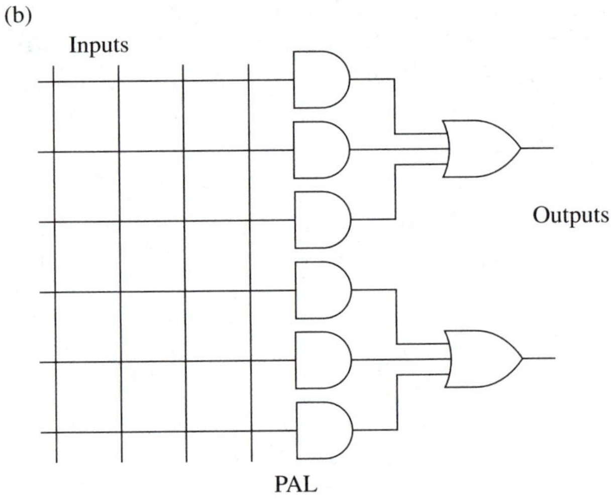

FIGURE 10–22
PLA and PAL representations

and outputs called Programmable Logic Arrays (PLAs). These are the two distinct variations in the structure of PLDs: PALs that have a programmable AND array with fixed OR connections or PLAs that have both a programmable AND and OR array (see Figure 10–22).

In PROM-type PLDs, fuseable links provide the logic circuit programmability. Nonvolatile memory cells can also be used as a way of implementing logical functions where each cell reflects the status of each fuseable link position. Both PALs and PLAs can configure each AND gate to provide an output for any combination of inputs to output a logical one for any truth table combination. The output OR gates for a PAL are fixed and have a maximum number of inputs that limit the AND gate combinations available to provide the desired logical output function. When using a PAL, it is important to select one that provides enough OR gate inputs to allow implementing the desired logic function. Programming the PAL is like programming a one-time programmable PROM, unless an EPROM version with an erase window is available.

The outputs from PALs can also be configured in a variety of different ways: active high logic, active low logic, complementary, programmable, and versatile. Higher-level PALs include the programmable and versatile output features by using Output Logic Macro Cells (OLMCs). OLMCs allow the output of the PAL to be programmed as active high or low, programmable or synchronous. Flip-flop "registered" outputs can also be selected.

Generic Array Logic (GAL) devices are similar to PALs, but they are electrically eraseable. They usually include OLMCs and the complete output programmability that they allow. Simple Programmable Logic Devices (SPLDs) include PLDs, PLAs, PALs, and GALs; the name is just a way of differentiating this group of devices from other, more complex devices such as CPLDs and FPGAs. SPLDs are on the low end of the PLD family of devices and include anywhere from 4 to 22 output cells that can replace a limited amount of discrete logic devices.

PALs were the first PLDs to be developed and marketed. The PAL16L8 is well known and has been in use for years; considered a mature product by some and by many others, retired. The structure of the part number system often used to identify PALs, GALs, and other PLDs is shown in Figure 10–23. The part number identifies directly the number of inputs and outputs, as well as the type of output. Note the variety of output types shown in Figure 10–23. The major variations in output configurations include those that simply determine the active logical state (L, H, P, and C designations), those that have "registered" outputs (R designations), or more complex variable output modules (V designations). PLDs with an "R" designation have registered outputs that include some type of flip-flop on the output to allow clocking the change in output to coincide with system timing requirements.

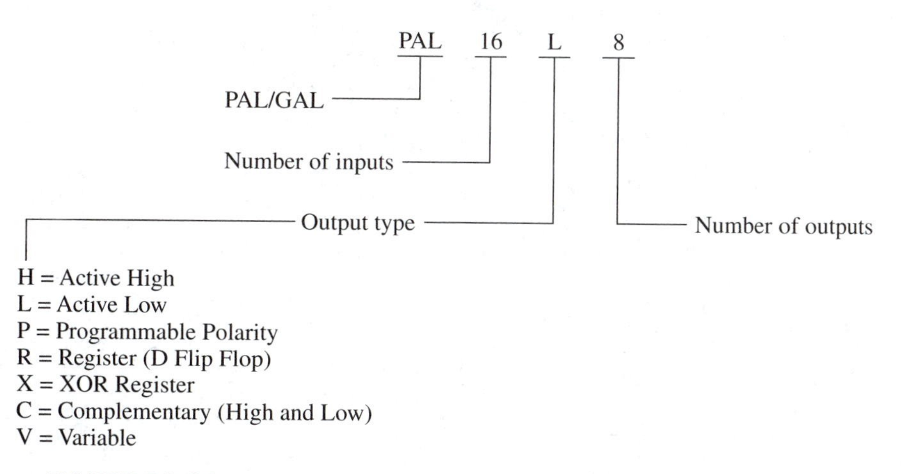

▲ **FIGURE 10–23**
PAL part numbering system

Variable or versatile architecture PLDs, denoted with a "V," have complex output module circuits.

Let's review the operation of the PAL16L8. It is designated as a PAL, and therefore only the inputs to the AND gates are programmable links. Figure 10–24 shows the logic connections to the PAL16L8. The part number scheme indicates that the

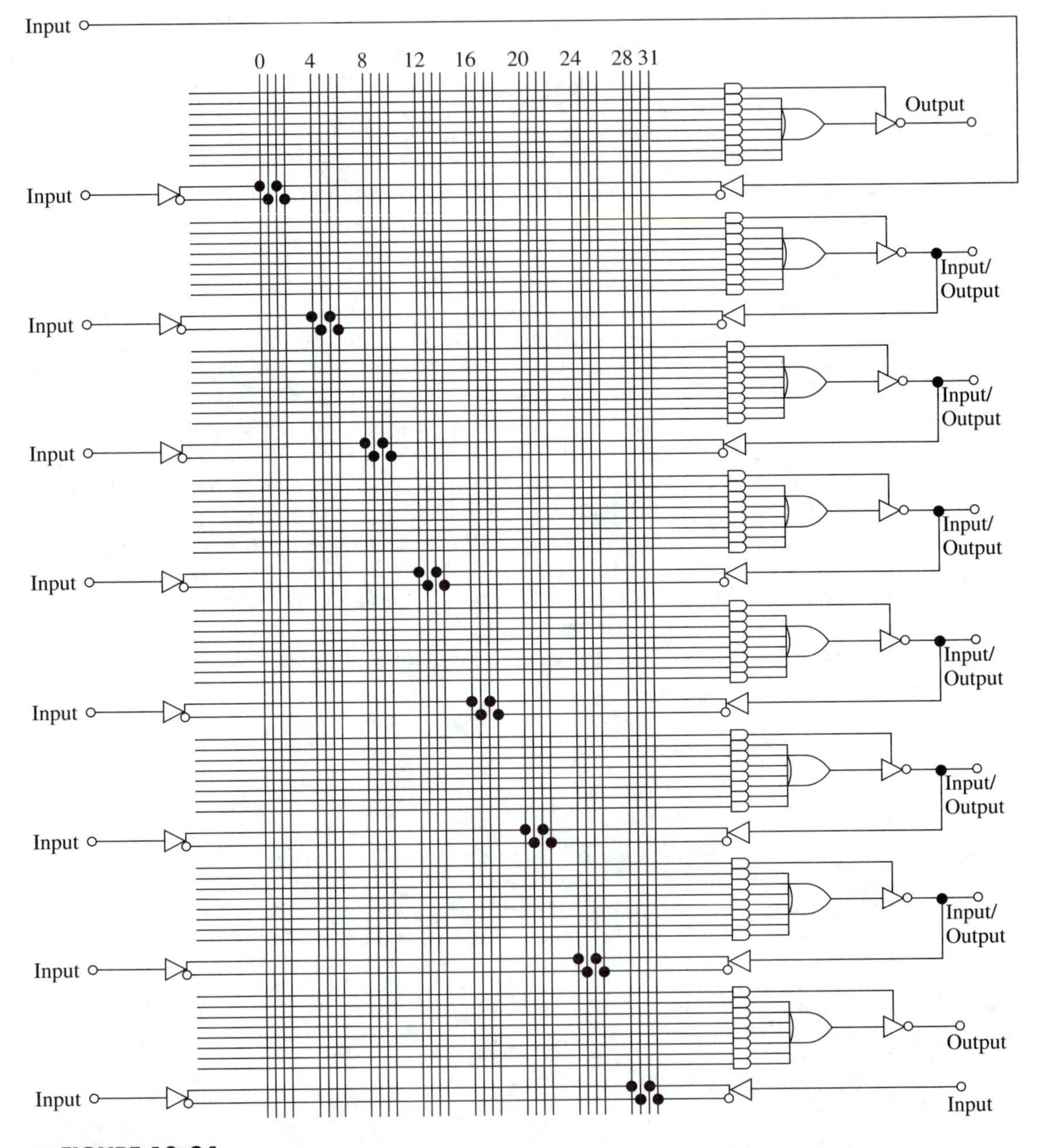

▲ FIGURE 10–24
PAL16L8 schematic diagram

PAL16L8 has 16 inputs with 8 outputs. While this is possible, 6 of the pins can be used as either inputs or outputs, so it is more correct to say that the PAL16L8 has a maximum of 16 inputs or 8 outputs. The PAL16L8 has 10 dedicated inputs and 2 dedicated outputs, with 6 pins that can be used as inputs or outputs.

The outputs from a PAL are generally tri-state so that some form of output enable is needed to connect the logic output to the output pin. The PAL16L8 has an inverter between the ORed product lines and the output pin that is enabled by a separate product line. The enable connection is active high. When disabled, the output reverts to its high impedance state. If all connections on the enable product line are blown, the inverter is always enabled because the inputs are all seen as high. If all connections are left in place, the AND gate always has both logical states for each variable, which causes the output to be low, disabling the inverter.

The GAL16V8 is a step up from the PAL16L8, because it is electrically erasable and it has versatile outputs, as denoted by the V located between the maximum inputs and maximum output designations. GALs can typically be erased and reprogrammed about 100 times and will retain their memory for at least 20 years. The GAL has 8 pins designated as inputs and 8 that are selectable as inputs or outputs. There are also 2 pins that can be dedicated inputs or clock and output enable pins, depending on the operating mode selected. With power supply and ground, the GAL16V8 requires 20 pins that accommodate a 20-pin DIP, a 20-pin PLCC, or a 20-pin SOIC package.

The key to the extensive output variability of PALs and GALs is the output logic macro cell (OLMC). The GAL16V8 includes eight OLMCs. Each can be selected to function globally in one of three modes: simple, registered, or complex. The functional diagram for the GAL16V8, configured for operation in the simple mode, is shown in Figure 10–25. When programming the GAL16V8, the operating mode of all eight OLMCs available on the device is accomplished by selecting the status of the global architecture cell. This cell consists of two programming bits called SYN and AC0. The bit combination selected by the operating modes is shown below:

SYN	AC0	Operating Mode
0	1	Simple Mode
1	0	Registered Mode
1	1	Complex Mode

GAL16V8—Simple Mode

When selected to function in the simple mode, six of the eight OLMCs of the GAL16V8 can be programmed to function as either a dedicated input or as an always active, combinatorial output. The remaining two OLMCs (pins 15 and 16) can only be dedicated outputs. Pins 1 and 11 become dedicated inputs to the AND array in this operating mode. Each cell designated as an output has a maximum of eight product terms with programmable polarity. There are two bits for each OLMC that determine whether it is an input or an output and the output polarity, XOR and AC1. The XCOR bit determines the output polarity for any cell designated an output—XCOR = 0 means the cell has an active low output and XCOR = 1 signifies active high

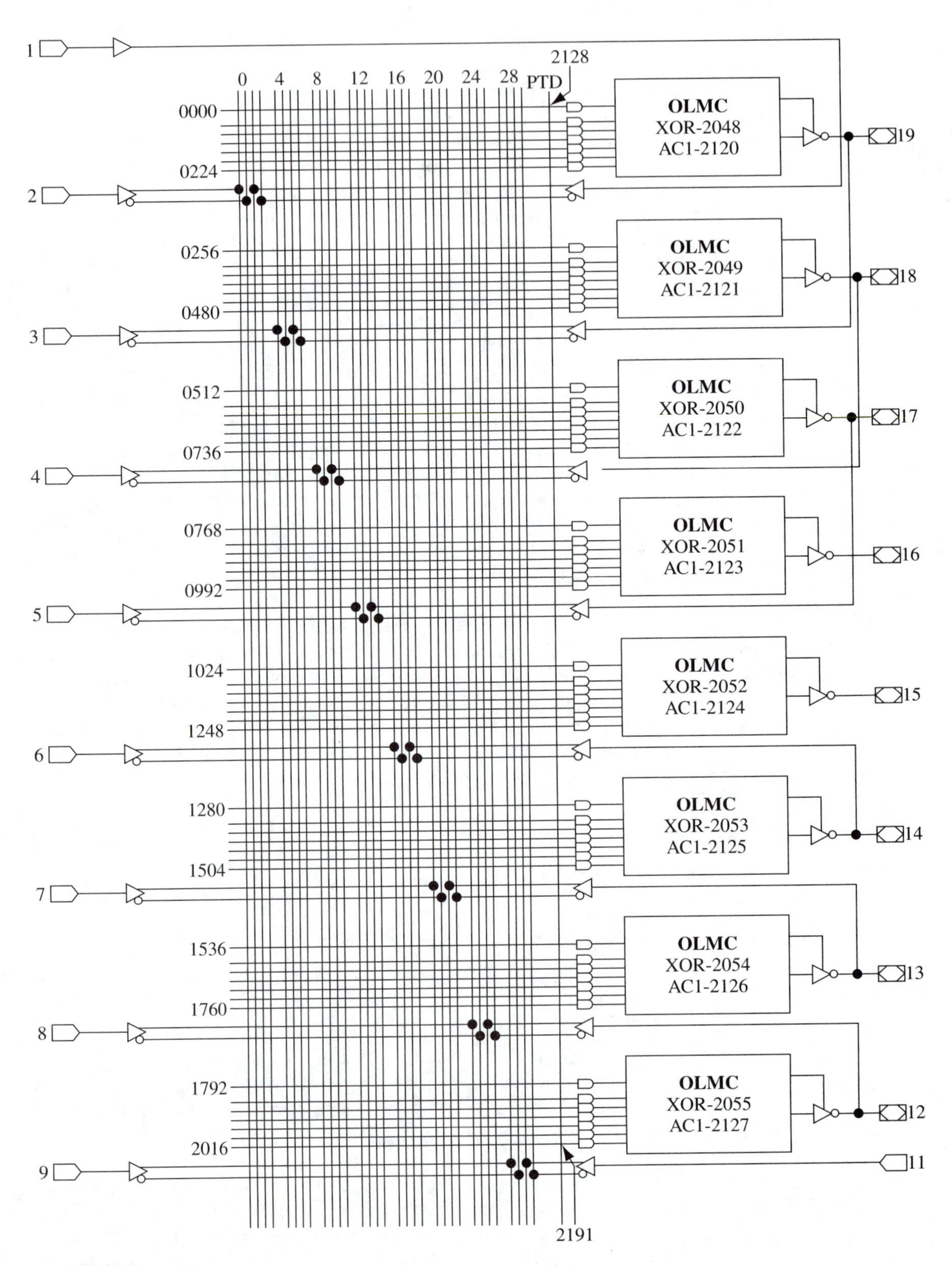

▲ **FIGURE 10–25**
GAL16V8 in simple output mode

operation. AC1 = 0 selects the OLMC to be an output while AC1 = 1 means input. The development software, as programmed by the circuit designer, configures the status of these bits for each OLMC.

Example 10–3

Set up the GAL16V8 to function in the simple mode. The three OLMCs associated with pins 12, 13, and 14 are to be configured as dedicated inputs. Pins 15 and 16 are to be dedicated as always active low outputs with no feedback, while pins 17, 18, and 19 will be always active high outputs that can be fed back to the inputs.

Solution

1. Select the Simple Mode of operation by making AC0 = 0, SYN = 1.
2. Set up the OLMCs as follows:

   OLMC pin 12—AC1 = 1, XOR = X = don't care these are inputs
   OLMC pin 13—AC1 = 1, XOR = X = don't care these are inputs
   OLMC pin 14—AC1 = 1, XOR = X = don't care these are inputs
   OLMC pin 15—AC1 = 0, XOR = 0
   OLMC pin 16—AC1 = 0, XOR = 0
   OLMC pin 17—AC1 = 0, XOR = 1
   OLMC pin 18—AC1 = 0, XOR = 1
   OLMC pin 19—AC1 = 0, XOR = 1

GAL16V8—Registered Mode

In the registered mode each OLMC can be configured either as a dedicated registered tri-state output or a combinational tri-state I/O pin. Registered operation simply means that a flip-flop register buffers the output. The polarity of the logic output is selectable for each output. In this mode pins 1 and 11 become fixed as the clock and output enable inputs, respectively, for the registered operation.

The same two bits, XOR and AC1, discussed for the simple operating mode select the variations of operation provided for in the registered mode. The XOR bit determines output polarity—0 = active low and 1 = active high. AC1 = 0 selects a tri-state registered output for a particular OLMC. AC1 = 1 selects a combinational tri-state output for the OLMC.

GAL16V8—Complex Mode

The complex mode allows six of the eight OLMCs (pins 13 through 18) to function as tri-state combinational I/O while the remaining two OLMCs associated with pins 12 and 19 operate as dedicated tri-state outputs. AC1 = 1 selects this configuration while the XOR bit defines the polarity of the outputs. Each OLMC in this mode has seven product terms per output as one of the product terms is always used as an enable output control. Pins 1 and 11 are always data inputs to the AND array.

Example 10–4

You are developing a logic circuit that includes mostly combinational logic with a synchronous output. You wish to implement the circuit with a GAL16V8. The general requirements for the circuit are as follows:

Inputs: eight logic level inputs, one clock input and an output enable input.

Outputs: one registered output and four combinational tri-state logic outputs. All outputs are active high.

Develop the global and local configuration bits for the GAL16V8.

Solution

1. To provide even one registered output, the GAL16V8 must be configured in the registered mode. Therefore AC0 = 1 and SYN = 0 define the registered mode of operation. The clock is connected to pin 1 and the output enable line to pin 11.
2. The OLMC connected to pin 19 is configured as a registered active high tri-state output—XOR = 1, AC1 = 0.
3. The OLMCs connected to pins 15, 16, 17, and 18 and are configured all in the same way—XOR = 1 and AC1 = 1.

As with all digital devices, the selection of PALs and GALs should include the consideration of speed (total propagation delay through the device), power consumption, package, and the overall grade (commercial, industrial, etc.). PALs and GALs are often used for the control of Direct Memory Access (DMA), state machine control, high-speed graphics processing, and simply replacing existing logic functions with smaller and cheaper alternatives.

Complex PLDs (CPLDs) include a number of PALs on one IC that can be connected together to provide complex logical functions. A CPLD can contain a few hundred macrocells. CPLDs are usually divided into logic blocks that include a portion of the macrocells. The logic blocks can then be connected through a switch matrix. The degree to which the logic blocks are connected varies from CPLD family to the manufacturer. One hundred percent connection is not always provided. Each of the macrocells includes an OLMC as discussed for the GAL and PAL devices.

An example CPLD is the Xilinx XC9500 family. This family of CPLDs includes anywhere from 36 to 288 macrocells. The product designations indicate the number of macrocells available on each device. For example, the XC9536 is on the low end of the product family with 36 macrocells. The XC9536 contains 800 usable gates and 36 registers. The general structure if the XC9500 family is shown in Figure 10–26.

Field Programmable Gate Arrays (FPGAs) are a class of PLDs that are significantly different from the devices we have discussed so far. They offer the highest logic capability. FPGAs have logic blocks or cells that are located in the center of groups of programmable interconnections. Around the outer edge of the chip are programmable I/O blocks. The general functional diagram for a FPGA is shown in Figure 10–27.

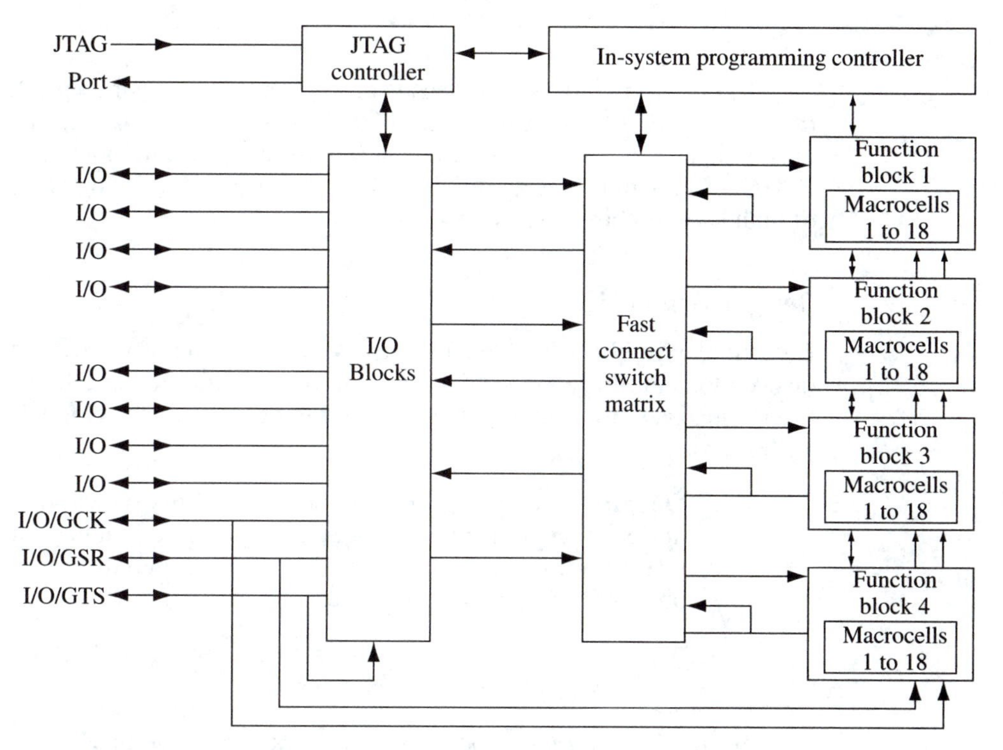

▲ **FIGURE 10–26**
XC9500 CPLD family

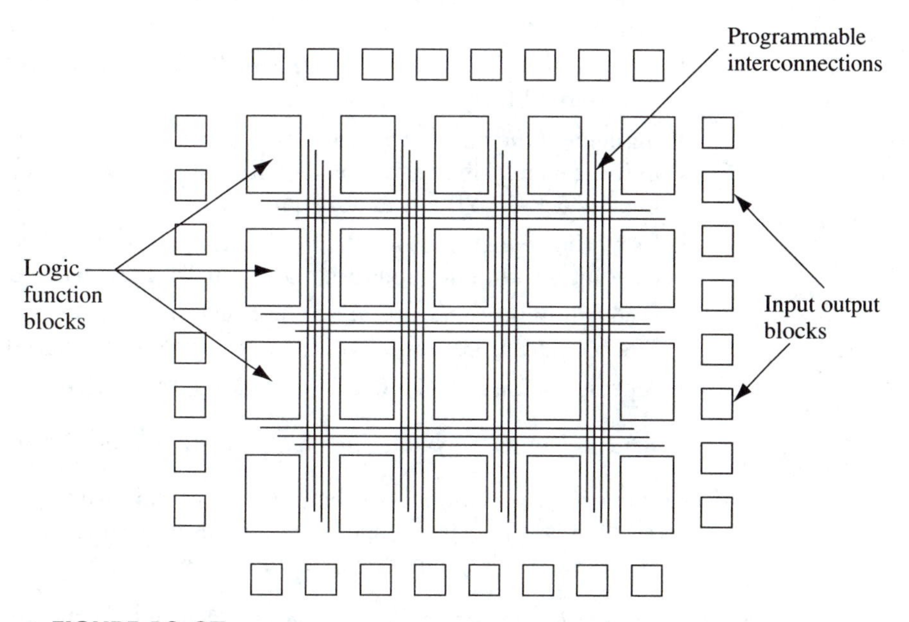

▲ **FIGURE 10–27**
Gate array functional diagram

There are two different classes of FPGAs that categorize the size and quantity of the logic blocks. Coarse-grained FPGAs contain relatively few large-sized logic blocks. FPGAs classified as fine grain have a larger number of smaller blocks. FPGAs offer a lot of design flexibility and are most often used in the development of high-level processor-based systems.

Designing with PLDs

Designing with PLDs involves the use of development software that translates the desired logical function into the program that will implement it. Let's use the Six Steps discussed in Chapter 1 to outline this process of developing logic circuits with PLDs.

Step One: *Research and gather information* about the various types of PLD devices that are available. Search out periodicals and books on the subject. Determine the manufacturers and accumulate data on the various product lines that are available.

Step Two: *Define the problem.* Develop a complete definition of the design problem in the form of a specification.

Step Three: *Plan the solution.* Develop a plan for completing the PLD design.

Here is a typical plan for PLD development:

1. Develop either a schematic or Boolean logic statement for the logic function required. Develop the logic design either as a schematic capture file or in hardware description language (ABEL, Versilog, or VHDL).
2. Review the requirements with the available devices and select the appropriate PLD for this application.
3. Implement the design with the development software that corresponds to the device being used. The development software will translate the logic design into the available hardware.
4. Next, the design rules for the device are checked against the translated solution. Modifications are made consistent with these design rules.
5. Logic circuits are partitioned and fitted into the available logic blocks. The blocks are allocated and the connections are routed.
6. A program file is created and loaded into the device.

Step Four: *Implement the plan.* The plan outlined in Step Three is completed.

Step Five: *Verify the design.* The performance of the resulting circuit is measured against the original specifications. Key performance target areas are function, timing, power, cost, and space.

Step Six: *Develop a conclusion.* What was learned in the process? What would be done differently next time?

Summary

In this chapter we have gathered the current information about digital logic design. We have reviewed the practical aspects and requirements for successful design implementation. With many logic families and devices from which to choose, the designer is faced with a maze of decisions and considerations: CMOS, TTL, BiCMOS or Low-voltage technology, and PLDs or discrete logic devices. These decisions are more easily made when reliability, cost, and availability considerations are made. Devices considered mature will dictate increasingly higher prices and be harder to procure. Those that represent the newest technology usually have the least field experience and may also have longer lead times, initially. Unless the latest technology is required to make the design feasible, the more prudent approach is to use those devices that are still in their growth phase, yet have been applied enough to develop a track record of performance in a variety of application areas.

The continued development of PLDs into CPLDs and FPGAs has moved logic design closer and closer to on-chip integrated circuit design. As data, address, and control buses become wider and more intricate, the designs and design tools have expanded to fulfill this need. But in the end the design will be judged as digital designs have always been judged: speed, accuracy, efficiency, reliability, size, and cost.

References

Dueck, R.K. *Digital Design with CPLD Applications and VHDL.* Albany, NY: Delmar, 2001.

Klietz, W. *Digital Electronics: A Practical Approach.* Upper Saddle River, NJ: Prentice Hall, 1999.

Tocci R. J., and Widmer, N. S. *Digital Systems: Principles and Applications.* Upper Saddle River, NJ: Prentice Hall, 2001.

Wakerly, J. F. *Digital Design: Principles and Practices.* Upper Saddle River, NJ: Prentice Hall, 2001.

Exercises

10–1 Why do TTL devices tend to consume more power than CMOS devices?

10–2 Why do CMOS devices tend to be slower than TTL devices?

10–3 Explain the difference between combinational, sequential, and bus-oriented logic devices.

10–4 List the general digital design considerations discussed in this chapter.

10–5 Describe the 74HC and 74HCT logic families. What is the primary difference between these two families?

10–6 List three special considerations for designing and working with CMOS devices that are different from other logic families.

10–7 What is the basic premise behind the development of BiCMOS technology?

10–8 What performance benefits are there for lowering both power supply voltages and logic levels of digital logic devices?

10–9 What are the design criteria for calculating the value of pull-up resistors in digital logic circuits?

10–10 What are the design criteria for calculating the value of pull-down resistors in digital logic circuits?

10–11 What is the primary purpose behind the development of the emitter coupled logic (ECL) family?

10–12 Explain what is meant by the term *current mode logic*. To what family of logic devices does it apply? How does it compare to a totem pole arrangement?

10–13 Why are the logic levels associated with ECL defined as negative voltages? What noise margin does ECL provide?

10–14 What is meant by the term *lead pitch*?

10–15 List the four different lead styles currently available for SMDs.

10–16 Explain the difference between a PAL and a PLA.

10–17 How does a GAL differ from a PAL?

10–18 Describe what is meant by the term SPLD. Describe what is meant by the term CPLD. Explain the differences between them.

10–19 When using a GAL, describe the purpose of the global selection bits. What functions do the local bits select?

10–20 Describe the structure of a FPGA. What are the different classes of FPGAs, and how do they differ?

Embedded System Design

Introduction

In the early 1970s the world's first microprocessor, the Intel 4004, was introduced. It was a four-bit microprocessor with a very limited instruction set, but it was unique in the fact that it was a central processing unit (CPU) contained on one integrated circuit. The 4004 was followed by its eight-bit counterpart, the 8008, and then the 8080. This began a long chain of expanded data/address buses and processor capabilities that continues today. At the time the 4004 was introduced, CPUs consisted of multilayer circuit boards that contained complicated and extensive circuits, including many combinational logic and sequential devices. These CPU boards were physically large and expensive to manufacture. Consequently, they were confined to applications that included medium-level minicomputer or large mainframe computer applications. The introduction of the microprocessor initiated an entirely new way of thinking about computer applications. Engineers and scientists began thinking about how to apply microprocessors to many products already available, and, more importantly, how to create new products that became practical with the availability of a one-chip CPU.

Today it is hard to find a product that does not include a microprocessor embedded within it. Microwave ovens, electric ranges, VCRs, stereos, and CD players all are likely to rely on a microprocessor as a central control device. The microprocessor has offered many new features, versatility, reduced cost, and smaller size to products available before its introduction. The personal computer, electronic games, Palm Pilots, and MP3 players are just a few examples of products that became possible with the availability of the microprocessor.

Embedded systems are systems in which a microcomputer or microprocessor is embedded as a controlling device. This chapter discusses the design considerations that are important when designing low- to medium-level embedded systems. Following are the specific topics covered:

- Design Considerations
- Central Processing Units
- Memory
- Serial Communications
- Parallel I/O and Special Functions
- Signal Conversion

11-1 Design Considerations

The design process always begins with preliminary research and gathering information about the design problem. Then the design problem is defined completely by design specifications. Because embedded systems include microprocessors, there should be two sets of specifications: one for hardware and one for software. With embedded systems there are many critical design decisions that must be made in the initial stages of the project. For example, an estimate must be made of the time it will take to process a certain amount of data, well before it is determined exactly how the data will be processed. These are the types of decisions that warrant serious consideration up front. It is much like beginning the construction of the foundation for a new building when you are not sure exactly how large the building must be to fulfill its intended purpose. In this case decisions are weighted more heavily toward overdesign; yet if the system is overdesigned, its cost and size might be unworkable.

In order to study the design problem and develop hardware and software specifications, the following analysis steps should be taken:

1. Develop an overall flowchart for the top level of processor operations.
2. Develop preliminary flowcharts for all key data-processing activities or any activity that is expected to take a significant amount of processor time.
3. Determine a realistic estimate of the quantity and type of instructions required to complete all software operations identified on the flow chart.
4. Identify the time constraints that are inherent in the design problem. Is this a real-time application for which data must be processed within a certain time period in order to be ready for the new data? If so, what is the required time period? In applications that are not real-time, what are the time requirements for performing the identified software operations?
5. Develop estimates for the amount and type of data memory required. Will the data memory requirements be volatile or nonvolatile?
6. Develop estimates for the amount and type of program memory required. Will the program memory requirements be volatile or nonvolatile?
7. Study the requirements for functions that will be programmable or selectable by the intended user of the device. Determine a preliminary process for providing this programmability.

8. Identify the quantity and type of input/output operations needed to support all operations.
9. Define the operational environment where the embedded system will be used: consumer, industrial, military, or hazardous atmosphere. These translate into ambient temperature, moisture proof, vibration, and shock requirements stated in the environmental specifications.
10. Complete the development of hardware and software design specifications.

The completion of these steps will result in enough data to allow consideration and completion of the following:

1. Develop a preliminary hardware block diagram for the embedded system.
2. Determine a range of microprocessors that provide the proper operational structure for the task at hand.
3. Identify a clock speed with significant safety factor for all operations.
4. Estimate the amount of data memory required, along with the access time and volatility requirements.
5. Estimate the amount of program memory required, as well as the access time and volatility.
6. Identify parallel and serial I/O operations and develop preliminary selections for the devices that will carry out these operations.

A summary of the analysis steps and the resulting design decisions is shown in Figure 11–1.

Embedded systems preliminary data ——► Embedded systems preliminary decisions

Embedded systems preliminary data	Embedded systems preliminary decisions
• Develop flowcharts and estimate the instruction cycles required.	• Determine the range of processor requirements.
• Identify system operational time requirements.	• Select the processor clock speed.
• Develop requirements for data and program memory.	• Specify the amount and type of data memory.
• Identify and develop preliminary operations for user-programmable functions.	• Specify the amount and type of program memory.
• Identify the quantity and type of I/O required.	• Select the I/O devices required.
• Define the operational environment.	• Specify the packaging requirements.
• Develop hardware and software specifications.	

▲ **FIGURE 11–1**
Preliminary considerations for embedded systems

Along with the selection of the CPU, memory and I/O, the embedded system will require at least consideration for the design or procurement of an adequate power supply and a system clock. (The design and application of power supplies is discussed in Chapter 6 and clock generators are covered in Chapter 8.) In many cases the I/O will be analog in nature and the signals may require amplification and scaling. (Chapter 7 covers the design and application of amplifiers and scaling circuits.)

EMI Immunity

EMI is always a prime consideration in any electronic circuit and is a significant source of failure in an embedded system. (The topics of EMI immunity and EMI emissions are discussed in general in Chapter 2.) In embedded systems there are some additional methods that can be utilized. As discussed in Chapter 2, it is always best to eliminate the source of the noise. If this is not possible, then shielding and grounding can be used to keep the noise signal from entering the operational circuit of the device in question. Once present in the circuit, filtering can be used to keep EMI noise from affecting sensitive circuits. While high levels of noise immunity can be achieved, there is always the possibility of noise with a large enough amplitude, at just the right frequency, which can affect the operation of a particular circuit.

In embedded systems there are additional steps that can be taken to minimize the effect of noise on the processor circuit. EMI noise can disrupt the processor operation by causing erroneous operations resulting from:

1. Control lines that are driven high or low to cause improper functions—for example, the processor performs a memory read instead of a write
2. Address lines that are driven to incorrect addresses, causing the processor to look for its next instruction at the wrong location
3. Data lines read incorrectly, resulting in the failure of the processor

These all can result in the processor accessing incorrect memory locations for instructions that are, most likely, completely invalid. One incorrect instruction can easily knock the CPU out of its proper program loop, causing it to become hopelessly lost. This can be seen on your personal computer when it ceases to respond to any key depressions and is hopelessly locked up until you press CTRL-ALT-DEL or restart the system.

Because they are usually developed for one specific purpose, embedded systems have a well-defined, top-level program loop. In this situation there are additional measures that can be taken to minimize the impact of any EMI. Let's say that a CPU is part of a data acquisition system that examines and processes an input signal 60 times a second. In order to keep up with the data flow, the data acquisition and processing must occur faster than 16.7 ms (1/60 cycles/sec). If the processor always performs this task in less than 10 ms, we can use this fact, with the addition of a watchdog timer to the CPU operation, to minimize the effect of any EMI. A watchdog timer is a separate timer that is reset every time the CPU

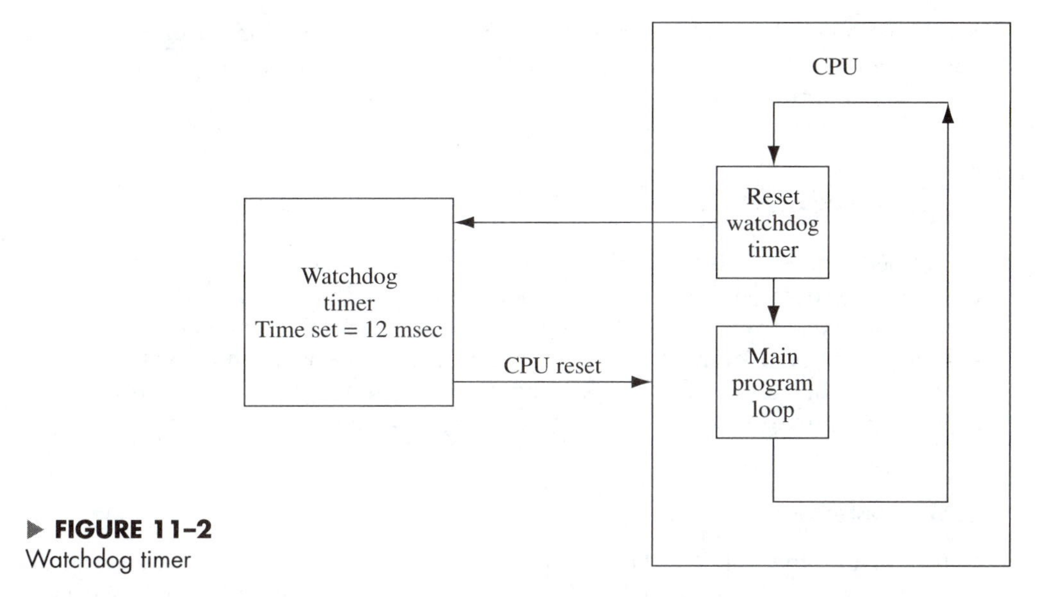

▶ **FIGURE 11–2**
Watchdog timer

starts in the main program loop. The timer is set to a time—in this case, let's say 12 ms. The timer counts down, starting at 12 ms, while the processor is performing its normal program loop operations. If the timer is allowed to count down to zero, this means that the CPU took much longer than its normal 10 ms to complete the main program loop and has likely been knocked out of its program loop. If the CPU completes its operations as usual, it resets the watchdog timer to 12 ms and the process is repeated. If the timer is allowed to count down to zero, its output resets the CPU to start over as if initially powered up (see Figure 11–2).

For this to be an acceptable solution, the process must be one in which the loss of one program loop operation is not critical to the system. It is also important that critical program and data memory locations are protected from corruption during the time that the CPU may be functioning outside its normal operating loop. If program memory locations are corrupted, they will have to be restored before proper operation can commence. Many systems that use watchdog timers use elaborate memory locks to prevent erroneously writing to program memory. As a last resort, correct program memory is verified and corrected after a watchdog timer-induced reset occurs.

Embedded systems can enjoy significant improvements in EMI noise immunity by using watchdog timers combined with methods that preclude writing to program and key data memory locations.

Ambient Temperature

Discussed in general in Chapter 2, ambient temperature is a serious consideration for embedded systems. Because embedded systems are inherently digital, variations of voltage levels with ambient temperature are not generally a concern. However, the large number of power-hungry components often included in embedded systems requires the maintenance of operating temperatures well below the absolute maximum levels for all devices. In order to accomplish this, elaborate heat sinks, cooling fans, package considerations, and judicious selection of components (speed vs. power) may all be necessary.

11–2 Central Processing Units

Selection of the CPU is a critical decision for any embedded system design. The CPU decodes instructions that are contained in memory, completes arithmetic and logic operations, transfers data to and from memory and I/O, and coordinates all system operations. There are many technical, manufacturing, and business issues that impact this decision. It is best to start with the technical issues, because if they are not met, the business issues become inconsequential.

CPU Architecture

Before discussing the different classes of CPUs, it is important to discuss some key CPU architecture variations. The first is the von-Neumann vs. the Harvard approach for transferring data between the CPU and program/data memory. The traditional method has been the von-Neumann approach, which provides one data bus for sending program and data information to and from the CPU. The Harvard method provides separate data busses for program and data memory CPU information transfers (see Figure 11–3). The von-Neumann approach provides much greater versatility and flexibility, but the fact that program and data memory transfers use the same data bus decreases CPU throughput for one operating cycle. Because of its separate program and data memory busses, the Harvard approach is set up to read the next instruction while it executes the current instruction.

Another key concept is called *pipelined data flow.* Non-pipelined data flow occurs when a CPU transfers data and only deals with one transfer at a time. Pipelined data flow breaks down the transfer process into steps that can occur at the same time. Therefore, pipelined data is transferred in concurrent steps that allow for data

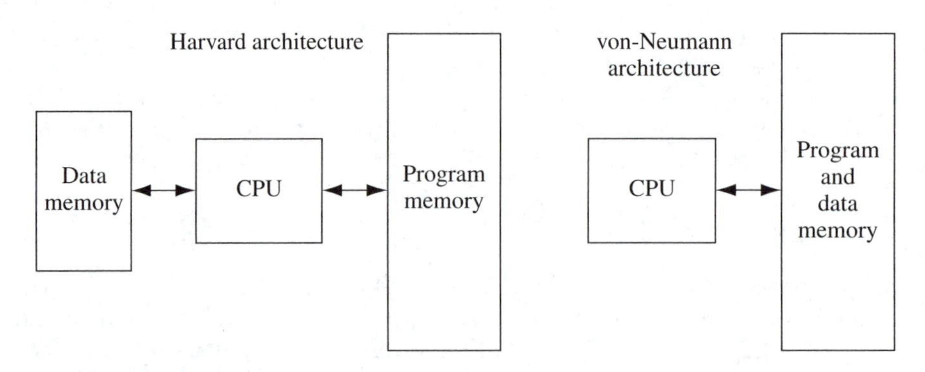

▲ FIGURE 11–3
Data memory transfers: von-Neumann vs. Harvard

to be sent and received at the same time. Pipelined data flow increases CPU throughput significantly but requires extra registers and limits flexibility.

There are two basic varieties of central processing units: the complex instruction set computer (CISC) or the reduced instruction set computer (RISC). These terms are general categories that are used to classify processor units. A CISC is a traditional CPU in which the instruction set is relatively extensive, variable, and can be located anywhere in memory. A CISC generally uses von-Neumann architecture and the data path flow is said to be non-pipelined. Most middle- to high-end CPUs are considered to be a CISC.

The RISC type of CPU features a reduced instruction set, Harvard architecture, and pipelined data flow, which all serve to increase throughput significantly for relatively simple and repeated operations. The reduced instruction set means that fewer bits are needed to decode the instruction. RISC CPUs often feature instructions that provide enough extra bits to include the address for a data transfer within the instruction itself. So not only is the instruction read from a separate data bus using a pipelined process (while other instructions are being executed), but all of the information about the transfer (the instruction plus the memory location of the data to be transferred) is included in the instruction itself.

RISC processors also feature a concept called *register file architecture.* Register file architecture allows for CPU registers to be addressed either directly (by their direct name or address location) or indirectly (as a relative address).

Defining CPU Requirements

The following questions help to define the requirements for the CPU and help to identify those CPUs that can potentially meet the application's requirements. A number of CPUs that meet these overall requirements should be identified. The final decision is made by selecting the CPU that best meets overall short and long-term objectives of the project or product.

1. Is the application best served with one CPU or can a number of distributed processors better meet the requirements? For applications in which data must be acquired, converted to digital, and processed in real time, it is often better to have a front-end processor that performs these tasks. This allows the main processor to handle the more central operations. Other examples are updating displays or printing data. These tasks are repetitive and time-consuming, and are often best left to a dedicated low-end processor.

2. What are the input and output data requirements? How many bits per data word? How many different types of data are there? How much calculation and sorting must be done to the data? How fast must the data be processed? After answering these questions, determine the ideal data bus size and clock speed for this application.

3. What is the complexity of the instruction set required to perform CPU operations? What is the estimated size of the application program? After answering these questions, determine the ideal address bus size for this application.

4. Review the instruction set for potential CPUs. Do the necessary instructions exist for the software development of the application? Are compilers and assemblers available for the CPU under consideration and the development software to be used? Is adequate documentation available?

5. Determine whether the application is a low-, medium-, or high-end application. There is no definite way to discriminate between these application levels other than to review the results of the previous questions combined with other facts about the application process. Low-end applications have less data and variability and therefore smaller programs that can be executed quickly, even with relatively slow clock speeds. High-end applications have a lot of data with much variety and complex long programs that require fast clock speeds to meet the speed requirements of the application. Medium applications fall somewhere in between. Later we will categorize some common CPUs into these same categories to aid in matching the processor to the application.

6. What type of memory will contain the program—RAM, ROM, EPROM, or EEROM? Will the program memory be maintained off-chip or must it be part of the CPU IC?

7. What type and how much memory will be required for read/write purposes? Will this memory be on- or off-chip?

8. What are the interrupt requirements of the application? Do the CPUs under consideration fulfill the basic requirements for interrupts?

9. What are the package requirements for this application? Is SMT or through-hole technology a requirement? Does the application require a microcomputer where all I/O, memory, and clock circuits are available on-chip or a microprocessor where they are provided off-chip?

10. Determine the power supply and environmental requirements for the CPU.

11. Determine the cost and size criteria for the CPU and select only those available CPUs that meet these criteria.

12. What is the degree of reliability for the CPU under consideration? Is it a tried and tested product or new and unproven? This is an important quality and reliability issue for the CPU.

13. Does the company have any experience or background with a CPU under consideration for this project? Identify this expertise and its relevance to the project in question.

14. What is the current lead-time for receipt of the CPU after it is ordered? What is the minimum order quantity? Is the CPU available from another source? The advantage of a second source is critical for the manufacturing and procurement departments. Single-source components must be managed much differently than those with two or more sources. Figure 11–4 shows a summary of the CPU requirements that must be identified to select the proper CPU.

Level of operation: Low-Medium-High

Variation of user program: Low-Medium-High

Program memory size and type

Data memory size and type

Quantity and type of I/O

Instruction set requirements: Simple-Complex

Package requirements

Power requirements

Environmental specifications

Reliability requirements

Cost/Leadtime/Minimum order quantity

▶ **FIGURE 11–4**
CPU requirements

Integrated-circuit central processing units can be segmented into four categories: 8-bit microcontrollers, 8-bit microprocessors, 16-bit microprocessors, and 32/64-bit microprocessors. Notice the distinction made between the term *microcontroller* and *microprocessor*. Microcontrollers are typically low-end microprocessors that include memory and I/O circuitry on-chip that make them fully functional without any peripheral ICs. Microprocessors are CPUs that require off-chip memory and peripheral ICs. Let's review the variety of microcontrollers and processors that are currently available.

8-bit Microcontrollers

PIC Microcontroller

Low-end processors are often called *microcontrollers*. The PIC microcontroller is manufactured and sold by Microchip Technology. PIC stands for *p*eripheral *i*nterface *c*ontroller. This name implies that its main purpose is to provide an interface with peripheral devices, so it is a prime candidate for performing this task in multiprocessor environments. The PIC microcontroller represents an entire product line of processors that are available with optional features in a variety of packages and memory types. The applications range from low end to medium levels. Options include serial ports, parallel ports, and analog-to-digital converters.

The PIC microcontroller features many aspects of the RISC processor structure that serve to provide high throughput for highly repetitious processes. These RISC oriented features are as follows:

- Harvard architecture
- Long, single-word instructions
- Single-cycle instructions
- Instruction pipelining
- Reduced instruction set
- Register file architecture

There are four distinct families of PIC microcontroller devices with distinguishing features:

PIC12CXXX: 12-bit instructions, 8-pin package, CMOS microcontroller, DC to 4 MHz operation for extremely low end applications

PIC16C5X: 12-bit instructions, CMOS low-end microcontroller, DC-20 MHz operation available with one-time programmable or ROM program memory

PIC14C000: 14-bit instructions, mixed signal microcontroller, DC to 20 MHz operation, one-time programmable program memory with built-in analog-to-digital and digital-to-analog converters; mid-range applications

PIC16CXXX: 14-bit instructions, CMOS mid-range microcontroller, DC to 20 MHz operation; variations include combinations of one-time programmable or ROM with optional analog and comparator functions

PIC17CXXX: 16-bit instructions, CMOS mid- to high-end microcontroller, DC to 25 MHz operation; variations include combinations of one-time programmable or ROM with optional mixed signal functions.

The discussion that follows focuses on the PIC16 and PIC 17 product lines. These are the mid-to-high range of the PIC product line and they are available with many options. The lower-level products merely represent a subset of the PIC16/17 product capabilities discussed here. The basic block diagram for the PIC16/17 products is shown in Figure 11–5.

Clock: The fastest clock speed for the PIC processor family is 25 MHz. With the reduced instruction set and other RISC features, the PIC executes most instructions in .2 μsec. The user determines the internally generated clock speed with the selection and connection of two capacitors and a crystal oscillator. If desired, the PIC can operate from an externally generated clock or a less-expensive RC network oscillator.

Program Memory: The program memory is contained internally on the PIC microcontroller. Up to 8K words are available. The size of the word varies with the PIC processor being used. The PIC 16 uses 14-bit instructions so its maximum program memory is 8K × 14 bits. The PIC 17 uses 16-bit instructions with a maximum program space of 8K × 16 bits. The program

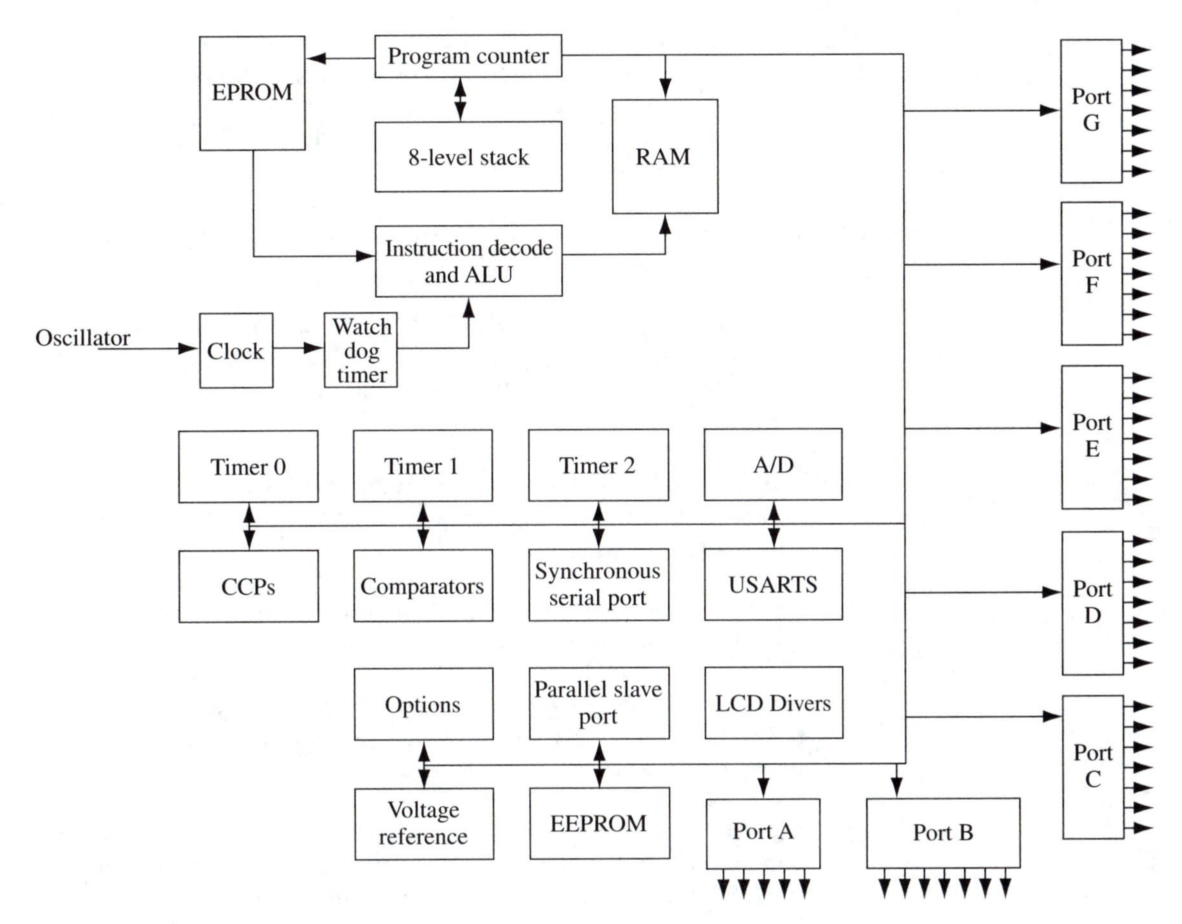

▲ **FIGURE 11–5**
PIC processor block diagram

memory can be one-time programmable (OTP) PROM, EPROM, or EEROM.

Instruction Set: The PIC processor uses a reduced instruction set that includes just 35 instructions, most of which operate in one machine cycle.

Data Memory: This is RAM that is available as a maximum of 256 bytes located on-chip. RAM is available in two banks, 0 or 1. Each bank includes 96 bytes of RAM plus 32, 8-bit special-function registers. The memory used for the stack to save addresses when calls are executed is separate and eight levels deep.

I/O Ports: The PIC features up to 56 I/O pins, most of which are individually configurable as inputs or outputs. The number of I/O pins available depends on the version PIC selected, including the package size. Some of the I/O pin functions are also multiplexed with special functions that are included on the chip. When these functions are being used, the I/O pins become dedicated to that function. These special functions include a compare/capture/pwm (CCP) module, a synchronous serial port (SSP), a universal serial asynchronous receiver transmitter (USART), and A/D and D/A converters.

Special Functions: The following describes the special functions optionally available on PIC microcontrollers:

Compare/Capture/Pulse Width Modulation Function (CCP): This is a variable-function block whose function is selected by bit values in the control registers. The CCP module can operate in one of the following ways:

Compare Mode: A 16-bit register associated with CCP function is compared to timer 1's register values. If they are equal, an output pin can be driven high or low.

Capture Mode: When an event triggers an input pin to the PIC, timer 1's value is captured and saved for further use.

Pulse Width Modulation (PWM): The CCP module operates to output a pulse on an output pin whose duty cycle and frequency depends on the value of registers in the PIC microcontroller.

Synchronous Serial Port (SSP): This port provides synchronous serial communications that can take the form of a serial peripheral interface (SPI) or to provide inter-IC communications (I^2C).

SPI: When operating in the SPI mode, the SSP allows 8 bits of data to be synchronously transmitted and received simultaneously. This is accomplished with the use of three I/O pins labeled serial data out (SDO), serial data in (SDI), and serial clock (SCK). This mode is ideal for communicating with one peripheral device.

I^2C: This is a standard two-wire serial communications protocol developed by Phillips Semiconductor that allows many serial IC devices to communicate with each other. I^2C can support communications ranging from 100 Kbps to 400 Kbps. Any device connected to the bus can initiate communications. The sending device becomes the master device and all listeners act as slave devices. Each device connected to the bus has a unique address. The PIC microcontroller implements the I^2C bus with two I/O pins labeled *SCL* for serial clock and *SDA* for serial data. These pins are multipurpose, open-drain outputs that are labeled RC3/SCK/SCL and RC4/SDI/SDA. This operating mode supports serial communications with many peripheral devices.

USART: The USART provides a second serial communications module that can operate either synchronously or asynchronously. Synchronous communications are half duplex while asynchronous communications are full duplex. The USART module utilizes two output I/O pins, transmit Tx and receive Rx.

Voltage Reference Module: This functional module provides a programmable voltage reference that is specified by values placed in the voltage reference control register. The output reference voltage connects to one I/O pin labeled V_{REF}.

Comparator Module: This module includes two analog comparators whose inputs can be connected to the analog input channels or V_{REF}. The comparator outputs can be connected to two separate I/O pins.

A/D Converter Module: Various PIC microcontrollers include A/D converters with either 8- or 10-bit capabilities. In each case the A/D converter can process any where from five to eight different analog input channels connected to the I/O pins. A variety of clock sources are possible. The internal voltage reference can be used to scale the analog signal range or an external reference can be utilized. The A/D module can be programmed to generate an interrupt when the A/D conversion process is completed.

LCD Driver: The PIC microcontroller can include an LCD driver module on certain versions of the product line (PIC16C9XX series) that is capable of driving up to 32 LCD segments.

Parallel Ports: Some of the PIC microcontrollers feature an 8-bit-wide Parallel Slave Port (PSP) that is multiplexed onto one of the device's normal I/O ports. Where available, this port is made functional by setting a control bit called *PSPMODE*. In this mode the PSP is asynchronously readable and writable from the external devices.

Watchdog Timer: PIC microcontrollers include an on-chip RC oscillator that can be used as a watchdog timer. A register is associated with the watchdog timer that serves to scale the timer setting. As discussed previously, watchdog timers provide a method of invoking a reset if the processor strays from the main program loop.

Timer 0: There are three timers available on PIC processors. Timer 0 is 8 bits wide and can be used to cause interrupts that are generated by dividing either the internal clock or an external clock connected to bit 4 of output port A. A prescaler register determines the amount to divide the input clock by. Timer 0 generates an interrupt when it overflows from FF Hex to 00 Hex.

Timer 1: This timer is a 16-bit timer/counter that is more versatile than Timer 0 or 2. It can be used as a timer in conjunction with the CCP module. When configured to work together, Timer 1 and the CCP module can drive an output pin at specific times without intervention by the processor.

This is ideal for applications such as starting A/D conversion at a precise time on a continuous basis. Timer 1 can also be used as an external event counter, both synchronously and asynchronously.

Timer 2: This is another 8-bit timer that is either used as the time base for the PWM function or as a separate 8-bit timer. Typical uses for Timer 2 are the generation of baud for serial communications. When used as a counter, Timer 2 can be set up to count the program loop time or other similar internal operations.

68HC11/12

The 68HC11/12 series is a good example of the next level up in processor complexity and capabilities. The 68HC11 was the first in this series of microprocessors and was introduced in 1985. The 68HC12 is the most recent version and was released for sale in 1997. The 68HC11/12 series fits the CISC processor model. It uses von-Neumman architecture, so program and data memory are accessed through the same data bus. Its instruction set is relatively large and the processing of instructions is not a pipelined process. The 68HC11 features approximately 145 instructions, while the 68HC12 offers an additional 40 instructions for an approximate total of 185.

All the 68HC11/12 products include a watchdog timer (WDT), a serial peripheral interface (SPI), an asynchronous serial communications interface (SCI), and a 16-bit timer. As with the PIC microcontroller, there are many varieties of the 68HC11/12 that include different amounts of memory and I/O as well as optional modules and features. Figure 11–6 shows some of these variations for the 68 HC11.

This processor can operate on instructions and data that are included in on-chip memory, or operations can be extended to multiplexed operations that provide access to external memory. This is called the *expanded mode*, which allows a selectable 8-bit or 16-bit data bus to be connected to ports C and D while a 16-bit address bus is provided through interconnections with ports A and B. Also, the 6 bits associated with port G can be used to extend memory addressing to a full 22 bits. Figure 11–7 shows the general block diagram for the 68HC12A4. The following describes the function of the various optional 68 HC11/12 modules:

Parallel I/O: For each of the parallel port I/O lines, there is a data register as well as a data direction register. The data direction register selects each

FIGURE 11–6 68HC11 microcontroller options

Part	ROM/EPROM	EEPROM	RAM	A/D	I/O Pins
68HC11A8	8 k	512	256	Yes	38
68HC711D3	4 k	0	192	No	32
68HC711E9	12 k	512	512	Yes	38
68HC711K4	24 k	640	768	Yes	62

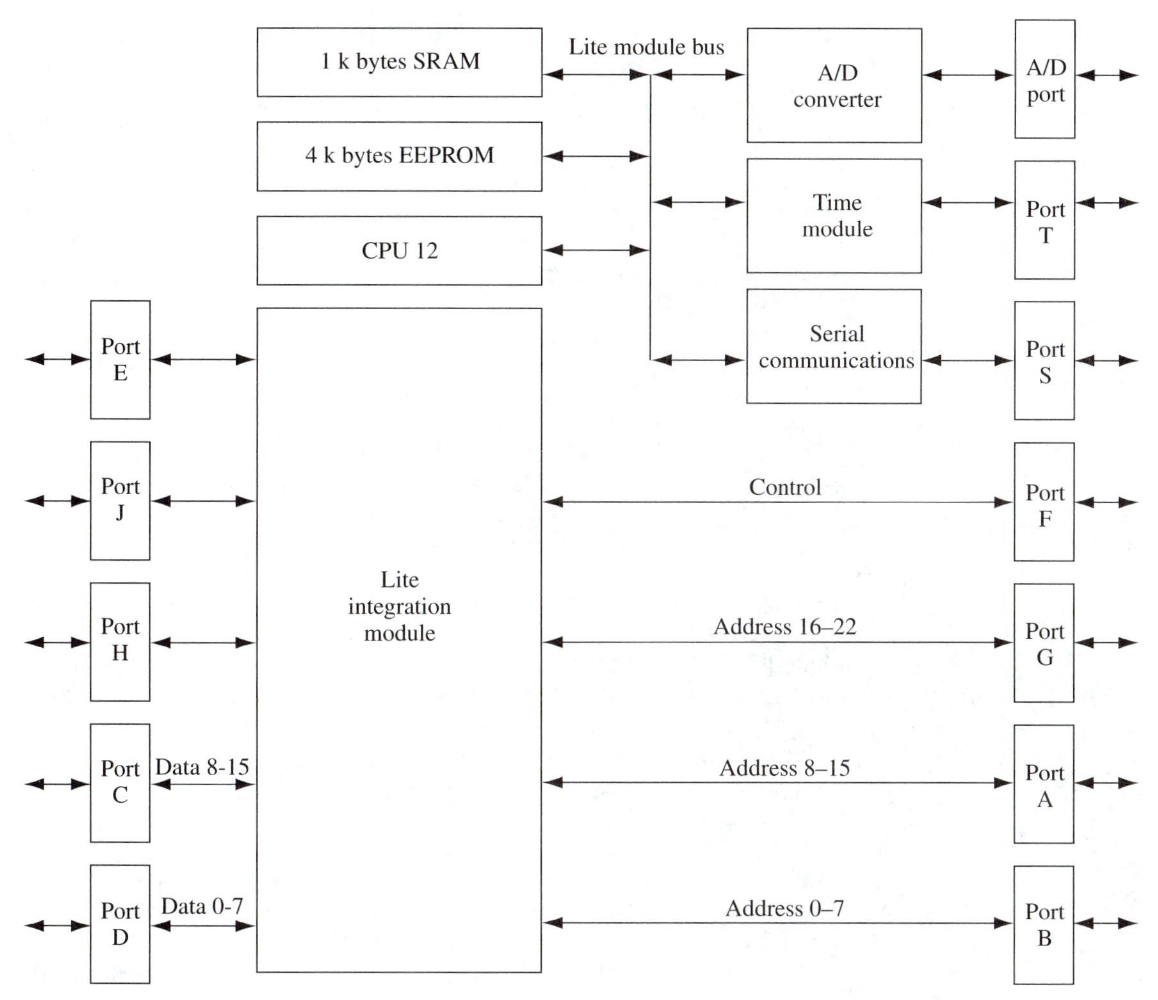

▲ **FIGURE 11–7**
68HC12 block diagram

individual I/O line as an input or output. When the output direction is selected, the data register contains information to be sent out. Otherwise the data register contains data that has been input. All of the ports are bidirectional except for the port for the A/D converter called *PORTAD*, which is input only. Many of the ports have alternate functions that they are intended to perform. For example, ports A and B are used as the data bus in the normal expanded mode, while ports C and D function as the address bus. If this mode is selected, these ports, which add up to 32 I/O lines, are not available as parallel I/O. Up to 85 I/O lines are available on the most complete versions of the product, with eight more lines available as analog inputs for the A/D module.

Serial Peripheral Interface (SPI): This is the synchronous serial interface for the 68HC11/12. It is used to communicate synchronously with other serial devices or processors through the serial data in (SDI) and the serial data out (SDO) pins. The serial clock I/O pin labeled *SCK* is the clock for the synchronous communications.

Serial Communication Interface (SCI): Some versions of the 68HC11/12 have two of these asynchronous serial communications ports (SCI0 and SCI1) while others have just one. These ports provide the proper start and stop bits and generate baud rate and parity information to support formal RS232 serial communications.

Analog to Digital Converter (ADC): This module is available on some versions of the 68HC11/12 series product line. The A/D module supports up to eight analog channels of 8-bit successive approximation type A/D conversion. On the 68HC11 product there are no interrupt capabilities, but the 68HC12 can generate an interrupt when the A/D conversion is complete.

Timer: The timer-generated features that are included in the 68HC11/12 family are all based on the operation of one free-running 16-bit up counter labeled *TCNT*. The timer operates in conjunction with a fairly complicated timer system that connects to port T. The timer module utilizes eight 16-bit capture/compare registers to initiate interrupts when the TCNT counter value is equal to the register value. The time for certain outside events can also be determined by capturing the event with port T's I/O pins.

The PIC and 68HC11/12 microcontrollers were covered with some detail because they represent two distinctively different and popular microcontroller operational philosophies. There are many other 8-bit microcontroller devices available, such as Intel's MCS51 family, the Zilog Z8 microcontroller, and the Atmel AVR series.

8-bit Microprocessors

Intel Corporation developed the first 8-bit microprocessor, the 8008. In 1974 it was replaced by the Intel 8080, which represented a significant improvement in performance. About the same time, Motorola introduced its 6800 8-bit microprocessor, Fairchild produced the F8, and Advanced Micro-Devices released the 6502. A few years afterward, the newly founded Zilog Corporation released its first product, called the Z80. For many years these five 8-bit microprocessors were the workhorses behind many computer and embedded system products and systems.

The 8-bit microcontroller has taken over many of the applications of 8-bit microprocessors while high-end requirements have pushed microprocessor development up to 16-, 32-, and now 64-bit systems. Once the only microprocessor product available, now the 8-bit microprocessor market is one with few players that covers the middle ground between 8-bit microcontrollers and higher bit processors.

Z80 Microprocessor

The Z80 was developed and released in the late 1970s. It was developed by Zilog Corporation, a new venture company at the time. The Z80 utilized the same basic instruction set as the Intel 8080, with some additional powerful instructions and features. The Z80 could be used for existing 8080 applications and offered many programming improvements. After the Z80 was released, Zilog developed many peripheral ICs that supported the Z80 and increased its popularity even further. Consequently, the Z80 is one of just a few still available from the era when the 8-bit microprocessor was king.

The Z80 is available in 40-pin DIP or 44-pin PLCC packages. It features an architecture that was common at the time of its introduction. An external 16-bit address bus, 8-bit data bus, and numerous control bus signals. Dependent on the particular model, the operating clock frequencies range anywhere from 2.5 to 8 MHz. The input/output diagram for the Z80 is shown in Figure 11–8.

The Z80 includes six general-purpose registers in addition to its accumulator and flag registers. There is a mirror set of all eight registers (general purpose registers plus accumulator and flags) that cannot be directly accessed by the programmer but can serve as an easy way to save the contents of registers during the execution of interrupt service routines or other subroutine calls. The instruction set includes 158 different types of instructions that are categorized into the following groups: data transfer, logic, arithmetic, bit manipulation, and branch and control operations.

In order to develop a functional embedded system using the Z80, external program and data memory, and serial and parallel I/O may be needed, as well as data conversion or counter/timer peripherals. The Z80 appears as a very simple device today when compared to the microcontrollers and high-end microprocessors that have been developed since its initial release.

FIGURE 11–8
Z80 I/O diagram

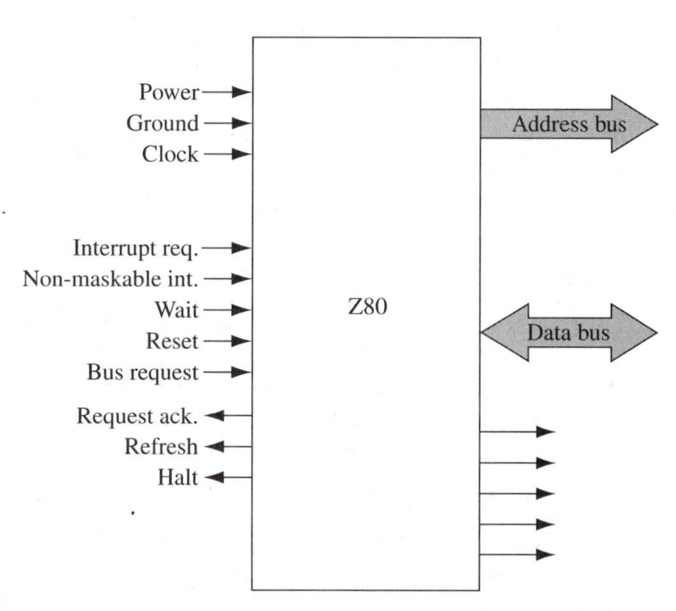

Zilog has enhanced the Z80 with numerous product enhancements such as the Z180. The Z180 is an 8-bit microprocessor built on Z80 architecture with an advanced processor core that improves performance by 33%. The Z180 also includes on-chip two asynchronous serial ports and an ISA bus interface.

Rabbit 2000

Another 8-bit microprocessor is the Rabbit 2000, which was developed by Rabbit Semiconductor Corporation. The Rabbit 2000 is a modern 8-bit microprocessor that possesses a unique combination of features. The Rabbit 2000 was developed from the basic Z80/Z180 architecture with improvements that offer higher performance. It possesses all of the capabilities that would classify it a microcontroller except for on-chip memory. It includes on-chip parallel and serial I/O as well as timers. The Rabbit 2000 is packaged in a 100-pin rectangular PQFP surface-mount package. Following are most of the Rabbit 2000's features:

- Clock speeds up to 32 MHz
- 40 pins for parallel I/O
- Four asynchronous serial I/O ports, two of which can function as synchronous serial ports
- Two sets of timers: one that includes five 8-bit reloadable down counters and a second group that has a 10-bit free running counter with two 10-bit match registers. These timers are typically used for providing baud rate clocks for the serial ports and for generating interrupts.
- Slave port that allows the Rabbit 2000 to be used as an intelligent peripheral device
- 20-bit address bus that can access up to 1 Mb of memory
- Enhanced Z180 instruction set
- Battery-backed-up internal real-time clock
- Watchdog timer
- Unique sleep mode with a current draw of about 200 µA
- Software developed using Dynamic C development software

Figure 11–9 shows the block diagram for the Rabbit 2000 microprocessor.

16/32/64-bit Microprocessors

Just as integrated circuit designers were continuing to find ways to pack more circuitry and features into smaller microcontrollers, they applied these same innovations to high-end processors as the race to increase processor speed and throughput was just beginning. The result has been the continuous development and application of the following strategies:

1. Increase circuit densities by increasing process resolution and exploring unique geometries.
2. Increase power efficiency.

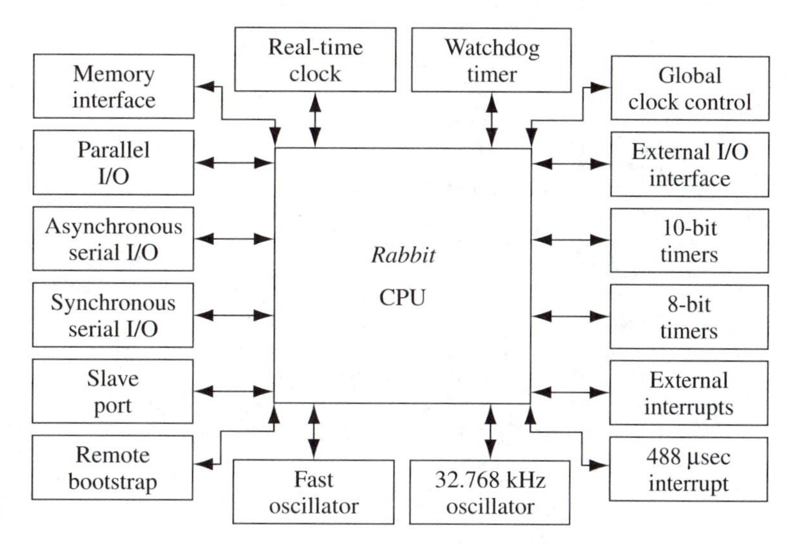

FIGURE 11–9
Rabbit 2000 microprocessor

3. Increase clock speed by further minimization of inherent capacitance levels and signal lead lengths.

4. Increase throughput with the use of pipelined processes and increasing the size of parallel data busses.

These strategies have led to the ongoing release of improved high-performance 16/32/64-bit processors, starting with the 8086 and culminating most recently with the Intel Pentium IV. Intel released the 8086 in 1978 as a significant improvement in performance over the microprocessors available at the time. It could address 1 Mb of memory and its data bus and internal registers were all 16 bits wide. The 8086 also featured pipelined processing. The ongoing improvements that began with the 8086 have led to 32- and then 64-bit data busses, 36-bit address busses (can address 64 Gb of memory) and clock speeds up to 2 GHz. Package sizes have now expanded to 478 pins. The Pentium IV represents the latest in microprocessor development with 64-bit data, 36-bit address, a 2 GHz clock, and 478 electrical connections. It requires a large heat sink attached to the chip to dissipate power. Figure 11–10 shows the general trend in microprocessor advances over the last 25 years. Most of these processors are aimed at the personal computer market or for other high-end computer applications.

11–3 Memory

Semiconductor memories have undergone developments similar and as significant as those employed in microprocessor technology. The densities of memory cells have increased substantially and power efficiency has also been improved. Pipelined processes are used to increase effective access times and wider data

Intel microprocessor developments

Processor	CPU clock	Internal register size	Data I/O bus width	Memory address width	Maximum memory	Number of transistors
8088	100M	16-bit	8-bit	20-bit	1M	29,000
8086	100M	16-bit	16-bit	20-bit	1M	29,000
286	100M	16-bit	16-bit	24-bit	16M	134,000
386SX	100M	32-bit	16-bit	24-bit	16M	275,000
486SX	100M	32-bit	32-bit	32-bit	4G	1,185,000
486DX2	200M	32-bit	32-bit	32-bit	4G	1,100,000
486DX4	300M	32-bit	32-bit	32-bit	4G	1,600,000
Pentium	300M	32-bit	64-bit	36-bit	64G	5,500,000
Pentium II	500M	32-bit	64-bit	36-bit	64G	7,500,000
Pentium III	1G	32-bit	64-bit	36-bit	64G	28,000,000
Pentium IV	2G	32-bit	64-bit	36-bit	64G	42,000,000

▲ **FIGURE 11–10**
Microprocessor development

words can be stored in one location. Semiconductor memories are categorized as either volatile (data is lost after the power is removed) or nonvolatile (data is retained when the device is not powered).

Volatile Memory

Volatile semiconductor memory is usually called *random access memory* (*RAM*). The term *random access* actually means that each location is accessible with the same access time. This name came about because older volatile memory was called either *RAM* or *sequential access memory* (*SAM*). Sequential access memory is an older style of memory in which the access time for each location depended on its location in the memory. All currently available volatile semiconductor memory is RAM, although the distinction *random access* is no longer important. Volatile memory is sometimes called *read-write memory* (*RWM*), which would be a better name for it today.

There are two distinct categories of RAM: static RAM (SRAM) or dynamic RAM (DRAM). Static RAM retains its data as long as power is applied. Dynamic RAM retains data while power is applied if each location is refreshed every few milliseconds. This refresh is necessary because the charge that is stored in the dynamic memory cells leaks out and must be refreshed.

DRAMs

DRAMs are MOS-type devices that provide high chip densities and power efficiency with average access times. This means that more memory cells can be located on a single DRAM memory device than any other type of memory. Information that is stored in a DRAM is actually stored in a small MOS capacitor. The charge can leak out of the capacitor over a small amount of time so the DRAM must be refreshed every few milliseconds. This refresh requirement complicates the use of DRAMs because circuitry must be included on the DRAM chip itself to implement refresh, or it must be done with external circuitry. In spite of the refresh requirement, DRAMs are still the most popular choice where large amounts of memory must be provided in a small space with less power and the lowest cost. DRAMs typically have four times the density of SRAMs. Consequently, DRAMs are most often used as the main memory in personal computers.

DRAMs contain such large capacities that the number of address lines required to address them would increase the package size to an unpractical size. To resolve this, DRAMs share or multiplex their address lines. Usually, each address line is multiplexed into two different address bits and the data for each is sent and latched into the DRAM during different parts of memory transfer cycle.

Because the inherent advantage of the DRAM is capacity, power efficiency, and low cost, the challenge for DRAM manufacturers is to develop DRAMs that can keep up with the ever-increasing clock speeds offered by computer manufacturers. Many innovations are being developed that help to solve this problem:

Fast Page Mode DRAMs (FPM DRAMs): These provide faster access to data that is on the same page.

Synchronous DRAMs (SDRAMs): SDRAMs were developed in the early 1990s. These memories offer a simpler synchronous interface between other system components and an internal process that is similar to the pipelined approach discussed for processors. SDRAMs internally have multiple banks of DRAMs that can perform operations simultaneously. All the address and control inputs are sampled synchronously on the same rising edge of the clock input.

Double Data Rate SDRAMs (DDR SDRAMs): This is an improved SDRAM that performs memory read or write operations on both the rising and falling edges of the synchronous clock.

SRAMs

Static RAMs are read/write volatile memory devices that retain data as long as power is applied. SRAMs can be fabricated with bipolar, MOS, or BiCMOS processes and the access speeds of the SRAM are directly related to the type of process used. Fast SRAMs are often used in personal computers to support operations that require very fast access time. The SRAMs used in these applications are called *cache memory*.

The older standard type of SRAMs are now referred to as *asynchronous SRAMs* because SRAM manufacturers have now developed a new type of synchronous SRAM (SSRAMs). The SSRAM is similar to what was done to speed up SDRAMs discussed previously. SSRAMs use a synchronous clock to set up the memory read/write operations into a pipelined process in which many operations are performed at the same time.

Nonvolatile Memory

Most embedded systems require a large amount of memory that is maintained when the power is removed. Even a personal computer maintains its BIOS program, with instructions of what to do when the power is first turned on, in nonvolatile memory. The program memory for embedded systems is usually maintained in ROM, PROM, EPROM, EEPROM, or Flash ROM. In addition, embedded systems usually employ many user-selected parameters that must be maintained when the power is shut off. These parameters must be retained in some type of nonvolatile memory.

Read Only Memory (ROM): This type of memory is programmed by the IC manufacturer per the user's program. The user pays a masking charge of a few thousand dollars to set up the proprietary program. ROMs are much cheaper in volume than other nonvolatile memories but the data cannot be changed once programmed. Masking charges preclude their use in low volume or with new programs that expect some change.

Programmable Read Only Memories (PROMS): These are a type of ROM in which the user can program the PROM using a device called a *PROM programmer*. The PROM programmer blows fuseable links inside the PROM to store the information permanently. Once programmed, the PROM functions just as a ROM; it retains the data under all conditions and it cannot be modified. PROMs provide low-volume users with an affordable alternative to the ROM because there is no masking charge. However, PROMs are significantly more expensive than ROMs. Another name for a PROM is "one time programmable" or OTP.

Eraseable Programmable Read Only Memories (EPROMs): similar to PROMs, EPROMs are programmed by the user with an EPROM programmer. The EPROM programmer functions much differently than the PROM programmer. An EPROM is programmed by the application of a voltage to each cell where a binary 1 is to be stored. The charge used to program the cell remains after the voltage is removed because there is no path for discharge. The EPROM includes a window that is positioned over the semiconductor chip. An EPROM can be erased by exposing it to ultra violet light. The ultraviolet light causes current to flow through a photo transistor, allowing the programming charge to dissipate, erasing all the cells in the EPROM. Because the EPROM can be erased, it provides a significant advantage over the PROM. However, in order to erase an EPROM, it must be removed from the circuit. EPROMs also cost more than ROMs or PROMs.

Electrically Eraseable Read Only Memories (EEROMs): EEROMs have the same basic structure as an EPROM except that they have a small oxide layer above the drain of the MOSFET memory cell. When 21 V are applied between the MOSFET's gate and drain, a charge flows into the gate that can program the cell. This charge is retained even after power is removed from the device. The voltage applied to a cell can be reversed to erase the cell. EEROMs offer ease in programming and erasing. Both can be accomplished in-circuit and only the cell to be modified need be erased and reprogrammed. EEROMs offer significant advantages over all other types of nonvolatile memories. However, they are much more expensive than other nonvolatile memory types and memory density is also much lower.

Flash Memory: Flash memory is a newer style of EEROM that provides erasure of large sections of memory or the entire memory at one time. True EEROMs can program and erase a particular word or byte in the memory. Flash memory can program individual words but must erase large blocks of memory at one time. While flash memory is not as versatile as EEROMs, there is a benefit to the bulk erasure requirement of flash memory. In the EEPROM, two MOS transistors are required to maintain the data and erase each memory cell. Flash memories only require one transistor; a second transistor is used to erase many cells. Consequently, flash memories can achieve a much higher memory density when compared to EEROMs and they are available at a much lower cost.

There are two basic varieties of flash memory devices: NOR and NAND architectures. The NOR-type flash memory is named because it resembles NOR logic. NOR-type flash memories provide fast read times and slow write/erase cycles. NAND-type flash memories, whose architecture resembles the NAND gate, exhibit the opposite features—slow read times with fast write/erase cycles.

Direct Memory Access (DMA)

In many embedded systems it is necessary to transmit large amounts of data from peripheral devices to memory and vice versa. In a basic microprocessor-based system, this data must be read in through the processor. To accomplish this, the processor must fetch an instruction to read in the data, fetch an instruction on where to send it, and then fetch another instruction to repeat the process for the next location. For large amounts of contiguous data, this represents a significant amount of processing time. If the task of moving the data is repetitious, or if the amount of data is large and located together in memory, then the application of a DMA controller may offer significant savings in processing time.

The DMA concept, as the name implies, is to give direct access to memory locations without having to pass data through the processor. A DMA controller IC is a specialized chip that is designed to accomplish this task. A DMA controller connects to the system control, address and data busses. Additional inputs to the DMA are the DMA Request and Hold Acknowledge; the outputs are DMA Acknowledge and Hold Request. Their function and interconnection is as follows:

DMA Request Input: Connects to external I/O devices and is used to request DMA transfers

Hold Request Output: Connects to the system processor and requests permission from the processor to use the system busses. The Hold Request line must remain high for the entire time that the DMA is using the system bus.

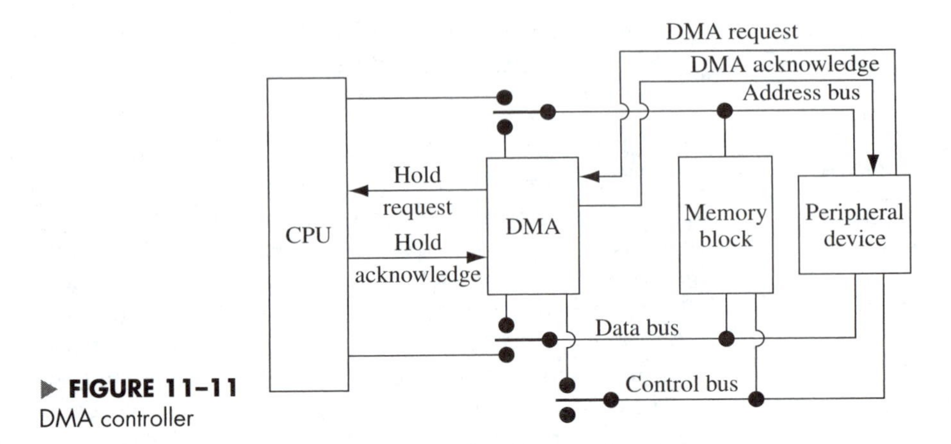

FIGURE 11–11
DMA controller

Hold Acknowledge Input: Connects to the system processor line that is used to grant permission to the DMA to take over the system bus. When the processor is ready to relinquish control of the bus, it drives this pin high.

DMA Acknowledge Output: Connects to the peripheral device that requested the DMA transfer. When the DMA takes over control of the system bus, the DMA signals the peripheral to begin transmission by driving this pin high.

A DMA chip can have multiple channels for controlling the transfer of data. It will have a DMA Request and DMA Acknowledge pin for each channel, but only one Hold Request output and Hold Acknowledge input (see Figure 11–11).

When a DMA transfer is to be performed, the device that initiates the transfer activates the DMA Request input on the DMA chip. The DMA follows by forcing its Hold Request line high. The processor receives the Hold Request and, after finishing the current instruction, drives the Hold acknowledge high, indicating to the DMA that it can proceed. The DMA follows by driving the DMA Acknowledge line high, which signals the peripheral device to begin the transfer.

Each DMA channel is programmed by the processor to set up its data transfer task before the DMA Request is made. The initial data required is twofold: the starting memory location and the number of locations to be transferred. Each DMA channel also has a control word location written to by the processor that dictates the type of operation to be performed and the address of the device to which or from which the data is being transferred. DMA operations can consist of transfers from I/O to memory, memory to I/O, or memory to memory.

The 8237 is a specific DMA controller that was used in the early personal computers. It includes four DMA channels and can transfer up to 64k bytes of data. The 40-pin DIP version of the 8237 is shown in Figure 11–12. Note that there are only eight address lines available on the 8237. These eight lines will support only 256 memory transfer operations. The additional eight bits required to support 64k data transfers are multiplexed through the data bus. The ADSTB line is used to indicate that the high byte of the address bus is present at the data bus pins. An external chip must be provided that will latch the high order address byte. A 74LS373 is often used for this purpose because of its tri-state outputs.

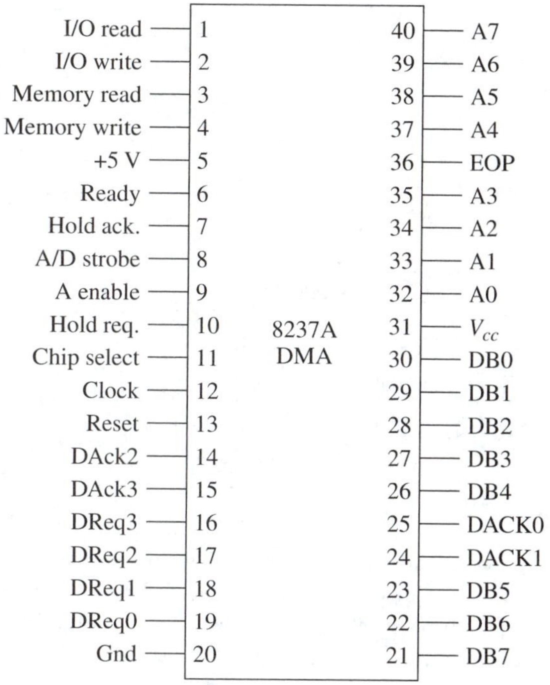

Signal	Pin		Pin	Signal
I/O read	1		40	A7
I/O write	2		39	A6
Memory read	3		38	A5
Memory write	4		37	A4
+5 V	5		36	EOP
Ready	6		35	A3
Hold ack.	7		34	A2
A/D strobe	8		33	A1
A enable	9		32	A0
Hold req.	10	8237A	31	V_{cc}
Chip select	11	DMA	30	DB0
Clock	12		29	DB1
Reset	13		28	DB2
DAck2	14		27	DB3
DAck3	15		26	DB4
DReq3	16		25	DACK0
DReq2	17		24	DACK1
DReq1	18		23	DB5
DReq0	19		22	DB6
Gnd	20		21	DB7

FIGURE 11–12
8237 DMA

11–4 Serial Communications

Many of the microcontrollers discussed in Section 11–2 included serial communications ports on-chip. Most microprocessors do not include serial communications ports on-chip because of all the circuitry required to process massive amounts of data quickly. Many serial communications ports are available as stand-alone ICs that interface with the microprocessor as an I/O port and handle all serial communications. This saves space on the processor chip and gives the user the flexibility to decide which and if any serial communications peripherals are needed.

All serial communications require a method for synchronizing the transmission and receipt of the information. There are two types of serial communications: synchronous and asynchronous. Synchronous communication requires a separate hardware clock line to synchronize communications between the transmitter and receiver. Asynchronous serial communications embed the clock within the data being sent.

Synchronous I/O ports are often called a *serial-peripheral-interface (SPI)*, as seen with many of the microcontrollers discussed in Section 11–2. Asynchronous serial ports are usually called *universal-asynchronous-receiver-transmitters* (*UARTs*). Most serial communications ICs are completely programmable and can function either as synchronous or asynchronous serial peripherals. These are often called *universal-serial-asynchronous-receiver-transmitters* (*USARTs*).

The Zilog Z85C30 is a prime example of a modern serial communications peripheral often called a *serial communications controller* (*SCC*). It is fabricated with low-power CMOS technology and features two independent full-duplex

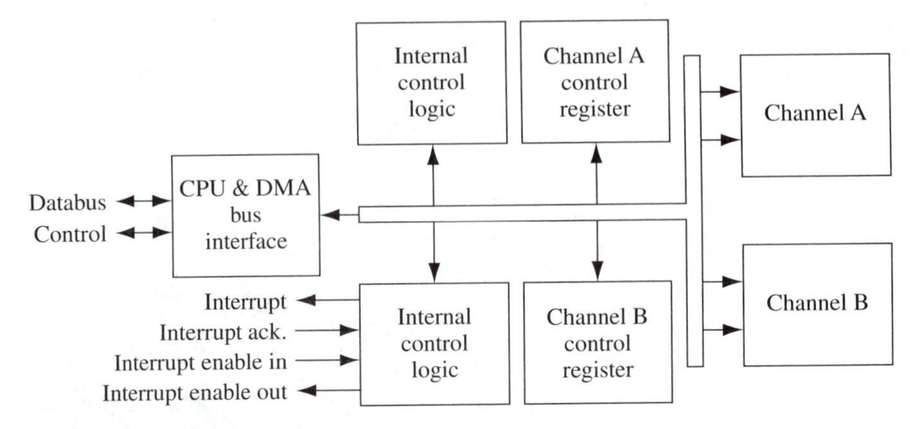

▲ **FIGURE 11–13**
SCC functional block diagram

(can send and receive at the same time) serial ports. These ports can be programmed to operate synchronously or asynchronously at transmission rates of up to 4 Mbits/Sec. They can encode data in NRZ, NRZI or FM formats and also provide for error checking. Each of the ports included in the Z85C30 can connect to separate crystals to generate the baud rate. The block diagram for the Z85C30 is shown in Figure 11–13.

11–5 ▶ Parallel I/O and Special Functions

Programmable parallel I/O devices are also available for connection to high-end microprocessors or to expand microcontroller parallel I/O capabilities. The 8255 IC is one of the oldest and still popular devices used for this purpose. The 8255 connects to microprocessor data and control busses. The 8255 can be connected to function as memory-mapped I/O (I/O that is located at unused memory addresses) or as peripheral I/O (I/O addresses). It includes three 8-bit parallel I/O ports. Port A can be programmed so that all 8-bits are inputs, outputs, or bi-directional. Port B can function with all 8-bits as inputs or outputs. The bi-directional capability is not provided for ports B or C. Port C is the same as port B except that it can be broken down into two 4-bit nibbles, and each nibble can be used as an input or an output.

The 8255 is connected to the processor data and control busses. The software programs ports A, B, and C in the 8255 to function as required by the program. The processor then writes the appropriate data to the ports, which present the data to their output pins. Any 8255 pins that are selected as inputs receive data that is input to them. The 8255 signals the microprocessor via the interrupt request line that it has data to be read. The interrupt priority status of this particular I/O will determine how quickly the processor will respond to answer this request. Eventually the processor will read the data that the 8255 is holding in one of its ports. The block diagram for the 8255 is shown in Figure 11–14.

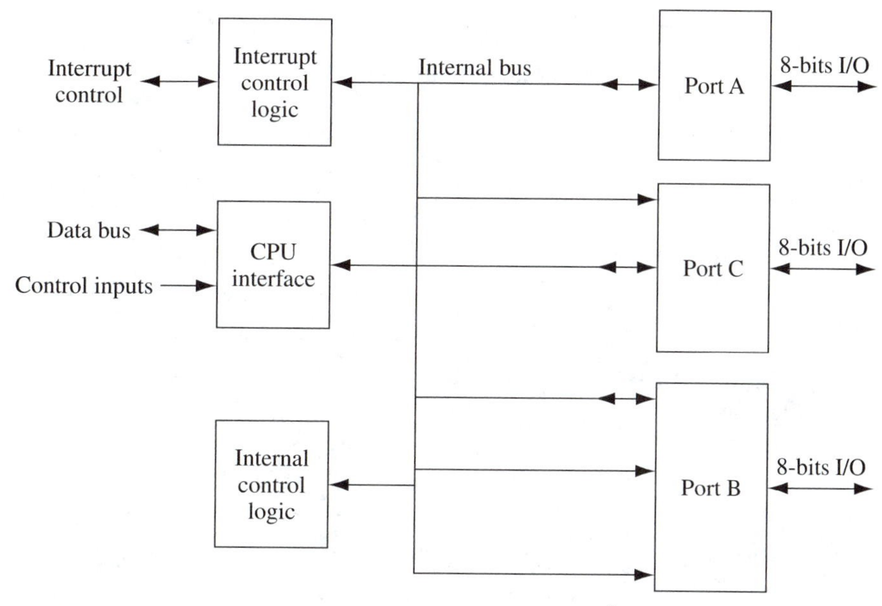

▲ **FIGURE 11–14**
8255 block diagram

Counter-Timer Circuits

Many of the microcontrollers discussed in Section 11–2 also included counter-timer modules on-chip. Counter-timers are useful in creating software time delays and in generating baud rates for serial communications. These devices are also available as standalone ICs for connection to microprocessors.

Counter-timers are simply counters that divide down an input signal by a scaling factor. The range of the scaling factor depends on the size of the counter. The 8-bit counters can divide the input frequencies over a range of 1 to 256, while 16-bit counters have a scaling range of 1 to 65,536. The output of the counter-timer usually connects to some output device and supplies an output frequency as determined by the input frequency divided by the scaling factor programmed into the counter-timer IC.

The Intel 8253 is a counter-timer IC that was utilized on the initial IBM PC computers. It includes three 16-bit counter-timer functional blocks that must be programmed separately. The block diagram for the 8253 is shown in Figure 11–15. Each counter in the 8253 has an input clock, a gate, and an output. Once programmed and selected, each counter divides down the input clock signal by the scaling factor programmed into the counter-timer scaling register. The gate signal is used as an external hardware inhibit for any one of the three counter-timers.

The Zilog Z8536 CIO, called a *counter-timer and parallel I/O unit,* is a newer device that combines the functions of the counter-timer and parallel I/O all on one IC. The Z8536 includes two independent 8-bit, bi-directional I/O ports, a 4-bit special purpose I/O port and three independent 16-bit counter-timer modules. The block diagram for the Z8536 CIO is shown in Figure 11–16.

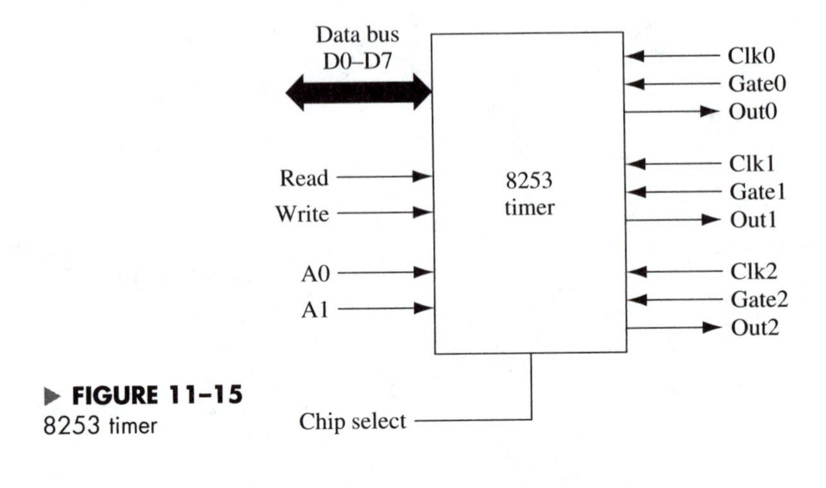

▶ **FIGURE 11–15**
8253 timer

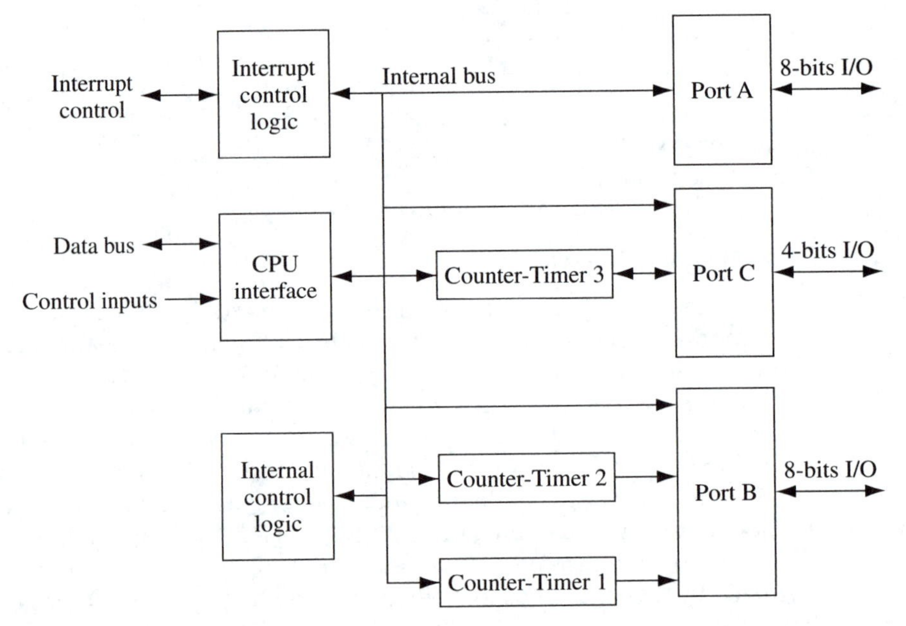

▲ **FIGURE 11–16**
Z8536 CIO block diagram

As the diagram shows, Ports B and C can be used to provide external access to the counter-timer modules. Port A does not interface with the counter-timers. Ports A and B can be configured to function together as one 16-bit I/O port. Each of the three ports includes pattern recognition logic that can be used to generate an interrupt when a certain pattern is input to the port. The three 16-bit counter-timers are all identical down counters. As many as four I/O lines can be dedicated to each of the counter-timers to provide an input, gate, trigger, and output. Each counter-timer can output either pulse, one-shot, or square-wave outputs. The Z8536 is available in a 40-pin DIP package or a 44-pin PLCC package.

Real-time Clock

In many applications it is desirable to maintain and record real time on an ongoing basis. Real-time clock ICs have been developed for use with microprocessors or microcontrollers that do not include this function on-chip. Most real-time clocks are stand-alone devices that require only DC power and an external crystal to operate. If the real-time clock is to operate when the rest of the system is powered down, a battery backup must also be connected. Real-time clocks maintain the date and time for all personal computers as well as a host of other embedded systems.

Real-time clocks like the 3285 RTC shown in Figure 11–17 simply count and maintain a calendar and time based upon a crystal input. The 3285 utilizes a 32.768 kHz oscillator. One cycle of this clock represents 30 μsec. Real-time clocks connect to the microprocessor data and address bus so that the microprocessor can program and read the current time value. The real-time clock can also act as an alarm clock for the processor by generating an interrupt to initiate real-time operations. The calendars for most real-time clocks are set up for a hundred years and feature automatic correction for leap year and daylight savings time changes. The 3285 RTC also provides a programmable square-wave output that can be used as a synchronous clock source for external devices.

11–6 ▶ Signal Conversion

Signal conversion circuits are needed for any embedded system in which there are analog inputs or outputs. Many of the microcontrollers discussed in Section 11–2 include analog-to-digital (A/D) converters on-chip that convert analog voltages to digital signals as inputs to the microcontroller. Digital-to-analog (D/A) converters convert digital signals back to analog signals to drive devices that require analog voltages. Much development effort has been applied to the development of faster conversion devices with higher levels of resolution.

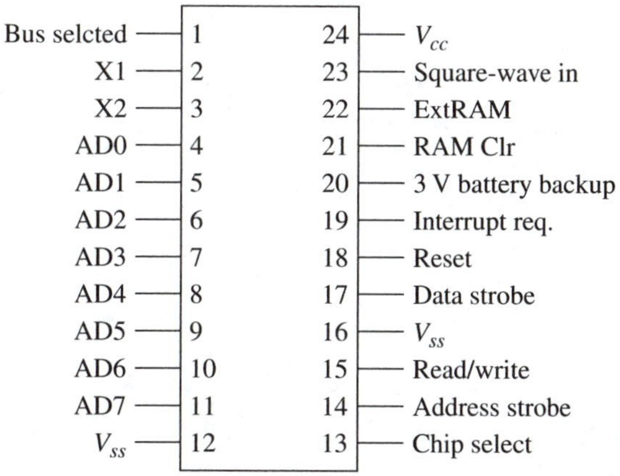

▶ FIGURE 11–17
3285 real-time clock

Signal	Pin	Pin	Signal
Bus selcted	1	24	V_{cc}
X1	2	23	Square-wave in
X2	3	22	ExtRAM
AD0	4	21	RAM Clr
AD1	5	20	3 V battery backup
AD2	6	19	Interrupt req.
AD3	7	18	Reset
AD4	8	17	Data strobe
AD5	9	16	V_{ss}
AD6	10	15	Read/write
AD7	11	14	Address strobe
V_{ss}	12	13	Chip select

A/D Converters

A/D converters are used whenever an analog signal must be converted to a digital signal for processing by some digital system. The key functional requirements for any A/D converter are its resolution, conversion time, and voltage range. Other important details are the power supply requirements, power efficiency, package size, and cost.

There are many different types of A/D converters but they all essentially perform the same basic operations: the input signal passes through a low pass filter; and a sample and hold circuit holds the sampled value, which is then converted to digital bits that represent the analog voltage sampled. The low pass filter is designed to filter out any signal that is higher than the highest frequency expected as an input signal. Sample and hold circuits serve to take a snapshot of the input signal and hold it constant while the conversion process takes place. The conversion process is where the many different types of converters are created. The most common types of A/D converters are successive approximation, integrating, and flash converters.

Successive Approximation A/D Converters

Successive approximation converters test each bit of a digital signal pattern, starting with the most significant bit down to the least significant bit, in order to develop the digital number that represents the input analog signal. The block diagram for an 8-bit successive approximation A/D is shown in Figure 11–18. This figure shows the control logic, D/A converter, and comparator that are the key component in the successive approximation converter. A sample conversion is requested by setting an input sample request signal to the A/D converter. The input is sampled, held, and converted to a representative digital signal. The A/D converter first sets the MSB of the digital word—in this case, D7. The D/A converter converts this digital number (10000000) to analog and it is compared to the sampled analog value being converted. If the sampled input signal is greater than the D/A output, then D7 will remain a 1; otherwise, it becomes a 0. Next the A/D control logic loads the result, D7 = 0 or D7 = 1, and sets D6 = 1. The process is repeated using the comparator to indicate whether the D/A output is greater than or equal to the sampled input signal. The converter continues through all 8-bits, testing the resulting D/A output

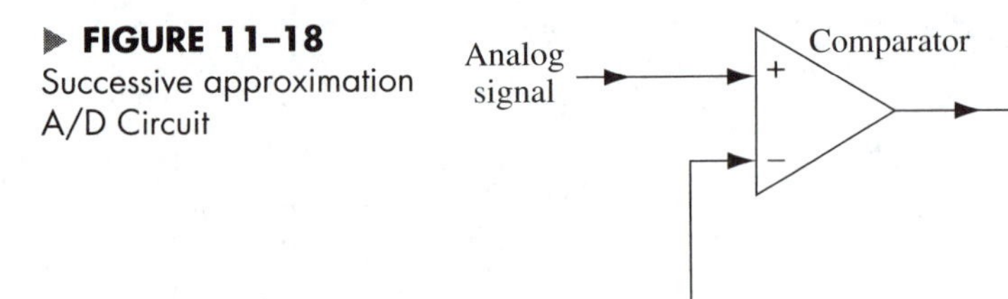

FIGURE 11–18 Successive approximation A/D Circuit

while compiling the digital word that most closely represents the amplitude of the sampled input. The number of comparisons to be made is determined by the number of bits included in the output digital word. In this case the 8-bit word will require that eight comparisons be performed.

Successive approximation converters are reasonably fast, require relatively few components, and are very versatile. Their conversion time and resolution are directly related to the number of bits, which determines the number of comparisons that must be made. Obviously, the conversion time increases with the number of bits that are output because the number of comparisons increases.

The ADC0841 is an example of a successive approximation converter that is designed to interface easily with a microprocessor. The chip pin out and block diagram are shown in Figure 11–19. The digital outputs of the ADC0841 are tristate so they can connect directly to the processor's data bus. The processor can initiate a conversion by toggling both the chip select (CS) and read (WR) lines. When the converter is complete it signals the processor by pulling the interrupt request line (INTR) low. The processor can then read the digital outputs by pulling the read (RD) line low. The analog input connects to the V_{IN+} and V_{IN-} connections and V_{REF} determines the range of the analog signal.

Integrating A/D Converters

Integrating converters are useful for conversion of lower frequency or DC types of signals. This is because they suffer from very slow conversion times, which means that they will not be able to accurately represent a higher-frequency waveform. Consequently, this type of converter is used extensively in digital multimeters and in instrumentation circuits for slowly changing signals, such as temperature. There are some variations in the types of integrating converters. Single- and dual-slope are the two most popular.

The block diagram for the single-slope converter is shown in Figure 11–20. These converters begin conversion by integrating a voltage reference signal with

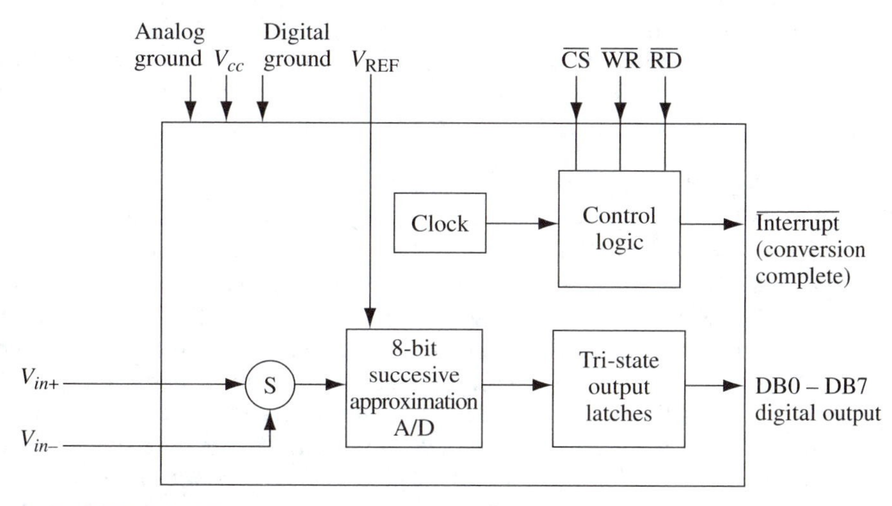

▲ FIGURE 11–19
ADC0841 A/D converter

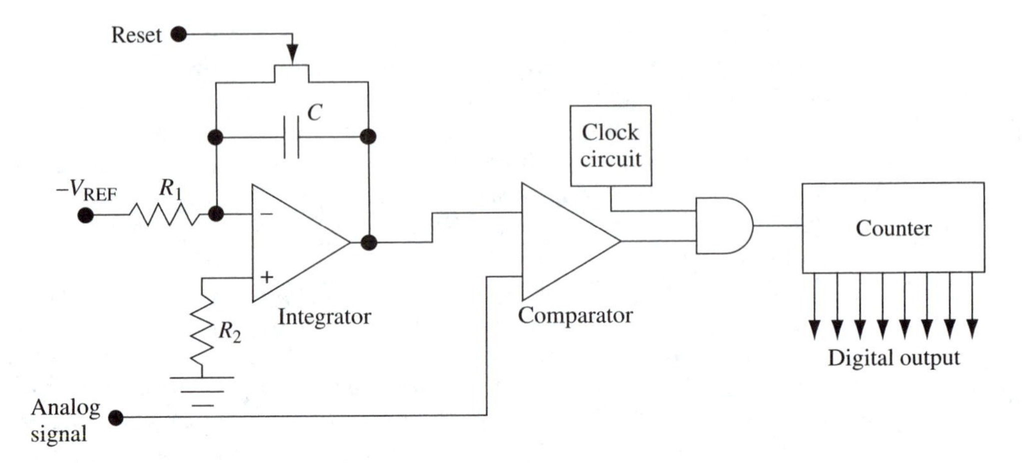

▲ **FIGURE 11–20**
Single-slope A/D converter

an op amp integrator circuit. At the same time, a clock is started that counts while the integration proceeds. The output of the integrator is compared to the analog signal being converted. When the integrator output equals the input signal, the comparator switches, stopping the counter and the conversion process. The counter value is the digital representation of the analog input. The single-slope converter is sensitive to changes in the reference voltage, the integrating capacitor, and the clock circuit.

A dual-slope integrator block diagram is shown in Figure 11–21. It improves on the performance of the single-slope converter significantly. The dual-slope converter initiates conversion by starting a counter and integrating the analog signal for a period of time such that the counter counts through all states. When this happens, the integrator is connected to the reference signal, and the counter starts again until the integrator output goes back to 0 V. At this point the comparator switches, signaling the end of conversion. The end result of this process is that the digital value is proportional to the two cycles of integration, and it is independent of the RC values of the circuit as well as clock variations.

Flash A/D Converters

Flash A/D converters are the fastest A/D converters available. The block diagram for a Flash converter is shown in Figure 11–22. In a flash converter the input analog signal is connected to comparators that represent each combination of digital outputs. For an 8-bit flash converter, there are 255 (2^{n-1}) comparators. The reference inputs to these comparators are the analog voltages that are coincident with the combinations of digital outputs. The internal analog reference voltages are created by dividing down the reference voltage with 255 equal resistor values. The flash converter is fast because all the comparisons take place in parallel at the same time. The comparator outputs are fed to a decoder that interprets them and outputs the appropriate digital word. Flash converters approach conversion times of 20 ns and are practically limited to 8 bits.

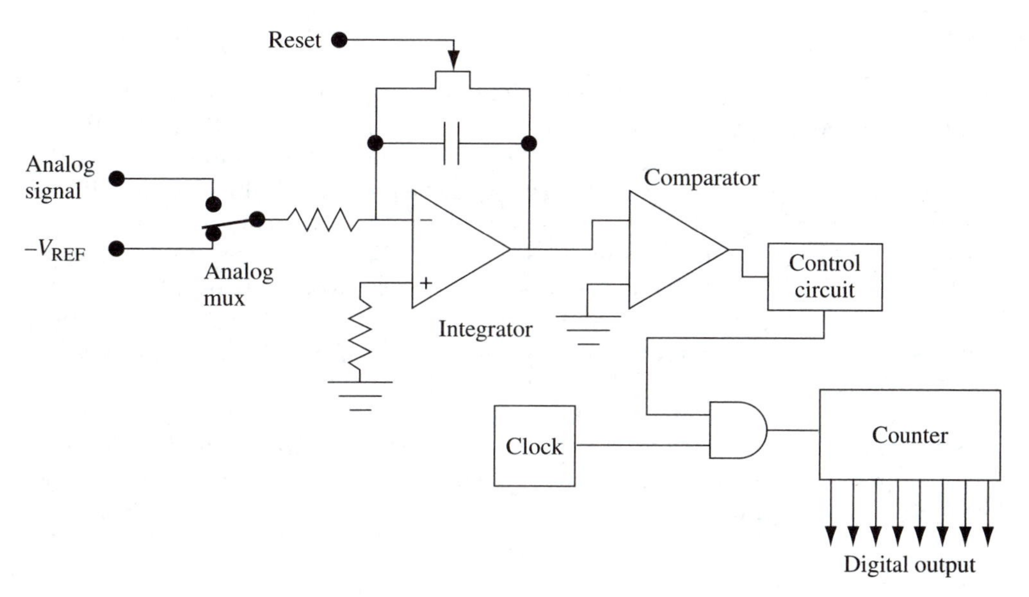

▲ **FIGURE 11–21**
Dual-slope A/D converter

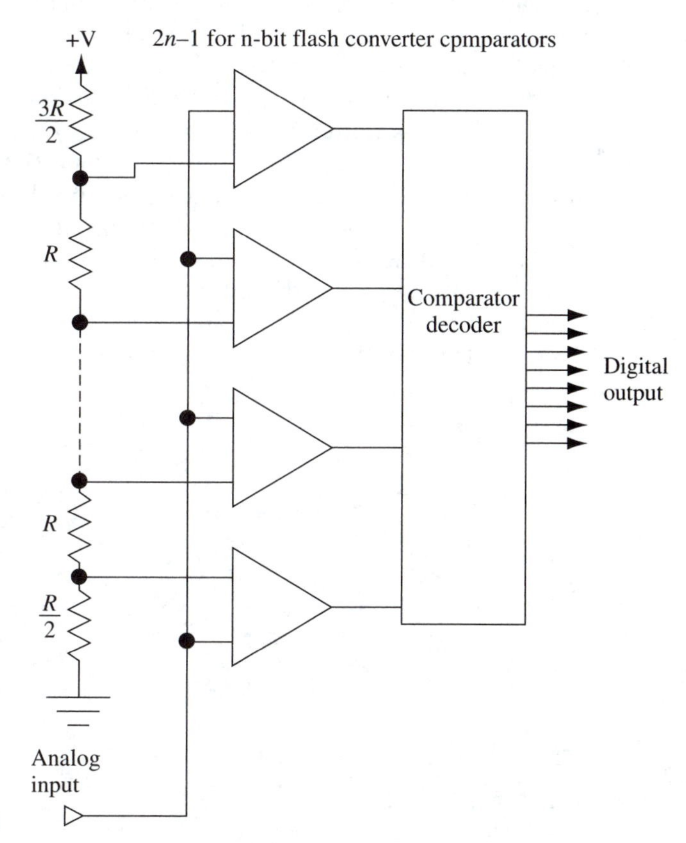

▶ **FIGURE 11–22**
Flash A/D converter block diagram

D/A converters

D/A converters take an input digital number and output the corresponding analog output voltage. The most popular type of D/A converter is called the R-2R D/A circuit. Figure 11–23 shows the functional block diagram for the R-2R circuit. The name of the circuit comes from the use of just two resistor values, R and 2R. The result is that the digital bits that connect the various 2R resistor segments are weighted according to their labeled bit position. The op amp output becomes the net sum of all these weights and represents the desired analog output.

The accuracy of the D/A converter is tied to the weighting resistors tolerances and the op amp and the reference voltage variations. The speed of the D/A is determined by the speed of the analog switches that switch in the resistors and the overall op amp frequency response. The performance of a D/A is typically noted on the data sheet by the settling time required after conversion and the slew rate of the op amp. The block diagram and pin out for the ever popular DAC08 D/A converter is shown in Figure 11–24.

Summary

Think about all of the appliances and devices that you use every day. Begin with the thermostat that automatically warms the house before you get out of bed and the alarm clock that stirs you from your sleep. Perhaps you listen to an MP3 player while you take a morning walk. The microwave oven, electric range, television, satellite TV, or cable selector are all devices you will likely use before heading out for the day. Then perhaps you will drive your car while listening to the car radio or use your cell phone. While at work in an office, you use a personal computer, advanced calculators, fax machines, copy machines, and possibly computer projection devices. Working in more specialized fields almost always involves the use of some specialized equipment or devices. While you are at work, your VCR records your favorite television program for later viewing and your answering machine records messages from important callers. If you need cash for shopping, you access an ATM; then when you spend that cash, the product is scanned with a bar

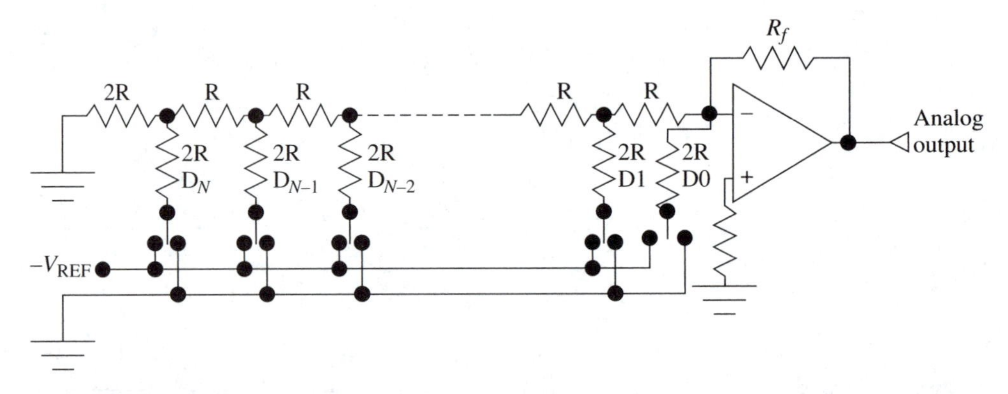

▲ **FIGURE 11–23**
R-2R D/A converter block diagram

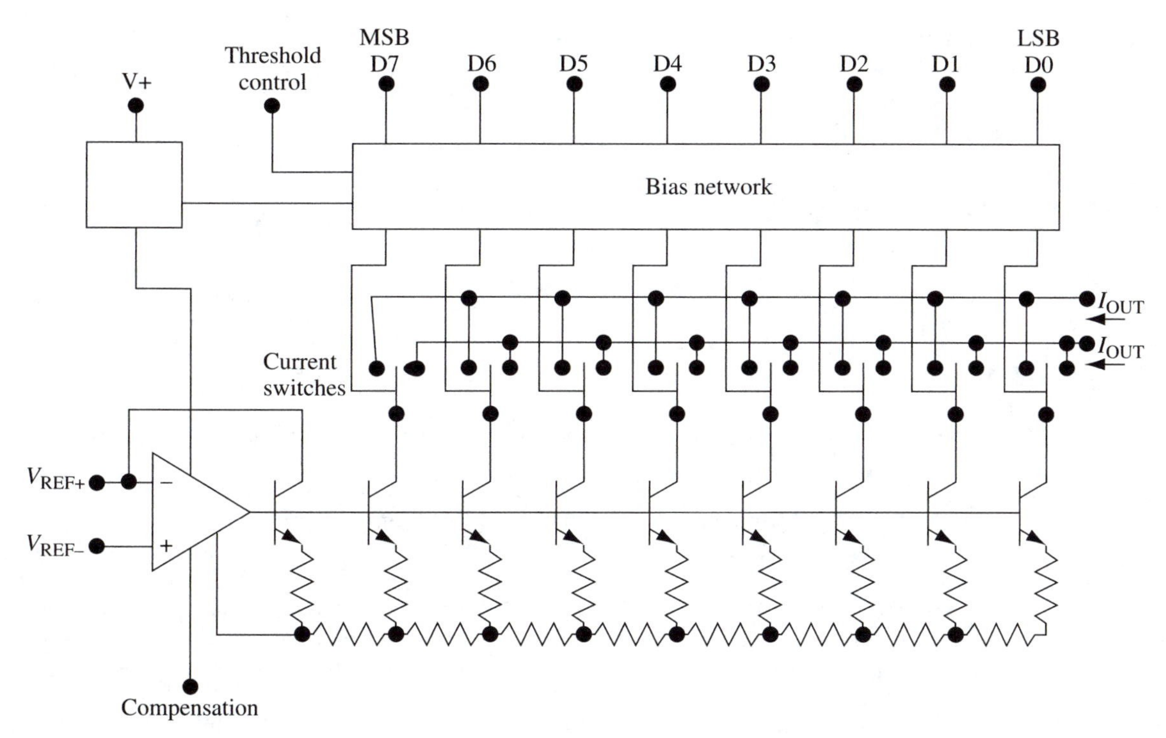

▲ **FIGURE 11–24**
DAC08 Digital-to-analog converter block diagram

code reader, processed on a terminal, and printed on a printer. These devices are just some of the examples of embedded systems. A microcontroller or microprocessor is inside controlling the device as it has been programmed to operate.

Embedded systems products must be both reliable and easy to use. Accomplishing this goal while maintaining a competitive price is a goal that's accomplished only with thorough planning, a high level of innovation, attention to detail, and teamwork. The raised bar of performance for embedded systems demands that they are easily set up and used. They must be impervious to electrical noise while not radiating excess EMI. They must withstand vibration and temperature variations while using little power and operating at high speeds—all of this while the average product development cycle is made shorter and shorter. Embedded system design usually requires an extended team of specialized hardware and software professionals. These goals can only be achieved by managing the project on a concurrent engineering basis.

References

Gaonkar, R. S. *The Z80 Microprocessor.* Upper Saddle River, NJ: Prentice Hall, 2001.

Haskell, R. E. *Design of Embedded Systems Using the 68HC12/11 Microcontrollers.* Upper Saddle River, NJ: Prentice Hall, 2000.

Mazidi, M. A., Mazidi, J. *The 80X86 IBM PC and Compatible Computers: Design and Interfacing of IBM PC; PS and Compatible Computers.* Upper Saddle River, NJ: Prentice Hall, 2000.

Peatman, J. B. *Design with PIC Microcontrollers.* Upper Saddle River, NJ: Prentice Hall, 1997.

Stanley, W. D. *Operational Amplifiers with Linear Integrated Circuits.* Upper Saddle River, NJ: Prentice Hall, 1994.

Exercises

11–1 List the key design decisions to be made before starting the design of an embedded system.

11–2 What additional steps can and should be taken to promote noise immunity in embedded systems?

11–3 What two primary processor architectures provide uniquely different schemes for transferring program and data information from memory to the processor? Describe each of these architectures and highlight the difference between them.

11–4 List the primary characteristics of an RISC processor system.

11–5 List the primary characteristics of a CISC processor system.

11–6 Explain what is meant by the term *pipelined processing.* What is the net effect of pipelined processing on system performance?

11–7 List three examples of microcontrollers.

11–8 List the major categories of volatile memory.

11–9 List the key differences between DRAMs and SRAMs.

11–10 List the various types of improved DRAMs and discuss the advantage of each.

11–11 Explain the meaning of the term *cache RAM.* What type of RAM is usually used as cache RAM in a computer system?

11–12 What enhancements are used in the development of memories called *Static Synchronous RAMs* (*SSRAMs*) and what are the benefits?

11–13 List the major categories of nonvolatile memory.

11–14 Explain the primary differences between EEROMs and Flash memory.

11–15 List the two different types of Flash memory and the characteristics of each.

11–16 What is the purpose of a DMA controller? Explain the basic steps of a DMA data transfer.

11–17 List and explain the two major categories of serial communications ports.

11–18 Describe the basic functions included on a counter-timer IC and list some typical applications.

11–19 List the three different types of A/D converters and explain the operation of each.

11–20 What is the fastest type of A/D converter available and what characteristics are the basis for its speed?

12 Telecommunications and Fiber Optics

Introduction

In the early stages of the electronics revolution, people became aware of new discoveries by their application in audio and video products, transistor radios, and solid-state television. After the development of the microprocessor, the application of state-of-the-art electronics to all types of products became more visible. All along, some advances in electronics had been applied to telecommunications products. However, the monopolistic environment that existed in the industry at that time kept many innovations in the laboratory. During this time, the telecommunications industry enjoyed stable growth. This would all change as the breakup of AT&T became law and many new market opportunities emerged. The telecommunications revolution is the result of the following contributing factors:

- The breakup of AT&T and the technical ramifications of the legal directive
- The development of the microprocessor and its application in many telecommunications products
- The development of a computer network that provided for communications between various research universities and defense programs. This network evolved into the Internet.
- The rise of many other communications protocols, which improved the efficient use of the existing telecommunications infrastructure
- The development of the personal computer and the ensuing information society
- The development of fiber optics as a new vehicle for transmitting telecommunications information
- The development of a variety of wireless communications technologies

These developments positioned many companies to compete in this rapidly growing and profitable business. The capabilities made possible by the technology created new demand that provided paths of opportunity. In the background, the Internet and the protocol TCP/IP were becoming a reality. During the last two decades, the telecommunications industry expanded at such a rate it was considered, on its own, a revolution.

Today, high-speed computerized routers, bridges, and switches all send packets of data to their assigned destinations through the quickest path available, either as pulses of light or electricity. In this chapter we will look at modern telecommunications systems and how new electronics and fiber-optic technology have been applied to make it all work. The specific topics of this chapter are as follows:

- Defining Telecommunications
- Telecommunications System Design Considerations
- Modulation
- The Public Switched Telephone Network
- Modems
- Error Detection and Correction
- Protocols
- Computer Networks
- The Internet
- Fiber Optics in Telecommunication

12–1 ▶ Defining Telecommunications

Telecommunications can be defined simply as the communication of information from one location to another. Typically, the two locations are some distance apart. Initially, the information to be transmitted is electrical in nature and its form is either analog or digital. The most common methods for transmitting this information are as electrical signals transmitted on a wire cable, light information sent down an optical fiber, or wireless radio signals transmitted through the atmosphere and space. This broad definition of telecommunications includes radio transmissions, television, telephone lines, and computer networks.

Because the two locations communicating on a telecommunication network are some distance apart, it is always desirable for the communication medium to require as few connections as possible. This is because each connection requires more electrical or fiber cable or radio frequencies to complete the connection. This also requires the information to be sent in serial verses parallel form. To make the system even more efficient, the ability to share the transmitting medium with many users is a significant advantage. A basic telecommunications system is shown in Figure 12–1. It includes inputting information, encoding the information, and transmitting the encoded signal down the communication medium. On the other end the receiver accepts the information, decodes it, and converts it back to its original form. The signal is encoded so that many users can transmit information

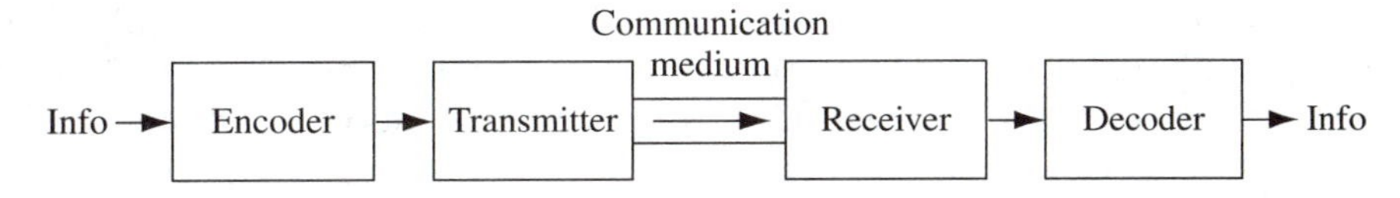

▲ **FIGURE 12–1**
Telecommunications system block diagram

over the communicating medium at the same time. Encoding provides a way of identifying the signal so that it can be decoded on the other end.

Typical residential telephone system communications today consist of landline telephone, microwave, or cellular telephone communications. Television signals are transmitted either as local wireless signals, via cable TV, or as directed satellite communications. Audio radio information is transmitted either as a local wireless signal or on cable TV mediums. Computer communications can be performed through the telephone system by using a modem, through dedicated local area networks (LANs) or metropolitan area networks, (MANs) or the Internet.

The performance of any method of telecommunication is always based upon the amount of information that can be sent; how far it can travel; the cost of installation, maintenance, and transmission; and the quality of the information that is received. The requirements for quality largely depend on the type of information that is being transmitted. Voice telecommunications are the least demanding, while audio, video, multimedia and computer data require higher levels of accuracy and quality.

12–2 ▶ Telecommunications System Design Considerations

The design considerations for telecommunication systems are closely tied to the performance measures already mentioned in the last section—how much data can travel how far, and at what cost and quality level. The basic types of transmission mediums to be compared are conductive electrical cable, fiber-optic cable, and wireless communication.

Bandwidth

The various methods of information encoding are discussed in the next section. However, each encoding scheme uses a carrier frequency of some type. In each case either an analog or digital communication channel is encoded onto a carrier frequency. The next channel is encoded on the next available carrier frequency. The frequency bandwidth required for a communication channel depends on the type of signal being transmitted by that channel. Digital signals require significantly more bandwidth than analog signals. The amount of information that can be carried over the communications medium is determined by its overall frequency bandwidth. The bandwidth capability of the medium determines how many communication channels can be transmitted at the same time. In telephony, this converts to the number of voice channels that can be transmitted at the same time, which equates to the number of phone calls that can take place simultaneously on the same phone line. The greater the bandwidth of the medium, the more data can be transmitted simultaneously over the medium.

Each communication medium has a bandwidth limitation just as any electronic circuit. The physical basis for this limitation differs for the medium in question. For electrically conductive communication cable, the sources of these limitations are usually linked to the inductive and capacitive properties that are inherently part of the conductor. Conductive communication cables offer the lowest bandwidth because of their inductive and capacitive properties. Fiber-optic cable and wireless communications offer much higher bandwidths because neither have limiting factors as significant as the affect of inductance and capacitance on electrical signals.

Attenuation

Attenuation is a measure of the loss of signal that occurs as the signal passes through a component. In telecommunications, attenuation measures the signal loss over a length of the communications medium. Attenuation experienced on a transmission line is usually measured in dB per unit length. Again, the physical basis for attenuation differs significantly for the communication mediums. In electrically conductive cable, attenuation results from the resistance and inductance of the cable. The higher the signal frequency, the greater the loss due to the inductive reactance (X_L) of the cable ($X_L = 2\pi fL$). Wireless communications offer the next highest attenuation while fiber-optic cable offers the lowest. Attenuation is important because it determines how far a signal can be transmitted down a certain medium before regeneration of the signal is required.

Quality

The perceived quality level of a transmitted signal depends greatly on the transmission mode (analog or digital) and the type of signal being transmitted (voice, audio, video, data). Voice telephone communications have the lowest quality requirements while computer data has the highest. In voice communications the quality level can be determined by the following criteria:

1. Amplitude level: Can the voice be heard?
2. Clarity: Is the voice clear and distinct without the presence of background noise?
3. Recognition: Can the voice be recognized as that of the caller?

Audio and video signal quality is usually a function of transmission method. Electromagnetically transmitted radio and television signals are always subject to noise and interference. Routine audio and video signals transmitted digitally are usually subject to the same quality requirements as voice communications.

Critical computer communications have only one measure of signal quality: they require that the data be received accurately. Because perfect transmission and reception cannot be guaranteed, error detection and correction circuits and algorithms are required for important computer data communications.

Cost

The cost of sending information includes the cost of installing telecommunication facilities and the cost of operation and maintenance. This varies greatly for the differing telecommunication methods. Fiber-optic cable requires more elaborate equipment to complete splices when compared to electrical wire. Expensive and more intricate laser light sources are used as fiber-optic transmitters in many telecommunications applications. However, fiber optics offers much less signal attenuation, so the need for amplifiers or repeaters to rejuvenate the signal is much less than the quantity required for electrical signals. Also, the increased bandwidth of fiber optics means that each fiber can carry more information. The least costly method is selected to resolve the telecommunication problem. Figure 12–2 summarizes the design considerations for telecommunication system design.

12–3 ▶ Modulation

Encoding is defined as the process of placing information on what is called a *carrier.* The carrier for this book is the paper on which it is printed. The encoding format is the English language printed on the pages. When you speak, the carrier is the air around you. Your vocal cords modulate a signal onto the air, which carries the signal to whom you are speaking. Your listener's ear decodes the signal and hopefully understands your message. In telecommunications, the carrier is an electromagnetic waveform. The electromagnetic waveform takes the form of either

Bandwidth: How much data can be transmitted?

Attenuation: How far can the data be transmitted without rejuvination?

Quality: How accurate is the received signal when compared to the transmitted signal?

Cost: What are the installation, operation, and maintenance costs?

▲ **FIGURE 12–2**
Telecommunications design considerations

an electrical waveform transmitted down a wire, electromagnetic radio waves, or as light passing through fiber-optic cable.

The purpose of encoding is to identify the signal uniquely, as it is transmitted through a medium where many other signals travel. The best example of this is radio signals transmitted from a radio station. The radio broadcast is encoded on the frequency assigned to the station, either as an amplitude modulation (AM) or frequency modulation (FM) encoded signal. The desired station is selected by tuning in the carrier signal frequency of the station on your radio dial. Your radio receiver blocks out all other frequencies and decodes the signal from the carrier signal.

Amplitude Modulation

Amplitude modulation and frequency modulation are two common ways to encode analog signals. Amplitude modulation simply changes the amplitude of the carrier frequency as the amplitude of the signal waveform changes. Amplitude modulation is usually used to encode signals within the audio frequency range (20–20 kHz). The carrier frequencies utilized for AM encoding are in the middle of the radio frequency ranges, typically 300 kHz to 3 MHz. The modulation index, m_a, is the relative magnitude of the peak voltage of the modulated signal, V_m compared to the peak carrier signal amplitude, V_c. Figure 12–3 shows an example of an input signal that is amplitude modulated onto a carrier signal.

$$m_a = V_m/V_c \tag{12–1}$$

AM transmissions are relatively simple to complete, but they suffer from sensitivity to noise. Noise is easily picked up and added to the AM transmitted signal, but it is difficult to detect and remove. If the amplitude of the noise is large enough and it is within the frequency band of the receiver, it has to be considered as part of the modulated signal. AM also suffers from power inefficiency. This is because the information being transmitted is actually contained in the portion of the signal above and below the carrier signal amplitude. These are called the *upper* and *lower sidebands*. The power used to transmit all of the carrier frequency is wasted because it contains no information.

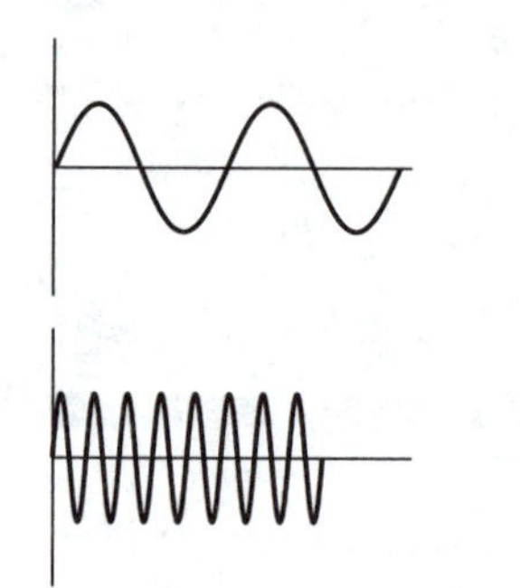
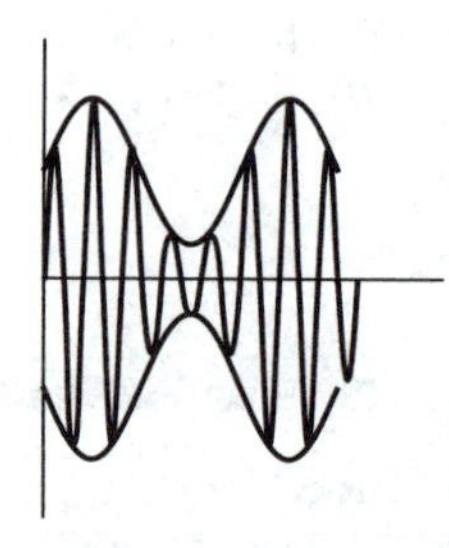

FIGURE 12–3
Amplitude modulation

Frequency Modulation

Frequency modulation offers a significant improvement over both of the inherent weaknesses experienced with the AM system. FM systems encode the signal waveform onto a carrier signal by varying the frequency of the carrier waveform as the amplitude of the modulating signal. The difference of the carrier frequency from the unmodulated value is proportional to the amplitude of the modulated signal. The largest frequency variation induced by the amplitude of the modulated signal, δ determines the modulation index, m_F for the FM system. The modulation index m_F is given by the formula:

$$m_F = \delta / f_m \quad (12\text{–}2)$$

where f_m = the frequency of the modulated signal
and δ = frequency deviation/amplitude

Following are the key performance differences between AM and FM:

1. Noise usually couples onto the modulated carrier waveform by increasing its amplitude. Because the amplitude of the carrier signal does not contain information about the input signal, noise is easily detected and rejected.
2. The encoded information is included as part of the primary carrier waveform; therefore, the power used to transmit the combined FM signal is used more efficiently.
3. The frequency band needed for FM is obviously greater than AM because the signal inherently varies around the center carrier frequency.

The commercial FM broadcast band is from 88 to 108 MHz. An example of FM is shown in Figure 12–4.

Pulse Code Modulation

Both AM and FM are considered to be analog encoding methods. Pulse code modulation (PCM) is essentially a digital method for encoding an analog signal and is the preferred method of telecommunications. In order to encode an analog waveform with PCM, the input signal is sampled by a sample-and-hold circuit, which connects to an analog-to-digital converter (A/D). The A/D develops a digital number that represents the amplitude of the analog waveform at the instant it was sampled.

▶ FIGURE 12–4
Frequency modulation

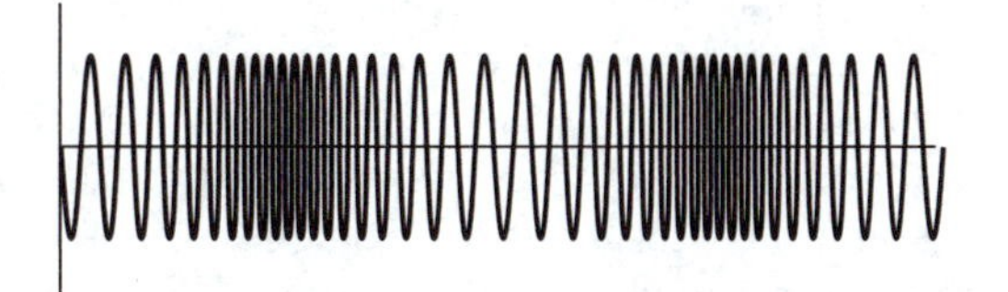

The number of bits provided by the A/D converter determines its ability to break down the input signal into small pieces. The resolution of the A/D, $R_{A/D} = 2^n$, where n is the number of A/D bits. An 8-bit A/D can break down an input signal into 256 pieces. The bit pattern representing the amplitude of the input signal is converted to serial pulses and is ready to be encoded further and then transmitted over the communication medium.

The other half of the PCM process is the method by which the pulses are sent over the communication medium. The communication of digital signals requires that the transmitter and receiver be synchronized. This is because the amplitude of the received signal is used to determine whether the received signal is a digital 1 or a 0 at a specific time. The synchronization signal indicates when the receiver should sample the transmission. Synchronization can be provided with a separate clock transmission or a clock signal can be encoded within the information to be transmitted. For most telecommunications applications the requirement for a separate clock is not practical because an additional communication line would be required. Therefore, a self-clocking encoding method for PCM signals is highly desirable. Self-clocking encoding means that the clock signal is embedded in the transmission signal. This is accomplished by causing a transition of some type (0 to 1 or 1 to 0) to occur during each clock cycle. The transition is used as the synchronizing clock. Self-clocking also resolves the problem of nonchanging voltage levels resulting from strings of 1s and 0s that occur with many encoding schemes. Certain encoding schemes require a wider-channel bandwidth than others because of the method used to include the clock in the transmission. When selecting the encoding scheme, the important considerations are the required channel bandwidth, whether the encoder is unipolar or bipolar, and the expected bit error rate (BER).

Unipolar signals are those that have only one polarity: the signal is either a peak high level or zero. Bipolar signals utilize two polarities: the signal is either a peak positive value or a peak negative value. Unipolar signals are simpler and easier to develop and encode. However, long strings of high or low levels affect the net DC level seen by the receiver. Receivers use the DC level to help develop a threshold for defining the high and low logic levels. Because the DC component varies with the nature of the data being transmitted, recovering the unipolar signal is more difficult. Bipolar signals offer a constant zero DC component, which promotes the establishment of a consistent and accurate logical threshold. In addition, bipolar signals can have twice the differential between logical highs and lows as unipolar signals.

Non-Return-to-Zero Encoding

The various serial encoding schemes define a binary 1 and 0 differently. Figure 12–5 shows examples of various encoding schemes using the same data and clock segments. Unipolar Non-Return-to-Zero (NRZ) encoding is referred to as standard digital encoding because a binary 1 is defined simply as the high logic level and a binary 0 equals the low logic level. This is the same definition used in standard logic representations. Bipolar NRZ defines a binary 1 as a positive logic

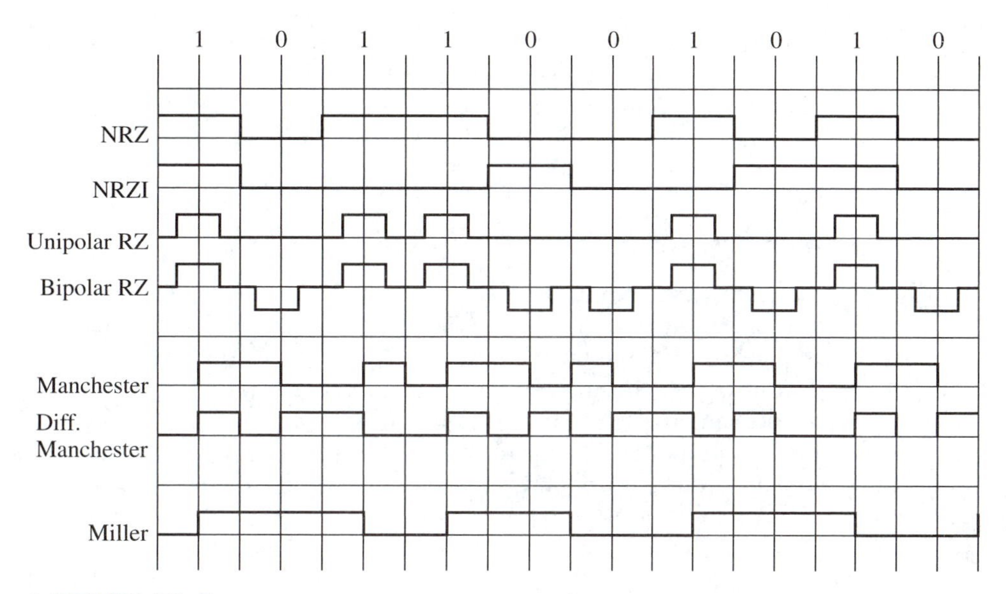

▲ **FIGURE 12–5**
Digital serial encoding schemes

voltage and a binary 0 as a negative logic voltage. NRZ code does not include a clock signal within and therefore requires a separate hardware clock signal. Notice that with long strings of binary 1s or 0s the NRZ does not change for many clock segments.

Non-Return-to Zero-Inverted Encoding

Non-Return-to Zero-Inverted (NRZI) encoding defines a binary 0 as any transition at the beginning of the clock cycle; a binary 1 equates to no transition. NRZI eliminates part of the problem experienced with NRZ code; long periods of low logic levels for a string of binary zeros. However, continuous binary 1s still result in a string of nonchanging transmission levels. NRZI is a differential encoding technique because its definition of binary states is determined by voltage transitions as compared to voltage levels. NRZI also does not include a clock within the transmission. Unipolar NRZI and bipolar NRZI signals are the same except that a negative voltage level represents a logical 0 (see Figure 12–5).

Return-to Zero Encoding

Return-to-Zero (RZ) code results in significant variations when applied to unipolar or bipolar signals. RZ code splits the clock cycle into two half periods. Unipolar RZ defines a binary 1 as a transition to a high level in the first half of the clock signal with a transition to low level in the second half. A binary 0 is interpreted when there is no transition. Note that unipolar RZ results in strings of nonchanging signals with successive binary 0s. Bipolar RZ results are quite different because of the negative logic level used for a binary 0. Bipolar RZ still defines a binary 1 the same as unipolar RZ; however, a binary 0 requires a transition from 0 V to the negative logic level and back to 0. The resulting encoded bipolar RZ signal changes state during each clock period and is therefore considered to be self-clocking.

Manchester Encoding

Manchester encoding is a self-clocking code that is used in many LAN systems and is specified for Ethernet LANs. A binary 1 is defined as a transition from 0 to 1 in the middle of the clock period. Binary 0 is interpreted as a 1 to 0 transition in the middle of the clock period. The transition in the middle of the clock period is used as the synchronizing clock. Bipolar Manchester encoding results in the same waveform as unipolar Manchester encoding except for the different levels seen for the low logic level (unipolar 0 = 0 V, bipolar 0 = negative voltage level).

Differential Manchester Encoding

Differential Manchester encoding is used in many token ring computer networks. The definition of a binary 1 is no transition at the beginning of the clock period. A binary 0 is read when a transition does occur at the beginning of the clock cycle. Differential Manchester is self-clocking and unipolar and bipolar waveforms look the same.

Miller Encoding

Miller encoding is a unique scheme that defines a binary 1 as any transition that occurs during the middle of the clock cycle. A binary 0 is interpreted when there is no transition at the middle of the clock period. In order to keep the signal changing for a string of binary 0s, Miller encoding inserts a transition at the end of the clock cycle if the previous bit was a zero. The result is a self-clocking waveform that changes at least every other clock cycle. The resulting clock is one-half the frequency of other encoding schemes, so Miller encoding requires less overall bandwidth to transmit a given amount of data.

Data Rate vs. Signal Rate

It should be noted that whenever an encoding scheme defines a binary 1 and 0 that requires more than one transition, the actual data or number of bits transmitted is less than the number of transitions. The data rate (bit rate) of a communication link is defined as the number of bits transmitted per unit time, usually stated in bits per second. The signal or baud rate is defined simply as the frequency of the signal transitions. The baud rate equals the bit rate for encoding schemes such as NRZ, NRZI, and Miller, because each transition equates to a bit being transmitted.

Manchester, differential Manchester, and RZ encoding require two transitions for each bit of data transmitted. Therefore, the actual bit rates for these codes are one-half of the signal or baud rate.

Binary Data Codes

In addition to the serial encoding, there is the issue of which binary code the data represents. Transmitted digital data can represent any one of the following codes:

Standard binary code

Binary coded decimal (BCD)

ASCII—Standard Keyboard Character

Baudot—older code that was used for Telex machines

Excess-3 Code—number code similar to BCD

EBCDIC—Extended Binary-Coded Decimal Interchange Code, used extensively in IBM mainframe computers

The most common digital codes transmitted are standard binary code for voice communication and ASCII for text information.

Bandwidth Requirements for PCM vs. Analog

As discussed earlier, AM and FM are considered to be analog methods of encoding information onto a carrier. Since FM requires more bandwidth than AM, let's use the bandwidth requirements of FM encoding to compare the bandwidth needs for PCM. For most telecommunications applications, we are concerned with the transmission of voice information. Since the very beginning, the telecommunications industry has used a bandwidth of 4000 Hz for voice signals. FM encoding therefore requires 4000 Hz of bandwidth for each voice channel transmitted down a transmission line or a wireless link. This means that the carrier frequencies must be at least 4000 Hz apart.

The bandwidth requirements for PCM encoding of voice information are dependent upon the A/D converter used to convert the analog data to digital and the rate at which the data is sampled. The standard used in the telecommunications industry is 8-bit A/D conversion at a sample rate of 8000 samples per second. 8-bit A/D conversion offers reasonable resolution of the signal amplitude for voice signals. The sample rate of 8000 samples per second is determined by communication theory that calls for sample rates of twice the expected highest signal frequency (2×4000 Hz). If an 8-bit A/D converter samples an analog voice signal at a rate of 8000 samples per second, it will create 64,000 bits of data for every second of the voice signal. Therefore the bandwidth required for PCM of voice signals is 64 kHz, which is one PCM voice channel.

Example 12–1

Let's take an example of a telephone cable that has a maximum bandwidth of 200,000 Hz. Determine the location of the carrier frequencies and the number of voice channels for both FM and PCM encoding. Compare the results.

Solution

1. The bandwidth requirements for FM = 4000 Hz = 1 voice channel

   200,000 Hz bandwidth of the telephone line/4000 Hz/voice channel = 50 voice channels

2. The voice channel carrier frequencies would start at 2000 Hz and be spaced every 4000 Hz up to 198 kHz (2000 Hz, 6000 Hz, 10,000 Hz, and so on).
3. The bandwidth requirements for PCM = 64 kHz = 1 voice channel

   200,000 Hz bandwidth of the telephone line/64,000 Hz/voice channel = 3.125 = 3 voice channels

4. The voice channel carrier frequencies would be 32 kHz, 96 kHz, and 160 kHz.
5. The results of these calculations indicate that the telephone cable in question can carry either 50 FM-encoded voice channels or 3 PCM-encoded voice channels at the same time. The bandwidth requirements for digital PCM encoding are 16 times those needed for analog FM encoding.

This example indicates that the bandwidth requirements are significantly greater for digitized signals compared to analog signals when they are transmitted over a communications medium. Why, then, has digital PCM become the preferred method of telecommunication? The answer to this question has many facets.

From the study of digital systems, we know that they possess inherent noise immunity that comes from their differential input and output voltage levels. Therefore, digital system noise immunity offers a significant advantage over any analog encoding scheme. When analog and digital signals are transmitted over a telecommunication line, the signal is attenuated over the long distance of the wire. After a certain distance, the signal must be rejuvenated in order to travel further and be recognized at its final destination. Analog signals are rejuvenated with amplifiers. The problem is that any noise picked up on an analog signal is also amplified. As the analog signal progresses, its signal to noise ratio actually becomes smaller, and the quality of the signal decreases.

Digital signals also experience attenuation and must be rejuvenated. The information encoded onto the digital signal is contained in the bits (whether a bit is a 1 or a 0). Therefore, digital signals can be reconstructed if the proper level for the signal can be interpreted. Devices called *repeaters* reconstruct digital signals. Repeaters sample the signal much like a receiver, determine the binary level, and output a reconstructed signal at that binary level.

Because telecommunications now includes the communication of much digital computer information, voice telecommunication systems must also be capable

Advantages of Digital Encoding

- Inherent noise immunity
- Ability to reconstruct the signal
- Need to transmit digital data
- Error detection and correction
- Digital signals are easily transmitted with fiber optics

Advantages of Analog Encoding

- Lower bandwidth requirements

FIGURE 12–6
Advantages of digital vs. analog encoding

of transmitting digital signals. If noise is picked up by a digital signal, error detection schemes can be used to detect the error and correct it.

One last benefit of digital encoding is that digital signals are very easy to transmit using light. When the light is on, it represents a binary 1; off, a binary 0. Fiber optics readily accommodates digital signals. The increased bandwidth capability of fiber optics more than makes up for the higher bandwidth requirements of digital signals.

A summary of the advantages of digital vs. analog encoding is shown in Figure 12–6.

Digital Transmission Levels

Bipolar digital signals are the most popular for use in computer networks because of the enhanced noise immunity created by the greater difference between logical 1 and 0. The older RS-232 standard requires bipolar digital signals and defines a logical 1 as –3 to –15 V and a logical 0 as +3 to +15 V. The newer RS-422 and RS-485 standards use lower voltage levels and define a logical 1 as –2 to –6 V and logical 0 as +2 to +6 V.

12–4 The Public Switched Telephone Network

The Public Switched Telephone Network (PSTN) represents the telecommunications connection for most of our voice telephone, cell phone, and computer communication. It begins with a connection of twisted pair cable that runs from our homes to a location called the *central office* (*CO*). The connections made to the CO are called the *local loop*. If a phone call is made to another party that is also connected directly to the same CO, the phone call is connected through the CO. Otherwise, the call must be passed on to another CO. The signals transmitted to the central office are most often analog signals with a bandwidth of 300 to 3400 Hz. If it is a long distance call, it will most likely be converted to a digital signal and then

Class 1 — Regional center
Class 2 — Sectional center
Class 3 — Primary center
Class 4 — Toll center
Class 5 — End office

▶ **FIGURE 12–7**
Switching exchange classifications

transmitted on a trunk line to the CO closest to the final destination. Then it is converted back to analog and received at the destination phone number. There are five classes of offices that develop a hierarchy of telephone switches within the PSTN. Figure 12–7 shows these classes of switching exchanges.

Today, computerized switches route phone calls from one exchange to the other. These switches determine the shortest route for a call, depending on the available circuits. The criteria used is to have the call travel the shortest route and use the smallest number of switching centers.

Trunks

Trunks provide the interconnection between the various switching exchanges. Trunks can take the form of twisted-pair, coax cable, microwave or satellite links, or fiber-optic cable.

Twisted-pair trunks are used for short distances with bandwidths under 1 MHz. Twisted-pair cable used for this purpose suffers from crosstalk and high signal attenuation per unit of distance. Coax cable is used for longer distances at frequencies higher than 1 MHz and can be buried underground or protected for underwater (submarine) applications.

Microwave radio links are often used instead of coax cable trunks. These links can carry several thousand voice channels over long distances with repeater stations located every 20 to 30 miles. Microwave links require a line-of-sight transmission and must be located high above any possible obstructions. Satellite communications are essentially microwave links with repeater stations located in space. These satellites are in orbit at approximately 22,300 miles in space above the equator. At that distance the satellite is said to have a geosynchronous orbit, which means the satellite maintains a fixed relative position to the same 40% segment of Earth. These satellites allow line-of-sight tracking 24 hours a day.

Most long-distance telecommunications are carried over fiber-optic trunk lines. (Fiber optic telecommunication is discussed in Section 12–8.) Fiber optics offers exceptional bandwidth capabilities because optical fiber does not possess the inductance and capacitance that limits the bandwidth of twisted pair and coax cable. This means that fiber-optic signals can be switched at much higher frequencies than electrical signals carried over conductive mediums. Fiber-optic fiber is also very small so many fibers can be placed in a fiber-optic trunk line. Fiber optics offers increased bandwidth in much smaller and lighter cables.

Multiplexing

Trunk lines must carry many voice channels of telecommunication signals. Modulation is applied to the signal to provide distinction for one voice channel from another. When analog signals are sent over a trunk, the method used to multiplex voice channels on the medium is called *frequency division multiplexing* (*FDM*). (This is the same as frequency modulation as described in Section 12–3.) The number of voice channels that can be transmitted over the trunk is determined by the bandwidth of the signals that can be transmitted on the trunk. The trunk bandwidth is simply the highest possible transmission frequency minus the lowest. The number of voice channels possible equal the overall bandwidth of the trunk, divided by the bandwidth required for one voice channel (4000 Hz for analog voice signals). FDM can be used to send many voice channels of analog signals over twisted-pair cable, coax cable, microwave, or satellite trunks.

The preferred method used today is the multiplexing of digital PCM signals using a method called *time division multiplexing* (*TDM*). TDM splits a transmission in specific time slices, allowing one slice for each voice channel that is input. For example, let's review a 10-channel TDM line that carries 1 second of data for each channel at any given time. The TDM circuit places the first second of data for channel 1 followed by channel 2 and so on until the data for channel 10 is the last data transmitted for the first second of the transmission. As the data arrives at the end of the trunk, it is demultiplexed and separated back into channels 1 through 10. TDM provides an added delay to the transmission because each channel shares transmission time. However, if the overall transmission bandwidth is high enough, the delays experienced will not be discernible.

12–5 ▶ Modems

The local loop for most PSTN connections is made up of twisted-pair telephone lines designed to carry the analog signals that are output from a typical telephone set. In order for a computer to communicate over these lines, the digital signals transmitted by a computer must be converted to a digitally encoded, analog signal. A modem (short for *modulator-demodulator*) is a device designed to accomplish this. Modems convert and transmit digital data to an analog signal that contains the digital information. A modem on the receiving end reads the digitally encoded analog signal and converts it back to its original digital format. There are five commonly used methods for encoding digital information on an analog signal: amplitude shift keying, frequency shift keying, phase shift keying, differential phase shift keying, and quadrature amplitude modulation.

Amplitude Shift Keying (ASK)

ASK is a simple scheme that applies a frequency with a defined high-level amplitude to transmit a binary 1 and a low-level amplitude for a binary 0. The low level amplitude usually selected is 0 V. ASK by itself is usable at very low transmission rates because of its inherent sensitivity to noise and the low bandwidth signals that are possible. Figure 12–8 shows examples of modem encoding methods.

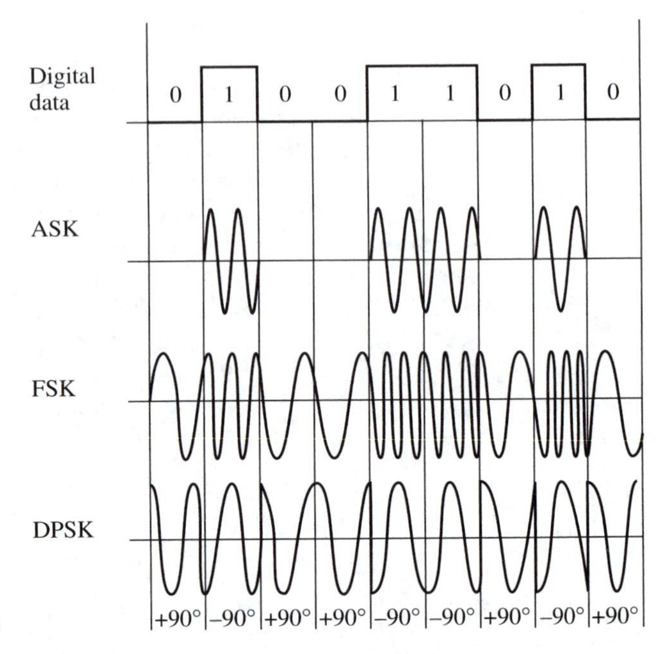

FIGURE 12–8
Modem modulation examples

Frequency Shift Keying (FSK)

FSK is FDM applied to digital signals. FSK uses an upper frequency to encode a binary 1 (called a *mark*) and a lower frequency to indicate a binary 0 (called a *space*). Recall that the bandwidth allotted for each analog voice channel is 4000 Hz. The actual transmitted frequency range is 300–3400 Hz. The frequencies used to implement FSK that equate to the mark and space must be within this 300–3400 Hz band. FSK is used for asynchronous communication at relatively low transmission rates with a maximum rate of 1800 bits per second (bps).

Phase Shift Keying (PSK)

PSK modulates the phase relationship of each succeeding portion of the waveform as compared to a reference waveform. A number of different phase shifts can be used. When only two phase shifts are encoded, this is called *binary phase shift keying.* The binary 1 waveform is 180 degrees out of phase with the binary 0 waveform. When more than two phase shifts are used, it is possible to encode more than one bit into each phase shift. For example, let's say that the following phase shifts are used to encode two bits:

0° phase shift—represents binary 0–0

90° phase shift—represents binary 0–1

180° phase shift—represents binary 1–0

270° phase shift—represents binary 1–1

This is possible because each phase shift is unique for all of the possible combinations of the two binary bits.

Differential Phase Shift Keying (DPSK)

DPSK is similar to the two-bit scheme just described for PSK. The primary difference between DPSK and PSK is the fact that the phase shift decoded is the difference between each succeeding waveform. Typically, DPSK is used to encode up to eight differential phases to achieve data rates up to 4800 bps. Each of the eight phase shifts encodes 3 bits of binary data, all the possible combinations for 3 bits.

Quadrature Amplitude Modulation (QAM)

QAM is an even more complex scheme that achieves yet higher data rates. QAM is a combination of the ASK and DPSK. QAM can encode more data bits into one waveform transition because it combines the phase shift information with differing amplitude levels. If we take a DPSK signal with eight phase shifts and then encode these signals with three different amplitudes, 24 different binary combinations are possible. Because 24 is not an incremental power of two and to allow maximum separation of the encoding requirements, the eight phase-shift and three amplitude QAM signal is structured as shown in Figure 12–9 to achieve data rates of 9600 bps.

▶ FIGURE 12–9
QAM phase diagram

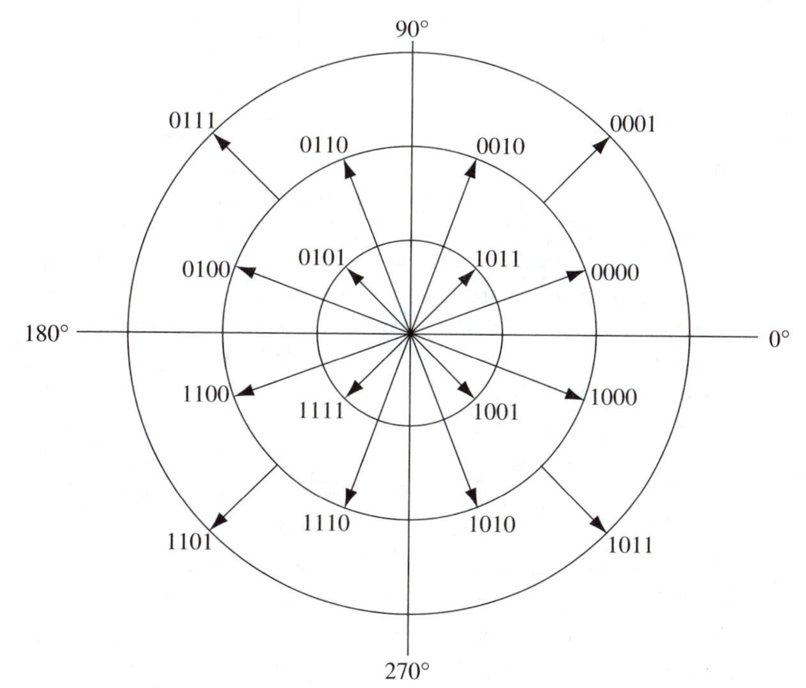

Modems were initially developed for teletype and telex communication over the Bell Telephone system. Consequently, Bell Telephone developed many of the older modem specifications. As the personal computer was introduced, the need for higher data rates was realized globally. The International Telecommunications Union (ITU) develops international standards for modems and other global telecommunications issues. Their specifications are issued as ITU-TS (Technical Specifications). There are many ITU-TS documents that technically define modem communication. In general, communication over the existing analog twisted-pair telephone wire is limited to 56 kbps. However, it is well recognized that 56 kbps significantly stretches the capabilities of twisted-pair telephone cable. The ITU-TS V .34 is the specification for most often used data rate for home personal computers. Data rates on the order of 28,800 and 33,600 are possible with QAM encoding schemes that use a greater number of amplitude levels and differential phase angles to encode more bits into each signal transition.

Cable Modem

Another type of modem is called the *cable modem*. This type of modem utilizes the high bandwidth capability available from coaxial cable that is already in place as part of the cable TV (CATV) network. Cable modems perform the same function as telephone system modems described previously; however, they are capable of much higher data rates. A typical CATV subscriber enjoys data rates on the order of five to ten times the data rates achieved on telephone lines with the additional benefit of not blocking phone communications. Television signals are transmitted over separate frequency channels and are transparent to any modem communications.

ISDN Modems

In order to be competitive with CATV cable modems, the telephone companies have developed the integrated services digital network (ISDN) or digital subscriber lines (DSL). These are telephone lines that connect to a home that provide direct digital communications to the phone system. This system provides much higher bandwidth. It also serves to allow a separate channel of communication for digital computer data that can occur simultaneously with voice communication. The net effect is that digital subscriber lines provide increased data rates and the ability to separate voice and data communication so that they can occur simultaneously.

Transmission Modes

Whenever two serial devices are communicating, the mode of the transmission must be defined. There are three levels of transmission modes: simplex, half-duplex, and full-duplex. Simplex mode is communication that takes place in one direction. Half-duplex mode is the transmission and receipt of data shared on one medium.

Transmission and reception occurs alternately. Full-duplex operation allows both of the communicating devices to send and receive data at the same time. This occurs over two separate lines or by splitting the frequency of communication in half over one line.

12–6 ▶ Error Detection and Correction

The performance of telecommunications systems is truly amazing when you consider the amount of information that is transferred in any one-second period of time. The bit error rate (BER) provides a measurement of the number of incorrect bits per total correct bits sent in a transmission. The BER depends on many factors, including signal bandwidth, distance traveled, signal-to-noise ratio, and the transmission medium and environment. The acceptable BER depends upon the type of communication. A missing bit or two from a voice communication will not make a significant difference in the result of the transmission. Computer data transfers of your bank account information are likely to require a greater level of accuracy and security. The BER is a measure of how accurately a transmission medium is transmitting data. There are many systems that offer extremely low BERs, but none is perfect. Therefore, whenever critical data is being transferred that must be received correctly, a means of detecting and correcting transmission errors is needed.

The oldest and most simple method of detecting errors is the parity method. Parity is an error detection scheme that adds one bit, called a *parity bit,* to any bit pattern being transferred. There are two types of parity detection: even parity and odd parity. In even parity, the parity bit is selected to make the entire transmitted bit pattern have an even number of 1s. Odd parity selects the parity bit so that the bit pattern possesses an odd number of 1s. When parity error detection is used, the sending and receiving device must be set up to operate on the same parity mode, even or odd parity.

In a system that uses parity, the transmitting device attaches a parity bit according to the parity mode that is selected. Let's say that odd parity is selected and that a 7-bit ASCII number of 40 HEX is being transmitted. Because the data 40 HEX transfers to the binary bit pattern 100 0000, the parity bit is made a 0 so that the transmitted number includes an odd number of ones. The bit pattern transmitted becomes 0100 0000, or an 8-bit binary number that still translates to a 40 HEX. On the other end, the receiver examines the received data using odd parity error detection. The receiver expects that all data received will have an odd number of binary ones; otherwise the transmission must be in error. When an error is detected, the receiver requests that the data be retransmitted.

Parity error detection can be implemented easily. The only overhead added to the transmission is the added parity bit and the small time required to add and transmit it and check for correct parity on the receiving end. Its shortcoming is that it can only detect errors with one or an odd number of bit changes in a transmission. An even number of bit changes cannot be detected for either even or odd parity detection. Parity error detection was developed to detect single-bit errors, and it performs this task very well. Higher levels of error detection require more sophisticated methods and greater overhead.

Cyclic Redundancy Checks (CRC)

The CRC concept provides a much better method of error detection and correction and is commonly used in block data transfers. CRC achieves performance levels of 99.9%. CRC functions by calculating a binary number that reflects the value of the data included in a block transfer. This number, called a *block check character* (*BCC*) is added to the data to be transmitted. The BCC represents the remainder of the binary division of the data by a CRC polynomial generator. The receiver performs the same division on the block of data received. The result of the receiver calculation should yield a zero remainder. Otherwise, the data is incorrect and must be resent or corrected by analyzing the remainder calculated by the receiver.

The key to the CRC process is the method used to calculate the BCC. Cyclic codes prescribe a specific number of BCC bits for a certain message size. CRC codes are also called *polynomial codes* because they view a bit string as a polynomial where the coefficients of each polynomial term are the 1s and 0s contained in the bit string. International standards have established 8-, 12-, 16-, and 32-bit BCCs, which are called CRC-8, CRC-12, CRC-16, and CRC-32, respectively. The size of the BCC corresponds to the size of the block being transferred; for example, CRC-8 is used to protect the 5-byte header on ATM data blocks.

The BCC is calculated by dividing the entire block of data to be transferred by what is called the *polynomial generator*. Polynomial generators are defined for each CRC standard. The standard for CRC-16 is $X^{16} + X^{15} + X^2 + 1$, which equals the binary number 1100000000000101. The BCC is generated by dividing the binary representation of the entire data block by the polynomial generator. The quotient resulting from the division is discarded and the BCC becomes the remainder of the binary division, which is added onto the data block. The resulting data plus BCC represents a binary number that is evenly divided by the polynomial generator. The receiver can verify the data transmitted by dividing the entire data block (including the BCC) by the polynomial generator. If the result is zero (no remainder), then the data is accepted. Otherwise, any remainder indicates that there is an error within the block of data. CRC-16 develops a 16-bit BCC. This concept can detect any odd number of bit errors as well as bursts of the number of BCC bits + 1 (17 bits for CRC-16).

Checksum

Checksum is another method used to detect errors in binary data. Checksums are often used to verify the content of programs and other variable data blocks. Checksums are also used in the transfer of data, usually to verify large blocks of transmitted data. There are a variety of different checksum methods that are used.

Single Precision Checksum

Single precision checksum is the most common. It results from the simple addition of succeeding words of data while ignoring any overflow of the MSB. The addition can be performed on any number of data words. After the last data word is added, the remaining word is the single precision checksum result. Any device receiving or reading the successive data words calculates its own single precision checksum for the data received or read. If the checksum values are the same, then the data is assumed to be correct. Single precision checksums are deficient because they allow one particular error to pass undetected. If the MSB of the data word is erroneously always a 1, the result is always the same as any other combination of 1s and 0s in the MSB positions. To test this theory, take the following hexadecimal numbers and add them, keeping only the least significant eight bits:

42H + 2DH + 17H + 34H = 6AH = single precision checksum

Now make the MSB of all of the previous numbers a 1, add the resulting numbers, and then compute the single precision checksum:

C2H + ADH + 97H + B4H = 26A, the single precision checksum = 6AH

Single precision checksums allow this constant high MSB error to exist because the result of the overflow operation of the MSBs are excluded from the result.

Residue Checksum

The residue checksum resolves the constant high MSB problem just described for the single precision checksum. Residue checksum is calculated just as the single precision checksum except that the carry from the addition of the MSB column is wrapped around and added onto the LSB. This step corrects for the MSB locked high problem because the carry information resulting from the MSB additions is included in the ongoing sum instead of being ignored. Every high MSB will impact the result of the checksum process uniquely. Calculating the residue checksum of the two hexadecimal numbers 87 H + BA H would follow like this:

87 H + BA H = 41 H with a carry
the carry is added to the 41 H to develop the residue checksum of 42 H

Double Precision Checksum

Double precision checksums double the width of the data words being added. With this method the carry-overs from the MSB additions are included in the extended width sum of the data. Consequently, this method also corrects for the MSB locked high problem as well as improving the performance of the checksum error detectors. Any carry that results from the overflow of the double-wide sum is ignored. The double precision checksum is calculated as follows:

D3 H + C5 H + F4 H + AB H = 27A H

It is not until the checksum total for a group of data exceeds the hexadecimal number FFFF H that the carry operation is ignored.

Error Correction

When errors are detected while transmitting information, there are two obvious alternatives: resend the information or correct the error. Resending the information is called *automatic repeat request* (*ARQ*), while correcting the error is known as forward error correction (FEC). The factors that dictate the use of ARQ or FEC are the size of the data block being transmitted, accuracy requirements for the data, the amount of time required to retransmit, and the error detection method being utilized.

The use of the ARQ method requires that the transmitter and receiver maintain bi-directional communications. The transmitter sends a block of data and the receiver either acknowledges the receipt of the correct information (no errors detected in the block) with an acknowledge (ACK), or replies with an ARQ indicating that an error was detected in the block of data.

The FEC method is often used when there exists only one-way communication between the transmitter and receiver. In this situation retransmission is not an option. FEC requires the use of more elaborate schemes that provide not only a method of error detection but a way of locating and correcting the error. This is called *redundant error-correction encoding.* These encoding methods insert redundant bits into a block of data to be transmitted. The location of these bits are pre-determined and made known to the transmitter and the receiver. The most commonly used redundant error correction code is called the *Hamming code.* A discussion of the Hamming code and other redundant error correction codes is beyond the scope of this book.

12–7 ▶ Protocols

Whenever two devices are communicating, there must be a set of rules that coordinate this communication. These rules are called *protocols* and they define the process of communication between devices. The simplest protocols are used to define the communication between two devices that interface directly over a dedicated connection.

For example, communication between a personal computer and the printer connected to it. This communication can be either serial or parallel format. The software driver for the printer used by the computer makes sure that the information being transmitted is in a code and format that will be understood by the printer. The protocol for this communication could be as follows:

1. The computer sends a message to the printer indicating that it has data to be sent to the printer.
2. The printer replies that it is ready and also sends back an identification number that specifies its manufacturer and model number.
3. The computer uses the appropriate software driver to encode and format the data, sending a block of data to the printer.
4. The printer either acknowledges the receipt of the data (ACK) or requests a retransmission (ARQ).

When communication involves two computers that are located across the country and connected by telephone lines and the Internet, the protocols required are much more complex. This is because this type of communication includes many layers of communication requiring many layers of protocols. In order to standardize global communications, the International Standards Organization (ISO) works with other global communication committees to establish standards for protocols. These studies have evolved into seven layers of protocols called the Open System Interconnect (OSI) model. The seven layers of the OSI are as follows:

1. Physical. This is the lowest of the layers and it defines the specific electrical and mechanical details of the interface and protocol between two devices.
2. Data Link. The data link layer defines the format of the data being sent, which includes the data code (ASCII, BCD, etc.), the framing characters (header and trailer), and the error detection/correction scheme.
3. Network. This layer determines the method for sending data throughout a communications network. The network can be the PSTN or a small LAN.
4. Transport. The transport layer deals with the communication over the entire path between the transmitter and the receiver. It promotes the efficient and reliable transmission of information.
5. Session. This layer manages the log-in and log-out procedures when the communication is part of a defined session.
6. Presentation. This deals with the conversion of data being transmitted from one type of code over to another.
7. Application. This layer defines the type of application software that the communication is related to. E-mail is a common application used on most PCs. If someone sends out an e-mail, the application layer information included with the transmitted data indicates to the receiving computer that the data is an e-mail message.

The seven-layer OSI model includes all the possible layers currently available in telecommunications. However, just five layers—application, transport, network, data-link, and physical—apply to the majority of computer communications. The session and presentation layers apply only to more specialized telecommunications applications. The transport, network, data-link, and physical layers all define hardware aspects of the communication while the application, session, and presentation layers deal only with software.

Let's discuss the application of the five layers of the OSI model to a typical e-mail communication between two people. A person named John@ABC.com writes a message to Sue@XYZ.edu, sending the message as e-mail to her given Internet address. John's computer is a home computer connected to the PSTN and he uses the Internet service provider called ABC. Sue is a college student at XYZ university, and her computer is connected directly to the school's network, which connects to the Internet.

When John writes the message as an e-mail, the application layer protocol used is called the *simple mail transfer protocol* (*SMTP*). The SMTP message is given a code number equal to 25, called the *port number*, which identifies it as an

SMTP transmission. The computer John is using sends out the message over the telephone connection to his computer with the SMTP port number 25 attached, along with other information about the application. The transport layer determines the information about the actual physical location of the address where the message is being sent, which in this case is XYZ university. It also attaches the physical address of the sender.

Next, the message is sent from John's computer through the PSTN to John's ISP ABC.com. Along the way, this message may be transmitted over a variety of communication links. Most likely the message will be sent through the local loop to the closest central office as an analog signal. Then it might be converted to a digital signal and sent over a trunk line to the central office closest to the ISP, ABC.com. ABC.com most likely retains a high bandwidth link with the central office because of the high volume of transmissions that they handle. Whatever path the message travels, the network layer of OSI is used to format and control the transmission from one end of the branch to the other. In other words the network layer identifies the method of communication from end-to-end of a particular branch of the path.

The data-link layer identifies the type of data being sent as well as the use of any error detection and correction schemes to be used with the transmission. The data-link layer would indicate that the data included in the e-mail is ASCII characters with parity error detection. The physical layer identifies the actual physical connection that exists over each step of the transmission, starting with the telephone wire to the central office, the trunk connection from John's central office to the central office closest to John's ISP, and so on.

The five OSI levels discussed serve to guide John's e-mail along the proper path (see Figure 12–10). Each level attaches information to the original message that is important to the transmission of the message. First is the message itself, which was written as an e-mail. The message plus its application (SMTP) are then readied for transmission and the transport layer adds the actual physical destination address. The data-link layer identifies the type of data and error detection/correction, while the network layer specifies details about the branch of the connection through which the message is currently passing. The physical level is where the signal is actually transformed to have the correct physical properties for transmission through a medium. The signal may be converted from FDM-modulated analog signals to TDM digital signals to fiber optics over the course of its transmission.

FIGURE 12–10
The OSI layers

1. Physical
2. Data-link
3. Network
4. Transport
5. Application

12–8 ▶ Computer Networks

The enhanced capabilities of computers and telecommunications systems have been applied to allow the interconnection of computers in ways that would not seem possible. Computer networks are the interconnection of computers for the purpose of communication to share information and peripherals (printers, ISPs, etc.). Computer networks are called *local area networks* (*LANs*) when the system components are located in relative close proximity to each other within a building or a group of buildings. Broader networks located within an area of approximately 100 miles are called *metropolitan area networks* (*MANs*). The method of communication between computers and peripherals is largely determined by the network topology, the method used to interconnect them. The most common network topologies are star, bus, and token ring.

Star Topology

The star topology was one of the first topologies developed and forms the basis for the existing PSTN. A diagram for the star topology is shown in Figure 12–11. A star topology network consists of a number of devices that connect to a central device that resides in the center of the star configuration. All communications must pass through the central node: data destined for a device within the star network as well as data to be sent to devices outside the network. The star configuration is the best alternative when most communication comes from the central node to some other network. When most of the communication is between devices located within the star topology, communication is limited by the overburdened central node. The function of the central node is critical to the operation of the star network; however, the malfunction of just one device within the star network does not render the entire network nonfunctional.

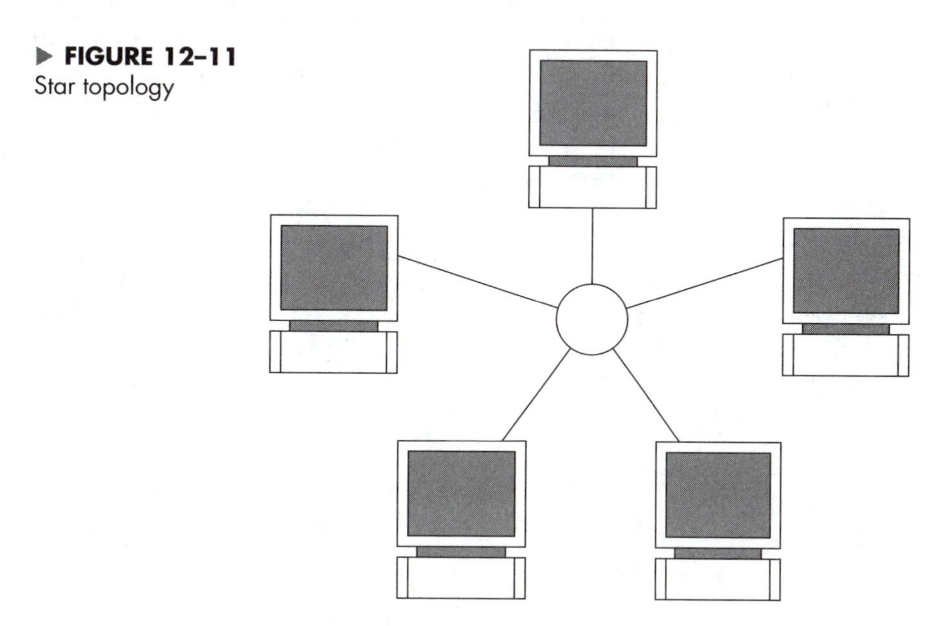

▶ **FIGURE 12–11**
Star topology

Bus Topology

A bus network is a very simple and commonly used topology for interconnecting computing devices. The bus topology consists of a common group of connections that exist between each device connected to the network (see Figure 12–12). These connections are referred to as the *network's backbone.* Each device connected to the bus can communicate with any other device on the bus; however, only one communication can be sent at one time. Therefore, the number of computers connected to the bus largely determines network performance. Bus topologies are often referred to as passive, because the devices connected to the bus simply listen to bus communications to determine if they should receive the information. They have no responsibility for passing data onto other devices. If one computer fails, the rest of the network continues to operate.

When a device wishes to communicate on the bus, it listens for a quiet period and then proceeds to transmit data. If by chance another device transmits at exactly the same time, there is a collision of the data. The transmissions are garbled and both devices, sensing the collision, retreat while selecting a random number that is counted down before attempting a re-transmission.

Token Ring Topology

The token ring topology is one in which the devices are connected serially in a ring, as shown in Figure 12–13. Communication commences only in one direction around the ring and each device is responsible for passing on the transmission. For this reason the token ring topology is classified as an active network. A data word called a *token* is passed around the ring in the direction of communication. When a device possesses the token, it has the opportunity to transmit data on the network. It modifies the token by adding the address of the device it is transmitting to and the data being sent. The data is sent around the ring until it reaches its intended destination, where the receiver acknowledges receipt by sending an acknowledge signal back to the sender. The failure of just one device in the token ring renders the entire network nonfunctional.

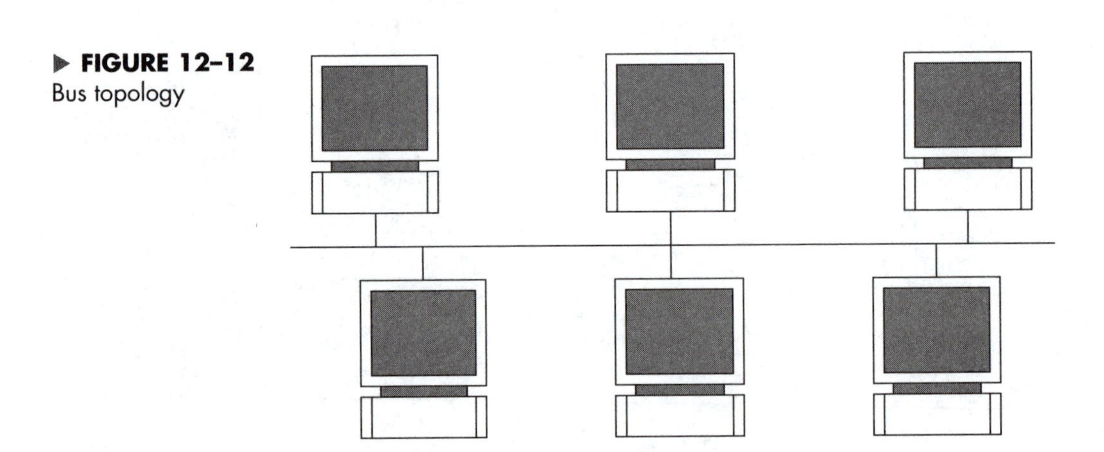

▶ FIGURE 12–12
Bus topology

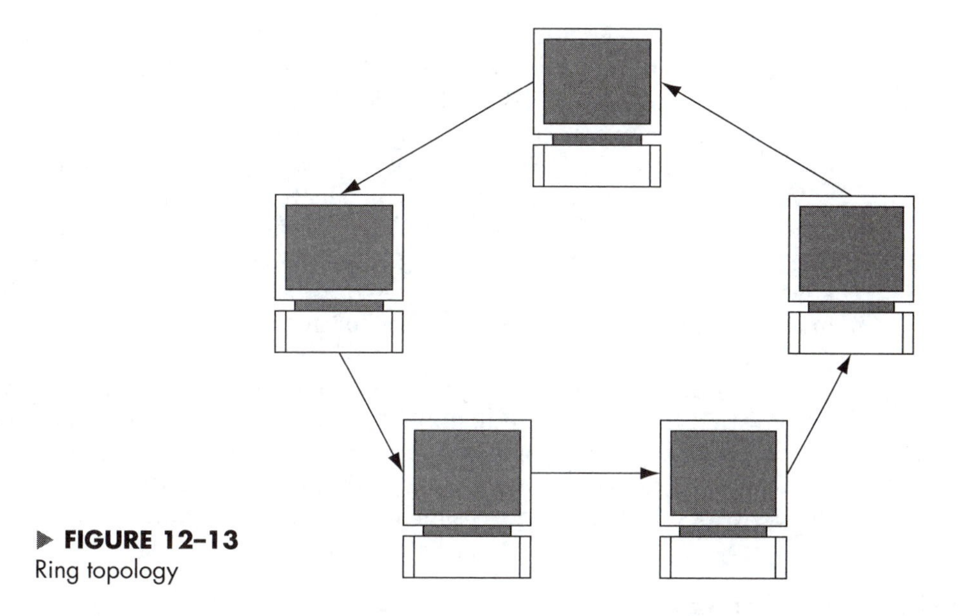

▶ **FIGURE 12–13**
Ring topology

Ethernet Technology

Ethernet is an example of a protocol that is widely used today. Ethernet can be implemented in either a bus or star topology and is defined by the IEEE specification number 802.3. The Ethernet standard essentially fits into the data-link and physical layers of the OSI model, defining the physical connection between the network components as well as the method for communication between all devices connected to the network. There are a number of Ethernet variations that specify the interconnection cable, transmission codes, network topology, and data rates.

10 Base 2 Ethernet is a commonly used network connection for businesses and schools. It uses thin coax cable, has a maximum data rate of 10 Mbps, and uses a bus topology and Manchester encoding. Another Ethernet version, called 10 Base T, utilizes twisted-pair cable wired in a star configuration that also has a 10 Mbps data rate and uses Manchester encoding. 100 Base T is an extended data rate version of 10 Base T with a 100 Mbps data rate. It is wired in a star configuration with twisted-pair cable but uses a more efficient encoding scheme called 4B5B.

Gigabit Ethernet is the latest version of the Ethernet standard that pushes data rates up to 1 Gbps. It is used as a backbone for many larger networks to eliminate the bottleneck in communications that occurs when various devices from one branch of a network communicate with other branches of the network. Gigabit Ethernet uses a star topology just the 10 Base T and 100 Base T Ethernet.

Network Devices

Three basic devices are used to interconnect computers that are part of a network: hubs, bridges, and routers. These devices are similar in nature because they all accept an input and provide a number of outputs. However, each function at a different layer of the OSI protocol hierarchy.

Hubs operate at the physical layer and are essentially repeaters or level-changing devices. The frames or bit patterns input to a hub are the same as those sent to all of the hub's outputs. The hub makes no decisions about the data transmitted and does not route the data to a particular output port. It simply passes on the information. The purpose of the hub may be to rejuvenate the signal or to change the physical output such as conversion from electrical to optical signals.

Bridges are devices that route data according to the link layer destination address of the frame transmitted. Bridges are used to interconnect networks of any kind. The networks can be of the same or different topologies. Data input to the bridge is routed to the link layer address of the network that is connected to a particular bridge output terminal.

Routers are switches that perform the same function as a bridge, yet they operate within the network, functioning at the network layer of the OSI protocol. Routers connect devices that are all part of the same network. Data input to the router is processed and routed to the output connection that will send the data to the network destination address included in the data frame.

12–9 ▶ The Internet

The Internet has evolved as an important communications device over the last 20 years. Also called the World Wide Web (WWW), the Internet can be defined as a global group of interconnected networks that all use the network layer protocol called, appropriately, the *Internet Protocol* (*IP*). Some of the networks that the Internet comprises are publicly owned while some are owned privately. Special-access networks provide connections to the Internet through companies called *Internet service providers* (*ISPs*). A unique aspect of the Internet involved the development of the Web page. A Web page is simply a document that can be published on the Internet that contains information as well as links to other Web pages. When a Web page is published, it is maintained as a file in the Internet network memory and can be read by anyone with Internet access. Links from the main Web page to other Web pages can be controlled with passwords and other security methods.

Another aspect of the Internet that is now part of many computer networks is an innovation in the way information is transmitted, called *packet switching*. In order to define packet switching, let's first define what is called a *connection* between communicating devices. A connection is a maintained continuous path for communication between two devices. The best example of a connection is a telephone call. Telephone calls all develop a maintained and ongoing connection between two parties until one party hangs up.

A packet is a small group of data that is sent out together that represents only part of the information to be transmitted. Each packet travels toward its intended network address destination separately along the shortest and quickest route available at the time of transmission. Therefore, it is entirely possible for the packets to arrive out of order. The packets include information that indicates the order of transmission, so the receiving device is able to reassemble the data in the correct order. The use of packets allows for the optimum utilization of the existing transmission mediums. If traffic is busy down one path, then the packet is diverted around the bottleneck over the next shortest and quickest path.

When dedicated connections exist between two devices, the communication is very secure and reliable; the transmitter can send data and the receiver can respond directly with an acknowledgement or request a retransmission. With packet switching there are two approaches to the transport level program: connection-oriented protocols and connectionless protocols. Connection-oriented protocols are used for packet-switched information where reliability is desired. In other words, we want to make sure that the data reached its destination and that it is received accurately. Connection-oriented protocols establish handshaking much the same as maintained connections, even though the data packets transmitted may travel completely different paths. The connection-oriented protocol used on the Internet is called the *transmission control protocol* (*TCP*). The cost of reliability is realized with the overhead needed to setup the handshaking between the two communicating devices. This overhead slows down TCP significantly.

Connectionless protocols send out packets of data blindly and hope the data is received accurately. The connectionless protocol used for Internet communication is called *user datagram protocol* (*UDP*). UDP is a simple and very fast protocol because it does not carry the overhead of establishing a virtual connection.

Communication over a network can utilize a maintained connection, connection-oriented packet communications or packets sent on a connectionless basis.

The Internet uses only five layers of the seven-layer OSI model described earlier in this section. The five layers used are application, transport, network, data-link, and physical. There are a number of application programs that are used for Internet communication but the most common are the hypertext transfer protocol (HTTP), simple mail transfer protocol (SMTP), and the file transfer protocol (FTP).

HTTP is the application layer protocol used to transmit Internet information that is published as Web pages. All Web pages and other Internet resources have a distinct uniform resource locator (URL), which is the Internet location of the resource. FTP is the file transfer protocol used by the Internet, and SMTP is the application protocol used for e-mail. The next OSI layer is the transport layer. The Internet transport protocol is either TCP or UDP. TCP is used for Web pages and other communications that require reliable transmission. UDP is used when higher data rates are necessary and the reliability of the data is not critical. Such applications include real-time audio, video, or multimedia communications.

The network layer protocol for Internet communications is called the Internet protocol (IP). The IP protocol works together with either TCP or UDP to complete the protocol needed for Internet communication. The IP protocol by itself establishes a connectionless link between the sender and receiver and relies on TCP to establish reliability. The data-link and physical layers vary for each branch of the packet's journey and are invisible to the sending and receiving devices. Figure 12–14 shows the five layers of protocols used for Internet communication.

12–10 ▶ Fiber Optics in Telecommunications

Nothing has had a larger impact on the nature of the telecommunications industry than developments in the field of fiber optics. Fiber optics uses light to transmit information instead of electrical signals. The primary advantages of fiber optics result from two facts:

1. Physical
2. Data-link
3. Network IP
4. Transport TCP/UDP
5. Application HTTP/FTP/SMTP

▶ **FIGURE 12–14**
Layers of OSI for Internet applications

1. Light can be switched on and off at a much higher frequency than electrical signals.
2. Light signals are attenuated only slightly (as compared to electrical signals) as they travel down the highly pure optical fiber available today.

Consequently, light signals that are sent over fiber-optic cable can be transmitted at much higher frequencies and can travel much further before requiring rejuvenation. A common misnomer is the thought that fiber optics provides more capacity because light travels faster than electrical signals. Electrical signals travel almost as fast as light does down a glass fiber, and the difference is not significant enough to warrant the capacity increases experienced with fiber optics. Fiber optics offers increased capacity simply because light can be switched on and off much faster than electrical signals.

Fiber-optic communications are implemented with a light source, either an LED or a laser that is connected to a fiber cable. The data to be transmitted is sent to the light source that switches according to the data stream input. On the receiving end of the fiber-optic cable, an optical sensor, usually either a PIN photodiode or an avalanche photodiode (APD), converts the light signal back to an electrical output signal.

The fiber cable itself consists of pure glass that includes two sections: an inner core with an outer cladding. The core has a higher index of refraction than the cladding, which means that light injected into the core is reflected off the border between the core and the cladding. The light is reflected according to a principle called *total internal reflection.* This results when the light is incident to the core-cladding border by an angle greater than what is called the *critical angle.* The critical angle is determined by the difference between the index of refraction of the core and the cladding. Light that is incident at angles less than the critical angle passes on through the cladding and is lost (see Figure12–15).

The increased bandwidth and attenuation performance of fiber optics happens because fiber cable lacks limitations such as the resistance, inductance, and capacitance effects on electrical signals as they travel down a conductive cable. All electrical cables offer resistance to current flow—the smaller the cable, the larger the resistance. Also, the inherent inductance and capacitance of wire limits the highest frequency that can be transmitted on the cable.

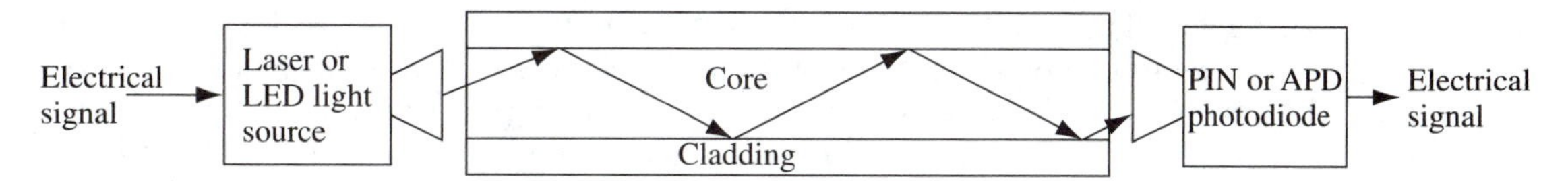

▲ **FIGURE 12–15**
Fiber-optic communication

A summary of the advantages of fiber optics is as follows:

- High bandwidth
- Low attenuation
- Small physical size
- Light in weight
- Noise immunity—insensitive to electrical noise
- Secure—does not radiate or easily tapped into
- Increased safety in hazardous atmosphere due to absence of potential for electrical spark

Fiber-optic cables do attenuate light signals as they pass through the cable. There are also frequency limitations due to properties of the cable and the way light is transmitted down the fiber. The attenuation of light signals is due to impurities that are present in the fiber that absorb the light instead of allowing it to pass through. The amount of attenuation is determined by the amount of impurities present and the size of the core.

The frequency limitations of fiber are caused by a property called *dispersion.* In general, dispersion results when the light injected into the fiber reaches the end of the fiber at different times. There are two distinctly different types of dispersion: the light can take different paths, which is called *modal dispersion* (modes = paths in fiber-optic terminology), or the light has a range of wavelengths that all travel at different speeds, which is called *chromatic dispersion* (see Figure 12–16).

Modal dispersion is the most significant bandwidth limitation affecting basic fiber-optic cable, called *step-index multimode cable.* Step-index multimode cable is the lowest performing grade of fiber cable. It suffers from limited bandwidth due to its large core size. This results in many paths (modes) for the light to travel down, therefore causing more dispersion of the light signal. As the input pulses

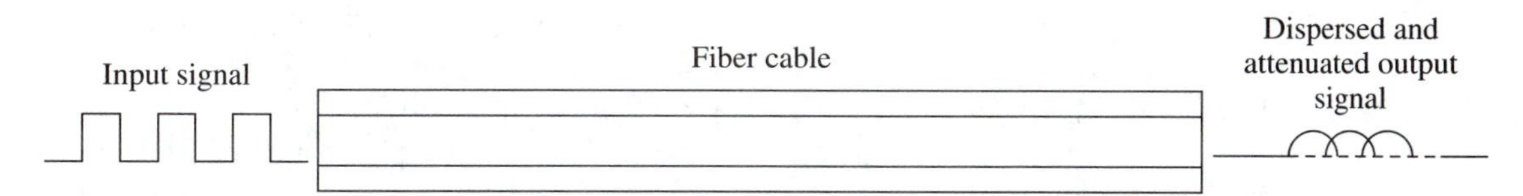

▲ **FIGURE 12–16**
Dispersion in fiber-optic cables

are dispersed on the other end of the fiber, the receiver cannot detect the 1-to-0 transition that indicates a binary low signal. Graded-index multimode cable improves the performance of step-index multimode fiber by graduating the index of refraction difference between the core and the cladding instead of the step-change in the index of refraction seen on step-index fiber. This gradual change in the index of refraction slows down the light that travels the shortest path through the fiber, thereby reducing modal dispersion significantly.

Single-mode fiber has been developed that, for all practical purposes, eliminates modal dispersion. Single-mode fiber has a core size that is much smaller than multimode cable, which effectively limits the number of paths to one. With only one path to travel, the light reaches its destination at the same time. With modal dispersion virtually eliminated, chromatic dispersion becomes the bandwidth-limiting factor for single-mode fiber. The level of chromatic dispersion is highly dependent on how wavelength consistent the light signal is.

Ideally, light transmitted down a fiber cable is just one wavelength. In practice, the spectral width (range of wavelengths) created by light sources varies significantly between the various types of light sources. LEDs have a spectral width of 30 nanometers while laser light sources can achieve narrow spectral widths of 2 to 3 nm. Because chromatic dispersion is based upon the varying speeds that wavelengths travel through fiber, it stands to reason that the wider the spectral width of the light source, the greater the level of chromatic dispersion. Therefore, the amount of chromatic dispersion experienced with single-mode fiber is highly dependent on the light source used (LED or Laser). The effect of using a wide spectral width light source (an LED) on the bandwidth of single-mode fiber is so dramatic that single-mode fiber is only used with laser light sources.

The applications of fiber optics can be separated into three categories: high bandwidth telecommunications applications, high-speed network backbones, and routine network applications. Telecommunications applications for fiber are usually trunk connections that require a very high bandwidth and throughput with low attenuation. Single-mode fiber cable and laser light sources are used almost exclusively in telecommunications applications. High-speed network backbones such as gigabit Ethernet can benefit from the application of fiber optics. These also use single-mode fiber cable with laser light sources. General application networks that interconnect with fiber cable will use graded-index multimode fiber and LED light sources because the bandwidth requirements are not as severe.

Summary

Telecommunications is a complicated area for electronic applications that now includes computer communication, voice, audio, video, multimedia, and wireless communications. Each application includes many levels of protocols and codes all symbolized with an ever-increasing multitude of acronyms. Yet there is a feeling that we are still just beginning to explore the many possibilities. The bandwidth potentials of fiber optics are yet to be fully utilized. Video conferencing and other multimedia communications offer more potential for change at a time when it seems like we have already experienced so much change. Sometime in the future, fiber-optic cable will connect every home to a global network that will offer all the potentials for telecommunications.

The growth opportunities for telecommunications are promising and the need for creative and knowledgeable technical professionals continues to be high. The telecommunications industry requires people with a range of knowledge that includes digital electronics, microprocessors, software development, fiber optics, general telecommunications, and business operations. The work can range from the design of telecommunications equipment (routers, hubs, bridges, telephone switches, peripherals, etc.), to laying out communications networks and trunks, to managing operations, field-service, or new construction. The skill levels required are high and the rewards are many.

References

Hioki, W. *Telecommunications*, 3rd ed. Upper Saddle River, NJ: Prentice Hall, 1998.

Kurose, J., and Ross, K. *Computer Networking.* New York: Addison Wesley, 2001.

Exercises

12–1 Explain how the bandwidth capability of a communication link determines the amount of information that can be transmitted at one time.

12–2 Explain the process of amplitude modulation, frequency modulation, and pulse code modulation.

12–3 Explain the difference between unipolar and bipolar digital telecommunications signals.

12–4 Draw the waveform that represents the digital bit stream 1100101101 when encoded with Manchester encoding. Repeat for Miller encoding.

12–5 Explain the difference between data rate and signal rate.

12–6 A communication link operates at a bit rate of 100 kbps. If NRZ code is used, what is the baud rate? What is the baud rate if RZ code is used?

12–7 When comparing the bandwidth required for digital PCM signals verses analog FDM signals, which type of signal requires the most bandwidth? Explain the reason for the large difference in bandwidth requirements.

12–8 What advantages do digital PCM signals offer over analog FDM signals?

12–9 Explain the purpose of multiplexing and the differences between time division multiplexing (TDM) and frequency division multiplexing (FDM).

12–10 What is the purpose of a modem? List the five different types of analog telephone line modems discussed in this chapter.

12–11 Explain the performance and functional differences between cable modems and telephone line modems.

12–12 Describe the function of simplex, half-duplex, and full-duplex serial transmission modes.

12–13 Define the term *bit error rate* (*BER*). What are the factors that determine the BER for a telecommunication system?

12–14 Explain the parity method of error detection. What types of parity detection are there? What types of errors can and cannot be detected with this method?

12–15 Explain the operation of CRC error detection. What performance level does CRC error checking achieve?

12–16 Explain the general process of checksum error detection methods and list the types of checksum error detection discussed in this chapter.

12–17 What is the limitation of single precision checksum error detection? How do residue checksums and double precision checksums overcome this problem?

12–18 When errors are detected in a transmission, what methods are available to correct them?

12–19 Explain the purpose behind each of the five layers included in the five-layer protocol used for Internet communications.

12–20 List and describe the three types of bus topologies discussed in this chapter.

12–21 Explain the difference between a hub, a bridge, and a router.

12–22 By what process does light travel down a fiber-optic cable?

12–23 Explain why fiber optics offers superior bandwidth performance when compared to communication using electrically conductive wire.

12–24 What are the three different types of fiber-optic cable? Which offers the lowest and highest bandwidth performance?

12–25 What characteristics of fiber-optic cable limits it bandwidth performance?

12–26 Describe what is meant by dispersion in fiber-optic cable. List and describe each type of dispersion discussed in this chapter.

Appendix A Component Reference Information

A–1 Resistors

Resistors limit the flow of current and divide voltage in electrical circuits. If the current through a resistor is known, the voltage drop across it can be calculated using Ohm's Law ($V = I \times R$). Resistors are not frequency dependent, so they have the same resistance value for AC and DC voltage.

Resistor color codes are shown in Figure A–1. Included in Figure A–1 are the four-band code used for 5%, 10% and 20% tolerance resistors and the five-band color code used for 1% tolerance resistors. Figure A–2 shows the standard value resistors that are available at different tolerance values.

A–2 Capacitors

Capacitors store electrical charge on plates separated by a dielectric material. Capacitors oppose any changes in the voltage across them by giving up previously stored charge if the voltage is decreasing or storing more charge if the voltage across them is increasing. Current never actually flows through a capacitor (except a small leakage current), but with AC voltages it appears to because the current is constantly changing direction. In AC circuits, the capacitor is continually charging and discharging as the current flows in one direction, and then it reverses. Capacitive Reactance, X_c, is the impedance that the capacitor offers resisting the flow of AC current and is equal to:

$$X_C = 1/(2\pi f C)$$

where f = frequency
C = Capacitance value

Resistor Color Codes

0 = Black	7 = Violet	Gold Multiplier = 0.1
1 = Brown	8 = Gray	Silver Multiplier = 0.01
2 = Red	9 = White	
3 = Orange	1% = Red	
4 = Yellow	5% = Gold	
5 = Green	10% = Silver	
6 = Blue	20% = No Color	

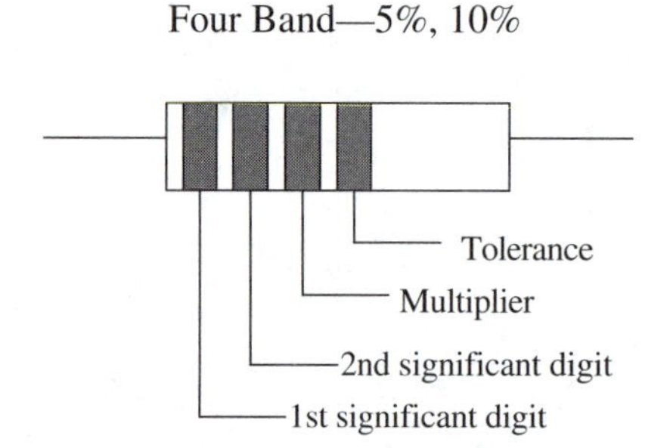

Five Band—1%

Tolerance
Multiplier
3rd significant digit
2nd significant digit
1st significant digit

▲ **FIGURE A–1**
Resistor color codes

As frequency increases, the Capacitive Reactance decreases. For DC circuits then, $f = 0$ so the Capacitive Reactance is theoretically infinite. In DC circuits capacitors completely block DC current flow (again, except for leakage) and charge up to the DC voltage. The capacitance value determines its ability to store charge. The larger the value, the more space it has to store charge. It makes sense that capacitance is added when you connect two capacitors in parallel. Conversely, it follows that two capacitors in series will react like two resistors in parallel and can be calculated by the formula, $C_t = C_1 \times C_2/(C_1 + C_2)$. This is because the effective plate area is reduced and the distance between the plates is increased.

Capacitors are rated by their value, dielectric type, and the voltage they can withstand. These values are either stamped on the capacitor or color codes are used. The standard capacitor values available use the same multipliers as for 10% tolerance resistors. Figure A–3 shows the standard capacitor values. It is important to note that not all of the values are available for a particular dielectric type or a specific manufacturer. For example, aluminum electrolytic capacitors are usually available only in multiples of 10, 22, 33, 47, and 68. Figure A–3 also shows the voltage ratings that are usually available for capacitors.

Capacitor color code locations are dependent on the type of dielectric used. The location and meaning of the color codes are shown in Figure A–4.

Standard Resistance Values
(the following values are available in powers of ten)

0.1% 0.25% 0.5%	1%	2% 5%	10%
10.0	10.0	10	10
10.1			
10.2	10.2		
10.4			
10.5	10.5		
10.6			
10.7	10.7		
10.9			
11.0	11.0	11	
11.1			
11.3	11.3		
11.4			
11.5	11.5		
11.7			
11.8	11.8		
12.0		12	12
12.1	12.1		
12.3			
12.4	12.4		
12.6			
12.7	12.7		
12.9			
13.0	13.0	13	
13.2			
13.3	13.3		
13.5			
13.7	13.7		
13.8			
14.0	14.0		
14.2			
14.3	14.3		
145			
14.7	14.7		
14.9			
15.0	15.0	15	15
15.2			
15.4	15.4		
15.6			
15.8	15.8		
16.0		16	
16.2	16.2		

0.1% 0.25% 0.5%	1%	2% 5%	10%
16.4			
16.5	16.5		
16.7			
16.9	16.9		
17.2			
17.4	17.4		
17.6			
17.8	17.8		
18.0		18	18
18.2	18.2		
18.4			
18.7	18.7		
18.9			
19.1	19.1		
19.3			
19.6	19.6		
19.8			
20.0	20.0	20	
20.3			
20.5	20.5		
20.8			
21.0	21.0		
21.3			
21.5	21.5		
21.8			
22.1	22.1		
22.3			
22.6	22.6		
22.9			
23.2	23.2		
23.4			
23.7	23.7		
24.0		24	
24.3	24.3		
24.6			
24.9	24.9		
25.2			
25.5	25.5		
25.8			
26.1	26.1		
26.4			

0.1% 0.25% 0.5%	1%	2% 5%	10%
26.7	26.7	27	27
27.1			
27.4	27.4		
27.7			
28.0	28.0		
28.4			
28.7	28.7		
29.1			
29.4	29.4		
29.8			
30.1	30.1	30	
30.5			
30.9	30.9		
31.2			
31.6	31.6		
32.0			
32.4	32.4		
32.8			
33.2	33.2	33	33
33.6			
34.0	34.0		
34.4			
34.8	34.8		
35.2			
35.7	35.7		
36.1		36	
36.5	36.5		
37.0			
37.4	37.4		
37.9			
38.3	38.3		
38.8			
39.2	39.2	39	39
39.7			
40.2	40.2		
40.7			
41.2	41.2		
41.7			
42.2	42.2		
42.7			
43.2	43.2		

0.1% 0.25% 0.5%	1%	2% 5%	10%
43.7		43	
44.2	44.2		
44.8			
45.3	45.3		
45.9			
46.4	46.4		
47.0		47	47
47.5	47.5		
48.1			
48.7	48.7		
49.3			
49.9	49.9		
50.5			
51.1	51.1	51	
51.7			
52.3	52.3		
53.0			
53.6	53.6		
54.2			
54.9	54.9		
55.6			
56.2	56.2	56	56
56.9			
57.6	57.6		
58.3			
59.0	59.0		
59.7			
60.4	60.4		
61.2			
61.9	61.9	62	
62.6			
63.4	63.4		
64.2			
64.9	64.9		
65.7			
66.5	66.5		
67.3			
68.1	68.1	68	68
69.0			
69.8	69.8		
70.6			

0.1% 0.25% 0.5%	1%	2% 5%	10%
71.5	71.5		
72.3			
73.2	73.2		
74.1			
75.0	75.0	75	
75.9			
76.8	76.8		
77.7			
78.7	78.7		
79.6			
80.6	80.6		
81.6			
82.5	82.5	82	82
83.5			
84.5	84.5		
85.6			
86.6	86.6		
87.6			
88.7	88.7		
89.8			
90.9	90.9	91	
92.0			
93.1	93.1		
94.2			
95.3	95.3		
96.5			
97.6	97.6		
98.8			

▲ **FIGURE A–2**
Standard resistor values

Standard Capacitance Values							
Picofarads	Picofarads	Microfarads	Picofarads	Microfarads	Microfarads	Microfarads	Microfarads
1.8	120		12,000	0.012	1.0	120	1800
2.2	180		15,000	0.015	1.2	180	2200
2.7	220		18,000	0.018	1.8	220	2700
3.3	270		22,000	0.022	2.2	270	3300
3.9	330		27,000	0.027	2.7	330	3900
4.7	390		33,000	0.033	3.3	390	4700
5.0	470		39,000	0.039	3.9	470	5600
5.6	560		47,000	0.047	4.7	560	6800
6.8	680		56,000	0.056	5.6	680	8200
8.2	820		68,000	0.068	6.8	820	
10	1000	.001	82,000	0.082	8.2	100	
12	1200	.0012	100,000	0.1	10	120	
15	1500	.0015		0.12	12	150	
18	1800	.0018		0.15	15	180	
22	2200	.0022		0.18	18	220	
27	2700	.0027		0.22	22	270	
33	3300	.0033		0.27	27	330	
39	3900	.0039		0.33	33	390	
47	4700	.0047		0.39	39	470	
56	5600	.0056		0.47	47	560	
68	6800	.0068		0.56	56	680	
82	8200	.0082		0.68	68	820	
100	10,000	.01		0.82	82	1000	
					100	1200	

Standard Voltages	
6.3	250
10	400
16	450
35	500
50	1000
63	2000
100	3000

▲ **FIGURE A–3**
Standard capacitor values

Capacitor color code designations are the same as for resistors:

0 = Black	7 = Violet	Gold Multiplier = 0.1
1 = Brown	8 = Gray	Silver Multiplier = 0.01
2 = Red	9 = White	
3 = Orange	1% = Red	Band 1 = 1st digit
4 = Yellow	5% = Gold	Band 2 = 2nd digit
5 = Green	10% = Silver	Band 3 = Multiplier
6 = Blue	20% = No Color	Band 4 = Tolerance
		Band 5 = Temperature coefficient or voltage rating

The location and meaning can vary with the manufacturer but generally the following are used for color bands:

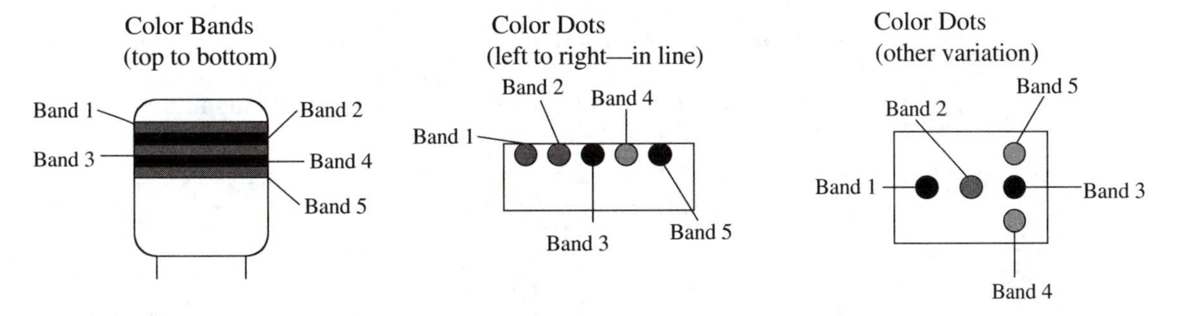

▲ **FIGURE A-4**
Capacitor color codes

A-3 ▶ Inductors

An inductor is a coil of conductive material that stores electrical energy in the magnetic field created by the *changing* current passing through the inductor. Inductors oppose any change in the current passing through them. They accomplish this by giving up previously stored energy if the current is decreasing or storing more energy in the magnetic field if the current is increasing. In AC circuits the inductors continually store and release energy as the current increases, reaches a maximum, then declines and eventually flows in the other direction. Inductive Reactance, X_L, is the impedance that the inductor offers to resist the flow of current and is equal to:

$$X_L = 2\pi f L$$

where f = frequency and L = Inductance

As frequency increases the inductive reactance, X_L increases. This is exactly opposite of the way capacitors function (X_C decreases as frequency increases). For DC circuits, $f = 0$ so inductive reactance is = 0. In DC circuits inductors are a steady state short circuit. The inductance value, L, is given in units called *henrys* and determines the inductor's ability to store energy. Practical ranges of inductance are millihenries (mH) or microhenries (μH). The larger the inductance value, the greater the capacity to store energy. Inductors are similar to resistors in that the equivalent of two inductors in series is simply the total of the two inductances. The equivalent of two inductors in parallel can be calculated by:

$$L_1 = L_1 \times L_2/(L_1 + L_2)$$

It is impossible to have a pure inductive component because of the resistance of the conductor used to make up the coil. The quality factor Q is a measure of this fact and is a ratio of the inductive reactance over the DC resistance:

$$Q = X_L/R_{DC}$$

Inductor values are given in henries. Inductors are rated by their value, tolerance, and maximum current. Inductor values are determined from the same standard number set as resistors and capacitors. Available inductance values range from nanohenries, microhenries, and millihenries. The nominal inductance values available are shown in Figure A–5. The actual values available depend upon the magnetic material used, the type of package, and the inductor manufacturer.

Standard Inductance Values					
Nanohenries	Nanohenries	Microhenries	Nanohenries	Microhenries	Microhenries
1.0	120	0.12	12,000	12	1000
1.2	180	0.18	15,000	15	1200
1.8	220	0.22	18,000	18	1800
2.2	270	0.27	22,000	22	2200
2.7	330	0.33	27,000	27	2700
3.3	390	0.39	33,000	33	3300
3.9	470	0.47	39,000	39	3900
4.7	560	0.56	47,000	47	4700
5.6	680	0.68	56,000	56	5600
6.8	820	0.82	68,000	68	6800
8.2	1000	1	82,000	82	8200
10	1200	1.2	100,000	100	
12	1500	1.5		120	
15	1800	1.8		150	
18	2200	2.2		180	
22	2700	2.7		220	
27	3300	3.3		270	
33	3900	3.9		330	
39	4700	4.7		390	
47	5600	5.6		470	
56	6800	6.8		560	
68	8200	8.2		680	
82	10,000	10		820	
100					

▲ **FIGURE A–5**
Standard inductance values

A–4 ▶ Diodes

There are many different types of diodes designed for use in specific applications. Each type of diode will be discussed along with the key parameters involved in its selection. The available package designations are shown in Figure A–6.

Rectifier Diodes

The primary purpose of rectifier diodes is simply the conduction of current in one direction, most often for the purpose of converting AC voltage to DC voltage. Generally, the key parameters for rectifier diodes are the forward-biased current and the peak inverse voltage. The switching speed and reverse-bias leakage are also important.

Zener Diodes

Zener diodes are commonly used to regulate or clamp voltages and as such are connected reverse-biased. The key parameters are the zener voltage, zener current, and power rating. The tolerance and variation of the zener voltage and zener current with temperature are also important considerations. The most significant parameter for their selection is the zener voltage.

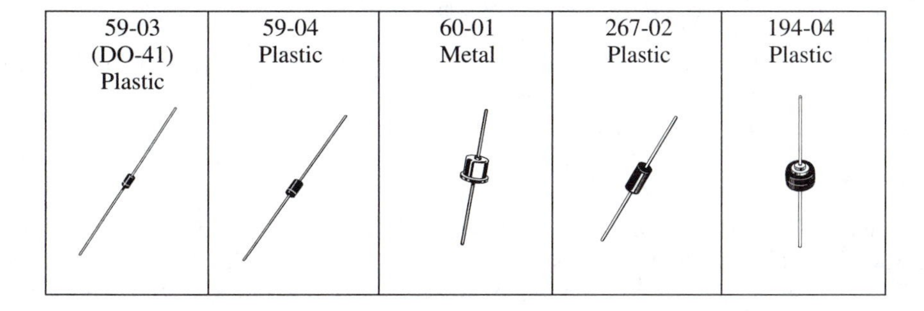

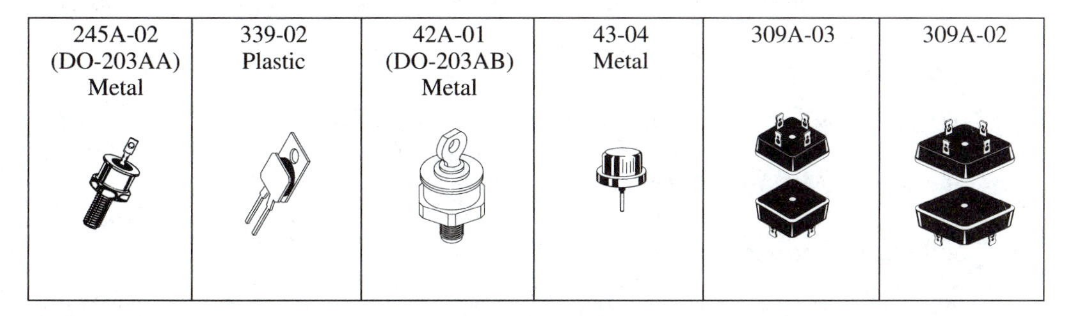

▲ **FIGURE A–6**
Diode packages (*Copyright of Semiconductor Component Industries, LLC. Used by permission*)

Shottky Diodes

These diodes feature a very high switching speed and are consequently used in high frequency and fast-switching applications. The key parameter in their selection is the speed at which the diode goes from the forward to reverse-biased operation.

Light-emitting Diodes

Light-emitting diodes are used as indicators or displays or to transmit information optically. As indicators, the LEDs are usually used in individual packages. As displays, 7 segment or 5×7 dot matrix LEDs are common. LEDs are used to transmit optical information down glass and plastic fiber in fiber-optic applications. The key selection parameters are:

1. The wavelength or color of the output light
2. The intensity of the output light
3. The switching speed
4. Efficiency (the amount of output light intensity per the input current)

Photodiodes

The photodiode is operated reverse-biased. When light is applied to the reverse-biased photodiode the leakage current increases. The key operational parameters for selecting a photodiode are:

1. The wavelength of maximum sensitivity
2. The sensitivity—this is the amount of reverse current per the input optical power

A–5 ▶ Transistors

Transistor selection involves first the choosing of either bipolar (BJTs) or unipolar (FETs) devices. BJTs are characterized by faster switching speeds and poorer power efficiency. FETs are generally slower but draw very little current and are therefore very efficient. The available transistor packages are shown in Figure A–7.

Bipolar Junction Transistor (BJT)

BJTs are generally used in small- to medium-power applications. The key parameters for selection are the maximum collector to emitter voltage, the maximum collector current, and the DC current gain, β.

Field Effect Transistor (FET)

The FET is also used in small- to medium-power applications where power efficiency is more important than speed. High-power FETs are called MOSFETs and are discussed next. The key FET parameters are the drain-to-source and the gate-

TO-39 or TO-205AD
TO-72 or TO-206AF
TO-92 or TO-226AA
TO-92 or TO-226AE
SOT-23 or TO-236AB

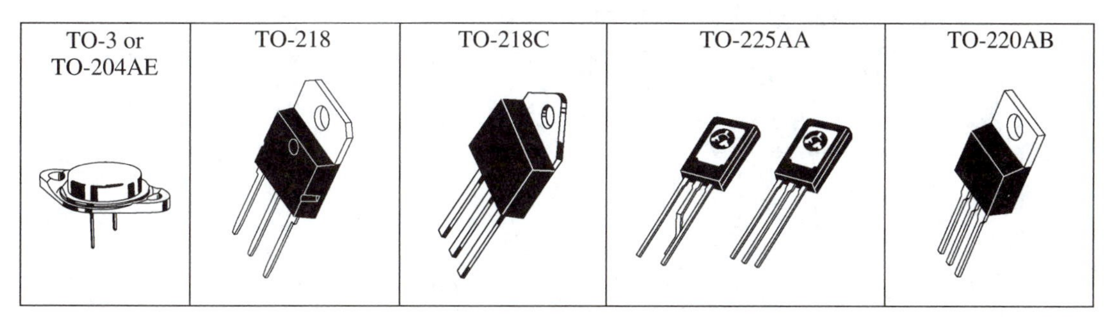

▲ **FIGURE A–7**
Transistor packages (*Copyright of Semiconductor Component Industries, LLC. Used by permission*)

to-source voltages, the drain to source current and the forward transconductance. The forward transconductance is the change in drain-to-source current per the change in gate-to-source voltage. This is similar to the current gain, β, in BJTs.

Metal Oxide Semiconductor FET (MOSFET)

The primary application for MOSFETs are as power transistors called *Power MOSFETs*. The Power MOSFET is capable of high currents because of the very small voltage drop across the drain to source. The key MOSFET parameters are the drain-to-source and the gate-to-source voltages, the drain-to-source current, and the forward transconductance.

A–6 ▶ Integrated Circuits

There are a wide variety of analog, digital, and hybrid (analog/digital) integrated circuits available. Their selection will involve all of the usual parameters—input and output voltage and current, power supply voltage, temperature coefficients, and frequency/speed capabilities. The different integrated circuit packages available are shown in Figure A–8 and A–9.

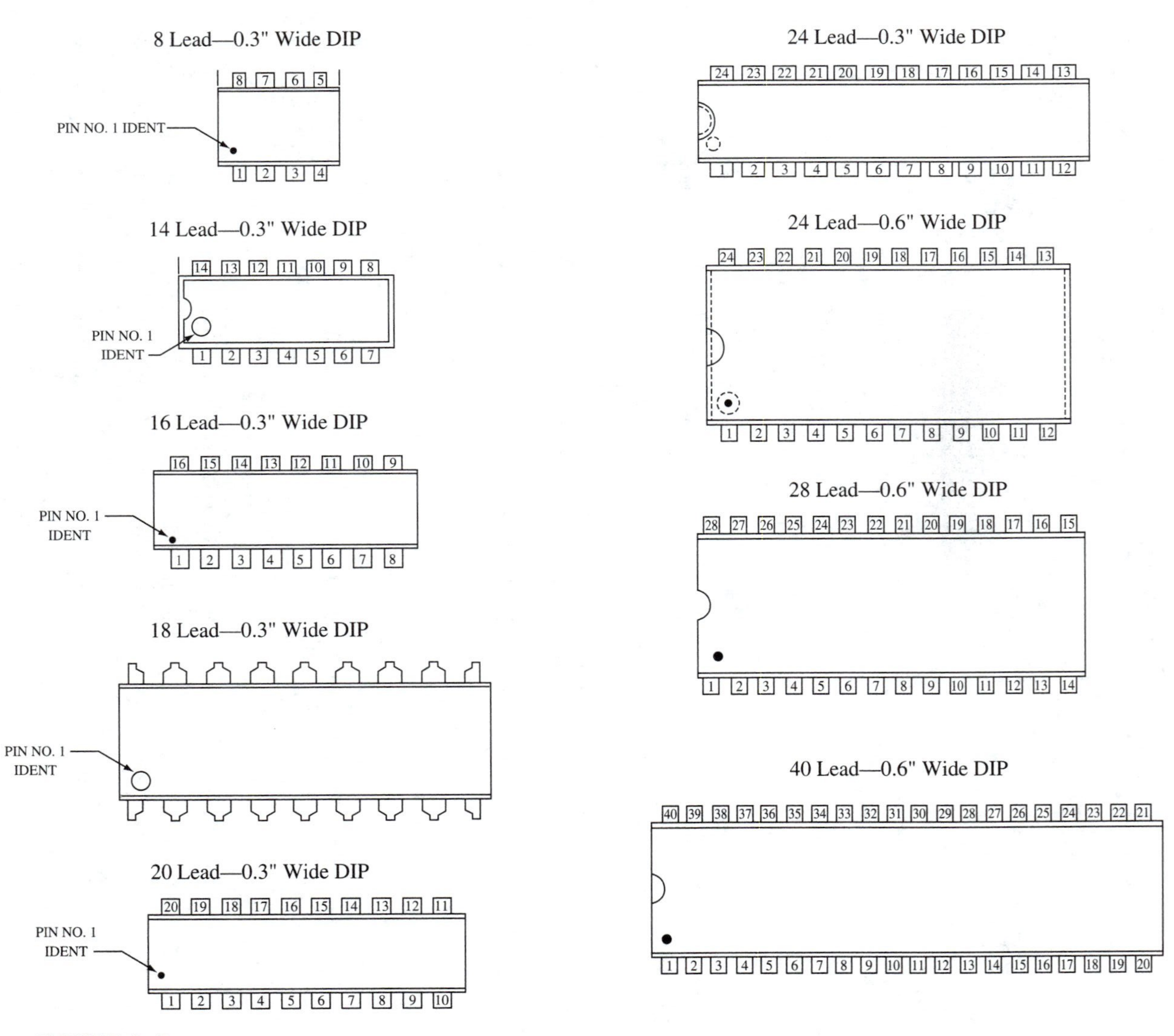

▲ **FIGURE A–8**
Integrated circuit packages

20-lead PLCC Package

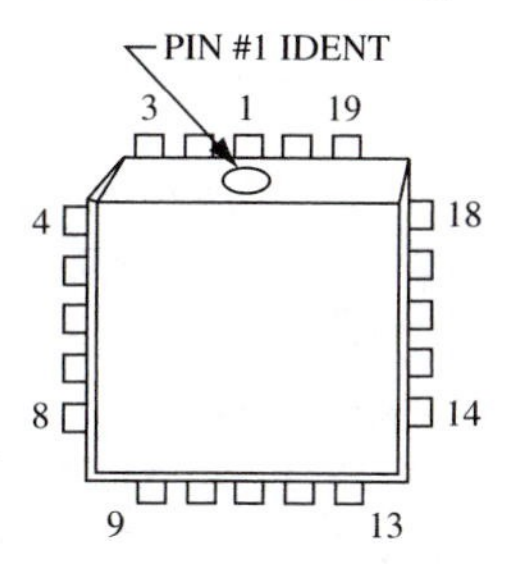

44-lead Quad Flat Package

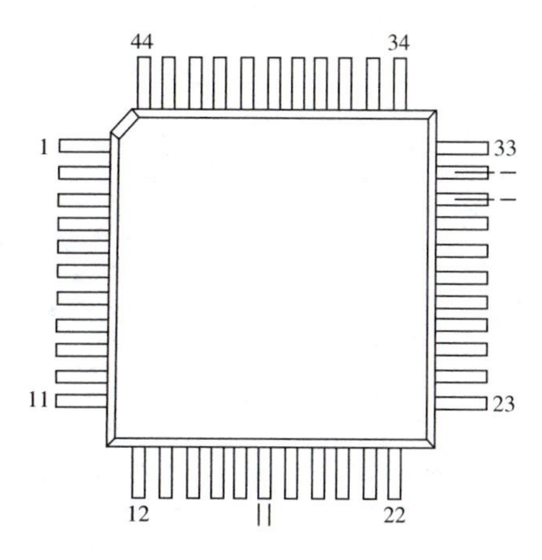

28-lead PLCC Package

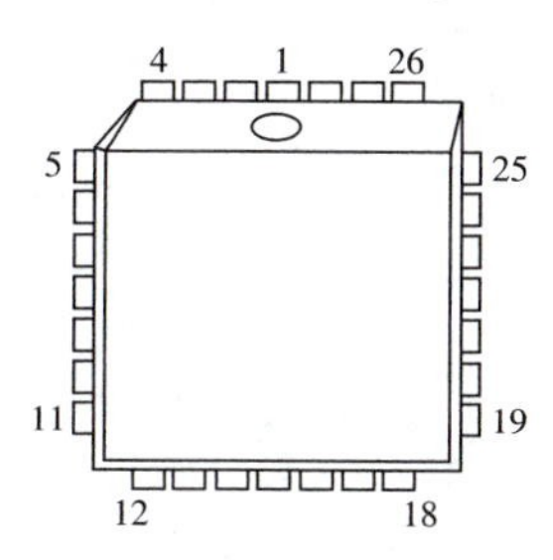

44-lead Cerquad Package

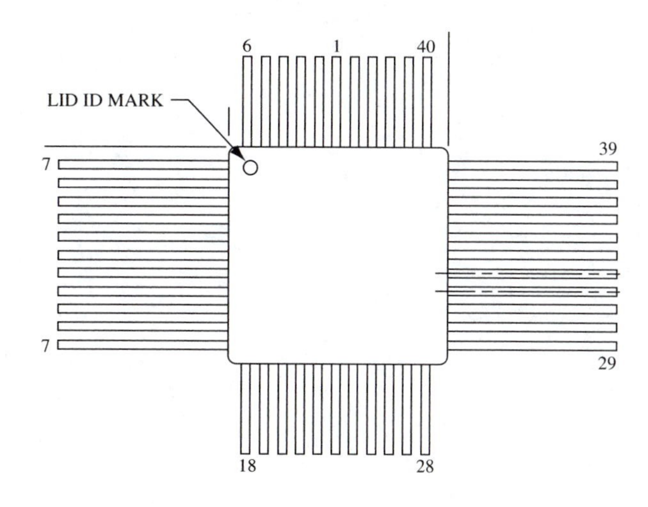

44-lead PLCC Package

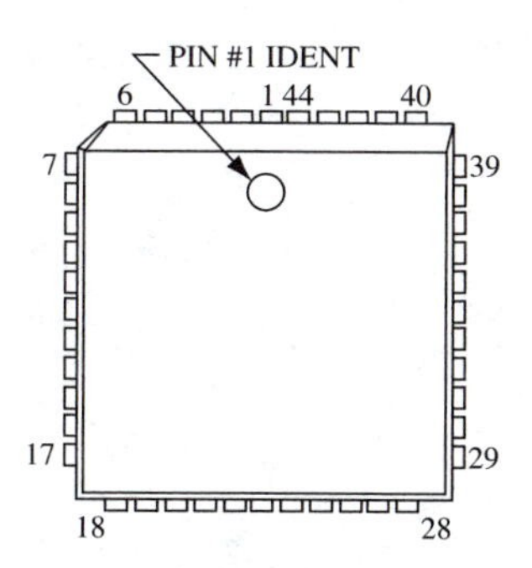

▲ **FIGURE A–9**
Integrated circuit packages

Index